lonely planet

TRIP BUILDER

CONTENTS

PREVIOUS SPREAD: VOLCANIC AEOLIAN ISLANDS
RIGHT: CANNES AND THE CÔTE D'AZUR
NEXT SPREAD: SENGLEA, MALTA; THE OSLO-BERGEN TRAIN

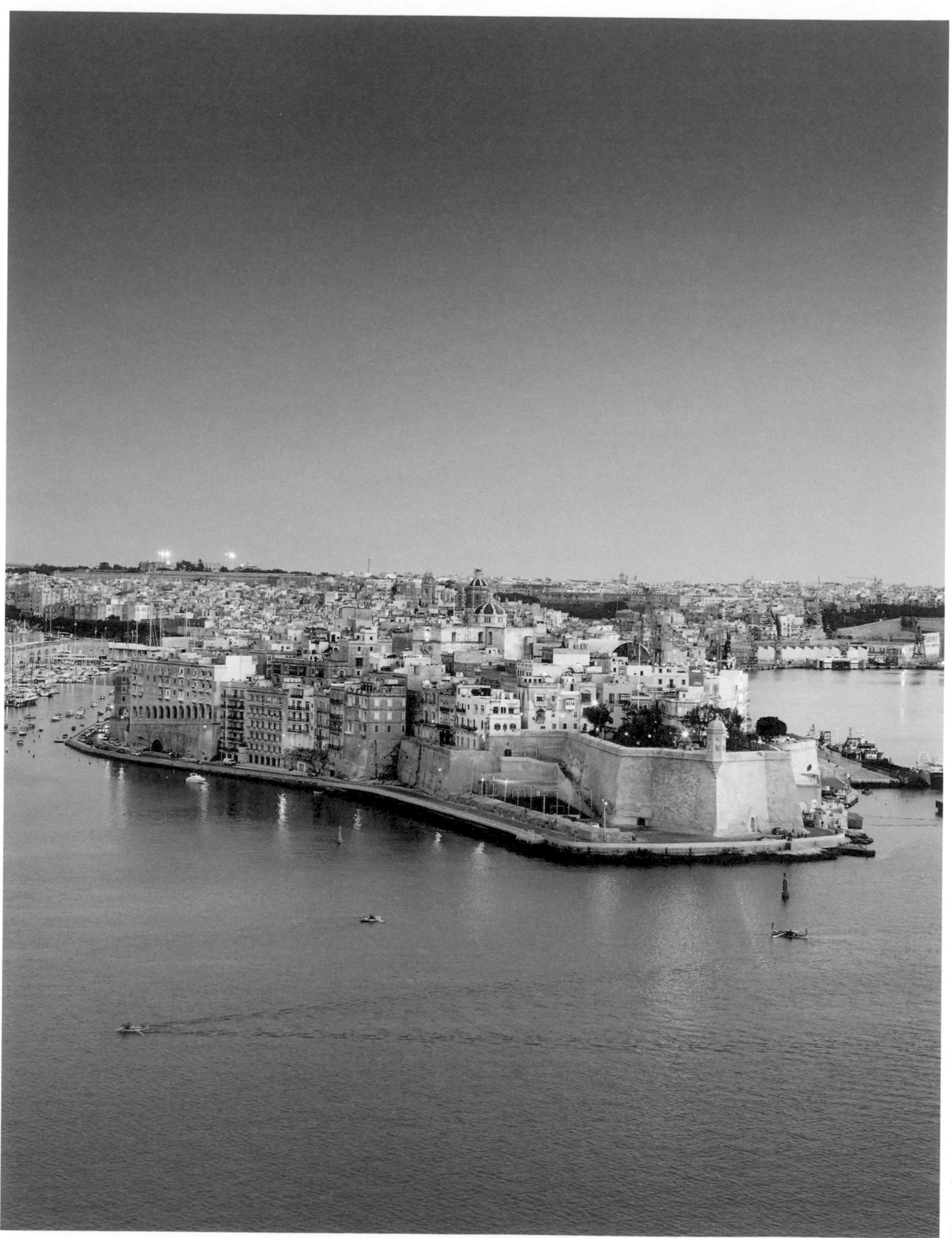

INTRODUCTION

When the German polymath Goethe journeyed across Italy in the 18th century, he wrote in his journal: 'I have spent the day well just looking and looking. It is the same in art as in life. The deeper one penetrates, the broader grows the view.'

The same idea is true today: travel broadens the mind. But although travelling around our world is easier than ever before, with flights to every corner of the planet, ferries plying seas, highways crossing countries and, in some locations at least, a high-speed rail network connecting people and places, we are perhaps more spoiled for choice than ever. And maybe more intimidated by the complexity of the options available. It's difficult to decide where you wish to go next. Where do you start to make a plan for the trip? And how do you ensure that the route is set up as sustainably as possible?

This book aims to simplify the whole process. Lonely Planet's *Trip Builder* presents hundreds of ideas for itineraries in every region of the globe, all designed by our expert travel writers, who have lived and researched Lonely Planet's travel guides for more than 45 years. These itineraries are their distillation of what you should see and do around the world, from remote corners of the Pacific to pulsating cities. Of course, most of us can't take many months to travel on a grand tour. So these itineraries range in duration from a few days, for a long weekend exploring a city or a wine region, to a few weeks voyaging through a country or continent by train.

The itineraries are also classified according to the themes of the trip: if you prefer museums and galleries to beaches and bars, or hiking trails and campfires to dining in swanky restaurants, the themes highlighted at the start of every itinerary will tell you what to expect. In the contents section for each chapter we've also provided the duration of the trip and the modes of transport required.

Transport choices are an important part of travel today. We want to know what impact our travel has on the planet and how to mitigate it. Each trip displays the carbon cost of the journey: many trips use trains, but we also include self-driving itineraries, boat trips and several cycle routes. The data helps you make an informed choice about how to manage the carbon footprint of your trip. All of the transport connections and timings are explained in detail, with the aim of making the planning of your next adventure as easy and enjoyable as possible.

THEMES FOR DREAM TRIPS

Active

Drink

Food

Sustainable

Adventure

Culture

Art

Architecture

History

Family

Solo

Short

Beach

Winter

Wellness

Wildlife

SOUTHERN EUROPE

From Aegean island-hopping and priceless art in Florence to swimming in the gin-clear Med or ticking off Baroque beauties and Moorish palaces, Southern Europe promises pleasures aplenty. Navigate the ferries, footpaths, backroads and bike trails to discover ancient history, fabulous museums and some of the best food on the planet, from Spanish tapas and Italian slow-food delights to Porto's perfect ports.

11
AZORES

NORTH ATLANTIC OCEAN

20 MADEIRA

CANARY ISLANDS
25

SOUTHERN EUROPE
BALTIC SEA
BLACK SEA
CORSICA
ITALY
SARDINIA
GREECE
SPORADES
DODECANESE
RANEAN SEA
SICILY
AEGEAN SEA
MALTA
CRETE
CYPRUS
34
13
23
15
26
33
12
38
39
16
27
31
37
35
2
19
14
1
30
41
21
4
9

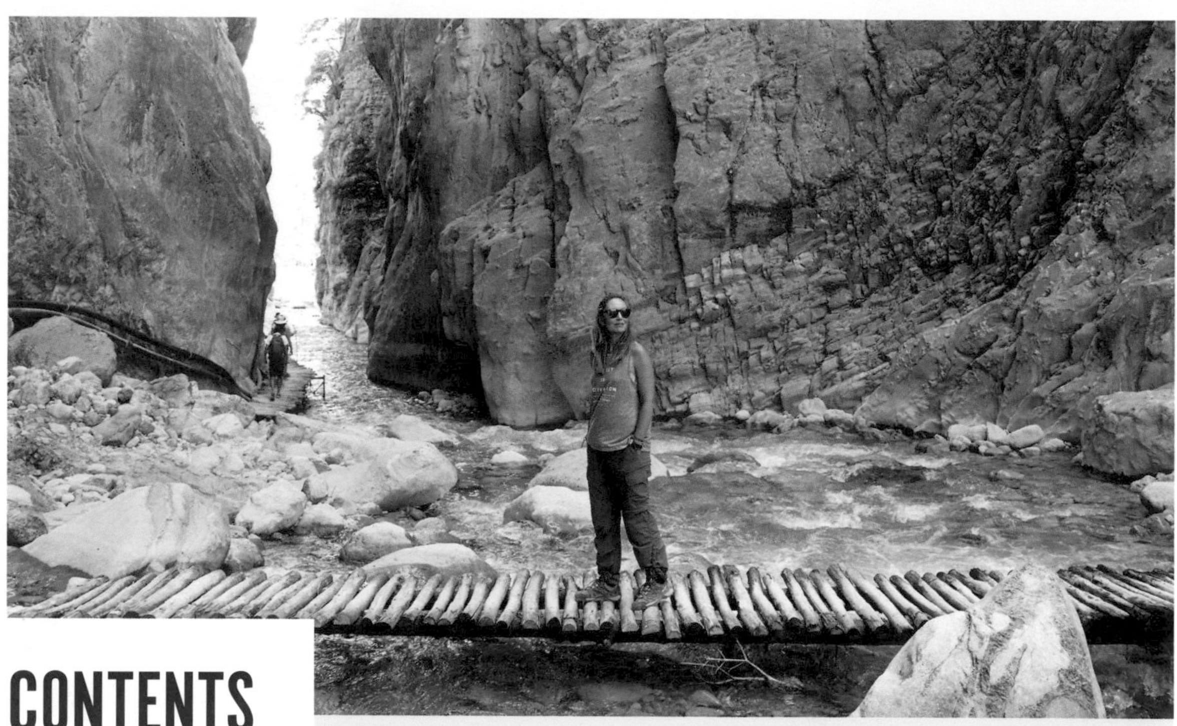

ABOVE: SAMARIA GORGE, CRETE

CONTENTS

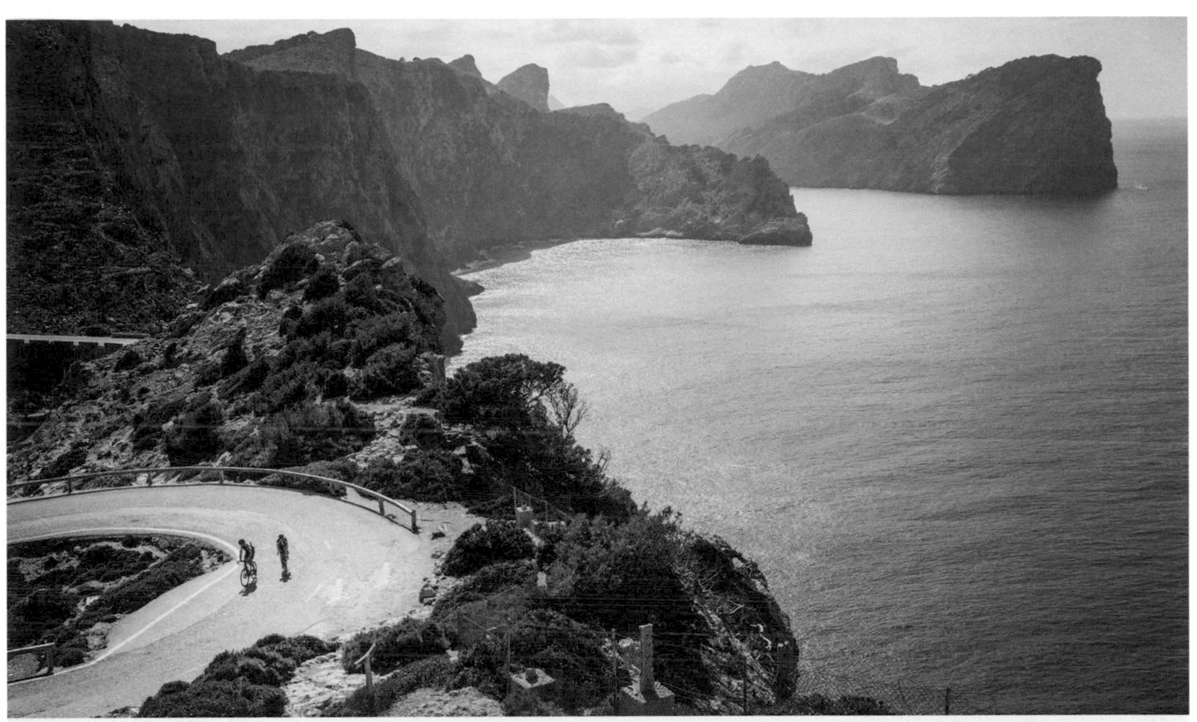

ABOVE: CAP DE FORMENTOR, MALLORCA

AN AEOLIAN ISLANDS ADVENTURE

Milazzo – Stromboli

Island-hop along Italy's string of volcanic outcrops stretching north from Sicily, clambering up smoking craters and swimming from peaceful beaches.

FACT BOX

Carbon (kg per person): 12
Distance (km): 109.5
Nights: 5-7
Budget: \$\$-\$\$\$
When: Apr-Sep

❶ Milazzo

Milazzo, a gritty industrial city with a surprisingly alluring old core, is the gateway to the Aeolians and sits near Sicily's northeastern tip. The ancient hilltop citadel and Borgo Antico (Old Town) are absorbing places to wander, followed by a traditional sundown *passeggiata* (stroll) along the waterfront, gelato in hand.

Take the ferry to Vulcano (several daily; 1hr 15min).

❷ Vulcano

Fumaroles still belch yellow-tinted, sulphurous steam from the rim of Vulcano's crater, an easy (though hot) hike on glittering silicate sand from the harbour at Porto di Levante. More fumaroles heat seawater at Spiaggia delle Acque Calde (Hot Water Beach), where you can also apply 'therapeutic' – and very smelly – mud to your skin.

Jump back on the ferry to Lípari (several daily; 20min).

❸ Lípari

The elegant main town on this largest and liveliest of the islands boasts a fascinating fortified old centre dominated by the Castello, an imposing 16th-century bastion with roots dating back to Neolithic times; it's now home to an archaeological museum. Bustling streets and alleys are lined with stores piled high with capers, sun-blushed tomatoes, chestnuts and olives, while quiet roads into the hinterland beg to be explored by bike.

Take the ferry to Salina (daily; 50min).

ABOVE: REFRESHING GRANITA ON SALINA

ABOVE: VIEW TO VULCANO FROM LÍPARI

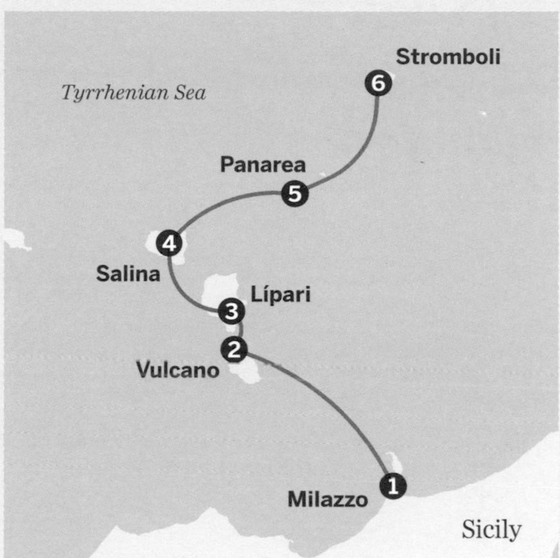

❹ Salina
A top pick for hiking, the twin extinct volcanoes on this most verdant island provide spectacular views to the other Aeolians, and across the vineyards producing Salina's renowned orangey Malvasia dessert wine.

Take the ferry to Panarea (several weekly; 1hr).

❺ Panarea
Smallest and most upmarket of the islands, Panarea is a sophisticated Mediterranean hideaway with just a few hamlets strung along its east coast. At the southern tip, the Bronze Age hut circles of Punta Milazzese line a promontory overlooking idyllic hidden coves.

Ferries sail to Stromboli several times a week (1hr 30min).

❻ Stromboli
Less an island, more a volcano ringed by shore, Stromboli's crater is the big attraction – climb at dusk (when safe) to witness rumbles, fumes and fiery outbursts. Beneath it spread lanes spangled with pink oleander leading to chic whitewashed restaurants.

Ferries sail to Milazzo (5hr 30min) and Naples (11hr 30min) several times a week.

ALTERNATIVES

Sidetrip: Filicudi and Alicudi
Two far flung Aeolian outposts lie west of the main chain. Just a handful of settlements dot Filicudi's flanks, but its coast, lapped by the blue, blue Med, is studded with gorgeous grottoes. Alicudi is smaller, steeper and more isolated still, a road-free speck where transport is by mule or boat or, ideally, foot – there's wonderful walking.

Ferries sail to Filicudi (1hr 30min) and Alicudi (2hr 30min) from Salina several times a week.

Extension: Taormina and Mt Etna
Taormina, eastern Sicily's premier resort, is a handy stop en route to Catania airport. Its spectacularly set Teatro Antico, the remains of a Greek amphitheatre scooped into the mountainside in the 3rd century BCE, sits above a string of fine beaches. Looming above is Mt Etna, its black lava slopes easily accessible by car or bus.

Frequent trains serve Taormina from Milazzo, changing in Messina (2hr 30min).

ANCIENT GREECE EXPLORER

Athens – Meteora

Not interested in the classical past? Rub shoulders with the ghosts of historical greats at some of Greece's most famous sites and you'll soon change your mind.

FACT BOX

Carbon (kg per person): 22
Distance (km): 775
Nights: 7
Budget: $$-$$$
When: Mar-Nov

❶ Athens

Athens is ground zero for 'the classics' in every sense. The city is dotted with small and large landmarks, all remnants of its long, often glorious, always interesting past. The remarkable National Archaeological Museum is the spot to get the best overview of all things ancient. From there head to the Acropolis, where the Parthenon and other millennia-old buildings form an awe-inspiring icon of Western civilisation. There are sweeping views across the city, too. Then knock your marble socks off at the Acropolis Museum, home to (some of) the Parthenon sculptures. Finally, browse the superb treasures from the Bronze Age up to WWII in the impeccable Benaki Museum of Greek Culture.

Take the bus for Delphi from Liossion Intercity Bus Station (2-3 daily; 3hr).

❷ Delphi

To the ancient Greeks, Delphi was the centre or 'navel' of the earth. This magical place, one of the most visited sites in all of Greece, has a superb archaeological site, the Sanctuary of Apollo, set among magnificent hillside scenery. Pilgrims from around the world used to voyage here to consult the oracle. You'll need at least a day to wander through

ABOVE: METEORA'S MONASTERIES

ABOVE: ATHENS FROM THE ACROPOLIS

both the site and the museum. The latter exhibits many extraordinary treasures – gifts presented by the petitioners who visited here – and is worth a visit just to see the Bronze Charioteer alone. Spend the night to avoid rushing to see everything in a single day.

Getting to Meteora is fiddly but not impossible – one daily bus service goes via Amfissa, Lamia and Trikala (4hr).

3 Meteora

The breathtaking Meteora region, at the northwestern edge of the plain of Thessaly, features towering rocky outcrops topped by Byzantine monasteries. Spend a day doing a circuit of these remarkable buildings (six of them are open to visitors), and allow an extra day to hike around the area on the former *monopati* (footpaths) that weave between the rock pinnacles. Head off on a hiking, climbing or kayaking adventure with Visit Meteora (visitmeteora.travel). They also offer walking tours, useful if you don't want to wander alone.

Buses return to Athens from Kalambaka, Meteora's main village (around 4 daily, via Trikala; 5hr). Alternatively, a daily train service heads from Kalambaka to the capital's Larissa station.

ALTERNATIVES

Extension: Ancient Corinth, Peloponnese

If you're still on a classical roll, venture west to Ancient Corinth, an evocative ruined Greek and Roman city: the Temple of Apollo here is one of the most complete structures of its kind in the world. The on-site museum provides excellent context.

A rented car is most convenient but buses run from Athens to modern Corinth, from where you can catch a local bus to the ancient city.

Extension: Epidauros, Peloponnese

Located 140km southwest of Athens, this ancient healing sanctuary dates from the late 4th century BCE and is home to one of the country's best-preserved ancient theatres – time your visit with the summer Epidauros Festival to see a performance.

A rental car is simplest, but you can catch a bus from Athens to Nafplio (hourly; 2hr 30min), and a bus from Nafplio to/from the site (1-4 daily; 45min).

CYCLE ASTURIAS' GREEN SPAIN

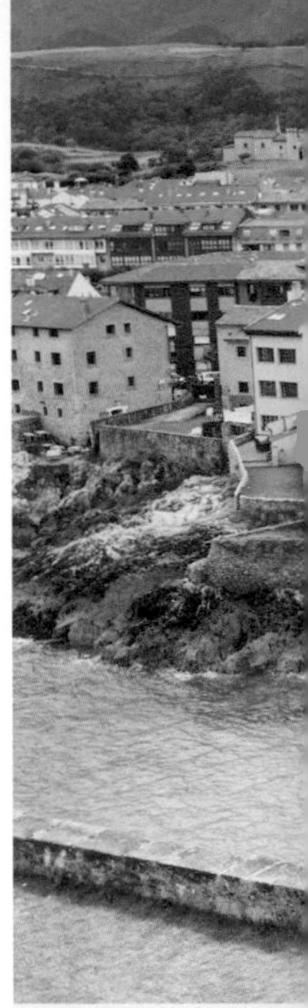

Oviedo – Gijón

Go green in Green Spain and take to two wheels, exploring Asturias' compelling cities, mountains, beaches and history by bike.

FACT BOX

Carbon (kg per person): 0kg
Distance (km): 240
Nights: 5–7
Budget: $$
When: May–Sep

❶ Oviedo

The Asturian capital is famous for the unique 9th-century pre-Romanesque churches in the hills around the city. In the centre, visit the Fine Arts Museum and try Asturias' signature dish, *fabada* stew.

Cycle 75km east.

❷ Cangas de Onís

Gateway to the magnificent Picos de Europa mountains, Cangas is also an historic place with a 'Roman' bridge and a 12th-century monastery turned luxury hotel. A steep 20km ride southeast is Covadonga, site of a famous Reconquista battle, a pilgrimage basilica and two picturesque lakes.

Cycle 30km east.

ABOVE: COOKING ASTURIAS' SIGNATURE *FABADA*

❸ Arenas de Cabrales

In the heart of the Picos, Arenas' pretty buildings are outshone by the views of the surrounding mountains. The main reasons to stop are to try the mouthnumbingly-strong *cabrales* cheese and to hike the wonderful, 22km-round-trip Ruta del Cares.

Ride 35km north.

❹ Llanes

The port of Llanes packs plenty of charm within its medieval walls. Your tough choices here are between a lazy lunch at a harbourside restaurant, strolling clifftop Paseo San Pedro or visiting one of the many local beaches: Torimbia is a stunner; Gulpiyuri is a surprise, located inland from the sea in the middle of a field.

Cycle 30km west.

❺ Ribadesella

It has a glorious swirl of beach, a colourful port and

ABOVE: LLANES

kayaking on its river, but Ribadesella's main claim to fame is the Tito Bustillo cave. In 1968, dozens of Palaeolithic paintings were discovered here, some around 33,000 years old. The Main Panel, depicting bison, deer and horses in black, red and violet, is the only one open to visitors.

Cycle 40km west.

6 Villaviciosa

Apple orchards lining the roads give a clue to Villaviciosa's popularity – *sidra* (cider). Not your usual fizzy cider, Asturian *sidra* is like strong apple juice, best drunk in a *sidrería* where it's poured from overhead into a huge glass shared by everyone. Down your *culín* (little arse) in one and be careful – it's surprisingly powerful.

Cycle 30km west.

7 Gijón

Asturias' largest city, Gijón is a seaside town that likes to have a good time. Laze on San Lorenzo beach, shop on Calle Corrida and visit the Roman remains, but save some energy for a lively night out in the *sidrerías* and bars of the old port neighbourhood, Cimadevilla. Buses and trains run from Gijón to many Spanish cities.

The regional airport, 40km west, has national and international flights.

ALTERNATIVES

Extension: Parque Natural de Somiedo

Somiedo offers your best chance of seeing the endangered Cantabrian brown bear. Making a slow comeback, they number 250-300, and can sometimes be seen wandering the park's four main valleys; other less elusive wildlife here include deer, boar and wolves. And while you're looking, there are trails through forests and past lakes to enjoy.

Somiedo is around 70km southwest of Oviedo.

Extension: The Camino de Santiago

If all that cycling has whetted your appetite for outdoor activities, consider walking one of the two branches of the Camino de Santiago pilgrimage route that go through Asturias. The 284km Coastal Route follows the shore, passing beaches and fishing villages. The 148km Primitive Route was the choice of the very first pilgrim (allegedly), King Alfonso II.

Get Camino info at santiago-compostela.net.

ROAD-TRIP CRETE'S WINE COUNTRY

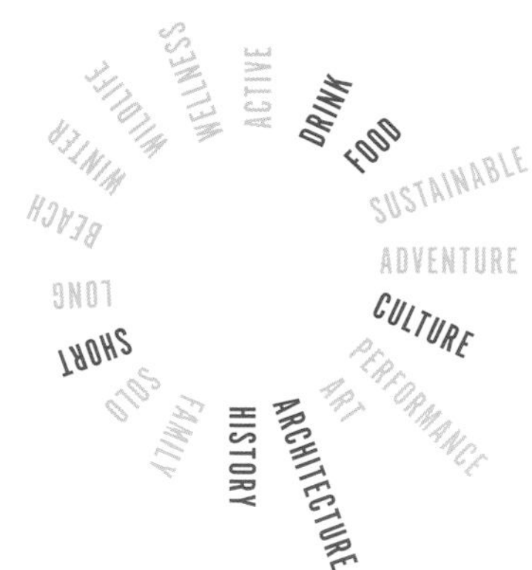

Iraklio – Arhanes

One of Greece's best wine regions unravels south of Crete's capital, Iraklio, and is easily visited by car (designated driver required).

FACT BOX

Carbon (kg per person): 10
Distance (km): 81
Nights: 2
Budget: $$
When: May, Jun, Sep, Oct

❶ Iraklio

Crete's capital doesn't usually evoke love at first sight, but its Old Town is worth a wander, especially to find one of the island's best restaurants, the outstanding Peskesi. The Heraklion Archaeological Museum, one of Greece's most important, is also a must for anyone heading to ancient Knossos.

Drive 11km southeast into the Peza and Arhanes wine regions.

❷ Koronekes

Grapes and olive oil often go hand in hand and this gorgeous olive estate is home to some 2000 trees pushing upwards of three centuries in age. They press olives the traditional way and it's a great spot for picking up some of the island's outstanding olive oil and other local products.

Head another 13km south towards Peza.

❸ Digenakis Winery

Call ahead to visit this modern winery, a striking architectural collage of concrete and glass that is unmissable from the highway. Digenakis' single-vineyard Kotsifali red is produced with grapes from 35-year-old vines, one of Crete's few to escape a devastating 1970s phylloxera outbreak.

Circle back north and drive 11km to Skalani.

❹ Skalani

Tiny Skalani is home to one of wine country's best restaurants, Elia & Diosmos, which is well worth the

ABOVE: MINOAN RHYTON, HERAKLION ARCHAEOLOGICAL MUSEUM

VINEYARDS NEAR IRAKLIO

detour for lunch. Chef Argyro Bardo does market-fresh Cretan dishes with a modern flair (that pork shank in wine, honey and citrus), paired with carefully curated local wines.

Return south for 17km to one of Crete's most visitor friendly wineries.

5 Lyrarakis Winery

Award-wining Lyrarakis is well-known for resurrecting nearly extinct varietals you aren't likely to find elsewhere, including whites like Dafni, Plyto and Melissaki. It was also the first to produce a single-vineyard Mandilari red; originally mocked as 'absurd', it went on to bestseller status.

Wrap things up 13km north in Arhanes.

6 Arhanes

On your way into the lovely town of Arhanes, stop off at Vathypetro, 5km south, and check out one of the oldest surviving wine presses in the world. In Arhanes itself, make your way to the dazzling main square. Dinner beckons at Bakaliko, where sommelier Giorgos Kteniadakis offers 70 wines by the glass and wonderful food with which to pair them.

Heraklion International Airport is 16km north of Arhanes.

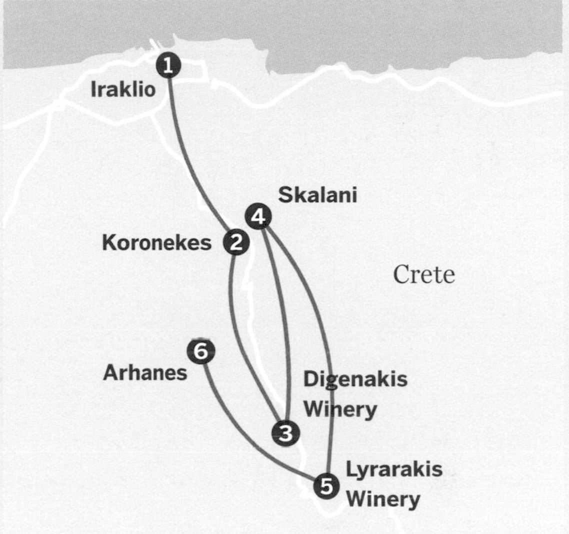

ALTERNATIVES

Day trip: Knossos

One of Greece's most important and well-preserved ruins lie at the gateway to wine country. The Palace of Knossos, the capital of Minoan Crete, is a fascinating, partially restored complex featuring colourful frescoes, grand staircases and commanding gateways. You can easily spend a day here on learning about this fascinating civilisation, which dominated southern Europe some 4000 years ago.

The ruins are 5km south of Iraklio.

Extension: Dafnes

If you'd like to tack on more wine tasting, the wineries of Dafnes are well worth exploring. Diamantakis Winery is one of Crete's most outstanding (its Petali Liatiko might be the island's best red), while Douloufakis, Silva and Idaia wineries are all just as recommendable. Earino is one of wine county's most scenic sleeps and best restaurants.

The Dafnes wineries are 22km southwest of Iraklio.

CÁDIZ FOR FOOD LOVERS

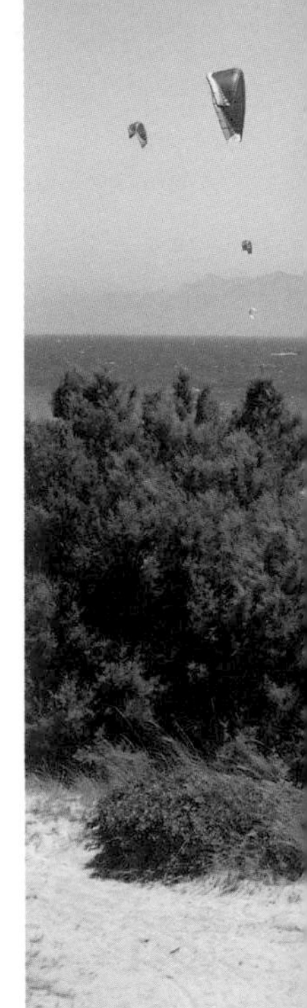

Jerez de la Frontera – Tarifa
Cádiz province, a breathtakingly beautiful pocket of southern Andalucía, has some of Spain's most outstanding gastronomy, all easily reached by bus.

FACT BOX

Carbon (kg per person): 6.4
Distance (km): 235
Nights: 7–10
Budget: $$-$$$
When: Mar-Oct

❶ Jerez de la Frontera

Soulful Jerez, queen of Andalucía's Sherry Triangle, is the perfect welcome to fabulous Cádiz province (with the region's only airport). Uncover elegant sherry *bodegas*, the Seville-inspired cathedral, the Almohad-era *alcázar*, some seriously original cuisine and feisty flamenco at the *tabancos* (sherry taverns).

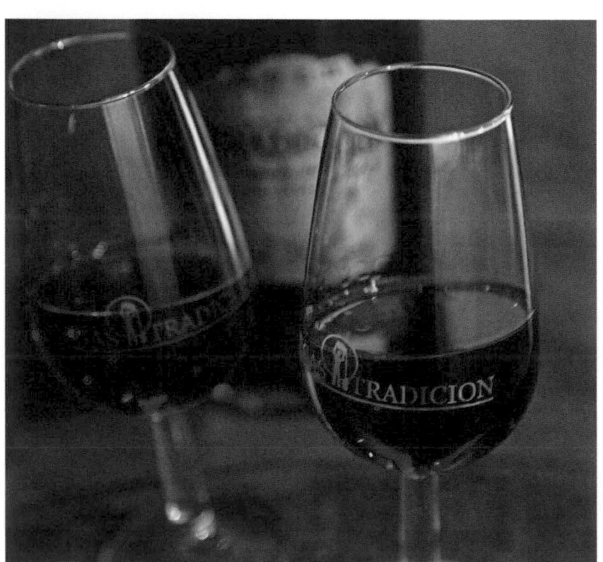

ABOVE: BODEGA TRADICION, JEREZ DE LA FRONTERA

Catch the bus to Sanlúcar (7-13 daily; 40min).

❷ Sanlúcar de Barrameda

Overlooking the Guadalquivir delta, understated Sanlúcar (Sherry Triangle town two) crafts its own one-of-a-kind sherry, Manzanilla, not to mention some of Cádiz's most superb seafood (the shrimp fritters are famous). Tour some *bodegas*, wander the old town, dine along waterside Bajo de Guía and day trip into Doñana's wetlands.

Take the bus to Cádiz (13 daily; 1hr).

❸ Cádiz

Founded by the Phoenicians, cheerful Cádiz is Europe's most ancient never-abandoned settlement, and its Carnaval is Spain's greatest. Three days suffice to enjoy its diverse barrios (seafood-heaven La Viña, flamenco-filled Santa María), its golden beaches and, most importantly, the top bars and restaurants for specialities like *papas aliñás* (tuna and potato salad) and *chicharrones* (fried pork).

Day-trip across the bay from Cádiz to El Puerto de Santa María by catamaran (8-16 daily; 30min).

❹ El Puerto de Santa María

Striking *bodegas* (including well-known Osborne),

ABOVE: TARIFA BEACH

white-sand beaches, excellent seafood and tapas and, if budget allows, a meal at Angel León's three-Michelin-star Aponiente await in the third Sherry Triangle town.

From Cádiz, take the bus to Vejer or La Barca de Vejer (5-6 daily; 1hr-1hr 30min).

5 Vejer de la Frontera

Is this Andalucía's most gorgeous white town? Indulge yourself for a few days in Vejer for the chic hotels, the richly layered Old Town and one of southern Spain's most innovative food scenes, with delicacies like *almadraba* (net-caught) tuna. Easy excursions take you to the Costa de la Luz's exquisite beaches and the tuna-tastic towns of Barbate and Zahara de los Atunes.

Take the bus to Tarifa (8 daily; 40min).

6 Tarifa

On mainland Europe's southernmost tip, easy-going Tarifa is Spain's kitesurf and windsurf capital, with miles of powder-soft beaches and an attractive Old Town. Food here is a deliciously creative fusion of Cádiz regional produce and flavours with international influences.

Tarifa has daily buses to Jerez, Málaga and Seville for airport connections.

ALTERNATIVES

Day trip: Parque Nacional de Doñana

Stretching around Europe's largest wetlands, this Unesco-listed national park is one of the most hauntingly beautiful natural spaces in Spain. Its rich biodiversity takes in wild boars, flocks of flamingos and even the endangered Iberian lynx.

Visitas Doñana runs boat-and-jeep trips from Sanlúcar, as does Viajes Doñana, which also arranges private jeep tours. Book ahead.

Extension: Arcos de la Frontera and the Sierra de Grazalema

With its evocative old quarter, turbulent history and wonderful tapas bars, clifftop Arcos is a classic Andalucian *pueblo blanco* (white town). To the east, Parque Natural Sierra de Grazalema has lovely hikes, villages and mountain gastronomy.

Frequent buses connect Arcos with Jerez (40min) and Cádiz (1hr), with connections to Grazalema via El Bosque.

CATALONIA – SURREAL ART TO SPANISH WINE AND WETLANDS

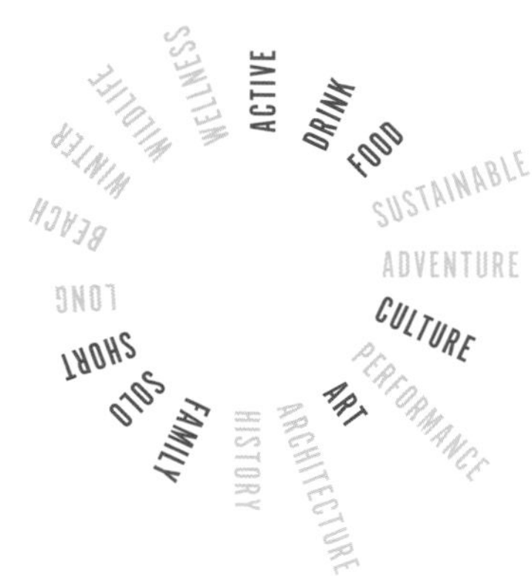

Cadaqués – Sant Carles de la Ràpita
Discover Dalí's surreal world in Cadaqués and Figueres, then bypass Barcelona and drive to Tarragona and Priorat's hilly wine country before exploring the Ebro Delta.

FACT BOX

Carbon (kg per person): 74
Distance (km): 591
Nights: 5
Budget: $$
When: Apr-Oct

❶ Cadaqués

Cadaqués is the perfect introduction to Catalonia, a hidden delight at the end of a coastal cul-de-sac whose sheltered, turquoise bay is part fishing village, part glamorous holiday resort. Forget the beach and join the locals in a traditional flat-bottomed boat in one of a dozen isolated creeks, only accessible from the sea. Then take a surreal turn with a visit to Salvador Dalí's home/museum, as delightfully eccentric as you'd expect.

Drive 25km west.

ABOVE: DALÍ MUSEUM, FIGUERES

❷ Figueres

Though he spent 52 years in Cadaqués, Dalí was born and chose to be buried in Figueres. His tomb is under the monumental museum bearing his name, which also contains an incredible collection of his works (the largest in the world) along with a theatre.

Continue south for 230km.

❸ Tarragona

With its Unesco World Heritage Roman amphitheatre, a resplendent 12th-century cathedral and walled medieval centre, Tarragona is very different from the rest of the Costa Daurada's 1970s beach resorts. After visiting the foodie paradise of Modernista Central Market, explore scenic Ruta del Cister, a wine and culture trail that passes Poblet monastery, then drive over to the historic Milmanda winery.

Leave Tarragona and head 50km inland.

❹ Priorat wine country

Picturesque Porrera, lively Falset and hillside hamlet

ABOVE: PRIORAT WINE COUNTRY

Gratallops form the heart of Priorat wine country, where artisan producers have long transformed the Grenache and Carignan grapes grown on spectacularly steep vineyards into some of the world's most sought-after vintages. Influential wineries to visit in and around Gratallops include the cathedral-like cellar of Alvaro Palacios, Clos de l'Obac and Clos Mogador. Top address to eat and drink is Falset's Hostal Sport, a cheap and cheerful diner where many winemakers drop in for lunch

From Gratallops drive 96km south towards the coast.

5 Sant Carles de la Ràpita

Driving into the delta where the Ebro River, Spain's longest, meets the Mediterranean resembles nothing else in Catalonia: immense, endless horizons of ever-moving sand and sea, lagoons and marshes, populated by hundreds of species of birds. At the edge of the delta, the fishing port of Santa Carles is the perfect base for exploring these protected wetlands as well as visiting traditional mussel and oyster farmers, salt harvesters and rice growers.

Barcelona and its international connections is a 190km drive northeast.

ALTERNATIVES

Sidetrip: Sant Sadurní d'Anoia

Catalonia's capital of cava is this festive town nestling in the Penedès hills, surrounded by over 100 wineries making the iconic sparkling wine. Take a memorable visit to one of the historic 19th-century Modernista cellars like Codorníu to witness the same production methods used for French champagne.

Sant Sadurni is on the route between Figueres and Tarragona.

Sidetrip: Montserrat

Catalonia's most popular attraction, this 11th-century Benedictine monastery perched 1200m atop a mountain range makes a stunning first impression. Pilgrims queue to see the holy Black Madonna; hikers to the summit from Monistrol avoid crowds and enjoy to-die-for views.

Montserrat makes an extra stop on the road between Figueres and Tarragona and can be combined with Sant Sadurní d'Anoia.

BALEARIC BLISS: THE OTHER IBIZA

Ibiza Town – Ses Salines

Tap into the White Isle's lesser-known corners, long history and laidback Balearic vibe on a circular road trip (with a little hiking, too).

FACT BOX

Carbon (kg per person): 17
Distance (km): 138
Nights: 7–10
Budget: $$$
When: Apr–Jun, Sep, Oct

❶ Ibiza Town

Originally settled by the Phoenicians, Ibiza's elegant capital challenges all the stereotypes. History-rich Ibiza Town centres on the Unesco-protected Dalt Vila: 16th-century defensive walls, steep cobbled streets, a Catalan Gothic cathedral and a crumbling Moorish castle. You'll dine wonderfully too!

Drive 16km northeast via Cala Llonga.

ABOVE: CALA D'EN SERRA BEACH

❷ Santa Eulària des Riu

The fascinating Puig de Missa (old quarter) of this attractive resort was once a key retreat from pirate attacks, and it's still guarded by a sparkling 1568 fortified church. Tempting tucked-away coves dot the coastline.

Continue 6km northeast.

❸ Sant Carles de Peralta

Loved for its hippie market, bohemian Sant Carles is also packed with Ibizan history, great restaurants and boutiques. The 18th-century church has a dark Civil War past, and among beaches nearby, secluded Aigües Blanques delivers the sunrises.

Head 15km north and west via Cala de Sant Vicent.

❹ Sant Joan de Labritja

Creative Sant Joan is one of northern Ibiza's main villages, with organic-focused restaurants and cafes, rustic-chic hotels and a weekend market.

Meander 8km north to Portinatx.

❺ North Coast

Ibiza's remote, off-grid northern coastline feels a world away from the island's party side; exploring here involves hikes and steep dirt tracks. Hidden-away spots include Portinatx lighthouse, ancient Cova des Culleram, and the

ABOVE: IBIZA TOWN

Port de Ses Caletes, Cala d'en Serra, Cala d'Aubarca, Es Portitxol and Caló des Moltons coves.

Continue 20km southwest.

6 Santa Gertrudis de Fruitera

At the island's heart, this charming village with a plaza and 18th-century church now bustles with boho-chic boutiques, antiques galleries and excellent restaurants.

Drive 28km west.

7 Stonehenge

Nicknamed Stonehenge by locals, this extraordinary clifftop creation by Australian artist Andrew Rogers consists of 13 bold basalt columns, one 10m high.

Head 15km south.

8 Es Vedrà

Magical at sunset, this mystical, off-limits islet is one of the Balearics' most entrancing sights.

Continue 23km east.

9 Ses Salines

Ibiza's shimmering salt pans – spectacular as the sun sinks – are protected by the 168-sq-km Unesco-listed Parc Natural de Ses Salines.

Drive 7km back to Ibiza Town.

ALTERNATIVES

Extension: Formentera

Spend a few nights on fabulously mellow, fiercely protected Formentera, off southeast Ibiza. Laze on the Caribbean like blonde-sand beaches, hit the *chiringuitos* and lighthouses, explore by bike or scooter or on foot, and wander the lively 'capital' Sant Francesc Xavier. The Trucador Peninsula, with its salt-white sands, is unmissable.

Frequent ferries run between Ibiza Town and Formentera (30min-1hr).

Diversion: Northern Villages

While roaming Ibiza's north coast, weave in some of the intriguing isolated villages just inland: Santa Agnès de Corona, Sant Mateu d'Aubarca, Sant Miquel de Balansat or Sant Llorenç de Balàfia. Most have traditional whitewashed homes, a central plaza with a restaurant or two and a gleaming fortress-church built several centuries ago.

All these villages are within an hour's drive of Sant Joan and Portinatx.

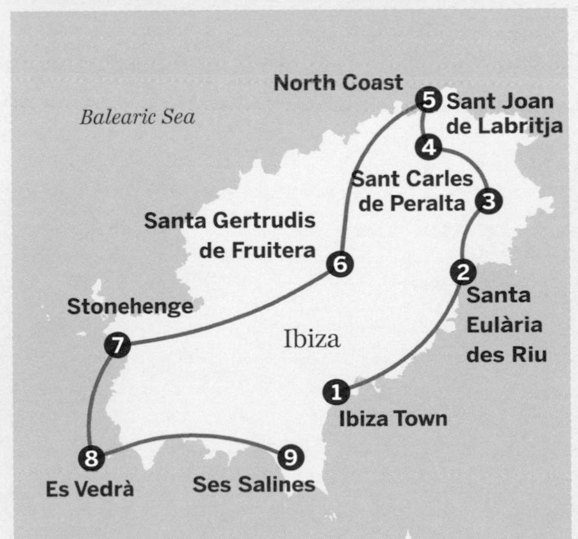

PORTUGAL'S DOURO VALLEY BY TRAIN

Porto – Pocinho

Drink in magnificent views and fine vintages on this train ride through the world's oldest demarcated wine region.

FACT BOX

Carbon (kg per person): 8
Distance (km): 185
Nights: 5-7
Budget: $$
When: Apr-Oct

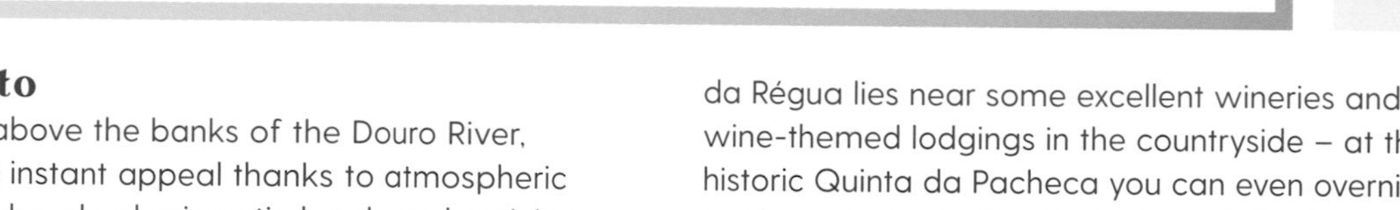

❶ Porto

Perched above the banks of the Douro River, Porto has instant appeal thanks to atmospheric neighbourhoods, charismatic locals and enticing food and drink scene. Grand architecture spans the centuries, from Baroque masterpieces like São Francisco Church to Rem Koolhaas' visionary contemporary concert hall. There are cast-iron markets, cutting-edge museums and relaxing waterfront spots in which to unwind after a day's exploring.

Walk across the lofty Dom Luis I Bridge and turn right.

❷ Vila Nova de Gaia

Historic port-wine lodges lie scatted among the streets facing the south bank of the Douro. At Espaço Porto Cruz you can take a virtual 360-degree wine journey over the Douro, then sample ports on the rooftop. Afterwards, tour the 300-year-old cellars at Taylor's and taste rare vintages at Graham's.

Take the train from Porto's São Bento Station to Peso da Régua (frequent; 2hr).

❸ Peso da Régua

Spread along the northern bank of the Douro, Peso da Régua lies near some excellent wineries and wine-themed lodgings in the countryside – at the historic Quinta da Pacheca you can even overnight inside a giant retrofitted wine barrel. On the riverfront, the Museu do Douro brings the valley's winemaking

ABOVE: ALFRESCO PORTO

ABOVE: DOURO VALLEY VINEYARDS

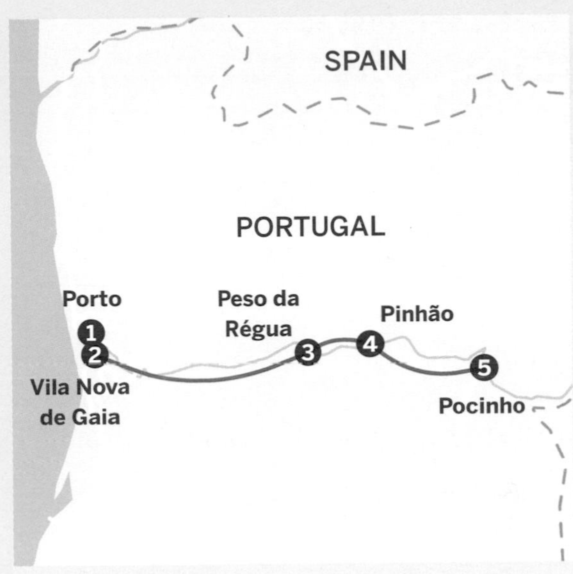

traditions to life with a fascinating collection of art and artefacts.

Take the train to Pinhão (3-5 daily; 30min).

4 Pinhão

Pretty Pinhão sits on a scenic bend of the Douro and the views here are sublime: towering terraced hillsides descend steeply to the riverbanks. Take in the backdrop on a stroll along the waterfront promenade, then head to one of the area's many wine estates. Vineyard and wine hotel Quinta Nova has excellent walking trails among the vines. Across the river, Quinta das Carvalhas has various tours and tastings, and you can ascend to the ridgeline for panoramic views over the countryside.

Continue on the train to Pocinho (1hr).

5 Pocinho

Although there isn't much to see in the tiny town of Pocinho, the train ride here is spectacular, taking you along the river's north bank then crossing to the south around Ferradosa. You'll follow sinewy bends in the Douro, enjoying majestic views of the lush vineyard-covered hillsides.

Take the train back to Porto (3-5 daily; 3hr 30min).

ALTERNATIVES

Extension: Vila Nova de Foz Côa

This sleepy whitewashed town lies near an astonishing collection of Paleolithic rock art in the nearby Rio Côa valley. Learn about the finds then see the wondrous carvings first-hand on a tour that departs from the museum and park entrance, 3km east of town.

From Pocinho, it's a 10min taxi ride to Vila Nova de Foz Côa (and a 15min walk to the park entrance).

Diversion: Lamego

The handsome town of Lamego is one of Portugal's top pilgrimage sites. A marvellous tile-covered staircase (with over 600 steps) leads up to the 18th-century church of Nossa Senhora dos Remédios. After making the scenic ascent (on knees for the faithful), wander through the Old Town, where you'll find traditional restaurants and a lively market.

Daily buses (or taxis) link Peso da Régua with Lamego (15min).

A ROAD TRIP AROUND CAPTIVATING CRETE

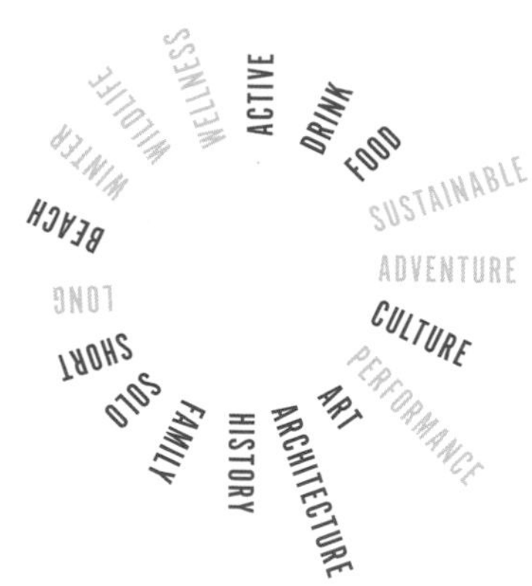

Iraklio – Elafonisi

Enjoy archaeological treasures, blissful beaches and heart-warming hospitality on a road trip around Crete's highlights.

FACT BOX

Carbon (kg per person): 40
Distance (km): 320
Nights: 7
Budget: $$
When: Apr-Jun, Sep, Oct

❶ Iraklio

The capital of Crete, Iraklio is home to a couple of the island's most significant sights: the incredible Heraklion Archaeological Museum and the nearby ruins of the Palace of Knossos, capital of Minoan Crete. The city's backstreets hide a lively cafe culture, along with craft-beer breweries and cool cocktail bars.

ABOVE: RETHYMNO TAVERNA

Drive 20km south into Iraklio Wine Country.

❷ Iraklio Wine Country

Vineyard-covered hills dot the countryside in Iraklio Wine Country. Stop off to visit a few of them and sample some of the best Cretan wines – many made from indigenous and nearly extinct Cretan grape varietals.

Drive west along the coast for 80km.

❸ Rethymno

Rethymno offers plenty of Mediterranean beauty with its Venetian-Ottoman quarter, 15th-century fortress and dazzling cerulean waters. Dine on fresh seafood at a harbourfront taverna, check out the views from the fortress and retreat to the old quarter's maze of lanes for the lively nightlife.

Make a return side-trip (40km each way) south to Preveli.

❹ Preveli

There are two excellent reasons to head out to the remote setting of Preveli: the historic Moni Preveli monastery and the paradisiacal Preveli Beach below it. The eye-pleasing monastery is perched high on the hill overlooking the glistening Libyan Sea, while

ABOVE: HANIA HARBOUR

the beach is backed by the craggy cliffs of the Kourtaliotiko Gorge and lined with palms.
Back in Rethymno, drive west for 60km to reach Hania.

5 Hania

Pretty Hania is Crete's most enchanting city, with a sublime Old Town and gorgeous Venetian harbour setting. Time is best spent getting lost in the maze of alleys crammed with restaurants, guesthouses, boutiques and bars, or strolling around the harbour, stopping for a drink or a meal and gazing out at the bobbing boats.
Drive 40km west along the coast road.

6 Elafonisi

With glassy turquoise waters and swirls of pretty pale-pink sands kissing the shore, the beauty of Elafonisi's beach is undeniable. The shallow lagoon attracts families, while naturalists head for the dunes beyond the main beach. There are a couple of nearby eateries and hotels; stay overnight to avoid the hordes of daytrippers.
Drive back to Hania from where there are international flights from its airport.

ALTERNATIVES

Day trip: Samaria Gorge

The 16km-long Samaria Gorge hike through one of Europe's longest canyons is on the must-do list for many visitors to Crete. It begins just south of Omalos at Xyloskalo and ends in the coastal village of Agia Roumeli, taking four to six hours to complete.
Early morning buses depart Hania for Xyloskalo (1hr). From Agia Roumeli, catch the ferry to Sougia or Hora Sfakion (40min) then a bus back to Hania (1hr 30min-2hr).

Diversion: Vaï Beach

A little slice of the Caribbean, Vaï Beach is a worthy detour to the Lasithi province on the eastern side of Crete for its shimmering water and soft sand backed by a large grove of swaying Cretan date palms. Its popularity means it does get crowded, however.
It's a 150km drive from Iraklio east to Vaï Beach.

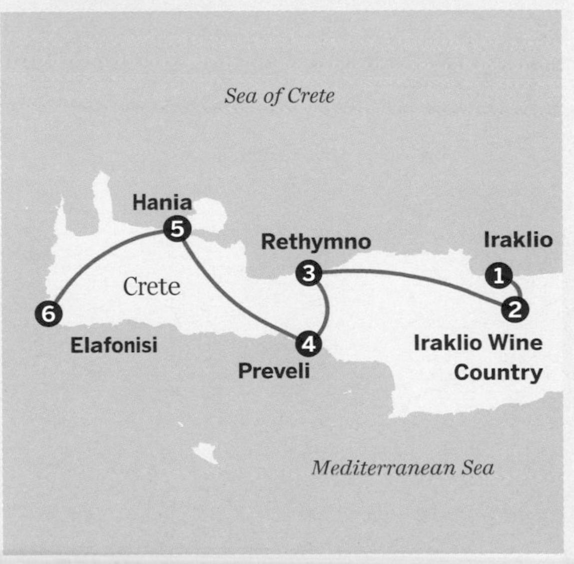

GRANADA AND THE ALPUJARRAS BY BUS

Granada – Eastern Alpujarras

Unravel Granada's Moorish riches, then escape by bus and on foot into Spain's rugged, trail-threaded Alpujarras mountains.

FACT BOX

Carbon (kg per person): 4.6
Distance (km): 170
Nights: 9–12
Budget: $$
When: Mar-Oct

❶ Granada

Overlooked by the splendid Alhambra palace-fortress, Granada is one of Spain's most beguiling cities. There are historical treasures around every corner, from the Alhambra's intricately tiled Palacios Nazaríes (book ahead) to the Albayzín's hidden *cármenes* (walled-in homes with gardens). Granada also has terrific tapas bars, outstanding street art, a thriving flamenco scene and some highly original places to stay.

ABOVE: ALHAMBRA, GRANADA

Take the bus to Capileira (3 daily; 2-3hr).

❷ Capileira

The highest of the three Poqueira Valley villages, Capileira showcases the Alpujarras' unique beauty: tightly built whitewashed houses, flat Berber-style roofs, narrow sloping alleys. Stay a couple of days, hiking to the villages of Bubión and Pampaneira and enjoying rich *alpujarreño* meals, traditional crafts and mountain panoramas.

From June to November, shuttle buses link Capileira with the Mirador de Trevélez lookout, from where it's a 5.1km hike to Mulhacén's summit (reserve ahead; reservatuvisita.es).

❸ Mulhacén

Rising to 3479m, Mulhacén, mainland Spain's tallest peak, is an irresistible adventure. You can be up and back in a day, or stay overnight at the 2500m-high Refugio Poqueira.

Take the bus (2 daily; 5-20min) or hike (6km) from Capileira to Pitres, La Tahá municipality's main village.

❹ La Tahá

This extraordinarily scenic valley sits just east of the

ABOVE: TRÉVELEZ

Poqueira area, dotted with tiny, tranquil villages, age-old walking paths and charming rural hotels for a few nights' stay (especially in Ferreirola).

Take the bus from Pitres to Trevélez (2-3 daily; 15-30min).

5 Trevélez

Some of Spain's most prized cured ham is produced in steeply stacked Trevélez, also the country's second-highest village at 1476m. Trevélez has some wonderful hiking trails, including a demanding route up Mulhacén (24km, 10-12hr round-trip).

Take the bus to Bérchules (2 daily, 1 continues to Yegen and Válor; 35min).

6 Eastern Alpujarras

The remote villages sprinkled near Ugíjar are some of the Alpujarras' most captivating (yet least visited): ancient churches, white-walled homes, wide-open mountain views. Follow in the footsteps of 20th-century British writer Gerald Brenan, who lived in sleepy Yegen, or explore historical Válor, peaceful Mairena (a great base), Cádiar and its wineries, and scenic Bérchules.

Jump on the bus from Válor back to Granada (2 daily; 3-4hr 15min), via Yegen and Bérchules.

ALTERNATIVES

Extension: Costa Tropical

Granada's 80km 'tropical' coastline is a delight: jagged cliffs tumble into turquoise bays and aged castles loom above buzzing seaside towns. Explore the hilltop old towns of Salobreña and Almuñécar (both with Moorish castles), La Herradura's water-sports scene, and lazy seafood lunches by the Med, perhaps at clothing-optional Cantarriján beach.

La Herradura, Almuñécar and Salobreña have good bus connections with Granada.

Extension: Cabo de Gata

Some of Andalucía's most beautiful, untouched beaches line the cliff-edged, desert-like coastline between Retamar and Agua Amarga, southeast of Almería. This is the protected Parque Natural Cabo de Gata-Níjar, where watersports, hidden coves, coastal walks and isolated villages abound.

Hire a car to reach and explore Cabo de Gata, 200km southeast of Granada.

ISLAND-HOPPING IN PORTUGAL'S AZORES

São Miguel – São Jorge – São Miguel
Island-hop by plane and ferry across an untouched volcanic archipelago where nature reigns supreme. Sample super-fresh fish and excellent wines along the way.

FACT BOX

Carbon (kg per person): 196
Distance (km): 1247
Nights: 14
Budget: €€
When: Apr-Oct

❶ São Miguel

São Miguel island is the Azores' main entry point, and while you'll return at the end of the trip, it's worth taking a night here at the start to rest post-flight. Ponta Delgada is the Azores' largest city, where you'll find excellent restaurants serving fresh fish and sushi.
Take one of the daily direct flights (80min) from Ponta Delgada to Flores.

ABOVE: PONTA DELGADA, SÃO MIGUEL

❷ Flores

Since there can be hiccups throughout this journey – the unpredictable weather can impact flights and ferries – it's best to start at the most remote point, Flores Island (the westernmost point of Europe) and work your way back east. The reward for making the effort is stunning, emerald-green nature, almost all to yourself – only about 4000 people live on the island. It's a Unesco Biosphere Reserve with pristine seas, deep valleys, mountain peaks, waterfalls and crater lakes to explore.
Regular flights (1hr) from Flores to Terceira run spring through autumn.

❸ Terceira

Terceira is the Azores' party island, with frequent festivals and events. It's a good place to tap into Azores culture, including the Unesco World Heritage-listed city of Angra do Heroísmo, the colourful *impérios* chapels and some of the local wineries. There are a number of good places to eat fresh tuna around the coast.
Daily direct flights (35min) connect Terceira to Pico year-round.

❹ Pico

The peak on Pico is the highest in Portugal (2351m to be exact) and many people visit the island in order to climb

ABOVE: FLORES

it – allow a full day, or longer if you want to sleep in the crater. After the hike, visit the island's many wineries, which produce unusual and delicious volcanic wines.

Daily ferries (30min) from Pico (Madalena or São Roque) head to Velas on São Jorge.

5 São Jorge

Gorgeous scenery, top-notch hiking and a feeling of remoteness are the main reasons to visit São Jorge, a long, skinny island with little infrastructure. Getting around is part of the adventure, as the roads are often quite steep and curvy.

Year-round daily flights link São Jorge to Ponta Delgada on São Miguel (2h 25min) with a stop in Terceira.

6 São Miguel

After the other islands, Ponta Delgada on São Miguel feels like a bustling metropolis. There are great restaurants and lovely hotels for unwinding at the end of an adventurous trip. There's also plenty to explore here, including crater lakes, waterfalls and a very warm swimming spot, heated by a hot spring in the Atlantic Ocean.

Flights connect Ponta Delgada with numerous European and North American destinations.

ALTERNATIVES

Extension: Corvo

In case Flores doesn't give you your fill of playing castaway on a remote island, Corvo is even more far-flung. The smallest and northernmost island in the Azores archipelago, it has fewer than 500 residents and nothing to stand in the way of your communion with nature.

Weekly ferry crossings (40min) sail from Santa Cruz on Flores to Vila Nova do Corvo.

Extension: Horta, Faial

Faial island doesn't have the same wow-factor as some of its neighbours, but sailing enthusiasts may want to check out the scene in its largest city, Horta – a prime stop on transatlantic crossings – where you're likely to find an international assortment of sailors.

Daily ferries connect Madalena on Pico to Horta on Faial.

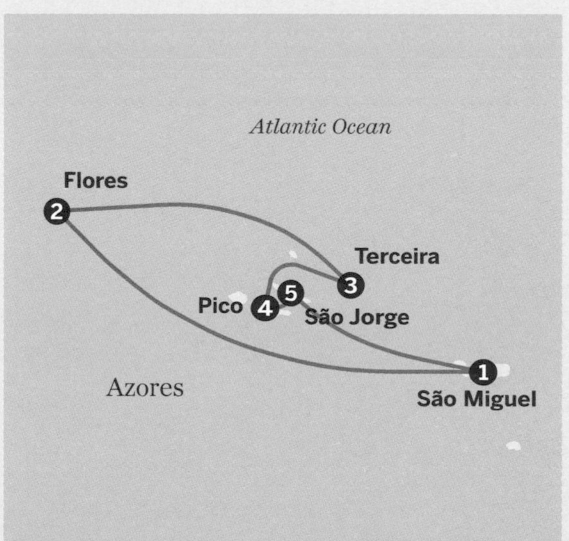

THE FRENCH AND ITALIAN RIVIERAS

Nice – Cinque Terre

Travel by rail through the charming coastal towns lining the Mediterranean's most glamorous shores, starting in France and ending in Italy.

FACT BOX

Carbon (kg per person): 13
Distance (km): 350
Nights: 10
Budget: $$$
When: May-Oct

❶ Nice, France

With characteristic Gallic swagger, Nice sprawls around a crescent-shaped bay with Provençal foothills as a backdrop. Follow in the footsteps of artists, aristocrats and aesthetes with a stroll on the Promenade des Anglais, a seafront walkway passing nodding palms and Belle Époque hotels.

Take the train to Menton (every 15-30min; 30min).

ABOVE: VERNAZZA, CINQUE TERRE

❷ Menton, France

Menton is the last French town before you hit Italian territory – and it makes for a marvellous finale, with pastel-hued houses rising over azure waters. The town's excellent Jean Cocteau Museum houses ceramics, drawings and cinematic works by the celebrated polymath.

Jump back on the train and cross the border to San Remo (hourly; 1hr).

❸ San Remo, Italy

Famous for its Art Nouveau casino, San Remo is often referred to as Italy's answer to Monte Carlo. Its appeal goes deeper than roulette, however. Wander the cobblestone alleyways of La Pigna, the Old Town, or step inside the seafront Russian Orthodox church, built at the behest of a visiting Tsarina.

Take the train to Genoa (hourly; 2hr).

❹ Genoa, Italy

Equal parts gritty and grand, Genoa represents a rare urban interlude on the Italian Riviera. The city once counted among the most powerful on the Mediterranean – explore the lavish *palazzi* of the Via Garibaldi to get a sense of Genoa's rich maritime history.

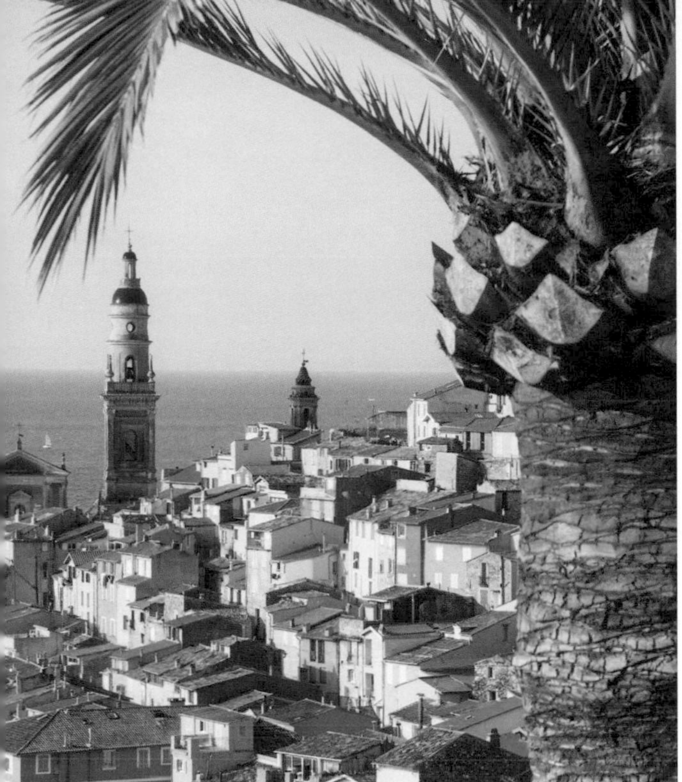
ABOVE: MENTON, FRANCE

Take a half-day sailing trip to Portofino – numerous ferry companies operate from Genoa harbour in summer.

5 Portofino, Italy

Cited as the most exclusive resort on the Mediterranean, Portofino hides on a wooded peninsula, a world away from the noise and sprawl of Genoa. Climb up to Castello Brown – a hilltop fortress commanding heavenly views over cypress trees, bell towers and moored yachts.

From Portofino you can reach the town of Santa Margherita Ligure on bus 82 (hourly; 20min) and then catch the train to Monterosso (hourly; 1hr).

6 Cinque Terre, Italy

The Cinque Terre are a quintet of fishing villages, poetically perched on near-vertical slopes. Undiscovered they are not – snatch moments of serenity walking the clifftop paths, such as the stretch of the Sentiero Azzurro between Monterosso and Corniglia.

From Corniglia, take the train to Pisa (hourly; 1hr 30min) whose airport has onward European connections.

ALTERNATIVES

Extension: Cannes, France

Cannes is known as Hollywood's little enclave on the Med: visit out of the festival season, and you won't need to be an A-lister to lounge on its sandy beaches or to promenade its quays. For a wilder taste of the Riviera, take the ferry to Île Sainte-Marguerite, a wooded offshore island with a fortress prison that once housed the Man in the Iron Mask.

Take the train to Cannes from Nice (every 15-30min; 30min).

Extension: Pisa, Italy

There's much more to Pisa than its wonky tower. Even so, expect vast queues on the Campo dei Miracoli (the 'field of miracles'), where the leaning structure rises over rolling green lawns and an ornate marble cathedral. Venture south into Pisa proper, and the crowds soon disperse – you'll find a likeable university town, with churches and galleries lining the banks of the Arno.

Hourly departures from Corniglia.

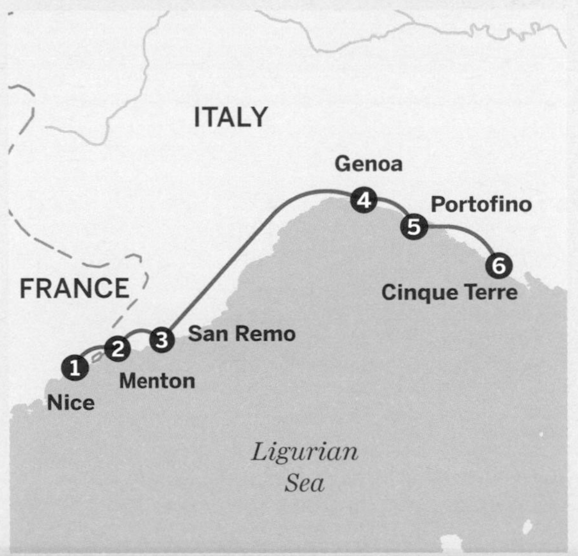

REV UP IN ITALY'S MOTOR VALLEY

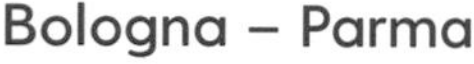

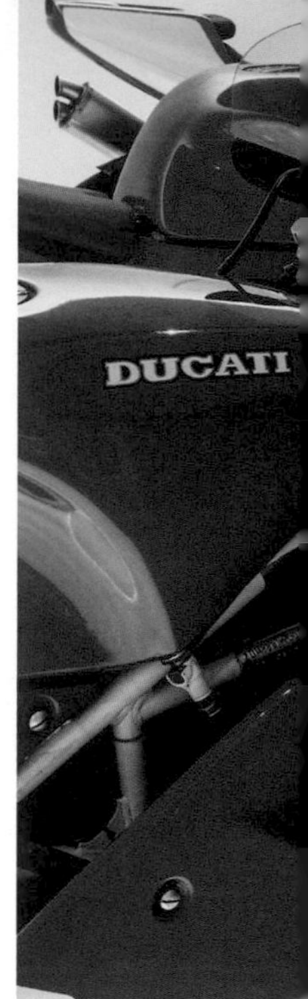

Bologna – Parma

Start your engine. Loosen your belt. Fast cars and food lure aficionados to Emilia-Romagna, a promised land of magnificent sports cars and gastronomic delights.

FACT BOX

Carbon (kg per person): 46kg
Distance (km): 340
Nights: 3
Budget: $$
When: Apr-Jun

1 Bologna

Emilia-Romagna's medieval capital is one of Italy's most wonderful places to eat – ragù, lasagna, tortellini and mortadella all hail from here (don't miss Trattoria da Me and All'Osteria Bottega). Between meals pay homage to exquisitely-designed motorcycles at Museo Ducati in Borgo Panigale, 6.5km northwest of downtown.

Take bus 97 from Bologna's Autostazione to Argelato (35min).

2 Museo Ferruccio Lamborghini

One of two Lamborghini museums in Motor Valley. Reserve ahead to peruse founder Ferruccio Lamborghini's personal collection of helicopters, tractors and legendary cars such as Miura SV and the Countach.

Return to Bologna on bus 97 and switch to bus 576 from the Autostazione to Sant'Agata Bolognese.

3 Mudetec

Lamborghini's Museum of Technologies (Mudetec) is the luxury sports car's principal museum and factory, chronicling history and innovation by decade with myriad cars on display. Book ahead for factory tours to see Lamborghinis custom-made by hand.

Return to Bologna on bus 576 and pick up a rental car for the 50km drive west on A1 to Maranello.

4 Maranello

Ferrari's world-class home is a pilgrimage-worthy essential for anyone with even a passing interest in high-end Italian sports cars. Factory tours are off limits if you don't own a Ferrari, but there's more than enough here to satisfy your need for speed.

Drive 24km northeast on SP16 to San Cesario sul Panaro.

ABOVE: MODENA

ABOVE: MUSEO DUCATI, BOLOGNA

5 Museo Horacio Pagani

There's no closer peek into the makings of outrageously expensive sports cars than a factory tour at Pagani, where just 40 hypercars per year are manufactured by hand for a cool €1.3-2.5 million each. Compared to Ferrari and Lamborghini, access here is astonishing. Book ahead through Modenatur (modenatur.it).

Drive 14km northwest on SS9 to Modena.

6 Modena

Another regional culinary capital, Modena is famous for Osteria Francescana, named the world's top restaurant. More modest meals at Ristorante da Danilo and Trattoria Ermes astound as well. Museo Enzo Ferrari and Maserati (book with Modenatur) are reachable on foot.

Continue 60km northwest on A1 to Parma.

7 Parma

Welcome to the home of the world's most famous cheese (Parmigiano-Reggiano, aka Parmesan) and ham (Prosciutto di Parma). Between indulging, see Parma's wonderful baptistery and Leonardo da Vinci's enthralling *La Scapigliata* inside Galleria Nazionale.

Head back to Bologna, 98km southeast, for international airport and train connections.

ALTERNATIVES

Extension: MAICC

Formula 1 fans can keep the engines running in Imola, 45km southeast of Bologna, home to Emilia-Romagna's F1 circuit. Brazilian racing legend Ayrton Senna was killed here in a 1994 accident – famed Brazilian street artist Eduardo Kobra's tribute to Senna adorns the facade of the track's hi-tech, multimedia MAICC (Museo Multimediale Autodromo di Imola Checco Costa).

Imola is served by trains from Bologna Centrale station (every 30min; 20min).

Diversion: Trattoria da Amerigo

No Emilia-Romagna restaurant better serves the region's outlandishly satisfying specialities than Michelin-starred Trattoria da Amerigo in Savigno, 33km southwest of Bologna. The late (and legendary) pasta maker Nonna Giuliana Vespucci's spirit is honoured by the top-quality tasting menus, often incorporating Savigno's coveted white truffles. Book a bed, too – you'll need it.

Savigno is most easily reached by car.

A TRIP THROUGH THE DODECANESE

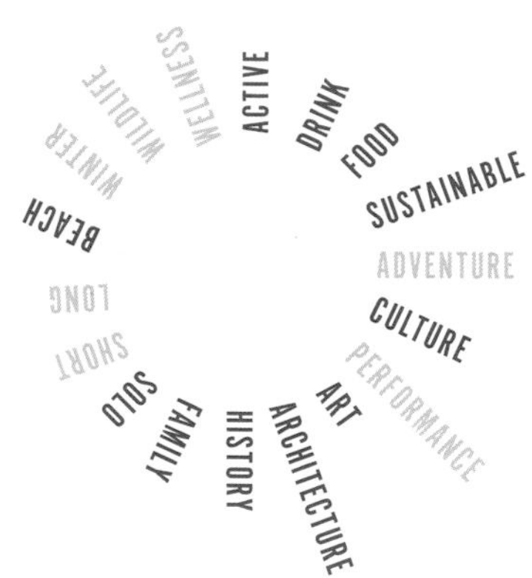

Leros – Kastellorizo

Greece's Dodecanese islands have it all, from hiking and birdwatching to chapels and shipwrecks, with each island's local dishes fuelling your explorations.

FACT BOX

Carbon (kg per person): 37
Distance (km): 330
Nights: 14
Budget: $$-$$$
When: May-Sep

❶ Leros

Food enthusiasts should head straight to Fish Restaurant Mylos on Leros, one of Greece's best seafood eating experiences. Don't miss the extraordinary village of Lakki too, one of the best-preserved Art Deco villages in the world. It was constructed as an Italian naval base by Mussolini.

Catch one of several daily ferries to Kalymnos (1-2hr).

ABOVE: RHODES TOWN, RHODES

❷ Kalymnos

Visitors to Kalymnos come for its excellent rock climbing, atmospheric low-key capital Pothia, and nature tours over herb-strewn peaks.

Jump on a ferry to Kos (daily; 30min).

❸ Kos

If you're a beachgoer, this is your 'happy (touristy) place': a great choice of sandy shores ring the island. To lose the crowds, head inland into the rugged Dikeos mountains to explore the lovely villages of Agios Dimitrios, Zia, Lagoudi and Pyli.

Several ferries link Kos with Nysiros weekly (1hr).

❹ Nysiros

This understated island sits on a volcanic faultline and the magnet for adventurers is the crater itself. But be aware – the fumaroles are still active. Nysiros is the place to sample top seafood too, served at any of the island's restaurants: Geusea Restaurant is a good place to start.

Take the ferry to Tilos (around 5 weekly; 1hr).

❺ Tilos

Head to tiny Tilos for peace, quiet, beaches, summer wildflowers – and plenty of walking. Around 54km of

ABOVE: KOS

scenic hiking trails cover the rocky mountainscape. A wonderful, if challenging, walk heads to and from the Agios Panteleimon monastery that's perched right on a cliff-edge.

Take the ferry to Rhodes (several a week; 3hr).

6 Rhodes

The largest of the Dodecanese is home to medieval history, beaches and tourist buzz. Take time to wander through Rhodes Town, an extraordinary mix of Byzantine, Turkish and Italian architecture, and be awed by the Archaeological Museum and the Palace of the Grand Master.

Given the limited ferries, the easiest way to get to Kastellorizo is by air (several flights a week; 40min).

7 Kastellorizo

This speck-in-the-ocean of an island is merely 2km from Turkey, yet it's the hardest of the Dodecanese to get to. So why make the effort? To do very little except revel in the beauty of the main waterfront lined with colourful mansions; to eat, drink, swim in the harbour – and repeat.

Fly to Rhodes or Athens (around three flights weekly).

ALTERNATIVES

Extension: Astypalea

Tiny, butterfly-shaped Astypalea island is known for its jaw-dropping villages, seafood taverns and Venetian *kastro* (castle). Find yourself a remote strip of sand, wander in the meadows and get lost in the maze of Pera Gialos' historic quarter.

You can fly to Astypalea from Rhodes, Leros, Kalymnos or Kos.

Extension: Symi

The highlight of this compact island is getting lost in the lanes lined with Italianate mansions that tumble down the cliffs to the harbour of Gialos, the main town. Be sure to journey across the pine-forested mountain range to visit the island's south, and wander through the historic Taxiarhou Mihail Panormiti monastery.

Ferries link Symi to Rhodes (daily; 50min).

LAKESIDE RAMBLE, ITALY

Milan – Borromean Islands

Begin this trip in Milan and then hop on trains and boats to lap up the enduring, beguiling beauty of the Italian Lakes and their shoreside towns.

FACT BOX

Carbon (kg per person): 16
Distance (km): 280
Nights: 7
Budget: $$$
When: Apr–Oct

❶ Milan

Art, fashion, food: ancient Milan wears many crowns. Italy's second-largest city is a monument to human achievement, from its jaw-dropping neo-Gothic Duomo (cathedral) to the da Vinci-designed Castello Sforzesco. After exploring Renaissance masterpieces, you can indulge in more 21st-century pursuits like window shopping in the glittering Galleria Vittorio Emanuele II, or sipping an *aperitivo* at a terrace cafe (try Terrazza Aperol).

Take the train from Stazione Centrale to Como (hourly; 1hr).

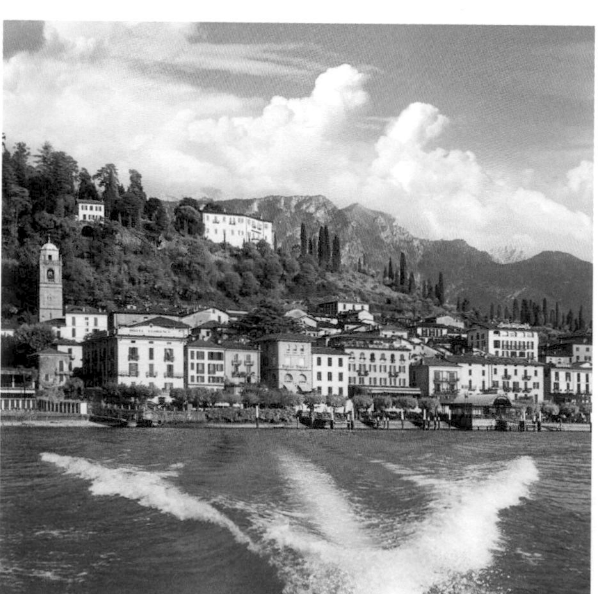

ABOVE: LAKE COMO

❷ Lake Como

Framed by steep, verdant hillsides, the elegant town of Como provides a suitable introduction to its eponymous lake's many charms – from waterfront strolls past grand villas to dining al fresco on a sun-dappled piazza. For a storybook view, take the funicular up to Brunate where mountains and water unfold before you.

Navigazione Lago di Como ferries run several times a day to Bellagio (45min).

❸ Bellagio

On a promontory jutting into the centre of Lake Como, Bellagio is a tranquil village of green-shuttered villas, rhododendron-filled gardens and elegant walkways along the shoreline. Days are spent exploring historic estates, like the Villa Serbelloni with its lavish gardens, followed by dinner and drinks at a waterfront restaurant.

ABOVE: BELLAGIO'S VILLA SERBELLONI, LAKE COMO

Take the ferry back to Como (45min), then a train via Milan to Stresa (2hr 10min).

4 Stresa

A Belle-Époque beauty on Lake Maggiore's west side, Stresa has a pleasant cobblestone centre and sun-dappled squares (like Piazza Marconi) perfect for taking in the lake's mountainous backdrop. Even better views await on Monte Mottarone. A 20-minute cable-car ride whisks visitors to its 1492m summit, where jagged mountains and distant lakes fill the horizon.

Frequent ferries and taxi boats run between Stresa and the three Borromean Islands (20min).

5 Borromean Islands

Owned by the noble Borromeo family since the 17th century, these three picturesque islands in Lake Maggiore are an easy day trip from Stresa. Highlights include Isola Madre's botanical gardens – among the oldest in Italy – and the art-filled Palazzo Borromeo on Isola Bella. Isola Superiore, aka Fisherman's Island, has fewer sights but is famed for its seafood restaurants.

Catch the ferry back to Stresa and hop on the train back to Milan (hourly; 1hr).

ALTERNATIVES

Extension: Orta San Giulio

Surrounded by thick woodlands, the lakeside village of Orta San Giulio has lovely architecture and narrow lanes, and is a short hop from idyllic San Giulio island. Just inland, a cobbled walk leads up to the Sacro Monte di San Francesco for lake views amid a series of Baroque, fresco-covered chapels.

SAF runs three daily buses from Stresa to Orta San Giulio in the summer (50min).

Diversion: Cannobio

Just south of the Swiss border, Cannobio has a delightfully quaint setting, complete with pastel-hued houses and a pedestrianised shore, making it a fine base for exploring the less trafficked end of Lake Maggiore. The public beach is ideal on warm days, and you can head off on great walks into the countryside, including a 3km stroll that follows the scenic Cannobino River.

Several daily ferries go from Stresa to Cannobio (2hr).

EXPLORING ITALY'S HEEL

Bari – Matera

Explore quieter backroads around the southern tip of Italy's heel to discover Puglia's compelling blend of history, art, culture and cuisine.

FACT BOX

Carbon (kg per person): 92
Distance (km): 507
Nights: 8-12
Budget: $$$
When: Apr-Jun

❶ Bari

The authenticity of a working fishing port infuses the capital of Puglia. The old town of Bari Vecchia includes Bari's 13th-century Romanesque cathedral, and the Basilica di San Nicola, containing the bones of the man who inspired the story of Santa Claus. Rent a bike to journey to the oceanfront bastion housing the city's archaeological museum before a lunch of fresh squid and oysters at the harbourfront seafood market.

Rent a Fiat 500 (a classic car for a classic road trip) and drive 151km from Bari to Lecce.

ABOVE: BASILICA DI SANTA CROCE, LECCE

❷ Lecce

Crafted in honey-coloured sandstone, Lecce's ornate and ostentatious churches, palaces and piazzas define the city's signature architectural style, *barocco leccese* (Lecce Baroque). Explore the beautiful Old Town, including the wildly flamboyant Basilica di Santa Croce, before pairing the authentic flavours of Puglia's *cucina povera* (poor person's cooking) with earthy wines from the southern Salento region.

Continue 50km along the quieter coastal road hugging the Adriatic Sea.

❸ Otranto

Otranto's 11th-century Norman cathedral features a stupendous 12th-century mosaic of the 'Tree of Life', covering the church's entire floor. The phantasmagorical design incorporates exotic flora and fauna, all balanced on two stylised elephants. And don't miss a guided tour to explore the storied underground vaults of the Castello Aragonese.

Cross the heel of Italy's boot and drive 54km to Gallipoli.

ABOVE: MATERA

❹ Gallipoli

On a promontory jutting into the Ionian Sea, white-washed Gallipoli ('beautiful city' from its Greek origins), is one of the Salento region's prettiest coastal towns. Stroll the oceanfront drive, and enjoy the town's famed red prawns served fresh from Gallipoli's fishing boats.

Continue 96km to Taranto via Manduria.

❺ Taranto

Taranto's past as the once-mighty Greek-Spartan colony of Taras is showcased at the prestigious Museo Archeologico Nazionale di Taranto (MArTA); displays include the world's most diverse collection of Greek terracotta art. Cross the bridge to Taranto's Città Vecchia and explore the leviathan Castello Aragonese.

Drive 90km into Basilicata and continue to Matera.

❻ Matera

One of the longest continuously inhabited settlements in the world, Matera's network of ancient *sassi* (cave dwellings) cut into the region's soft tufa rock is an urban landscape like no other. Crowned by the *duomo* (cathedral), the two labyrinthine neighbourhoods of Sasso Barisano and Sasso Caveoso are a fascinating destination recalling the confounding art of Escher.

Return 66km northeast to Bari.

ALTERNATIVES

Diversion: Capo Santa Maria di Leuca

Navigate Salento's southeastern edge to the remote tip of Italy's heel. Winds dancing across the Ionian Sea enliven cobalt waters, while stops en route include boat trips to the karst limestone formations of the Grotta Zinzulusa. Santa Maria di Leuca's soaring basilica and lighthouse punctuate the end of the road south.

Journey 53km down coastal roads from Otranto to Capo Santa Maria di Leuca.

Day trip: Ostuni

Along with Locorotondo and Martina Franca, Ostuni is one of Puglia's famed whitewashed hill towns. From the train station, take an Ape Calessino (motorcycle three-wheeler) tour through the narrow streets and lanes. Shop for local olive oil in artisan food shops, admire the 15th-century cathedral, and book ahead for a leisurely lunch at Osteria Ricanatti.

Frequent Trenitalia departures link Lecce's main station to Ostuni (1hr).

FINE FOOD AND ARCHITECTURE IN GALICIA

Santiago de Compostela – Ourense

Make a pilgrimage to Spain's gorgeously green northwest corner, with its beautiful architecture, dramatic coastline and tasty food.

FACT BOX

Carbon (kg per person): 95
Distance (km): 764
Nights: 14
Budget: $$
When: May-Sep

❶ Santiago de Compostela

As the goal of the Camino de Santiago pilgrimage path, the stone-paved streets of Galicia's enthralling capital have felt the tread of millions for over a millennium. Start exploring at the splendid cathedral and the lively Mercado de Abastos food market, then move on to graceful plazas, heaving tapas bars and excellent museums.

Drive southwest for 66km.

ABOVE: PONTEVEDRA

❷ Cambados, Rías Baixas

Welcome to the cheery waterside 'capital' of Albariño, Galicia's deliciously fruity signature white wine, produced near the calm Rías Baixas coast. Feast on terrific seafood, explore elegant *pazos* (country mansions) and visit outstanding wineries such as Gil Armada.

Continue southeast for 28km.

❸ Pontevedra

The charming provincial capital is perfect for an overnight stop, with a pedestrianised Old Town packed with slender alleys, plazas and tempting tapas spots.

Head 135km northwest.

❹ Costa da Morte

Base yourself in Fisterra, Muxía or Lires/Nemiña for a few nights on the remote, eerily beautiful Costa da Morte (Death Coast). Among the area's wild beaches, spectacular capes, stone-built fishing villages and isolated lighthouses, don't miss dramatic Cabo Fisterra, or walking part of the 200km Camiño dos Faros.

Drive 180km northeast to Cedeira.

❺ Rías Altas

The Rías Altas are a wild world of plunging cliffs, powder-soft beaches, horizon-reaching views and

ABOVE: SANTIAGO DE COMPOSTELA CATHEDRAL

Atlantic surf. Spend a day or two in laidback fishing port Cedeira, visiting rugged Cabo Ortegal, the clifftop Garita de Herbeira lookout, the awe-inspiring Bares Peninsula and Loiba's fabulously scenic 'Best Bench in the World'.
Drive 100km south.

6 Lugo

The brilliantly preserved Roman walls are low-key Lugo's Unesco-listed draw. The city also offers up a twisting-and-turning Old Town and some of Galicia's finest tapas.
Head south for 115km to Parada de Sil, a handy Ribeira Sacra base.

7 Ribeira Sacra

Hugging the Sil and Miño river canyons, this region's precipitous vineyards yield superb, up-and-coming wines. Narrow roads and wonderful hiking and cycling trails weave past ancient wineries, untouched woodlands, spine-tingling lookouts and medieval monasteries.
Drive 40km west.

8 Ourense

An overnight stop in ancient Ourense means unbeatable tapas, a maze-like old core and blissful soaks in the Miño-side thermal baths.
Drive back 100km to Santiago.

ALTERNATIVES

Day trip: Illas Cíes

Some of Galicia's dreamiest beaches are hidden away in the dramatically beautiful Parque Nacional de las Islas Atlánticas de Galicia. The three traffic-free Illas Cíes here make a wonderful day trip from the Cambados area.
Frequent boats (€16-19 return; 45min) depart from Vigo during Easter and from at least June to mid-September; boats also go from Baiona, Bueu and Cangas.

Extension: Picos de Europa

The soul-stirringly spectacular Picos de Europa mountains tempt with sparkling lakes, thrilling cable car rides, pungent cave-matured cheeses and adventure activities from river kayaking to high-altitude hikes.
From Cedeira or Lugo, it's a 300km drive east to Cangas de Onís in the foothills of the Asturian Picos, and a further 80km to Potes for the Cantabrian Picos.

DISCOVER THE BEST OF MOORISH SPAIN IN ANDALUCÍA

Málaga – Salobreña

Road-trip through 1300 years of Islamic heritage – and a few stellar museums – in the region that conquering North African Moors called 'al-Andulus'.

FACT BOX

Carbon (kg per person): 65
Distance (km): 521
Nights: 10-14
Budget: $$
When: year-round

❶ Málaga

The unprepossessing capital of the Costa del Sol might have just one fortress – the Alcazaba – which bears the signature architectural motifs of the conquering Moors, but its Picasso Museum, a Pompidou outpost, the wonderful Carmen Thyssen Museum and a quirky museum of glassware still make it well worth a few days. Shady botanical gardens, elegant tapas bars and seafood-filled *chiringuitos* (bars) on broad beaches offer lots of post-culture R&R.

Drive north for 161km.

❷ Córdoba

The candy-stripe arches of Córdoba's stand-out mosque-turned-cathedral, La Mezquita, are just one lovely feature in the city's mix of Islamic and Christian architecture – climb the 54m-high bell tower, once a minaret, for great views of this former Moorish capital. The well-preserved Old Town is a Unesco World Heritage Site: the Alcázar de los Reyes Cristianos, on the banks of the Guadalquivir River, is like a mini Alhambra, and the Roman Puente Romano is one of the most beautiful bridges in Spain.

Continue east for 108km.

❸ Jaén

Strategically placed Jaén makes an intriguing stop between Córdoba to its west and Granada to its south. The 10th-century mountain-top Castillo Santa Catalina is the obvious draw, while in the hillside neighbourhood of La Magdalena, once the

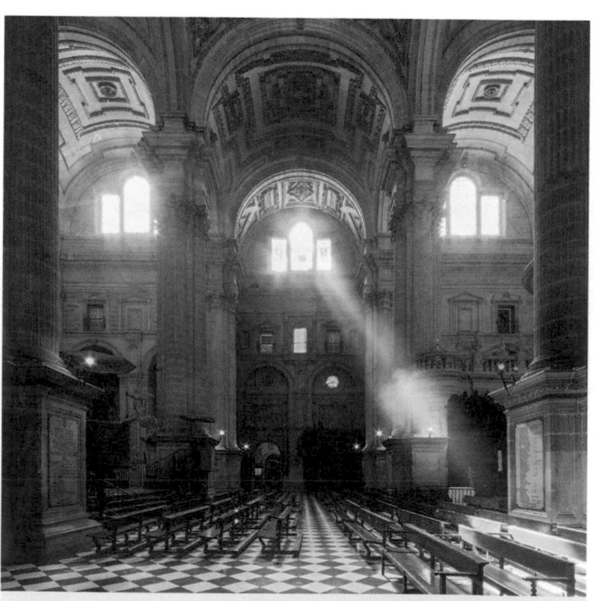

ABOVE: CATEDRAL DE LA ASUNCIÓN, JAÉN

ABOVE: LA MEZQUITA, CÓRDOBA

Moorish town, Spain's largest hammam is an 11th-century treasure concealed beneath the elegantly understated Renaissance Palacio de Villardompardo.

Head south for 94km.

4 Granada

Granada is without doubt the highlight of Moorish Andalucía. The Alhambra complex, built as a fortress in 889 CE and converted into a Sultan's palace in 1333, is as impressive as you would expect the most popular monument in Spain to be. But don't miss the winding hilltop Arab quarter of the Albayzín, where unexpected viewpoints (*miradores*) of the town below and snowy Sierra Nevada range above pop up between the cobbled streets like mirages.

Continue south for 68km.

5 Salobreña

The distinctive rock outcrop of this alluring *pueblo blanco* is visible for miles around, and with habitation stretching back 6000 years, it's a picture-perfect example of those Andalucían white towns that are a delight to just meander around. This one offers a 10th-century Moorish castle, a few Roman remains and a number of lovely little churches.

Drive 90km west along the coast road to Málaga.

ALTERNATIVES

Extension: Seville

In its monumental Gothic cathedral, located on the site of the 12th-century Almohad mosque, Seville boasts the biggest church in the world – officially. It has a certificate from the Guinness Book of Records to prove it. For a bird's-eye view head to the top of the Giralda tower and explore in detail the clever construction and elegant harmony of one of the world's loveliest minarets.

Seville is 200km northwest of Málaga.

Extension: Ronda

The dramatic setting above the Tajo de Ronda gorge makes the mountaintop eyrie of Ronda look like something out of *Game of Thrones*, right down to the three stunning bridges linking the city's new (15th-century) and old (Moorish) towns. The latter features a wealth of attractions, including Santa María la Mayor, the 13th-century San Sebastián minaret, and the Arab baths on the banks of the river.

Ronda is 100km west of Málaga.

FROM EUROPE TO ASIA BY RAIL

Athens – İstanbul

This train trip of a lifetime journeys between four great cities – Athens, Thessaloniki, Sofia and İstanbul – and two continents.

FACT BOX

Carbon (kg per person): 27
Distance (km): 1345
Nights: 5-6
Budget: €€
When: year-round

❶ Athens, Greece

The Greek capital boasts more claims to fame than most cities, being the birthplace of democracy, theatre, philosophy and much more. The marble columns of the famed Parthenon are visible across Athens, but it and the other structures of the Acropolis are best admired with a site visit – don't miss the Temple of Athena Nike at the top and the views across the city. Other essential stops include the ancient Agora, the splendid Acropolis Museum and the picturesque Plaka neighbourhood.

Intercity trains depart Larissa station heading to Thessaloniki (2 daily; 4hr 30min), passing Mt Olympus, home of the mythological Greek gods, on the way, and following the Aegean coast for the final stretch.

❷ Thessaloniki, Greece

Greece's second city sports Byzantine walls and churches, as well as a waterfront precinct packed with bars and restaurants and overlooked by the 15th-century White Tower, the city's most famous sight. Culture buffs will want to visit the Archaeological Museum and Museum of Byzantine Culture, and everyone will want to sample the local cuisine, often lauded as the best in the country.

Take the train to Sofia (1 daily in the morning; 8hr 30min).

ABOVE: MONASTIRAKI SQUARE AND THE ACROPOLIS, ATHENS

ABOVE: THESSALONIKI

❸ Sofia, Bulgaria

Called Serdika back when it was a colony of ancient Rome, Bulgaria's capital is a laidback city in which onion-domed Orthodox churches sit alongside Roman remains, Ottoman mosques and Red Army monuments. Most visitors gravitate towards the grandiose Aleksander Nevski Cathedral, but the city's greatest treasure is the World Heritage-listed Boyana Church, a repository of exquisite medieval murals.

Take the overnight İstanbul-Sofya Expresi train to Halkalı in İstanbul (13hr).

❹ İstanbul, Turkey

Straddling two continents, this magnificent metropolis has had many guises and names throughout history, from Byzantium to Constantinople and now İstanbul. The result is a place layered with mythology, history and culture. Explore the Byzantine and Ottoman monuments (especially the outstanding 1500-year-old Aya Sofya), museums and bazaars of historical Sultanahmet; sample the vibrant eating and drinking scenes in the 19th-century streets of Beyoğlu; and climb aboard ferries sailing the length of the Bosphorus Strait or across the Sea of Marmara to Anatolia in Asia.

İstanbul's international airport has global onward connections.

ALTERNATIVES

Extension: İstanbul's Asian shore

Taking a ferry trip between Europe and Asia is one of İstanbul's quintessential experiences. Views of the Old City's skyline are spectacular, as are those of the medieval Galata Tower in Beyoğlu and the suburb of Üsküdar, crowned by the Çamlıca Mosque.

Frequent commuter ferries sail from the docks in (European) Eminönü, Karaköy and Beşiktaş to (Asian) Üsküdar and Kadıköy.

Extension: Along the Bosphorus

Linking the Sea of Marmara with the Black Sea, the Bosphorus Strait is İstanbul's greatest natural asset. One shore is in Europe, the other in Asia – both are lined with Ottoman-era palaces, mansions and mosques that can be admired from the deck of ferries and cruise boats.

A tour boat operated by İstanbul Şehir Hatları departs from the Eminönü dock at 10.35am daily (2hr each way).

NATURE-LOVER'S MADEIRA

Funchal – Porto da Cruz

Lush and vertiginous, this Portuguese island offers fabulous walking and nature; thrill-seekers will love its hair-raising landing, one of the world's most dangerous.

FACT BOX

Carbon (kg per person): 17
Distance (km): 134
Nights: 7–10
Budget: $$
When: year-round

❶ Funchal

Cobbled lanes, decorative arts in the Frederico de Freitas House Museum and a lush Botanical Garden – plus the and the lively Lavradores market, where the black scabbard fish looks like a proverbial monster from the deep – make for a mesmerising mix in the island's main city.

ABOVE: LAVRADORES MARKET, FUNCHAL

Take the cable car to Monte.

❷ Monte

The journey up to Monte's expansive Monte Palace Tropical Garden, in a cable car rising almost 600m, is unmissable. The return is even better, with two agile runners, *carreiros*, in tyre-soled shoes deftly steering a wicker-basket toboggan around the hairpin bends of slick-as-ice hills.

From Funchal, rent a car and drive 20km west.

❸ Ribeira Brava and the south coast

West from Funchal, banana plantations and sugar-cane fields lead to the epic cliffs of the south coast's Ribeira Brava. Stop at the vertigo-inducing glass-floor platform of Cabo Girão and, just before Ribeira Brava, the beachside Fajã dos Padres restaurant, reached via a cable car and a walk through tropical fruit orchards.

Continue west for 26km to the Levada das 25 Fontes.

❹ Levada walks

Madeira's ancient man-made water channels were put to good use when more modern irrigation systems replaced them: turned into walking trails

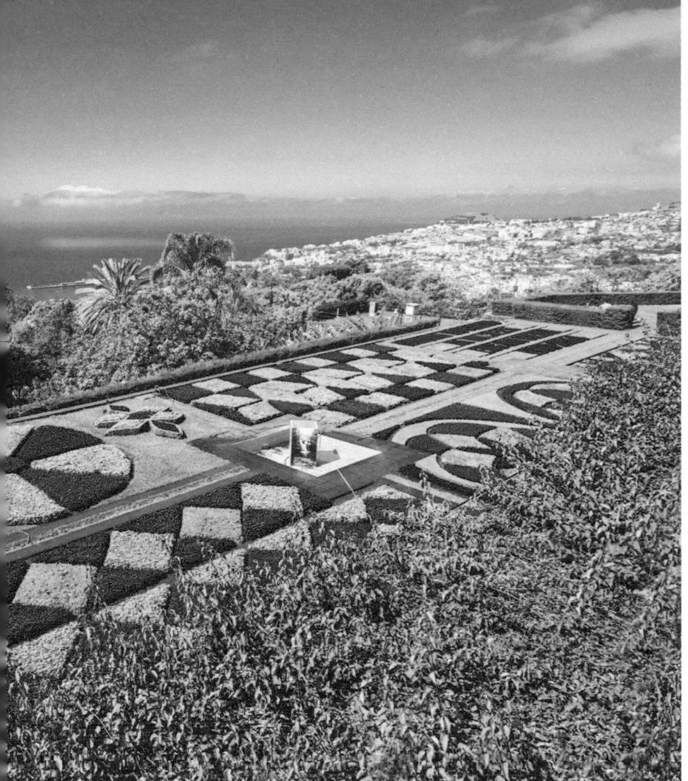

ABOVE: MADEIRA BOTANICAL GARDENS, FUNCHAL

across the whole interior. Hikes like the Levada do Alecrim and Levada das 25 Fontes offer forest paths to pools fed by waterfalls, while more strenuous routes lie in wait around the Pico Ruivo, at 1862m the island's highest peak.

From Levada das 25 Fontes, head north for 23km.

5 Porto Moniz to Sao Vicente

At the northwestern tip of Madeira, Porto Moniz has spectacular sights in store, notably the natural swimming pools formed by volcanic lava and the Bride's Veil waterfall. East along the untamed north coast lies São Vicente, starting point for walks to two quarries filled with five-million-year-old fossils.

From Sao Vicente, drive east for 51km.

6 Porto da Cruz

Pack your surfboard for a few hours at pretty Porto da Cruz, where the wonderful coastal scenery is matched by the island's best breakers and an impressive sandy beach at Banda d'Além, created by importing sand from Morocco. Sample local rum at the sugar-cane processing plant before heading to the airport or Funchal.

From Porto da Cruz, it's 14km to Madeira Airport.

ALTERNATIVES

Extension: Ribeiro Frio

From Funchal, a 48km-long road connecting the north and south coasts twists and winds its way though national parks. At the Ribeiro Frio, a short walk leads to the Vereda dos Balcões, with 880m high views over the surrounding laurel forest. It's one of Madeira's best lookout points; add in tame Madeiran chaffinches feeding from your hand and it's a highlight of any trip.

Ribeiro Frio is 17km north of Funchal.

Extension: Porto Santo island

Northeast of Madeira, Porto Santo offers a range of enjoyable peaks – chief being the Pico do Facho, at 516m – to be scaled for great Robinson Crusoe moments and ocean views. On the north coast, seek out underwater pillow lava formations at Zimbralinho's bay.

The ferry from Funchal to Porto Santo takes around 2hr (duration and frequency vary by season). Take your rental car, or use the island buses.

HIKING ACROSS THE TROÖDOS

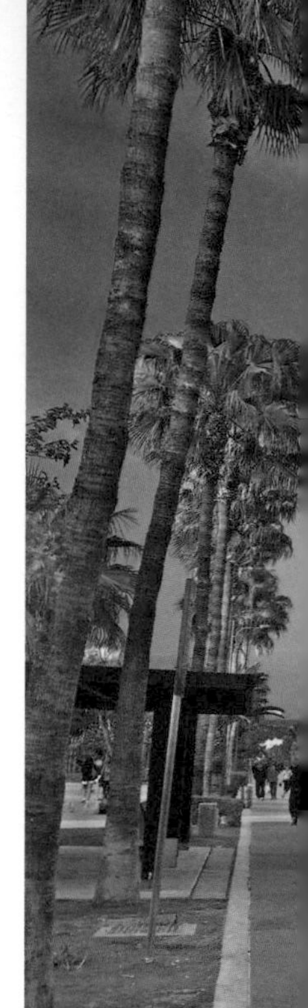

Lefkosia (Nicosia) – Lemesos (Limassol)
For the best of Cyprus, hike into the Troödos Mountains, where Byzantine monasteries and wine-producing medieval villages spill out of the landscape.

FACT BOX

Carbon (kg per person): 3.5
Distance (km): 140
Nights: 7
Budget: $$
When: Mar, Apr, Sep, Oct

❶ Lefkosia (Nicosia)

Start this mountain trip in Cyprus' atmospheric, divided capital, wandering backstreets dotted with venerable stone houses, Byzantine and Frankish churches and relics from the island's 1970s civil war. Pause for a slap-up Cypriot meal at Shiantris on Pericleous, then hit the bus stand early the next morning for the trip to Kakopetria.

Take the bus from Constanza Bastion bus stop to Kakopetria (hourly; 1hr 15min).

❷ Kakopetria

From the centre of this pretty village, hike 3km west to Agios Nikolaos tis Stegis, perhaps the loveliest of the Troödos' Unesco-listed Byzantine chapels. Inside the crudely-hewn stone church is a riot of 11th- to 14th-century murals. After the midday heat subsides, stroll north from Kakopetria to Galata, where the chapel of Panagia tis Podythou erupts with more mural colour from the 17th century.

Jump back on the bus to Kentriko Troödos (early morning; 30min).

❸ Kentriko Troödos

It's a short bus ride to the sprawling village of Kentriko Troödos, which serves as a hiking base in summer and, surprisingly, as a skiing centre in winter. Point your hiking boots towards the Artemis Trail, an invigorating 7km loop around 1952m Mt Olympus, Cyprus' highest peak, passing through pine-scented forests with swooping island views.

Walk 9km south to Pano Platres.

ABOVE: LEFKOSIA (NICOSIA)

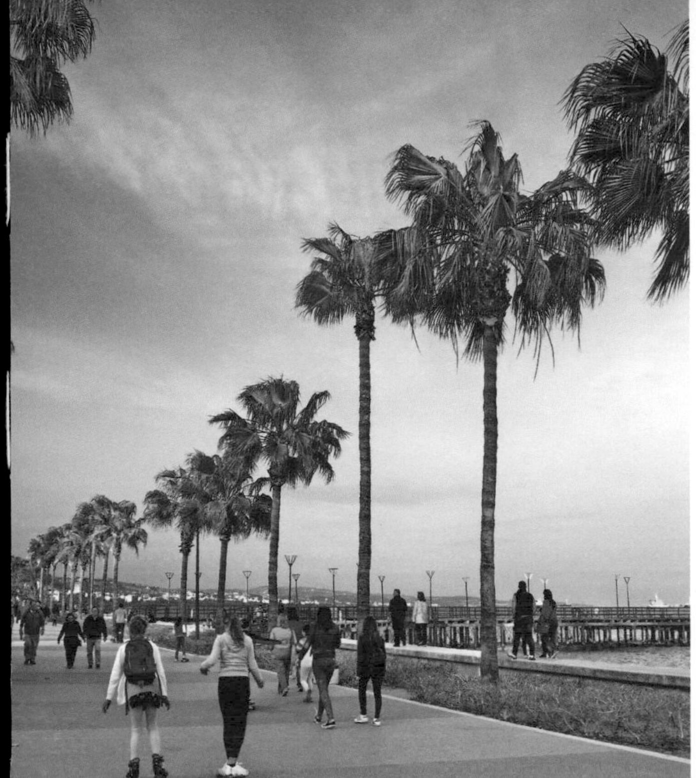

ABOVE: LEMESOS (LIMASSOL)

❹ Pano Platres

There's more serene forest walking around this British-developed hill station, with the added bonus of fine Cypriot country cooking; the *meze* spread at Mimi's Restaurant is a local legend. The charming villages of Foini, Perapedhi and Pouziaris are all worthy day-hike destinations.

Walk 8km southwest to Omodos.

❺ Omodos

The hike downhill to the village of Omodos generally follows quiet, sealed roads. The reward at the end is one of Cyprus' most perfect cobbled squares, plus the imposing Timiou Stavrou Monastery, and a string of small winemakers selling the sweet Commandaria wine that has been quaffed here since at least 800 BCE.

Take the bus to Lemesos (2-4 daily; 2hr).

❻ Lemesos (Limassol)

After the comparative cool of the hills, the coastal heat will hit you like a wall; cool off at one of the laidback beaches strung out west of Lemesos, accessible on the Paphos bus.

Lemesos (Limassol) is Cyprus' main flight hub, or buses zip back to Lefkosia (hourly; 1hr 15min).

ALTERNATIVES

Extension: Paphos

Why stop in Lemesos? The highway west provides easy access to a string of historic sites: the Crusader castle of Kolossi; the ruins of Kourion; the partly restored Sanctuary of Apollon Ylatis; and Petra tou Romiou, the rocky outcrop where Aphrodite reputedly emerged from the sea-foam.

Buses zip along the coast hourly, making it easy to visit Paphos, where beach parties rave within sight of more magnificent ruins.

Diversion: Pedoulas

With a few days to spare, extend the Mt Olympus walk from Troödos to Pedoulas, home to the church of Archangelos Michail, full of murals of sword-wielding angels from the 15th century, and a short walk from dainty Panagia tou Moutoulla, a 13th-century masterpiece adorned with stern-looking saints.

It's a 17km hike from Troödos to Pedoulas; to save backtracking, a daily bus runs on to Pano Platres.

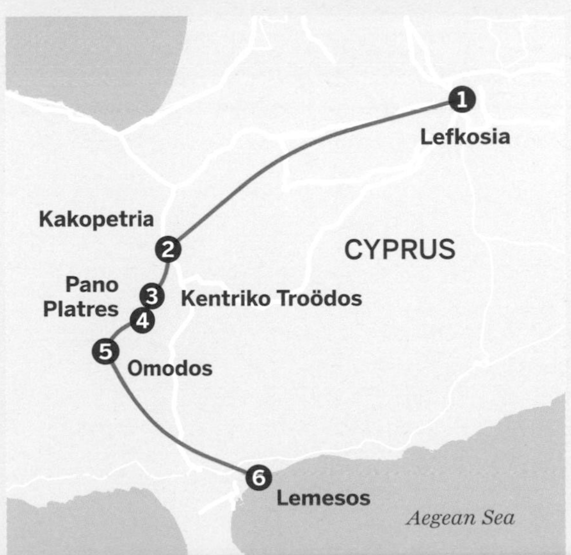

MALLORCA'S NORTH COAST

Pollença – Valldemossa

Escape Mallorca's resort crowds with this unforgettable road trip in the island's north, exploring the limestone heights of the Tramuntana range.

FACT BOX

Carbon (kg per person): 13
Distance (km): 103
Nights: 7
Budget: $$
When: Apr–Oct

❶ Pollença

Pollença, in the Tramuntana foothills 60km northeast of island capital Palma, is a honey-coloured hill town with a maze of alleys and cafe terraces in its Plaça Major to explore. For arresting views over the rooftops to the mountains beyond, climb the Calvari, 365 cypress-lined steps leading to an 18th-century chapel.

Drive 22km southwest, climbing steeply into the Tramuntana mountains.

ABOVE: TRAMUNTANA FOOTHILLS

❷ Lluc

For centuries, pilgrims have flocked to the monastery at Lluc for a glimpse of the 14th-century statue of the Virgin. The mountain setting is divine, as is the gilded church, with a design partially inspired by the work of Antoni Gaudí.

Continue 36km west, looking out for Mallorca's highest peak, 1445m Puig Major.

❸ Fornalutx

Fornalutx is a hill town to make your heart sing, with its jumble of shuttered, warm-stone houses and steep, flowery lanes rising above terraced citrus groves. Offering respite from the coastal crowds in summer, it's believed to have its origins as an Arab *alquería* (farming community).

Head 4km southwest.

❹ Sóller

Sóller is a vision of ochre-stone merchant houses,

ABOVE: DEIÀ

cupped in a lush valley of citrus groves and olive trees. Its beating heart is Plaça de la Constitució, with cafe terraces, vintage trams and the Baroque-meets-Modernist Església de Sant Bartomeu. The town's train station harbours galleries showing Picasso and Miró masterpieces.

Deià is an 11km drive west of Sóller.

5 Deià

Spilling down a cypress-stippled hillside with the sapphire-blue sea beyond, the village of Deià is astonishingly attractive. Once the beloved retreat of British poet Robert Graves, its landscapes still inspire towards the lyrical, with steep garden terraces dropping to citrus orchards and olive groves.

Drive 10km southwest to Valldemossa.

6 Valldemossa

Capped off by a Carthusian monastery and the high, forested peaks of the Tramuntana, Valldemossa is a wonder: it's a pleasure to stroll its cobbled alleys and fragrant gardens. Real Cartuja de Valldemossa monastery was a former home to kings, monks and 19th-century composer Frédéric Chopin and his lover, George Sand.

It's a 20km drive south to the capital, Palma.

ALTERNATIVES

Day trip: Cap de Formentor

From a distance, this cape's wind-buckled limestone peaks look like waves about to break. Buckle up for a white-knuckle ride along the peninsula, with hairpin bends revealing knife-edge cliffs and vertiginous sea views. Stop for a swim and picnic at Platja de Formentor's ribbon of pale, pine-fringed sand en route to the lighthouse-topped headland.

Formentor is a 23km drive northeast from Pollença.

Diversion: Sa Calobra

Mallorca's most hair-raising drive, the Ma-2141 twists 12km down to the sea in a series of loop-the-loops, carving its way precariously through a ravine to Sa Calobra. A training route for Tour de France pros, the road skirts breathtaking ridges as it unfurls to a white-pebble cove and the Torrent de Pareis, a rocky river gorge.

Turn off Ma10 coastal road just after Escorca, 7km west of Lluc.

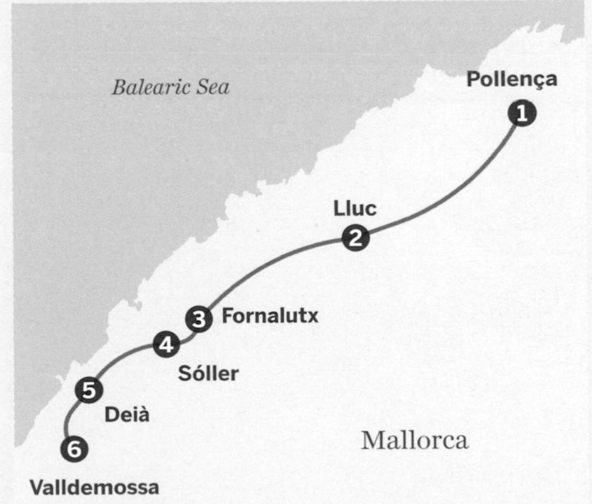

AN ADRIATIC ODYSSEY

Venice – Dubrovnik

Follow the Adriatic Sea south in the wake of the Venetian seafarers of old on a multi-country road trip through Italy and Croatia.

FACT BOX

Carbon (kg per person): 200
Distance (km): 1120
Nights: 8-12
Budget: ££
When: Apr-Oct

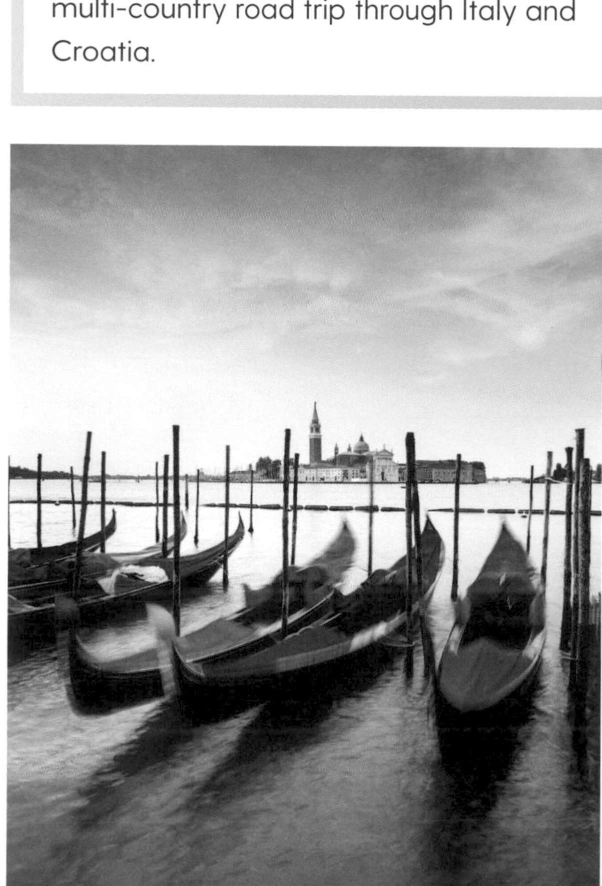

ABOVE: VENICE, ITALY

❶ Venice, Italy

Venice once ruled the waves from her watery perch at the top of Adriatic, and the legacy of that imperial wealth is still evident. Head to St Mark's Basilica to see artefacts plundered from the Siege of Constantinople, or to the Accademia for dusky canvases showing La Serenissima in her maritime heyday.

It's a 150km drive east to Trieste.

❷ Trieste, Italy

One of Italy's great unsung cities, Trieste was once the main seaport of the Austro-Hungarian Empire and still represents a confluence of Italian, Germanic and Slavic cultures. On the western edge of the city is Miramare Castle, a former residence of Austro-Hungarian royalty, overlooking a little marine reserve.

Contintue 110km south to Rovinj, first entering Slovenia before crossing the border into Croatia.

❸ Rovinj, Croatia

Rovinj sits on Croatia's Istrian Peninsula, but from some angles you could be forgiven for thinking it's a stray district of Venice. Climb up to the Bell Tower – a lookalike of the one in St Mark's Square – before wandering south from the harbour to Zlatni Rt, a forested cape where Aleppo pines rise over blue shallows.

ABOVE: ROVINJ, CROATIA

Korčula is a hefty 670km drive south (including a short ferry crossing) – consider splitting up the journey in Zadar, Trogir or the aptly named Split.

❹ Korčula, Croatia

A legend tells that Venice's most famous seafarer, Marco Polo, hailed from the island of Korčula. Quite how he found the inclination to leave his home island is a mystery: beyond the walls of the eponymous town is an idyllic hinterland of thick forests, mini vineyards and a coast strewn with pebbly coves.

Continue for a spectacular 170km along the coast.

❺ Dubrovnik, Croatia

The defining symbol of Dubrovnik is its city walls, originally medieval constructions that repelled sieges from the Venetian fleet, and later hit by shells during the Balkan Wars of the 1990s. Wander the length of the ramparts, or see them from a fresh perspective on the short ferry ride to the island of Lokrum, beyond the harbour.

Dubrovnik airport is 20km south of town, and has car drop-off facilities and European connections.

ALTERNATIVES

Extension: Mostar, Bosnia & Hercegovina

Mostar was devastated during the Balkan Wars, with its iconic Ottoman bridge destroyed in 1993. The rebuilt span has since become a symbol of the town's recovery – admire the bridge and then wander among the buzzing bazaars and cafes that line the Neretva River.

Mostar is a 140km drive north from Dubrovnik – stop to explore the village of Počitelj on the way, where minarets rise among cedar trees.

Extension: Kotor, Montenegro

The Gulf of Kotor sees the Adriatic surge inland, with little waves lapping against the towering karst cliffs of Montenegro. In a little nook at the far end you'll find the town of Kotor itself, a maze of medieval alleyways, guarded by ramparts and towers.

To get to Kotor, follow the coastal road 90km south from Dubrovnik.

PORT-WINE PILGRIMAGE IN PORTO

Cálem – Taylor's

Get the lowdown on this revered Portuguese tipple via a grape-to-glass walking tour around the historic port-wine lodges of Porto's Vila Nova de Gaia.

FACT BOX

Carbon (kg per person): 0
Distance (km): 1.1
Nights: 2
Budget: $$
When: Mar-Oct

❶ Cálem

Sitting across the Douro from central Porto, Cálem has been in the port-wine business since 1859, and its barrel-lined cellars are among the city's most atmospheric. Engaging all the senses, thirty-minute tours conclude with tastings of white and tawny ports; alternatively, upgrade to a tasting matched with Portuguese cheese, charcuterie or dark chocolate.

Kopke is just a minute's walk west of Cálem.

❷ Kopke Wine House

If you fancy something less formal than a tour, have a drink in this chicly contemporary tasting bar with views of Porto's iconic Ponte de Dom Luís I bridge. Dating back to 1683, Kopke is one of the world's oldest port-wine houses, so you can be guaranteed a seriously good aged tawny. Tastings can be paired with chocolate or olive oil.

Walk two minutes' west along the riverfront to Sandeman.

❸ Sandeman

Port-related paintings and memorabilia whisk you back to 1790 when the young Scotsman George Sandeman started dabbling in the port and sherry trade here. Guides dressed in black capes and sombreros lead forty-minute tours into the deep, dark cellars, finishing off with a two-port tasting.

Espaço Porto Cruz is on the opposite side of Largo de Miguel Bombarda.

❹ Espaço Porto Cruz

Get versed in port at Espaço Porto Cruz's exhibitions, shop and tasting bar. The Terrace Lounge 360° rooftop bar has sensational views of the medieval

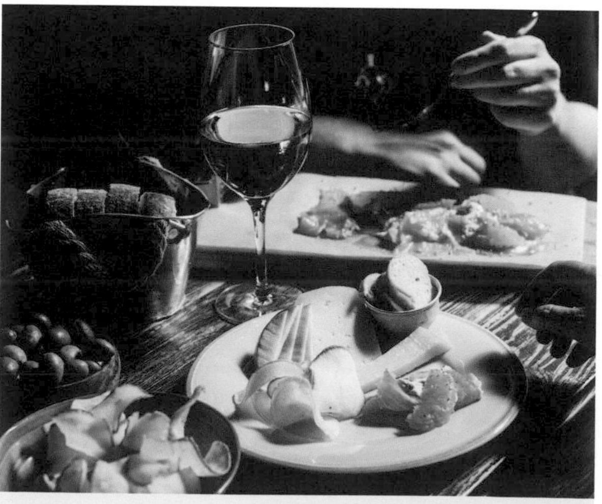

ABOVE: PERFECT PAIRINGS IN PORTO

ABOVE: PORTO'S RIBEIRA DISTRICT

Ribeira district spilling down to the river, while slick restaurant DeCastro Gaia puts creative riffs on Portuguese flavours.

It's a four-minute walk south via Rua Guilherme Gomes Fernandes to WOW.

5 WOW

Opened in 2020, Porto's World of Wine (WOW) is the city's shiny new wine district, with dreamy views of the skyline. The €105-million complex of wine-related experiences unites restored cellars, a chocolate factory, a wine school, a wine-themed gallery, gastronomy workshops, a cork-themed exhibition and bars and restaurants.

Walk three minutes' up Rua do Choupelo to Taylor's.

6 Taylor's

From its hilltop perch and landscaped gardens, British-run Taylor's has been making first-class port since 1692. Audioguide tours in 12 languages whizz through the 300-year-old cellars, ending with a two-port tasting. Private tours, port masterclasses and cocktail-making workshops can be arranged.

It's a 10 minute walk back to Cálem, downhill and along the river.

ALTERNATIVES

Extension: Graham's

With dress-circle views of Porto's medieval Ribeira district, Graham's has the scenic edge. British-founded in 1820, it offers tours of working cellars, a panoramic tasting room, the private-club-style Vintage Room for tastings of rare ports and tawnies, a shop and a museum. The Vinum restaurant pairs rich Trás-os-Montes dishes with Douro wines.

Graham's is a 20min walk west of Taylor's via Avenida de Ramos Pinto.

Extension: Ferreira

Celebrated for its Douro wines and ports since 1751, Ferreira is wholly (and unusually) Portuguese run. A guided tour of the cellars reveals some interesting facts about the founding family. Opt for the Premium Visit and you'll get to taste white, ruby and tawny ports.

Ferreira is a 12min walk from Taylor's (go downhill and turn left along the river).

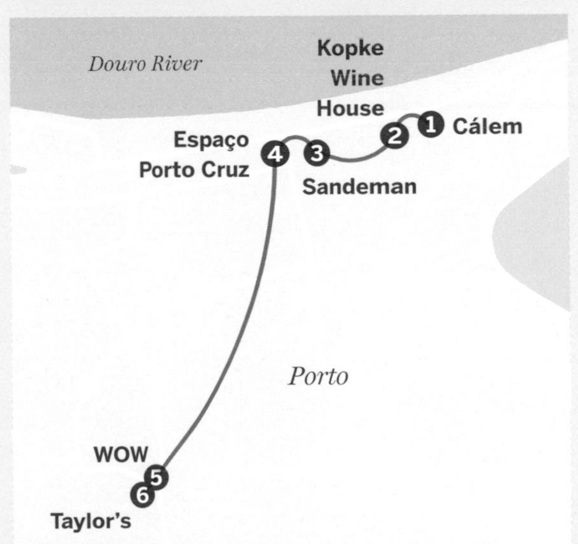

BELOW: THE DOURO RIVER FLOWS THROUGH PORTO

TOMAZ DO DOURO

OUTDOOR CANARIES: EL HIERRO & LA PALMA

El Hierro – La Palma

Wander off the beaten Canary Islands track to explore two unspoiled, outdoor-adventure-friendly islands.

FACT BOX

Carbon (kg per person): 56
Distance (km): 533
Nights: 10–14
Budget: $$
When: Mar–early Jul, Sep, Oct

❶ Eastern El Hierro

Base yourself in pretty little Tamaduste for a few days exploring eco-conscious El Hierro's eastern and northern coastlines. Visit the pint-sized capital Valverde, saltwater pools (La Caleta, Charco Azul), electric-green pine forests and stunning lookouts.

Drive 40km southwest.

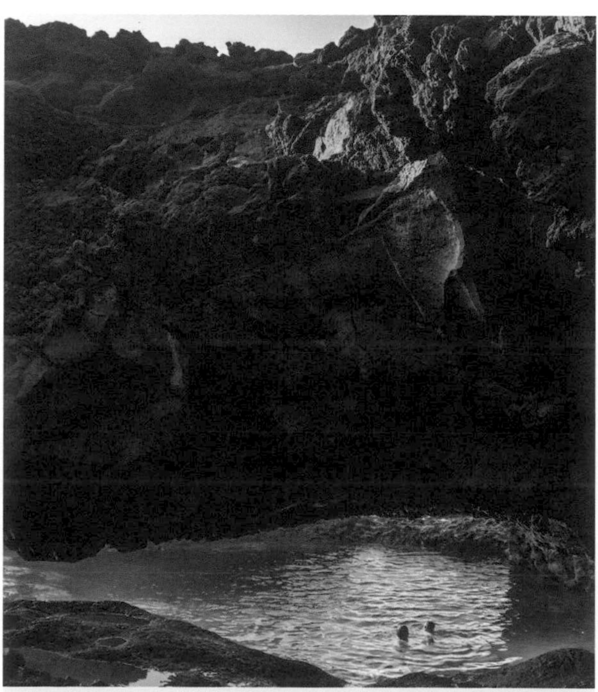

ABOVE: CHARCO AZUL, EL HIERRO

❷ La Restinga, El Hierro

Focused on the tranquil Mar de las Calmas marine reserve, this sunny southern fishing village is El Hierro's diving hub (good seafood restaurants too). Swimmers will love the twinkling Tacorón sea pools.

Drive 40km north.

❸ El Golfo, El Hierro

Spectacular El Golfo offers cascading cliffs, natural pools, sloping vineyards and brilliant hikes (try the Camino de Jinama). Stay in La Frontera or Las Puntas, seeking out wind-sculpted juniper trees, La Dehesa's 18th-century chapel, the far-flung Orchilla lighthouse and the Manrique-designed Mirador de la Peña.

From El Hierro's Puerto de la Estaca, ferries go to Tenerife's Los Cristianos (6 weekly; 2hr 45min), from where you catch another ferry to Santa Cruz de la Palma (1-3 daily; 2hr 30min-4hr).

❹ Santa Cruz, La Palma

One of the Canaries' loveliest capitals, seaside

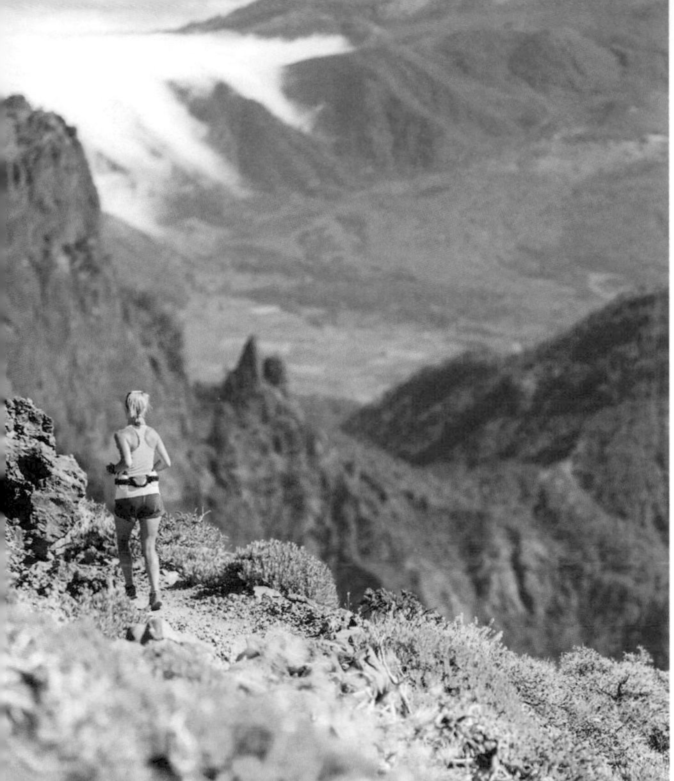
ABOVE: UPCOUNTRY LA PALMA

Santa Cruz has a mix of traditional Canarian houses, Renaissance architecture, fine museums, dazzling churches and wonderful cuisine.

Drive 25km north.

❺ Los Tilos, La Palma

With great day-hikes, rich wildlife and ancient laurisilva (laurel forest), this lush 140-sq-km rainforest in the island's north has been a protected Unesco Biosphere Reserve since 1983.

Head 40km southwest.

❻ Parque Nacional de la Caldera de Taburiente, La Palma

La Palma's heart is this unmissable natural paradise strung around the 8km-wide Caldera de Taburiente depression. Superb hikes include the clamber up 1854m Pico Bejenado, and there's stellar stargazing at the Roque de los Muchachos.

Drive 25km south.

❼ Fuencaliente, La Palma

In far-south La Palma, allow a day for Fuencaliente's eerie volcanic moonscapes, black-sand beaches and delectable wines.

Drive 27km north back to Santa Cruz.

ALTERNATIVES

Diversion: El Teide

Passing through Tenerife, you'll inevitably be tempted by Spain's tallest peak, El Teide (3718m). The World Heritage-listed volcanic national park sprawling around it has many view-laden trails.

The easiest way up El Teide is aboard the cable car. From the upper cable car station, it's an hour's hike to the summit.

Extension: La Gomera

Slow-paced La Gomera unfolds around the forests of Parque Nacional de Garajonay. Shimmering black-pebble beaches, palm-studded valleys, jagged volcanic cliffs, exhilarating hiking and the pastel-painted capital San Sebastián de la Gomera (where Christopher Columbus once stayed) add to the appeal.

Ferries run between San Sebastián and Tenerife's Los Cristianos (5-6 daily; 1hr).

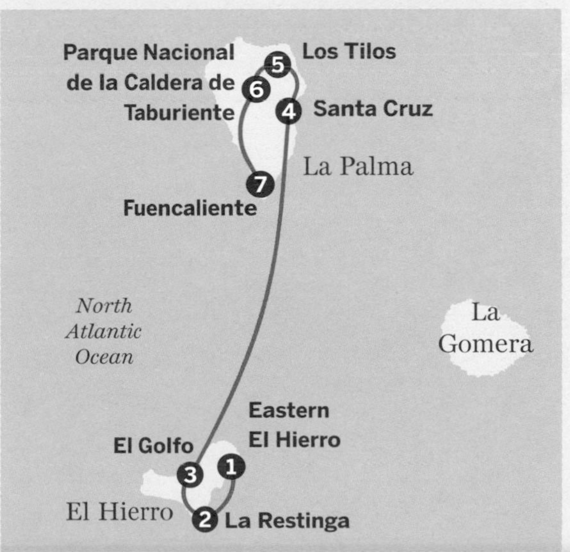

SLOW FOOD AND WINE THROUGH PIEDMONT AND LIGURIA

Turin – Genoa

Enjoy a leisurely train journey inspired by Italy's Slow Food movement, tasting Piedmont cuisine and wines before arriving at Genoa, gateway to Liguria's idyllic coast.

FACT BOX

Carbon (kg per person): 9
Distance (km): 230
Nights: 6
Budget: $
When: year-round

❶ Turin

Turin is a grand, regal city of wide avenues and monumental piazzas lined with Baroque and Rococo buildings. It found fame as modern Italy's first capital, yet is barely touched by tourism compared to the likes of Venice and Florence – despite diverse attractions ranging from world-class museums dedicated to Egyptology, cinema and automobiles, to the wonderfully anarchic Porta Portese flea market. Not forgetting the evening *aperitivo* ritual, when locals flock to sumptuous Belle Époque salons like Caffè Platti.

Local trains for Bra depart Turin's Lingotto station (hourly; 50min).

ABOVE: GENOA

❷ Bra

Sleepy Bra became globally famous when Carlo Petrini founded the Slow Food movement here in 1986. Initially a protest against unhealthy fast food, the movement evolved into today's worldwide organisation championing artisan, sustainable produce. The friendly town has tempting delis, wine bars and homecooking restaurants, and a vibrant farmers market that takes over the streets each Friday morning, while the biennial cheese festival draws tens of thousands of *formaggio* fans. Slow Food's respected University of Gastronomic Sciences is just outside Bra, in Castello di Pollenzo.

Take the train from Bra (hourly; 20min) to Alba.

❸ Alba

The train to Alba winds through Langhe's picturesque vine-clad hills, where several charming winemaking

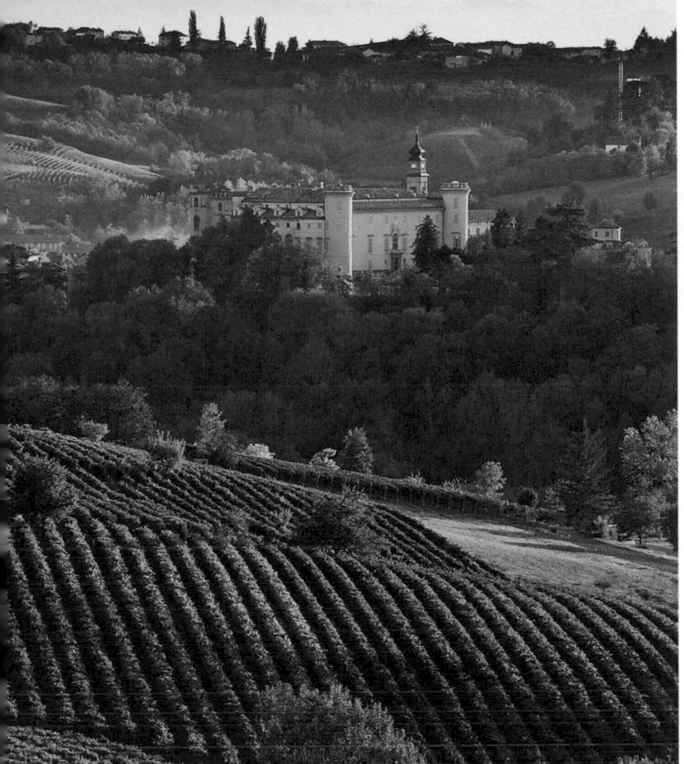
ABOVE: ASTI PROVINCE

villages – Serralunga, La Morra and world-renowned Barolo – are good places to break up the trip with wine tastings and hearty lunches of Piedmont's distinctive traditional cuisine. Alba itself is the gastronomic capital of precious white truffles, celebrated during a two-month fair each autumn.
Take bus 45 from Alba to Asti (55min).

4 Asti

The ancient Roman settlement of Asti has given its name to the world-famous Spumante sparkling wine: try it in the nearby medieval village of Canelli where it was invented in 1870. Asti hosts a prestigious antiques fair the last Sunday of each month.
Board the train from Asti for Genoa's Piazza Principe station (hourly; 1hr 30min).

5 Genoa

This ancient maritime port is an unexpected delight, well-off the tourist trail despite its glittering palaces, black-and-white striped Romanesque cathedral, jaw-dropping aquarium and local *osterie* serving delicious, affordable meals. From Genoa it's easy to explore the glamorous beach resorts of the Ligurian Riviera, especially millionaire's hideaway Portofino.
Take the train back to Turin (hourly; 2hr).

ALTERNATIVES

Extension: Sestriere

The gentle hills of Piedmont's Langhe wine country are quickly replaced by dramatic Alpine scenery as you head towards Sestriere. Located at 2000m above sea level, it has one of the world's most renowned ski resorts, but offers year-round attractions too including trekking and biking in summer.
Trains run from Turin's Porta Nuova station to Oulx station (1hr 15min), from where bus 285 runs to Sestriere (45min).

Extension: Cinque Terre

The vividly painted houses of Cinque Terre's five fishing villages tumble down steep cliffs to sandy beaches. A memorable walk through the national park and vineyards links all five, and though there's no road, the local train stops at each one.; don't miss Riomaggiore and Vernazza.
From Genoa's Nervi station take the regional train to La Spezia Centrale (1hr 15min), sometimes with a change Sestri Levante.

EAT YOUR WAY ROUND PUGLIA

Alberobello – Lecce

Puglia's *cucina povera* (poor cuisine) is a great reason to visit an Italian region also known for Baroque architecture, verdant hills, beguiling towns and brilliant beaches.

FACT BOX

Carbon (kg per person): 15
Distance (km): 118
Nights: 7-10
Budget: $$
When: year-round

❶ Alberobello

The Valle d'Itria's rolling fields epitomise the agricultural landscape that dominates Puglia, and in the heart of that landscape sits preposterously cute Alberobello. The town is chock-full of distinctive and unique thatch-topped limestone *trulli* – many now housing shops, cafes, bars and restaurants – and wandering among them is a wonderful way to spend a day. Break for a lunch of *orecchiette con cime di rapa* (pasta with turnip tops) in spring, or dishes made with local broad beans, wild onions, artichokes and chickpeas through the rest of the year.

Drive 36km southeast.

❷ Ostuni

The hilltop *città bianca* (white town) of Ostuni appears like a mirage out of the olive-filled plains. The steep main street, Via Cattedrale, will test your lungs and legs, but is worth the effort for the occasional expansive views to the plains or sea. The historic heart of Ostuni is a jumble of cobbled alleyways, arching porticos and graceful loggias. And everywhere, a dizzying range of antipasti waits to be sampled – try cured-pork specialities *capocollo* and *soppressata*.

Continue 43km southeast towards the coast.

ABOVE: *TRULLI*, ALBEROBELLO

ABOVE: BASILICA DI SANTA CROCE, LECCE

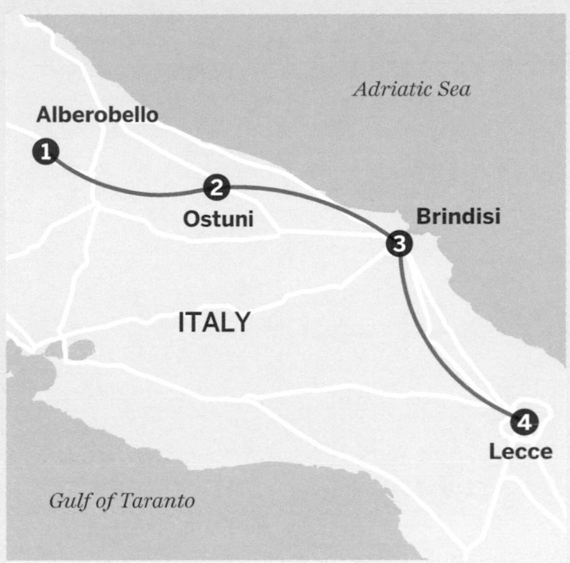

3 Brindisi

Historic port town Brindisi, gateway to the Salento region, brings together two of Italy's top food products: fish and olive oil. Along the palm-lined Corso Garibaldi and the seafront you'll be stopped in your tracks by displays of shellfish, spiny sea urchins, and heaps of all types of fresh fish. Try them grilled, roasted, in foil or fried, or go for local specialities such as *tiella barese* (a rice, potato and mussel flan) or octopus pickled in wine vinegar.

Drive 39km southeast.

4 Lecce

The Baroque fantasies of Salento's capital, Lecce, may leave you open-mouthed with astonishment in a city called, with justification, the 'Florence of the South' thanks to its mix of medieval alleyways, grand churches and imposing secular buildings. From the Piazza del Duomo take in the fanciful facade of the Basilica di Santa Croce, then promenade the Piazza Sant'Oronzo and wander streets filled with *pasticcerie* and *gelaterie* selling a tempting array of cakes and ice-cream that are as rich and indulgent as the city's architecture.

From Lecce, drive to Brindisi or Bari airports, both with international connections.

ALTERNATIVES

Extension: Gallipoli

Not for nothing is Gallipoli called the pearl of the Salento region. The historic centre, set on an island off the mainland, is a great base from which to explore the surrounding Salentine coast and Baroque towns like Nardò and Galatone. Diving is popular in the crystal-clear waters and at the natural park island of Sant'Andrea, underpinned by fantastical underwater caverns and canals.

Gallipoli is a 40km drive southwest of Lecce.

Extension: Bari

The narrow alleys and Italy's longest seafront in Old Bari are filled with the sights and sounds of everyday life in this food-focused region: women sit in the shade of their houses making *orecchiette* pasta; wineries sell the heady reds produced locally; and everywhere are churches and a wealth of historic buildings. Munch on fresh focaccia for fortification.

Bari is the arrival airport for Alberobello, 55km southeast.

PORTUGAL'S SUN-TOASTED SOUTHWEST

ACTIVE DRINK FOOD SUSTAINABLE ADVENTURE CULTURE PERFORMANCE ART ARCHITECTURE HISTORY FAMILY SOLO SHORT LONG BEACH WINTER WILDLIFE WELLNESS

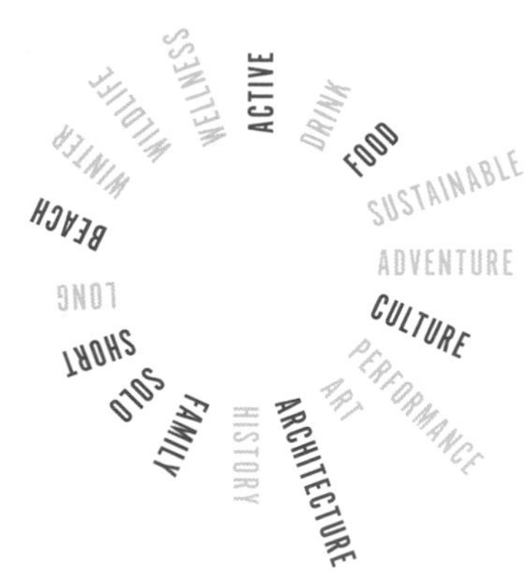

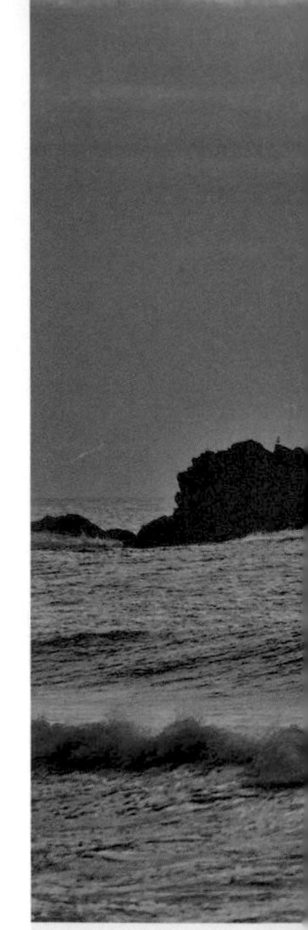

Lisbon – Sagres

The Algarve hogs the headlines, but this road trip along Portugal's other coast offers undeveloped beaches, picturesque ports and outstanding seafood.

FACT BOX

Carbon (kg per person): 48
Distance (km): 383
Nights: 6
Budget: $$
When: Mar, Jun, Sep, Oct

❶ Lisbon

Lisbon is the gateway to southern Portugal's duelling coasts; time is limited, so head to Belém to be awed by Unesco-listed sights like Torre de Belém and Mosteiro dos Jerónimos while munching on a world-famous *pastel de Belém* custard tart (or three).

Drive 49km south to Setúbal from where ferries cross the Sado River to Troia (7:30am to 10pm), then continue to Comporta, 18km further south.

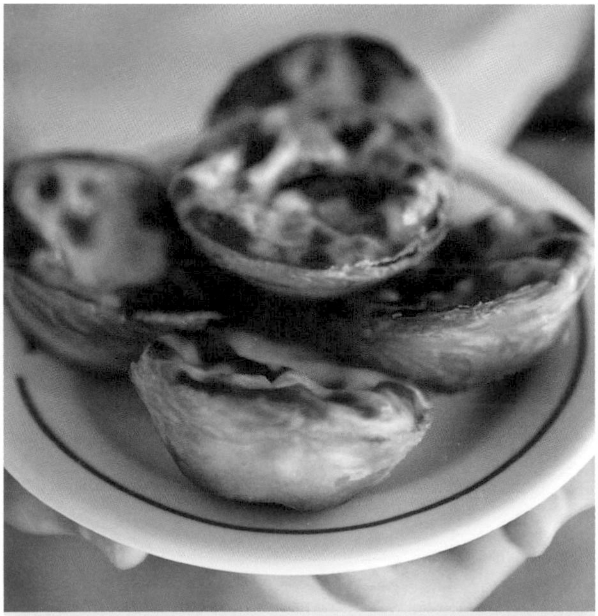

ABOVE: *PASTEL DE BELÉM*, LISBON

❷ Comporta

The hippie-chic Alentejan village of Comporta has drawn the attention of international fashionistas like Christian Louboutin and Philippe Starck for good reason: remarkable white-sand beaches – Tróia, Pego, Carvalhal – are surrounded by cork trees, sand dunes, rice fields and vineyards. *Perfeito!*

Drive 70km south.

❸ Porto Côvo

This archetypal traditional Portuguese fishing village of whitewashed cottages untangling across a perfect grid toward a fishing harbour feels stuck in simpler, far more wonderful times. Plant yourself on the idyllic Praia Grande (Big Beach), a postcard-perfect patch of sunburnt sands bound by natural rock formations.

Continue south for 18km.

❹ Vila Nova de Milfontes

A laidback Alentejan beach town, Vila Nova features don't-miss food (start at Tasca do Celso) along with

ABOVE: SAGRES SURF

an historic lighthouse, a 16th-century fortress and a cathedral. Golden sands unfold in all directions: Praia das Furnas and Praia da Franquia near town and, a little further afield, Praia do Malhão and Praia do Almograve.

Drive 26km south.

5 Zambujeira do Mar

Zambujeira do Mar is perched above rugged cliffs on the southern end of the Alentejo Coast; down below, cerulean seas roll onto floury sands backed by rugged cliffs. For a fishy feast, head to Barca Tranquitanas, a seafood restaurant where the owner's father used to feed the town's returning fisherman.

Head 82km further south.

6 Sagres

Cross into the less developed western Algarve. A pit-stop for fresh seafood at A Azenha do Mar restaurant is highly recommended – and only topped (perhaps) by the stunning technicolour sunsets on offer at Europe's most southwesterly point, the stunning Cabo de São Vicente. Sagres and its end-of-the-world feel lie in wait at the end of the road.

The nearest international airport is 120km east in Faro.

ALTERNATIVES

Diversion: Parque Natural do Sudoeste Alentejano e Costa Vicentina

This natural park preserves some 120km of Alentejo and Algarve coastline – including many of the stops in this itinerary – and exploring on foot affords the chance to see a wealth of flora and fauna, including rare fishing eagles (the only place on earth where they nest into seashore rocks) and marine otters. Two long-distance trails and 24 circular routes traverse the park.

Extension: Algarve coast

Continue east from Sagres along the world-famous Algarve coast, a favourite of holidaymakers for decades. Bigger towns like Lagos feature all the infrastructure of any major European resort, while smaller villages such as Tavira are a little more low-key. Beaches along this 155km stretch of coastline are usually more developed than the Alentenjo coast, but are often no less stunning.

AN ALENTEJO ADVENTURE

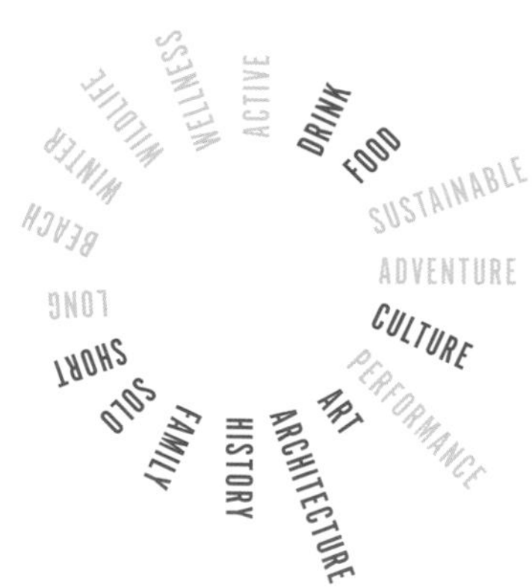

Évora – Mértola

Explore bewitching hilltop villages in Portugal's largest region, Alentejo, enjoying mighty castles, delightful town squares and lots of sweeping views.

FACT BOX

Carbon (kg per person): 45
Distance (km): 246
Nights: 3-6
Budget: $$-$$$
When: year-round

❶ Évora

Take time to stroll and soak up the atmosphere of Évora's narrow, winding lanes and the handsome main square, Praça do Giraldo. Don't miss the Capela dos Ossos (Bones Chapel) next to Igreja de São Fransisco (Church of St Francis). Swing by the main building of the historic Universidade de Évora before stopping at Rota dos Vinhos do Alentejo to get your head around Alentejo's Wine Route.

Drive 30km northeast.

ABOVE: ROMAN TEMPLE OF DIANA, ÉVORA

❷ Évoramonte

A side-trip on the way to Estremoz, Évoramonte is a tiny hamlet with a fabulous castle. Originally constructed in the early 14th century, it had to be rebuilt after a 1531 earthquake, and while there's not much to see inside it's worth heading to the roof for the spectacular views.

Back in your car, head 18km northeast.

❸ Estremoz

Estremoz is the least-touristy stop on this trip, a place whose agricultural roots can be felt, especially on Saturday, market day, when you can stock up on fruits and vegetables, plus local specialities including ewe-milk cheese. Wander through the pretty orange-tree-lined lanes of the lower town, before heading on foot to the upper town and the interesting Museu Municipal. For the best 360-degree views around? Enter the

ABOVE: ALENTEJO VINEYARD

Pousada de Rainha Santa Isabel and ask to access its roof, reached by several flights of narrow steps.
Drive 60km southeast.

4 Monsaraz

The most visited, and perhaps charming, of all the hilltop villages, Monsaraz' permanent population is nevertheless small. Wander the sleepy streets – the village is pedestrian-only and a sight in itself. Visit the castle for views and then grab a bite at one of the restaurants that overlook the walls and the Alentejan plains below.
Continue south for 140km.

5 Mértola

Stunning medieval Mértola is the place to head for more great views – it's set on a rocky spur, high above the Guadiana river – as well as for witnessing some extraordinary Moorish history. The town itself is a kind of open-air museum, so stroll the streets and discover a wealth of minarets and former mosques. And don't miss a climb to the castle to take in panoramic vistas.
From Mértola, it's 120km to Faro on the Algarve, from where there are international flights and train connections.

ALTERNATIVES

Extension: Megaliths and Dolmens

From Évora, it's worth taking the time to explore the megaliths and dolmens – ancient stone structures – that are dotted around the Alentejan landscape; most are well signed. The nearest site to Évora, and the most accessible, Cromeleque dos Almendres is also the largest megalithic monument on the Iberian Peninsula.
Allow half a day to drive around the sites.

Extension: Wine Route

Évora's Rota dos Vinhos tourist information centre has the gen on wine tastings and an Alentejan wine route from the city. A couple of good wineries are located outside the city walls, but you'll need a car for most of them. If you've only time for one, visit Herdade do Esporão. Reserve ahead for a world-class meal in its smart restaurant, plus birdwatching and other activities.
Herdade do Esporão is 44km southeast of Évora

EASTERN SICILY'S BAROQUE SPLENDOUR

Taormina – Ragusa

Dip in and out of sun-drenched piazzas, local trattorias, ornate churches and ancient ruins on this tour of Sicily's spectacular Baroque towns.

FACT BOX

Carbon (kg per person): 7.5
Distance (km): 200
Nights: 10
Budget: $$
When: May-Jul, Sep

1 Taormina

This mountainside Baroque belle has one of the world's most spectacularly sited ancient Greek theatres, with bonus views of Mt Etna. Luxury boutiques, local produce shops and bars selling aperol spritz and *arancini* (fried rice-balls) line Taormina's narrow streets. Take the cable car downhill to visit two crescents of sublime shingle beach, at Lido Mazzarò and Isola Bella.

Take the train to Syracuse (every 2hr; 2hr).

2 Syracuse

Jutting out into the Ionian Sea, Syracuse was once the largest settlement in the ancient world. The city ruins can still be toured at the Parco Archeologico Neapolis, but the show-stopper is Syracuse's lively, honey-coloured medieval centre with its grand Piazza del Duomo and seaview restaurants. Visit on market day and spend an afternoon on the tiny beach on the island of Ortygia.

Continue on the train to Noto (8 daily; 35min).

3 Noto

After a devastating earthquake in 1693, Noto was rebuilt as a Baroque beauty. At its heart is Corso Vittorio Emanuele – a promenade made for strolling, flanked by ornate churches and palaces. Pause at terrace cafes for *cannoli*, *gelato* or almond *granita*, and climb the bell tower of the Chiesa di San Carlo al Corso for wonderful views.

ABOVE: RAGUSA IBLA

ABOVE: DUOMO DI SAN PIETRO, MODICA

Take the bus from Noto's station to Modica (several daily; 1hr 30min).

4 Modica

This slow-food haven is renowned across Italy as a faithful producer of an Aztec chocolate recipe brought to Sicily during 17th-century Spanish rule. Artisan makers ply visitors with hot chocolate flavoured with cinnamon and orange peel, as well as chocolate *granita* and biscuits stuffed with chocolate, spices and minced beef; you can also tour Sicily's oldest chocolate factory, Dolceria Bonajuto. Walk it off with a steep climb up into medieval Modica Alta.
Take the train to Ragusa (every 2-3hr; 25min).

5 Ragusa

Like Unesco-listed Noto and Modica, Ragusa was rebuilt in exquisite Baroque style after the devastating earthquake that cleaved southeast Sicily's Val di Noto in the late 17th century. Clinging to a hillside as though it still fears for its life, Ragusa Ibla is the historic centre – a hauntingly beautiful, quiet maze of lanes that circle up to the sun-soaked Piazza Duomo and Neoclassical San Giorgio cathedral.
Buses run from Ragusa to Catania's international airport (every 1-3hr; 1hr 45min).

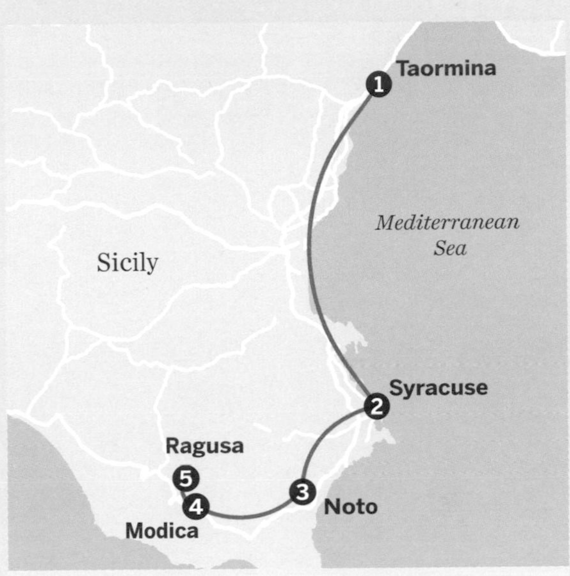

ALTERNATIVES

Extension: Mt Etna

The indomitable cone of Mt Etna appears like an unwavering mirage over Sicily's east coast. It's surprisingly accessible from Taormina, either on an organised day trip or via a helter-skelter drive in a hire car. Tours start with a cable car, then it's either a minibus to the highest accessible point, Torre del Filosfo, at 2920m, or you can hike the final 2km uphill. From there you walk the rim to peer into Etna's smoking, sulphur-tinged fumaroles.

Diversion: Chiaramonte Gulfi

To finish your trip on a gastronomic high, add a stop in this hilltop town north of Ragusa, nicknamed 'il Balcone della Sicilia' (Sicily's balcony). Italians flock here for a handful of famous local products: a much-prized olive oil and high-quality pork salami and sausages. Feast on them at lovely Ristorante Majore, just off Piazza Duomo.
Take the bus from Ragusa (4 daily; 50min).

A SALENTO SOJOURN IN SOUTHERN ITALY

Lecce – Porto Cesareo

Salento, the southern region of Puglia, is famous for its food, music, Baroque churches and wild and beautiful coastline, all best explored by car.

FACT BOX

Carbon (kg per person): 22.5
Distance (km): 184
Nights: 7
Budget: $$-$$$
When: Jun-Sep

❶ Lecce

University town Lecce is the capital of Salento, famous for its intricately carved Baroque facades and golden-stone buildings. During summer, the winding, narrow lanes of its historic centre get busy for the *passeggiata* (early evening stroll), when people eat ice cream, sip an *aperitivo* and browse the boutiques. The cathedral piazza is a beautiful space surrounded by honey-coloured 16th-century palaces, and there's a small Roman amphitheatre in the Piazza St Oronzo.

Drive 46km southeast.

ABOVE: BAROQUE ARCHITECTURE IN LECCE

❷ Otranto

Overlooking the kingfisher-blue Adriatic, the walled town of Otranto, topped by a 15th-century Aragonese castle, is a popular summer destination. Attacked multiple times by Ottoman pirates, the town has a grisly reminder of its tumultuous past in its cathedral – a chapel stacked with the bones of 813 Christians who (unsuccessfully) sought sanctuary here. On a happier note, the cathedral also has a remarkable floor mosaic depicting the tree of life.

Continue 30km south.

❸ Castro

Atop a 100m-high cliff, Castro has epic views over the Adriatic (on a clear day you can see Albania), plus a castle built by the Romans, rebuilt by the Normans, then rebuilt again by the Aragonese. The wind- and

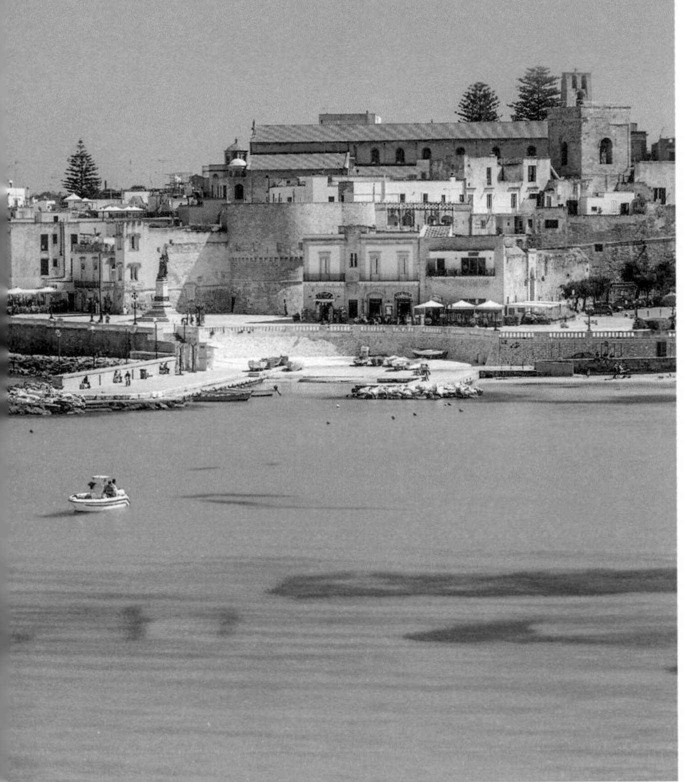

ABOVE: OTRANTO

sea-sculpted limestone coast is pocked with dramatic caves, most notably the cathedral-like Grotta Zinzulusa. For a dose of *dolce vita* (sweet life) take a boat trip from Castro along the coast.

It's a 48km drive to Puglia's Ionian coast.

4 Gallipoli

The walled town of Gallipoli (meaning 'beautiful city') sits on an almost-island promontory surrounded by 14th-century fortifications. There are superb sunset sea views from its rooftops and ramparts, and its maze of narrow lanes is full of picturesque corners and squares.

Drive 32km north.

5 Porto Cesareo

This stretch of Ionian coast has some of Puglia's loveliest white-sand beaches, and the gently shelving shores at Porto Cesareo are among the finest, overlooked by an ancient watchtower, a legacy of when the coast had to be defended against pirates. The beach gets very busy in July and August, but if you can make it here in June or September, you'll have it almost to yourself.

From Porto Cesareo it's a 28km drive back to Lecce.

ALTERNATIVES

Extension: Galatina

Only 29km south of Lecce, Galatina is worth a visit just for its astoundingly beautiful frescoed Basilica di Santa Caterina. Time it right and you'll also encounter its famous *taranta* folk dancing, special to this region and performed here every 29 June (the feast day of St Peter and St Paul).

Midway between Gallipoli (30km) and Lecce (29km) by car, Galatina is easily visited en route between the two.

Extension: Italy's heel

Take a trip from Castro to Santa Maria di Leuca, the point of Italy's heel, where there's a lighthouse and a monastery set high above cobalt-blue sea. More scenic still is the sandy Il Ciolo swimming cove, at the bottom of a canyon a few kilometres away; it's accessed via zigzagging steps and traversed by a graceful arched bridge.

Castro to Santa Maria di Leuca is 21km; Ponte Ciolo is 8km from Santa Maria.

PORTUGAL'S ROUTE 66: ROTA NACIONAL 2

Chaves – Faro

Take a top-to-bottom road trip along Portugal's Rota Nacional 2, sampling distinctive cuisine and sipping excellent wines along the way.

FACT BOX

Carbon (kg per person): 133.5
Distance (km): 800
Nights: 11
Budget: €€
When: Apr, May, Sep, Oct

❶ Chaves

Ninety minutes' drive from the international transportation hub of Porto, Chaves is the starting point for Rota Nacional 2, one of the longest roads in the world – a surprising fact in a small country like Portugal. The town has a history that goes back to the Roman Empire and there's much to explore, but be sure to leave time for nature, particularly the many thermal springs that dot the region.

Drive 92km south.

ABOVE: STRIPPED CORK-OAK TREE, ALENTEJO

❷ Douro Valley

Spend a couple of days taking in the gorgeous scenery of the Douro Valley, stopping to sample the wines produced in the vineyards which climb up steep hills on either side of the eponymous river. A boat tour offers an excellent, alternative way to discover the region.

Continue south for 76km.

❸ Viseu

The capital of the Centro district of Portugal, Viseu is also a cultural centre, with beautiful palaces, an important cathedral, the São João de Tarouca convent and the nationally acclaimed Grão Vasco museum, where you can take in an impressive collection of Renaissance art.

Head south for 309km.

❹ Portugal's schist villages

Take small detours to visit some of Portugal's most atmospheric and historic villages, such as Tondela, Penacova, Lousã and Sertã, all constructed from schist rock. In Abrantes, the largest municipality in the region, be sure to take in the castle and its sweeping views over the countryside.

Continue the drive south for about 76km.

ABOVE: DOURO VALLEY VINEYARD

❺ Upper Alentejo

In this agricultural region that's increasingly becoming known as the 'new Tuscany', make for a city like Mora or Montemor-O-Novo, or book a stay in a winery or countryside hotel. Plan to spend a couple days here, sampling the full-bodied red wines and hearty local gastronomy, with dishes like *açorda*, a garlicky bread soup, on many menus.

Back on the road continue south for 118km.

❻ Lower Alentejo

The Lower Antejo region is all about going deep into nature, with landscapes marked by gnarled olive trees and overhung by brilliant blue skies. Charming rural hotels are plentiful here, but during high season it's best to book in advance.

Drive south about 130km.

❼ Faro

The drive ends with a view over the Atlantic in Faro, capital of the Algarve. The old city has charming narrow streets and excellent fish restaurants, or you can head to the beaches of the Ria Formosa Natural Park and Ilha Deserta.

Many international flights depart from Faro's airport, along with buses to Spain.

ALTERNATIVES

Extension: Porto

Portugal's second city has an excellent restaurant scene and some of the country's best fashion boutiques. The historic centre, with narrow colourful buildings almost stacked one on top of the other, is a Unesco World Heritage Site.

Porto is a short drive (150km) from Chaves, and probably your entry point into Portugal anyway.

Extension: Tavira

No trip to Portugal is complete without some quality time on the beach. Tavira is a rare corner of the Algarve that isn't dominated by large resorts and still maintains its fishing village character. The fresh fish and seafood are a must.

The village is a short drive (38km on the A22) from Faro.

TUSCAN ART & ARCHITECTURE

Florence – Lucca

Head for central Italy to feast on art, architecture, fine food and old-world wines on a bus and train trip around Tuscany.

FACT BOX

Carbon (kg per person): 12
Distance (km): 333
Nights: 7
Budget: $$
When: year-round

❶ Florence

Utterly seductive and every bit as romantic as its reputation suggests, Florence wows with a warren of narrow cobbled streets lined with Renaissance *palazzi*, frescoed churches and world-class museums and galleries. You could spend a lifetime exploring, marvelling at art and architecture, gorging on delicious food, and watching the sun set over the River Arno and the glorious Duomo (cathedral).

ABOVE: DUOMO, FLORENCE

Frequent trains run from Florence's Santa Maria Novella station to Siena (1hr 30min).

❷ Siena

Siena's Gothic architecture remains remarkably untouched and its superb cathedral is matched by imposing churches, noblemen's houses and the glorious shell-shaped Piazza del Campo. This public square is surrounded by perfectly preserved medieval buildings and is the site of the world-famous horse-racing festival, Il Palio.

Jump on a bus to San Gimignano (every 3hr; 1hr).

❸ San Gimignano

A forest of medieval towers once dominated the walled town of San Gimignano. Today, only 14 remain, overlooking a rich collection of medieval *palazzi*, churches and squares and an impressive cathedral that's home to magnificent 14th-century frescoes. Climb the Torre Grossa for panoramic views.

Four buses daily serve Volterra, with a change in Colle di Val d'Elsa (1hr 30min).

❹ Volterra

A former Etruscan settlement perched on a rocky plateau, Volterra sits behind imposing ramparts,

ABOVE: SAN GIMIGNANO

its winding cobbled streets lined with medieval mansions. In addition to the handsome architecture you'll find a Roman theatre and the country's finest Etruscan museum.

Take the bus to Pontedera (every 4hr; 1hr 15min) from where frequent trains serve Pisa (15min).

5 Pisa

A trio of Romanesque wonders adorn Pisa's spectacular Campo dei Miracoli. Along with the renowned leaning tower, you'll find an exquisitely ornate cathedral and unusual round baptistry. Beyond the square, medieval lanes open into handsome squares, while a host of museums explore the history of this former maritime superpower.

Trains run every 30min to Lucca (25min).

6 Lucca

Narrow streets, handsome piazzas and medieval towers are enclosed by Renaissance walls in Lucca, the birthplace of Puccini. Spend your day wandering the Romanesque cathedral, examining the Byzantine mosaics of the Chiesa di San Frediano or soaking up the Rococo excess of the Palazzo Mansi.

Catch the train back to Florence (every 30min; 1hr 50min).

ALTERNATIVES

Extension: Montepulciano

Best known as the home of the renowned Vino Nobile, a classic Tuscan red wine, Montepulciano is a picturesque Tuscan town set on a volcanic ridge. Stroll streets lined with rustic stone houses for panoramic views of the Val di Chiana and Val d'Orcia, and the chance to sample the region's many artisan products and rich wine culture.

Five buses run daily between Siena and Montepulciano (1hr 45min).

Extension: Montalcino

High above the Val d'Orcia, hilltop Montalcino is the home of Brunello, one of Italy's best-loved and most expensive red wines. Montalcino's medieval town sits among picturesque vineyards; you can easily spend a day wandering the streets, walking the 14th-century ramparts, quaffing wine at local *enoteche* (wine cellars) or touring local vineyards.

Regular buses run between Siena and Montalcino, a journey of about 1hr 30min.

VENICE, GATEWAY TO PALLADIO'S VENETO

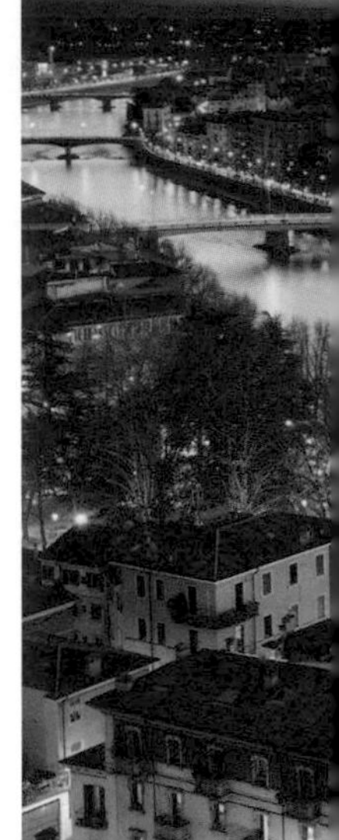

Venice – Verona

From romantic Venice, head west into the Veneto to discover masterworks by the great Italian architect Palladio, along with art, culture, distinctive cuisine and wines.

FACT BOX

Carbon (kg per person): 5
Distance (km): 120
Nights: 5
Budget: $$
When: year-round

❶ Venice

The *vaporetto* waterbus is ideal for exploring 'La Serenissima'. Take Line 1 along the Grand Canal, stopping off for a food tour of the Rialto market, then disembark at Piazza San Marco. From the nearby Doge's Palace, jump on Line 2 to San Giorgio Maggiore island, dominated by Palladio's majestic church. The *vaporetto* continues to the sleepy island of Giudecca, with another architectural masterpiece, Palladio's Il Redentore. From the Zattere pier, explore Dorsoduro neighbourhood's narrow alleyways, canals and welcoming *bacaro* winebars.

Take the train from Venice's Santa Lucia station to Padua (every 20min; 30min).

ABOVE: PRATO DELLA VALLE, PADUA

❷ Padua

The beating heart of venerable Padua is the immense Piazza dei Signori, with a seething morning market below the opulent 12th-century Palazzo della Ragione. Art lovers should make a beeline for Giotto's frescoes in the Scrovegni Chapel. The bucolic Brenta Canal links Padua with Venice and a waterside bike ride is perfect to explore the shady banks, lined with sumptuous summer villas built for Venetian nobility. Notable examples include Villa Pisani, with frescoes by Tiepolo, and Palladio's Villa Foscari.

There are frequent trains to Vicenza (24min).

❸ Vicenza

Prosperous Vicenza is dominated by Antonio Palladio, beginning with his immense Basilica Palladiana, today

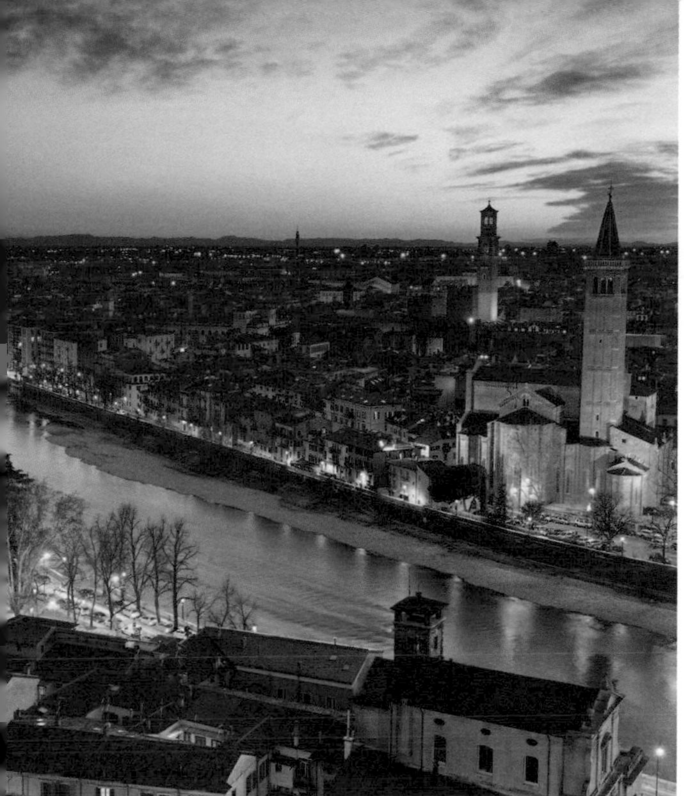

ABOVE: VERONA

a venue for blockbuster art shows. There's also the modern Museo Palladio; a day-long tour of 20 of his villas and churches in the surrounding countryside; and his final creation, Teatro Olimpico, where plays are still staged. This under-the-radar destination is also a goldmine of traditional family trattorias, cool lounge bars and innovative restaurants.

Take the local train to Verona Porta Nuova station (hourly; 40min).

4 Verona

Irresistible Verona offers something for everyone. Romantics visit the famous balcony from Shakespeare's *Romeo and Juliet*. Historians and music-lovers head to the breathtaking 2000-year-old Roman Arena, which comes alive when hosting operas like *Aida*. Chic shoppers appreciate designer wares, especially shoes. And everyone can tuck into hearty Veronese *cucina casalinga*, delicious and surprisingly affordable home-cooking, and local bubbly Prosecco. For the perfect final mix of culture and wine, take bus 21 to visit Villa Santa Sofia in nearby Valpolicella's vineyards, designed by Palladio in 1565 and today open for wine tastings.

Verona Villafranca airport has international flights.

ALTERNATIVES

Extension: Mantua

Mantua is an unspoiled medieval city, an historic cultural hub for music, art and literature. Explore the World Heritage centre, passing remarkable Baroque and Renaissance architecture – especially the sprawling Palazzo Ducale, home of the Dukes of Mantua. Mingle with locals over *aperitivi* in cosy arcade bars around Piazza delle Erbe.

Hourly trains run from Verona Porta Nuova station to Mantua (50min).

Extension: Bassano del Grappa

Beneath the towering Monte Grappa, the city of Bassano comes to life on Thursday and Saturday mornings during its colourful markets. The Municipal Museum has a good collection of fine art, but pride of place goes to Palladio's covered wooden bridge across the Brenta River. At the entrance, try Italy's most famous grappa at Grapperia Nardini.

Trains run to Bassano from Venice (every 30min; 90min).

SPORADES ISLAND-HOPPER

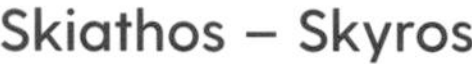

Skiathos – Skyros

Ferry-hop around the aquamarine Aegean between these pine-forested, beach-blessed Greek islands.

FACT BOX

Carbon (kg per person): 20.4
Distance (km): 254
Nights: 12–18
Budget: $$
When: late Apr to early Oct

❶ Skiathos

Anyone on the hunt for Greece's most perfect sands will love seductive Skiathos. Its 65 beaches range from pristine powdery blonde strands (hello Koukounaries!) to tiny pebble coves (hit the north coast). Stay three or four nights for snorkelling, diving, hiking, swimming and sailing, and seek out the former home of 19th-century Greek writer Alexandros Papadiamantis, as well as the 18th-century Evangelistrias Monastery (where the Greek flag was first raised) and the romantically ruined old capital, Kastro. Skiathos Town is a buzzy jumble of boutiques, boats, guesthouses and tavernas, and there's a handy beach bus.

Daily ferries run between Skiathos Town and Skopelos, stopping in Glossa/Loutraki (45min) and Skopelos Town (1hr).

ABOVE: SKYROS

❷ Skopelos

Despite stealing the show in *Mamma Mia!*, rugged Skopelos is all about wild natural beauty, arty vibes and a supremely laidback charm. Sparkling pebble beaches, peaceful hiking paths, endless olive groves, jagged coastal cliffs, divine cuisine and at least 40 monasteries and churches await here. Skopelos Town is a wonderfully scenic little port with terrific restaurants, watersports outfitters, creative arts-and-crafts boutiques and a lively music scene. Hiking, cycling and buses get you around.

Up to three ferries a day run from Glossa/Loutraki and Skopelos Town to Patitiri on Alonnisos (45min-1hr).

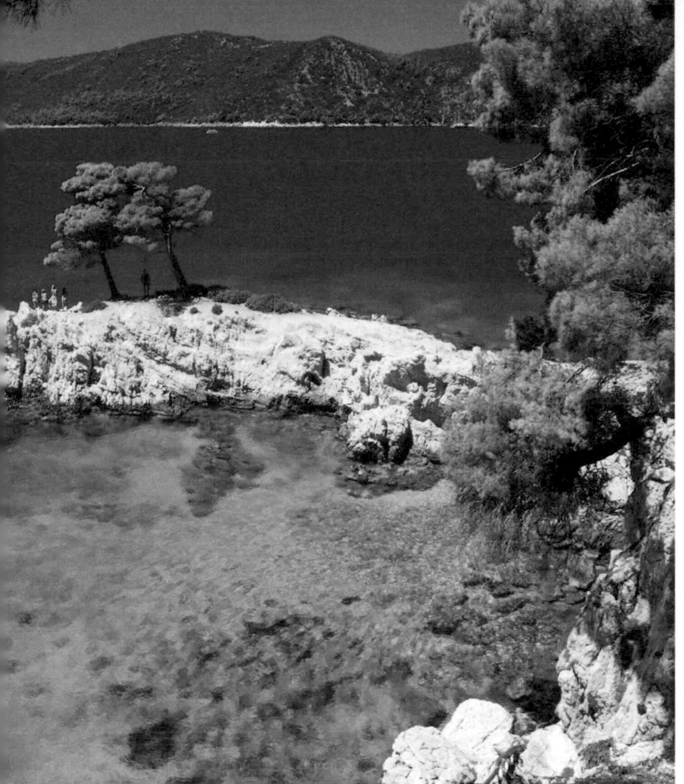

ABOVE: SKOPELOS

❸ Alonnisos

Distant, delightful Alonnisos remains the least-visited of the populated Sporades. It's absolutely worth a few nights, with its glittering pine-fringed bays, quiet walking trails, perfumed hillsides, creatively stylish accommodation, excellent diving and unmissable marine park. Explore Patitiri harbour, fairytale Old Alonnisos, tiny seaside Steni Vala and the east coast's untamed pebble beaches.

From mid-June to mid-September, three ferries a week link Alonnisos with Linaria on Skyros (5hr 15min); otherwise, change ferries in Evia.

❹ Skyros

So far south it feels more like the Cyclades, this arid, mellow and fascinating island is famed for its ancient crafts and endangered horses. Three or four nights give enough time for beach lazing, Molos' rock-hewn church, the Palamari Bronze Age ruins and sloping Skyros Town, with an imposing Byzantine-Venetian fortress, a brilliant folklore museum and some spectacular cuisine. Meeting island artisans and conservationists working to save the Skyrian horse is a unique highlight.

Skyros has several weekly flights to Athens for worldwide onward connections.

ALTERNATIVES

Day trip: National Marine Park of Alonnisos

One of Greece's most pristine natural spaces, this huge, strictly protected marine park (Europe's largest) is home to dolphins, turtles, whales, rare seabirds and Mediterranean monk seals. Take a boat trip between turquoise waves, sea caves and remote islets such as Kyra Panagia with its centuries-old monastery.

Full-day boat trips depart from Patitiri harbour on Alonnisos and Skopelos Town.

Diversion: Evia

Greece's second-largest island, Evia remains off the radar. Rent a car and explore pretty beaches, ancient Eretria, one-of-a-kind wineries, thermal baths and some epic hikes.

Ferries link Skopelos and Alonnisos with Kymi on Evia three times a week mid-Jun to mid-Sep; year-round ferries connect Kymi with Skyros. One to two daily ferries (May-Sep) link Skiathos and Skopelos with Evia's Mantoudi.

SPANISH BASQUE COUNTRY EXPLORER

Bilbao – San Juan de Gaztelugatxe

With its own culture, language and cuisine, the Basque Country stands apart from the rest of Spain. A road trip along its rugged coast feels like a step back in time.

FACT BOX

Carbon (kg per person): 47
Distance (km): 280
Nights: 7-10
Budget: $$$
When: May-Oct

❶ Bilbao

Welcome to the capital of the Basque Country. Bilbao is perhaps best known for the striking Guggenheim Museum Bilbao, designed by Frank Gehry. But there are plenty more architectural highlights to take in while strolling around the historic centre, stopping along the way in picturesque plazas and pavement bars where locals sip Txakoli, a sparkling white wine that's produced in the region.

Drive 100km east.

ABOVE: BASQUE *PINTXO*

❷ San Sebastián

There's so much to like about San Sebastián: beautiful beaches, elegant architecture and, of course, a food and wine scene to rival any in Europe. There are more places to eat well here than you could possibly cover in a single visit, from Michelin-starred fine dining to crowded *pintxo* bars serving local wines and bite-sized gourmet specialities. After dark, stroll along the gorgeously illuminated bayfront promenade and into the cobblestoned labyrinth of the Parte Vieja, or old city.

Head 25km west.

❸ Getaria

This charming port town is a pleasure to wander around or break for a leisurely lunch with wine pairings. Step into 15th-century churches and sample locally produced anchovies in the gourmet shops

ABOVE: SAN SEBASTIÁN

along the main street. One of its former residents became a star in the fashion world: stop into the Cristóbal Balenciaga Museoa to learn more about his life and legacy.

Continue 60km west.

4 Lekeitio

A traditional Basque fishing village with a scenic lighthouse and a busy working port, Lekeitio is a great place to soak up some local flavour. Look for the gilded Gothic altarpiece inside the grand basilica, take in panoramic ocean views from the hermitage of San Juan Talako, and have a seat in a pavement cafe for coffee or vermouth.

Drive 60km west.

5 San Juan de Gaztelugatxe

Photogenic San Juan de Gaztelugatxe – the name means 'Rock Castle' in Basque – has to be seen to be believed. A rocky islet in the Bay of Biscay, it's connected to the mainland by a man-made bridge. Hiking across it, and up to the hermitage perched on top of the island to ring the old church bell, is a rite of passage for travellers in the Basque Country. From here, it's an easy trip back to Bilbao.

Drive 35km south back to Bilbao.

ALTERNATIVES

Extension: Bermeo

You'll work up an appetite hiking at San Juan de Gaztelugatxe so detour to the sleepy fishing village of Bermeo for *pintxos* and wine overlooking the marina. Explore the steep streets around the old port, lined with colourfully painted houses. Dotted throughout town you'll notice a coat of arms featuring a whale: traditionally, Bermeo residents were whalers.

Bermeo is a 12km drive east of San Juan de Gaztelugatxe.

Extension: Mundaka

Surfers won't want to miss a side trip to Mundaka. The Basque port village is world-famous for having some of the longest waves in the world. A few surf schools and outfitters in town provide equipment, lessons and excursions. If you just want to watch from the shore, a number of charming outdoor cafes offer views of the sea.

Mundaka is a 38km drive northeast of Bilbao.

ROADTRIPPING SARDINIA'S EAST COAST

Cagliari – Orosei

Cliff-hanger views, dreamy beaches, Pisan towns and absorbing prehistory enthral on a drive along the east coast of this astoundingly scenic Italian island.

FACT BOX

Carbon (kg per person): 77
Distance (km): 617
Nights: 10
Budget: $$
When: Apr-Oct

❶ Cagliari

The Sardinian capital rises from the sea in a helter-skelter of gold-hued *palazzi*, domes, Pisan towers and pastel-painted facades. Devote a day or two to lounging in cafes in rocky hilltop citadel, Il Castello, and dipping into museums spotlighting archaeology and contemporary art.

Drive the snaking 60km coastal road east.

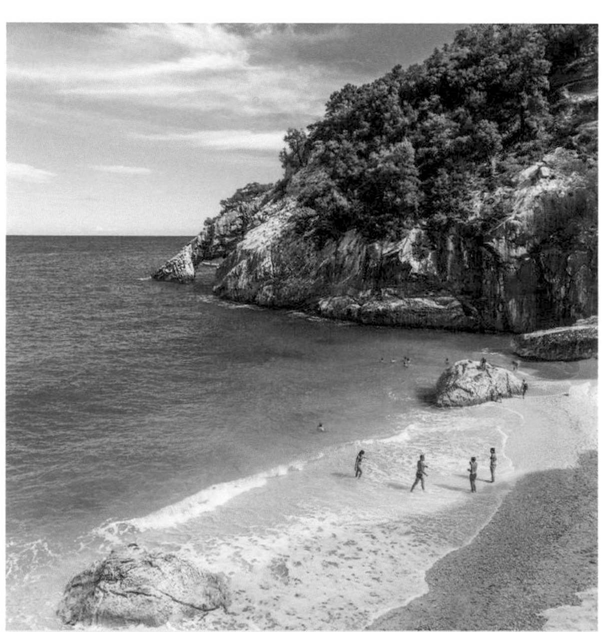

ABOVE: CALA GOLORITZE, NEAR OROSEI

❷ Villasimius

Once a fishing village and now a laidback resort, Villasimius is blessed with incredible frost-white beaches shelving into the glass-blue Med, with the forested mountains of the Sarrabus behind. It's also the springboard for Capo Carbonara, a protected marine park with secluded bays and excellent diving.

Drive 117km north on the spectacular, hairpin-bend-riddled road to Baunei.

❸ Baunei

Backed by the limestone cliffs, peaks and canyons of the Supramonte massif, this shepherds' village clings to a precipitous rocky ridge. Reached by steep switchbacks, its must-see is the otherworldly Golgo plateau. From here, descend on foot via an old mule trail through holm oak woods to the out-of-this-world bay of Cala Goloritzé.

Continue 52km north.

❹ Cala Gonone

Cala Gonone is at the heart of the astonishingly

ABOVE: CAGLIARI

beautiful Golfo di Orosei, where the high mountains of the Gennargentu collide with the sea, forming dramatic cliffs riven with wood-cloaked ravines and horseshoe-shaped bays pummelled by aquamarine water. Come to swim, climb, kayak, dive, hike or boat along the coast to coves like Cala Luna.

Back in the car, head 29km north.

5 Orosei

Orosei is surrounded by marble quarries and orchards. Once a Pisan port, it's now an atmospheric, refreshingly untouristy town, with stone houses, cobbled lanes and courtyards. A 5km ribbon of pale-gold sand lapped by topaz sea runs south, while sublime hidden bays like juniper-backed Cala Ginepro are a pebble's-throw to the north.

It's an 85km drive north of here to Olbia.

6 Olbia

Olbia is a cultured little city, with a fetching old town crammed with boutiques, wine bars and cafe-filled squares. Its showstopper is the Museo Archeologico, detailing Sardinian history in artefacts, from Bronze Age to Roman finds.

Taking the quickest route through the island's heart, it's a 274km drive back to Cagliari.

ALTERNATIVES

Extension: Costa Smeralda

Olbia is the gateway to the ritzy Costa Smeralda (Emerald Coast), so named for its brilliant green waters. The main hub is Porto Cervo, a fantasy village attracting the mega-rich and their yachts. The coast immediately south is dotted with gorgeous crescent-shaped bays, among them the Aga Khan's favourite, Portu Li Coggi.

Porto Cervo is a 30km drive north from Olbia.

Diversion: Gola Su Gorropu

The Ghenna Silana Pass is the trailhead for one of Sardinia's most spectacular day-hikes to Gola Su Gorropu, where limestone walls tower up to 500m. With sturdy shoes and sufficient water, you can wander some 8km into the boulder-strewn ravine without climbing gear. Allow four hours for the 16km return hike.

The trailhead is at Km 183 on the SS125, midway between Baunei and Cala Gonone.

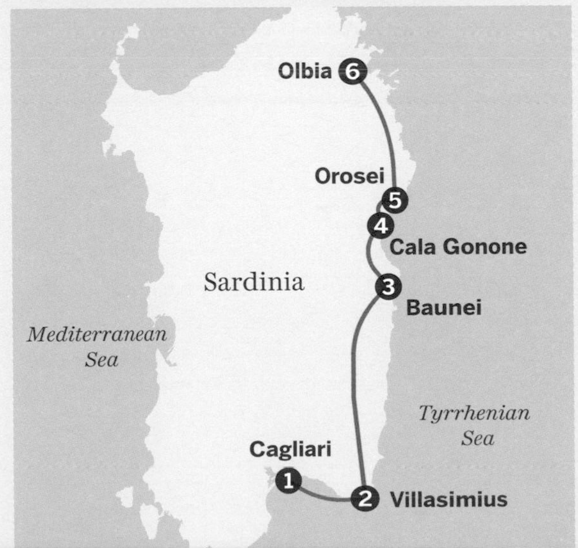

UNDISCOVERED ITALY IN UMBRIA

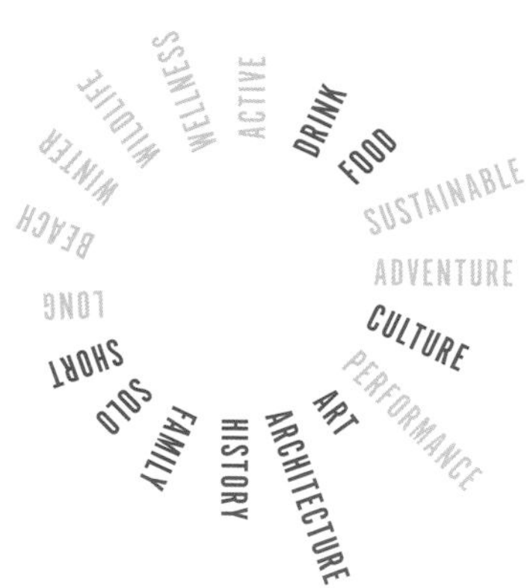

Perugia – Orvieto

Umbria delights with divine hill towns, rich history and delicious food on this epic Italian road trip.

FACT BOX

Carbon (kg per person): 26
Distance (km): 208
Nights: 5
Budget: $$
When: Apr-Oct

❶ Perugia

Sitting high above a valley where the River Tiber flows, Umbria's capital pulses with a lively cafe and food scene and party-loving student population. Cobbled alleys and arched stairways lead to ornate basilicas and grand piazzas in the *centro storico* (historic centre). Allow time to see Gothic Palazzo dei Priori's treasures and works by homegrown heroes Pinturicchio and Perugino at Galleria Nazionale dell'Umbria.

Drive 25km east to Assisi.

❷ Assisi

St Francis of Assisi, born here in 1181, made this strikingly beautiful town one of Italy's foremost pilgrimage destinations. Rising poetically above the plains, its pristinely preserved medieval centre and Unesco World Heritage Franciscan basilicas are by no means a secret: come in the early morning or late afternoon to experience them at their peaceful best.

Continue 12km south.

❸ Spello

Pretty Italian hill towns are ten-a-penny, but Spello really manages to stand out. Flowerpots and hanging baskets festoon the honey-coloured houses that line its

ABOVE: BASILICA DI SAN FRANCESCO, ASSISI

ABOVE: ASSISI

higgledy piggledy cobbled lanes, which are guarded by three Roman gates. Go for a saunter and linger over lunch and local wines at a family-run *osteria*.

Take a lovely 43km drive on backcountry lanes via humorously named Bastardo.

4 Todi

With warm-stone houses, *palazzi* and belfries pasted to a steep hillside, Todi is insanely photogenic. Its backstreets curl up to the *centro storico*, layered with Etruscan, Roman and medieval history. Slow the pace with a *passeggiata* (stroll) around medieval Piazza del Popolo.

Drive a highly scenic 48km west along the forest-rimmed shores of Lago di Corbara.

5 Orvieto

Sitting astride an extinct volcano, above vineyards, olive groves and cypress trees, Orvieto is stunning. The Gothic *duomo* (cathedral), one of Italy's most ravishing, dominates the historic centre, its black-and-white banded facade a lavish feast of frescoes and mosaics. Besides this there are medieval towers, churches, archaeological museums and a warren of underground passages and caves to discover.

Drive 80km northeast back to Perugia.

ALTERNATIVES

Day trip: Spoleto

A cinematically set hill-town at the foot of the Apennine mountains, Spoleto promises Roman arches, amphitheatre, archaeology museum, a Romanesque cathedral with mosaic frescoes, and the Ponte delle Torre, a 10-arch medieval bridge that leaps across a wooded gorge. The food is sensational, too: try *strangozzi* (literally 'priest strangler') pasta with tomato, garlic and chilli.

Spoleto is 46km east of Todi.

Diversion: Norcia

If you make it to Spoleto, continue east and the mountains become more rugged and the landscapes wilder on the edge of Parco Nazionale dei Monti Sibillini. Here, in the wildflower-freckled Valnerina valley, lies Norcia, a pleasing little town that sends gastronomes into raptures – it's prized for its *salumi* (cured meats) and *tartufo nero* (black truffle), available in local shops and restaurants.

Norcia is a 44km drive east of Spoleto.

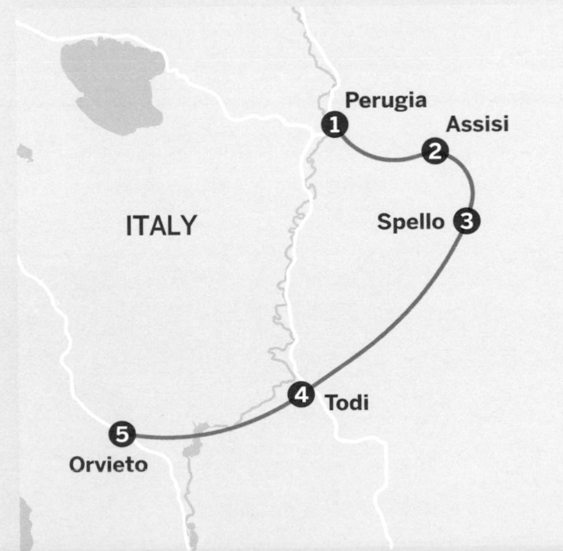

NAPLES AND THE AMALFI COAST

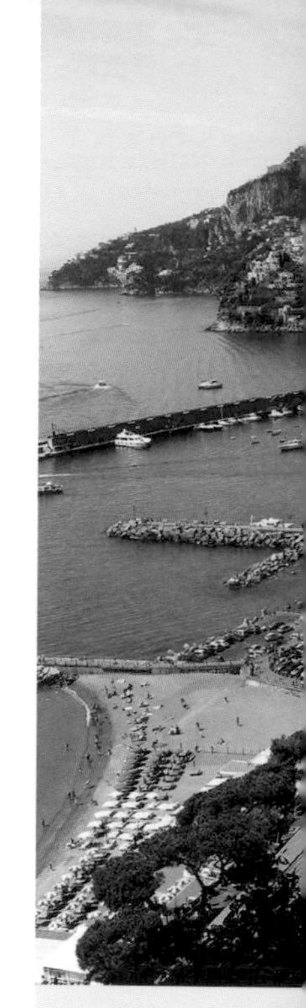

Naples – Amalfi Coast

Italy's iconic Amalfi region mixes perfect pizza with ancient archaeology, and a brooding volcano with one of the world's most glamorous islands.

FACT BOX

Carbon (kg per person): 5
Distance (km): 175
Nights: 7-10
Budget: $$
When: Apr-Jun, Sep, Oct

ABOVE: FOOD MARKETS IN NAPLES

❶ Naples

Naples (easily accessed via its international airport or a quick train ride from Rome) is sprawling, scruffy, edgy, arty, authentic – and delicious (the food that is). There's no tourist sheen, but plenty worth seeing. The Museo Archeologico Nazionale, with its Pompeiian finds, is a must. Quirkier are Napoli Sotterranea, an underground tour of the city's ancient aqueducts, burial chambers and air-raid shelters, or an offbeat walk in the neighbourhoods made famous by Elena Ferrante's *Neapolitan* novels.

Take the train to Sorrento (every 30min; 1hr).

❷ Sorrento

Cliff-top Sorrento, which stares out to Vesuvius over the Bay of Naples, is gateway to the Amalfi Coast, and a charming spot in itself with a handsome historic centre. A warm evening spent taking a *passeggiata* through Piazza Tasso is never wasted. The town has plenty of glitzy accommodation, but consider an *agriturismo* out of town, where you can stay with a local family, sleep amid fragrant lemon groves, kneed pizza dough in the *nonna's* home kitchen and have a go at making your own fresh-from-the-tree limoncello.

Take the ferry to Capri (around 10 daily; 25min).

ABOVE: AMALFI TOWN

❸ Capri

Touristy, yes. Crowded, yes. But, oh the glamour! This steep-sided isle, bursting from sapphire seas, has been beloved by all, from Roman Emperors to Hollywood A-listers. Walk (or ride the chairlift) to its summit, sail around its sides and look out from its flower-draped villas. And make sure to let an opera-belting boatman row you into the ethereal blue grotto – cheesy but worth every Euro.

Back in Sorrento take a bus (1-2 hourly; 1hr 40min) or ferry (1 daily, Apr-Oct; 1hr 20min) to Amalfi.

❹ Amalfi Coast

Spend a day – or a week – dipping in and out of this iconic coastline's colourful cliff-tumbling towns: pretty Amalfi and neighbouring whitewashed Atrani; refined hilltop Ravello; impossibly steep and snazzy Positano. Work up a thirst for an ice-cool *granita* by getting to the latter via the Sentiero degli Dei (Path of the Gods), a precipice-hugging 8km hike from Bomerano to Positano, with unmatchable views up the terraced hillsides and across the sparkling seas (buses run Amalfi-Bomerano to access the walk).

Return to Naples by bus (from Amalfi), by bus and train (via Sorrento) or ferry (in season).

ALTERNATIVES

Day trip: Pompeii

Nowhere offers greater insight into ancient Roman life than Pompeii. Thanks to Vesuvius blowing its top in 79 CE, the streets, villas, brothels, baths and unfortunate residents have been preserved under ash; walking around this extensive archaeological site is like time travel.

Take the train from Naples to Pompei Scavi-Villa dei Misteri Station (every 30min; 45min).

Extension: Ischia

Far less fancy than Capri, volcanic Ischia – the largest of the islands in the Bay of Naples – is a breath of fresh Italian air. Spend a few lazy days bar-hopping in lively Ischia Porto, inhaling the blooms at La Mortella (one of the country's loveliest botanical gardens) and lolling around in laidback car-free Sant'Angelo.

Frequent ferries run from Naples to Ischia (every 3hr; 1hr).

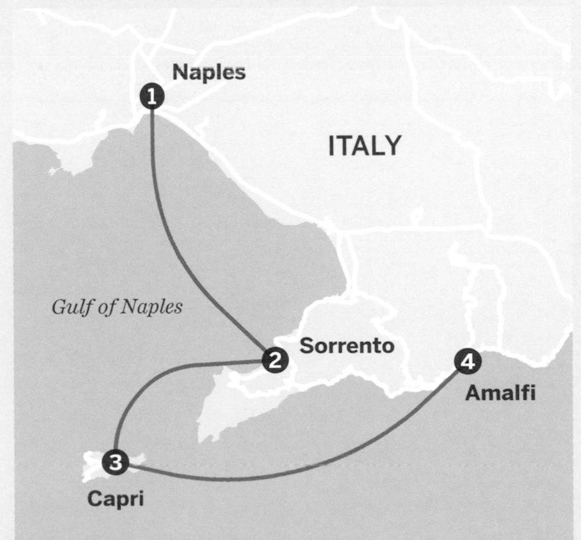

BIKE AND HIKE THE SIGHTS OF VALENCIA

Jardín del Turia – Malvarrosa

Capital of its eponymous region, the Spanish city of Valencia has culture, gastronomy, beaches, lagoons and even paddy fields – all easily explored on foot or by bike.

FACT BOX

Carbon (kg per person): 0
Distance (km): 30
Nights: 3-5
Budget: $$
When: year-round

❶ Jardín del Turia

Not many Spanish cities lend themselves to being explored by bike, but Valencia is an exception – in particular along the former Turia river bed, now turned into an 8km-long lung of beautiful parkland (the river was diverted after much flooding). You can rent a bike in Parque de Cabecera and cycle through the Jardín del Turia past the flowers, bridges and coffee shops dotted along the trails.

Head 1km south to the Ciutat Vella.

❷ Ciutat Vella

Slow down the pace and wheel your bike through the parks, squares and streets of Valencia's medieval Old Town (Ciutat Vella). The architecture spanning 700 years here is mesmerising, from the 12th-century cathedral containing two Goyas and the rude gargoyles leering down from the Lonja de la Seda (Silk Exchange) to the 13th-century cloisters of the Museo Convento del Carmen, now containing art exhibitions. Stops for ice-cold *horchata* and sweet *fartons* add to the appeal.

Continue southeast into neighbouring Eixample.

❸ Eixample

The delicate Art Nouveau ironwork and tiling of the Mercado de Colón and the soaring halls of the

ABOVE: VALENCIA'S OLD TOWN

ABOVE: CIUDAD DE LAS ARTES Y LAS CIENCIAS

Estación del Norte train station are just two highlights of the Modernista Eixample area.

Head east to rejoin the Jardín del Turia trails and cycle south to the Ciudad de las Artes y las Ciencias.

4 Ciudad de las Artes y las Ciencias

Conceived by the world-famous bridge designer and local boy Santiago Calatrava, this eye-popping arts and science park is home to, among many other things, L'Oceanogràfic, Europe's biggest marine park, where a meal at the Submarino restaurant feels like dining inside Jules Verne's *Twenty Thousand Leagues Under the Seas*. Outside, giant pools, fountains, terrace cafes and playgrounds create a playful recreation area.

Take the cycle path 5km northeast to Malvarossa.

5 Malvarrosa

Cycling along the Mediterranean promenade to the huge sandy stretches of Las Arenas and adjacent Malvarrosa is a delight. The beaches are stunners, and the paella houses that stretch across the back of Malvarrosa beach are the perfect stop for some lunchtime refuelling and people-watching.

Head inland and back through Jardín del Turia to drop the bike off in Parque de Cabecera.

ALTERNATIVES

Day trip: Xàtiva

The train ride to Xàtiva begins at Valencia's stunning Art Nouveau Estación del Norte and ends at this charmingly cinematic town under a castle-topped mountain. The particularly lovely Old Town is an intriguing mix of Moorish, Gothic Renaissance and 15th-century architecture, with some excellent examples of crossover in styles too, notably the cathedral and the old hospital.

Take the train from Estación del Norte (every 20min; 50min).

Day trip: Albufera

The extensive Albufera lagoon has been home to rice paddies for 600 years – take a boat trip to properly immerse yourself in them, then explore the surrounding coastline, including El Saler, the paella restaurants of El Palmar and the 5km-long La Devesa beach, whose rich ecosystem provides habitats for a wide variety of vegetation and wildlife.

Albufera is a 20km cycle ride south of Valencia.

STEP THROUGH MILLENNIA IN MALTA

Valletta and the Three Cities – Mellieha
From Neolithic complexes thought to be temples to Byzantine and even WWII sites, Malta has a fascinating history mapped onto its natural beauty.

FACT BOX

Carbon (kg per person): 13
Distance (km): 70
Nights: 7-10
Budget: $$
When: year-round

ABOVE: MDINA

❶ Valletta and the Three Cities

Malta's capital Valletta is an historical and cultural maze. Five centuries of architecture and the excellent National Museum of Archaeology nicely contextualise its past physically, while St John's Co-Cathedral, home to Caravaggio's *Beheading of St John the Baptist*, does the same spiritually. The Upper Barrakka Gardens offer a fantastic overview of the capital and the ancient Three Cities harbour towns, among which 16th-century Birgu adds more layers of history and culture to an already rich feast with its Collacchio area, Inquisitor's Palace and Malta at War Museum.

Drive 18km to Ħaġar Qim via Tarxien and the Blue Grotto.

❷ Tarxien to Ħaġar Qim and the Blue Grotto

At Tarxien, take in the first of Malta's prehistory sites at the Unesco-listed Ħal Saflieni Hypogeum, then be even more impressed at Mnajdra, where the megalithic complex Ħaġar Qim dates from 3600-2500 BCE (earlier than the Great Pyramid of Giza and Stonehenge). On the coast, the Blue Grotto is a big draw; visit it and other caves on a half-hour boat trip from tiny harbour of Wied iż-Żurrieq, set in a narrow inlet in the cliffs in the seaside village of Żurrieq.

ABOVE: VALLETTA

Drive 9km to Ħad-Dingli. From there, turn inland and follow signs 5km to Mdina.

3 Ħad-Dingli and Mdina

Ħad-Dingli, and in particular its cliffs, are a great spot from which to admire Malta's dramatic coastline and to watch (or join) the divers heading into waters that offer 40m visibility. Inland, the hilltop fortress town of Mdina is indeed medina-like, but the miniature old city boasts more than just Arab influences, with traces of a Jewish quarter, buildings from the medieval tenure of the Knights of the Order of St John, and even Roman mosaics in Domvs Romana.

Drive 12km to Mellieha.

4 Mellieha

The sweet town of Mellieha sits above a WWII air-raid shelter consisting of 500m of tunnels hand-hewn from solid rock. Nearby, the country's impressive screen history – *Midnight Express*, *Gladiator*, *The Count of Monte Cristo* and *Game of Thrones* were all filmed here – is playfully undermined by the set of the 1980 live-action movie *Popeye*, built on a beautiful little bay.

Drive 23km back to Valletta (or continue north to Cirkewwa for the 25min ferry ride to Gozo).

ALTERNATIVES

Extension: Gozo

Take the ferry to Gozo to explore Ggantija, Malta's earliest temple complex, and Neolithic Xagħra Stone Circle. At lovely Xwejni bay, people have been panning salt since medieval times. The hulking 20th-century neo-Gothic church Ta' Pinu, 5km southwest, looks decidedly out of place rising out of the dusty plains.

Car/passenger ferries to Gozo run from Cirkewwa or Valletta, with frequency and duration depending on the seasons.

Day trip: Comino

Tiny, rocky and car-free Comino, wedged between Malta and Gozo, makes a great day trip, particularly if you're a keen walker. Given its 2.5 by 1.5km size you can circumnavigate the entire island easily, and end with a dip in the Blue Lagoon, a sheltered cove whose colour has to be seen to be believed.

Regular ferries run to Comino from Cirkewwa on Malta and Mġarr Harbour on Gozo (both 25min).

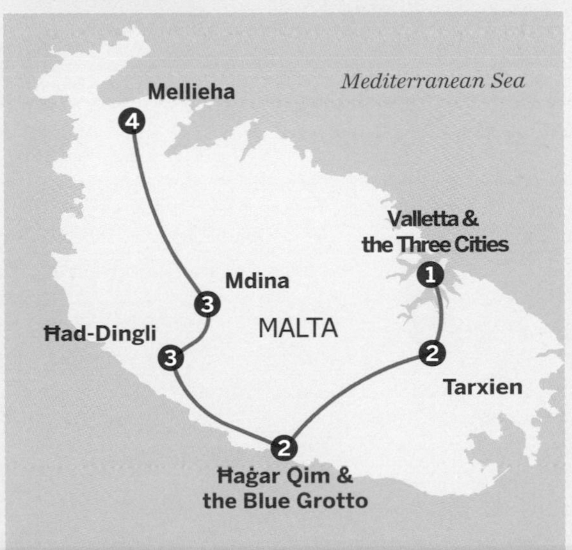

PILGRIMAGE THROUGH NORTHERN SPAIN

Pamplona – Santiago de Compostela
Follow in the footsteps of countless pilgrims on this train trip across northern Spain, taking in lesser-known cities, fine food and plenty of exquisite cathedrals.

FACT BOX
Carbon (kg per person): 14
Distance (km): 710
Nights: 7-10
Budget: $$
When: Easter-Sep

❶ Pamplona

Famous for July's wild Fiesta de San Fermín, with its daily *encierro* (running of the bulls), Navarre's capital is almost as appealing during the rest of the year thanks to its vast citadel, a compact and atmospheric centre, and city walls offering views down to the plains and up to the Pyrenees. Like every city on this tour, it features on the Camino de Santiago pilgrimage route, and the section between here and Burgos is a popular one, taking in monasteries, vineyards, towns and gorgeous mountain views.

Take the train to Burgos (2 daily; 2hr).

❷ Burgos

You might visit Burgos for its cathedral, an imposing, intricate and extraordinary Gothic marvel that's one of the finest in Spain – and which contains the body of El Cid, the legendary medieval hero of Spain's Reconquista. Or you might come for the *morcilla* (blood sausage), revered as the best in the country and offered in many local tapas bars. Either way, don't be put off by the rather soulless modern suburbs where the train station is located – Burgos' Old Town is an underrated jewel in the Spanish crown.

Continue on the train to León (3 daily; 2hr).

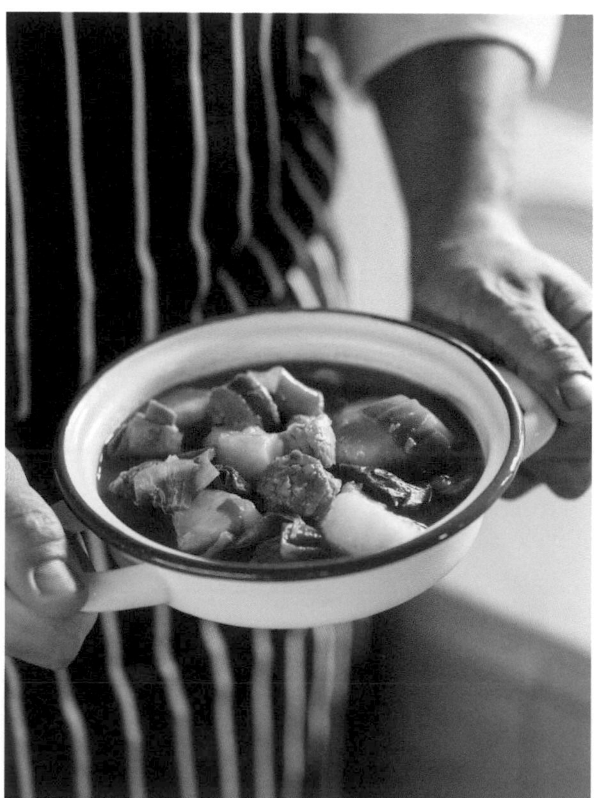

ABOVE: BASQUE TAPAS: *PATATAS A LA RIOJANA*

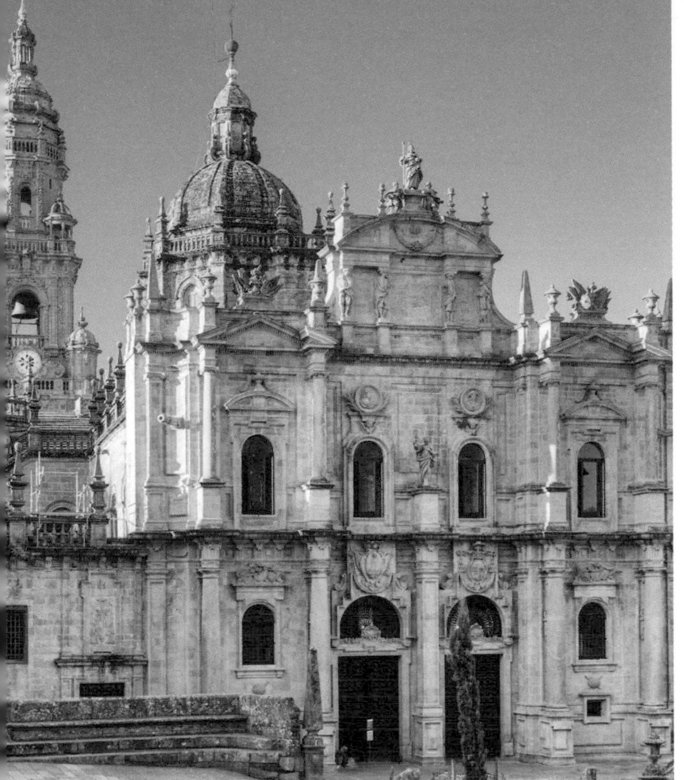
ABOVE: SANTIAGO DE COMPOSTELA CATHEDRAL

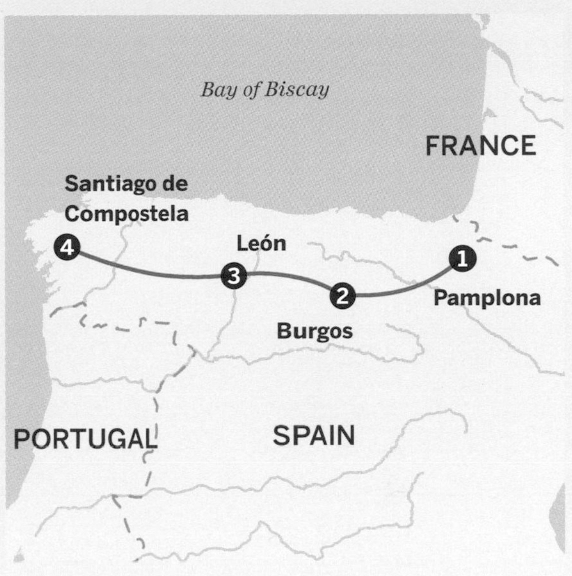

❸ León

Did we mention cathedrals? León's is truly magnificent, built when the city was still one of the driving forces in the reconquest of Spain, and its vast stained-glass windows create a magical light within. There's great food and drink here too, with a phenomenal (and meaty) selection of tapas bars and lively nightlife. Romanesque frescoes, modern art and one of Spain's most famous Semana Santa (Easter) celebrations complete the picture.

Take the train to Santiago de Compostela (2 daily; 4hr 30min).

❹ Santiago de Compostela

The end point of the Camino de Santiago and of your trip, Santiago de Compostela is something special. Its twisting lanes, arcaded streets and, you guessed it, wonderful cathedral astound, whatever the weather – this being Galicia, rain is a regular visitor. All that water feeds a lush landscape, and if you prise yourself away from the enchanting architecture you can chow down on some of Spain's best food and drink: succulent seafood, punchy green peppers from nearby Padrón and light, crisp white wines.

Santiago de Compostela's airport has international connections.

ALTERNATIVES

Extension: A Coruña

Jutting out from Galicia's northern coast, buzzing A Coruña is an underrated gem. Here you'll find beaches, cutting-edge Aquarium Finisterrae, and museums and parks, many accessible via a walkway circling the peninsula. The Tower of Hercules lighthouse offers great views, and the Old Town's alleys are packed with bars and restaurants.

Trains connect A Coruña with Santiago de Compostela (every 30min; 30min).

Extension: Fisterra

Most travellers end their pilgrimage in Santiago de Compostela, but the truly devout and/or enthusiastic continue to Fisterra (Finisterre). This is where, tradition has it, you strip naked, burn your clothes on the beach and rush into the Atlantic as a final act of absolution. Or just enjoy dramatic views along the Costa da Morte (Coast of Death), named for its treacherous ship-wrecking rocks.

Buses run to Fisterra from Santiago, or you could walk (80km).

NORTH
ATLANTIC
OCEAN

WESTERN EUROPE

Want world-class museums, architectural marvels and Roman-era treasures? Or ravishing national parks, quaint English gardens and off-grid beaches? See the best of Western Europe via train trips into the Alps, slow-boat canal journeys, riverside cycle routes or city walking tours – and don't forget the food and drink, from the freshest fish and chips to brilliant Belgian beers and world-beating French wines or Parisian patisserie.

WESTERN EUROPE
BALTIC SEA
NORTH SEA
SCOTLAND
IRELAND
WALES
ENGLAND
ISLES OF SCILLY
NETHERLANDS
BELGIUM
LUXEMBOURG
GERMANY
FRANCE
SWITZERLAND
AUSTRIA
MEDITERRANEAN SEA
BLACK SEA
AEGEAN SEA
29
18
34
5
6
2
14
33
38
41
11
12
7
40
17
35
15
27
4
28
10
3
26
30
31
1
13
43
22
21
24
25
16
9
36
23
41
19
20
32
37
39
8

CONTENTS

ABOVE: THE BLACK FOREST, GERMANY

ABOVE: ST MAWES, CORNWALL

ALSACE'S ROUTE DES VINS, FRANCE

Marlenheim – Colmar

Go slow on one of France's most evocative drives, with cellars, vineyard strolls and half-timbered villages straight out of a fairy tale.

FACT BOX

Carbon (kg per person): 20.5
Distance (km): 164
Nights: 7
Budget: $$
When: year-round

❶ Marlenheim

Vine-framed Marlenheim, 22km from Strasbourg and its international connections, is the wine route's northern gateway. Right in the half-timbered heart of things on Place du Kaufhaus, you can taste local Pinot, Riesling and Gewürztraminer wines at family-run Vins d'Alsace Mosbach, doing business for almost 500 years.

Drive 25km south.

ABOVE: COLMAR

❷ Obernai

Ringed by fortified walls, Obernai is a vision of half-timbered, vine-draped, alley-woven loveliness. On Place du Marché, the neo-Renaissance town hall and bell-topped Halle aux Blés (Corn Exchange) demand attention. Wander the ramparts or the 1.5km wine trail before lunch in a cosy beamed *winstub* (wine tavern) like Le Freiberg.

Continue 23km south.

❸ Dambach-la-Ville

The soaring summits of the Vosges mountains hover above walled Dambach-la-Ville, where the ancient gates are topped by stork nests. Frankstein grand cru vines encircle this town of gabled houses and wine cellars. Le Pressoir de Bacchus is a solid choice for local grub like *choucroute garnie* (sauerkraut with smoked meats).

Drive 24km south.

ABOVE: RIQUEWIHR

❹ Ribeauvillé
Ribeauvillé is a delight, with hilltop castle ruins, medieval towers and winding lanes stacked with candy-coloured, half-timbered houses. Get versed on Alsatian wine over a tasting at the ultra-contemporary Cave de Ribeauvillé, then lunch at the wonderfully rustic Pfifferhüs *winstub*.

Head south for 5km.

❺ Riquewihr
Riquewihr is pure Grimm's fairytale stuff, with mazy lanes, medieval ramparts, hidden courtyards and gingerbread houses that look good enough to eat. A pretty 2km trail leads through grand cru vineyards, stopping at wineries en route. Dinner at Michelin-starred La Table du Gourmet is the icing on the cake.

Drive 14km south.

❻ Colmar
The Alsatian capital is full of bridge-laced, canal-riven, half-timbered romance. Its illustrious past is recorded in museums celebrating local legends like Bartholdi (of Statue of Liberty fame), and treasures like the late-Gothic Isenheim Altarpiece, lodged in Dominican convent-turned-museum Musée Unterlinden.

Drive back to Strasbourg (73km).

ALTERNATIVES

Diversion: Château du Haut-Koenigsbourg
You'll never forget your first glimpse of the turrets and towers of Château du Haut-Koenigsbourg. Perched eyrie-like on a steep wooded hillside, the ramparts of this red-sandstone fantasy of a 900-year-old castle afford astounding views, which reach all the way to the Black Forest and Alps on clear days.

The castle is 17km southwest of Dambach-la-Ville.

Day trip: Strasbourg
With plenty of half-timbered buildings and a staggering Gothic cathedral crowning Unesco-listed centre, Grande Île, Strasbourg is a joy to explore. See big-hitters like the canals in Petite France, opulently Baroque Palais Rohan (hailed a 'Versailles in miniature') and contemporary art gallery MAMCS, but also devote time to the simple pleasures of the city's markets, patisseries and bistros.

Strasbourg is 22km east of Marlenheim.

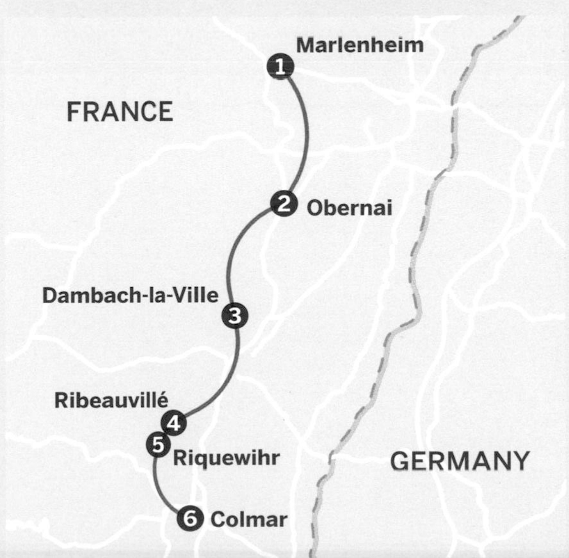

EMBRACE THE PAST IN ANCIENT IRELAND

Dublin – Glendalough

Take a scenic road trip to discover Ireland's most venerable historic sights, from Neolithic tombs to early Christian monasteries.

FACT BOX

Carbon (kg per person): 104
Distance (km): 831
Nights: 7-10
Budget: $$
When: May-Oct

❶ Dublin

Wander Dublin's Georgian streets for clues to the island's early history with a visit to the wide-ranging collection of Bronze and Iron Age artefacts at the National Museum, the extraordinary 8th-century *Book of Kells* at Trinity College, and to Dublinia for Viking history.

Drive 50km northwest.

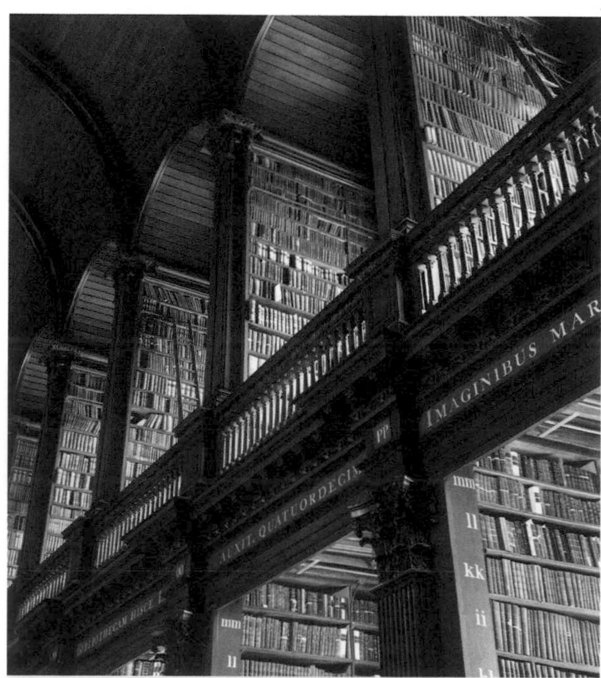

ABOVE: TRINITY COLLEGE LIBRARY, DUBLIN

❷ Newgrange

Ireland's finest passage tomb predates the Egyptian pyramids by six centuries and may have been a royal burial place or ceremonial site. Either way, its intricately carved stones and precise alignment with the winter sun prove its significance in Neolithic times.

Head 130km cross country.

❸ Clonmacnoise

Early churches, high crosses and round towers overlook the River Shannon at Clonmacnoise, once Ireland's most important ecclesiastical city. Founded by St Ciarán in 548 CE, it was soon attracting students from all over Europe.

Drive northwest for 130km.

❹ Knocknarea

This Neolithic stone cairn perched on a limestone plateau is thought to be the grave of legendary Queen Maeve. It takes about 45 minutes to climb

ABOVE: NEWGRANGE PASSAGE TOMB

but the views of the Atlantic and the flat-topped Benbulben (526m) are superb.

Continue 182km south.

5 Poulnabrone

Two massive stone uprights support a five-tonne capstone at Poulnabrone, an otherworldly grave marker dating back 5000 years. Once covered by an earthen mound, today it stands alone, eerily grand and evocative.

Continue 126km southeast.

6 Cashel

Rising starkly from the surrounding plain, the Rock of Cashel is crowned with a cluster of fortifications, a 13th-century cathedral and a fine 12th-century chapel.

Drive 160km northeast.

7 Glendalough

Two steely lakes nestle in a wooded valley at Glendalough, a tranquil spot whose splendour attracted St Kevin in 498 CE. His contemplative lifestyle drew legions of followers, making Glendalough the country's most significant monastic settlement by the 9th century.

Drive 53km north back to Dublin.

ALTERNATIVES

Diversion: Loughcrew

Little-visited and all the more atmospheric as a result, the 30 Stone Age cairns at Loughcrew are strewn across three hills that offer sweeping views of the surrounding countryside. Built around 3000 BCE, the largest is Cairn T, 35m in diameter and decorated with numerous carved stones.

The cairns are 53km west of Newgrange near Oldcastle.

Diversion: Gallarus Oratory

Several hours drive south of Poulnabrone but worth the slog for the exquisite craftmanship and glorious views, the Gallarus Oratory is a corbelled chapel in the shape of an upturned boat which dates back about 1200 years. For similar structures in an even more dramatic setting, take the boat to Skellig Michael.

Drive south through Limerick and Tralee to the Dingle Peninsula; the oratory is near Murreagh.

BERLIN'S HISTORIC BIG-HITTERS

Brandenburg Gate – East Side Gallery
This walking tour of Berlin lifts the lid on the German capital's epoch-defining role in 20th-century European history.

FACT BOX
Carbon (kg per person): 0
Distance (km): 13
Nights: 3
Budget: $$
When: year-round

❶ Brandenburg Gate

Anyone familiar with Ronald Reagan's 1987 speech imploring Gorbachev to tear down the Berlin Wall will recognise this Cold War landmark. The Neoclassical triumphal arch is most striking when illuminated at night. At the top is a quadriga, a horse-drawn chariot driven by the goddess of victory.

Walk 600m south along Ebertstrasse.

ABOVE: BRANDENBURG GATE

❷ Memorial to the Murdered Jews of Europe

Covering an area as big as two football pitches, this strikingly austere, maze-like memorial bears the imprint of American architect Peter Eisenman. With 2711 tomb-like concrete pillars of varying heights, it's a sombre and powerful testament to the murdered Jews of Europe.

Continue 1km south along Nazi nerve-centre Wilhelmstrasse.

❸ Topography of Terror

This moving exhibition spotlights the terror, persecution and brutality of Nazi Germany – from the rise of Hitler in 1933 to the postwar trials of SS officers – on the site where the Gestapo headquarters once stood. A walkable timeline, original documents, recordings and photos recall the horror.

Follow Wilhelmstrasse 1.3km south.

❹ Jewish Museum

American-Polish architect Daniel Libeskind put his indelible mark on this museum, which resembles a shattered Star of David with its zigzag design and titanium-zinc facade. The museum covers 2000 years

ABOVE: MEMORIAL TO THE MURDERED JEWS OF EUROPE

of Jewish-German life, leading visitors into concrete voids and along diagonal, intersecting axes.

Continue 1km north on Lindenstrasse.

5 Checkpoint Charlie

This former border checkpoint between East and West Berlin now has a replica hut, soldier's post and an open-air exhibition. For greater insight, the adjacent Mauermuseum compellingly narrates the rise and fall of the Wall and ingenious escape attempts in everything from submarines to hot-air balloons.

It's a 4km walk east via Köpenicker Strasse to the East Side Gallery (or a 30min bus ride).

6 East Side Gallery

Sitting along the Spree River in Friedrichshain, the longest remaining stretch of the Berlin Wall has been reborn as the 1.3km-long East Side Gallery, the world's largest open-air art collection. Its politically charged, tongue-in-cheek murals include Dmitri Vrubel's *Help Me To Survive This Deadly Love*, showing a passionate kiss between communist leaders Erich Honecker and Leonid Brezhnev.

Stroll 5km through central Berlin back to the Brandenburg Gate (or take the S-Bahn from Warschauer Strasse to Brandenburger Tor).

ALTERNATIVES

Extension: Stasimuseum

The real-life Big Brother, the Stasi were East Germany's secret police during Communist times. Lodged in the former grounds of the headquarters of the GDR Ministry for State Security, this museum dishes the dirt on their development, function and methods, most engrossingly with its collection of low-tech spying devices.

From the East Side Gallery, the museum is a 15min tram ride from Warschauer Strasse to Frankfurter Allee.

Diversion: DDR Museum

Spies next door, Skodas and sunburn on the Baltic – this time warp of a museum zooms in on everyday life in Communist East Germany. Designed like a prefab housing estate, it's a fascinating romp back in time: explore Soviet-era living rooms, rev the engine of a Trabi car and find out why nudist holidays were all the rage.

The DDR Museum is a 1.7km walk east of the Brandenburg Gate along Unter den Linden.

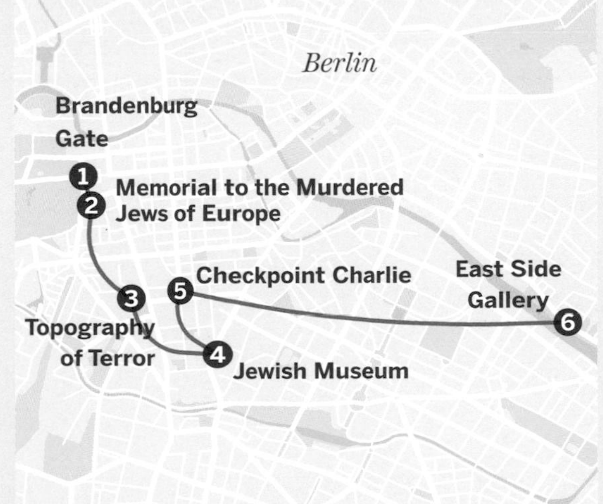

CHOCOLATE TOUR OF BRUSSELS

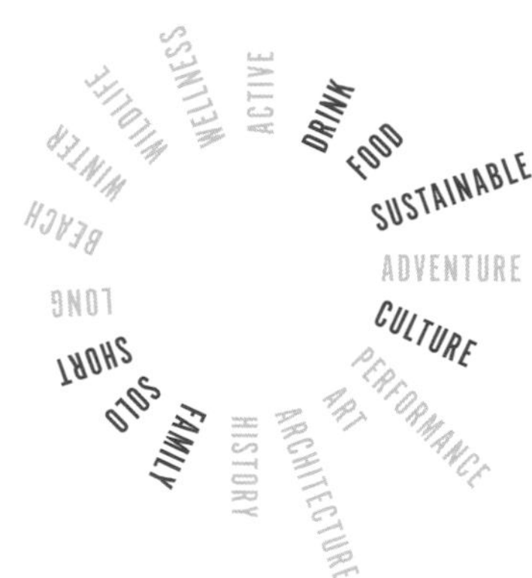

La Grande Place – Brasserie de la Senne
Discover the sights of Belgium's capital of chocolate on foot and by tram, tracking down the best pralines, truffles, bars, biscuits and hot chocolate.

FACT BOX

Carbon (kg per person): 0.3
Distance (km): 21
Nights: 2
Budget: $
When: year-round

❶ La Grand Place

Just a few minutes from Brussels' sumptuous Grand Place, arguably Europe's most beautiful square, lie some of the city's top chocolate addresses. Head southwest towards the famed *Mannekin Pis* statue to visit the newly opened Choco-Story, a brilliant interactive museum where your visit ends with a chocolate-making demonstration. Northeast of the Grand Place, in the gilded Art Nouveau Galerie de la Reine arcade, try a praline at Neuhaus, the chocolatier who invented this exquisite ganache-filled treat.

Walk southeast to Bozar Arts Centre (10min).

❷ Bozar Arts Centre

Housed in a landmark Art Deco building, Bozar hosts challenging exhibitions, concerts, cinema and installations. For post-art refuelling, Laurent Gerbaud's café-chocolatier is directly opposite, with to-die-for *chocolat chaud*, plus chocolate-making classes.

Walk southwest to Place Grand Sablon (10min).

❸ Place Grand Sablon

This majestic square is chocolate heaven, beginning with the gilded salons of distinguished royal suppliers Wittamer, famed for their *orangettes*: slivers of chocolate-covered confit orange. International shoppers will already know Leonidas and Godiva, while locals acclaim Atelier Sainte Catherine. Serious connoisseurs head for the flagship showroom and cafe of Pierre Marcolini, one of the world's leading chocolatiers.

From Petit Sablon tram stop, take tram 92 to Vanderkindere stop (30min).

ABOVE: BRUSSELS' ARTISANAL BEER AND CHOCOLATE

ABOVE: LA GRANDE PLACE

❹ Uccle

Next to the tram stop, funky chocolate store Mike & Becky specialises in single-plantation bars from sustainable producers in Belize, India and the Congo. Ten minutes' walk west through the chic residential Uccle neighbourhood, Jérôme Grimonpon's boutique cafe includes a glass-wallled bean-to-bar laboratory, where shoppers feast on chocolate cake while watching him invent innovative pralines and truffles.

Take tram 51 from Vanderkindere to Jubile (45min).

❺ Koekelberg

Fréderic Blondeel has created his own Willy Wonka factory where chocoholics witness the whole bean-to-bar production of his delicious products, passing through ingenious Heath Robinson-eque machinery that roasts, tempers and enrobes the beans.

Walk northeast to Brasserie de la Senne (25min).

❻ Brasserie de la Senne

Belgium is as famous for its beer as its chocolate, with the thriving artisan ale scene symbolised by Brasserie de la Senne. Their new state-of-the-art brewery offers tours and tastings and they even occasionally brew a seasonal chocolate stout.

A 25min walk gets you back to the Grand Place.

ALTERNATIVES

Extension: Hergé Museum

While Brussels has a brilliant Comic Strip museum, serious *bande déssinée* fans make a pilgrimage to the futuristic museum dedicated to Hergé, creator of *The Adventures of Tintin*. Original drawings and objects of the eponymous adventurer accompanied by Snowy and Captain Haddock enchant visitors of all ages.

Take the train from Bruxelles-Central to Louvain-la-Neuve (hourly; 50min).

Extension: Ghent

Though day-trippers flock to romantic Bruges, the equally beautiful medieval city of Ghent has transformed itself into a pioneering ethical destination where ecotourism is a way of life: vegetarian and sustainable diners; artisan brewers; a car-free centre; and green bike tours taking in classic sights like the ornate guild houses on Korenlei canal.

Regular trains run direct from Bruxelles-Central to Ghent Saint-Pierre station (35min).

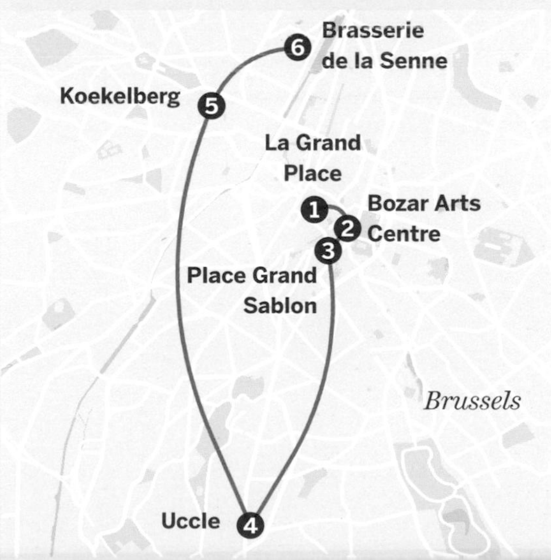

ENGLAND'S NORTHUMBERLAND COAST

Alnmouth – Lindisfarne

The Northumberland coast in northeast England has glorious sands, dramatic castles, quaint villages and seals – and much of it can be hiked.

FACT BOX

Carbon (kg per person): 2
Distance (km): 75
Nights: 7
Budget: $$
When: Jun-Sep

❶ Alnmouth

This pretty coastal village was an important grain port until a vicious storm in 1806 diverted the river and marooned the harbour. It's flanked by golden beaches and makes a good jumping-off point for hiking St Oswald's Way.

ABOVE: CHAPEL OF ST CUTHBERT, INNER FARNE

Take bus X18 to Alnwick (hourly Mon-Sat, every 2hr Sun; 10min), or hike the 7km.

❷ Alnwick Castle and Gardens

Alnwick's forest of turrets has a history dating back to the Norman Conquest and film-makers love it: you might recognise it from the *Harry Potter* movies. Learn to fly a broomstick in the same courtyard as Harry did, then marvel at Alnwick Gardens' water jets and treehouse restaurant.

Bus X18 heads to Craster via Alnmouth (3 daily; 55min).

❸ Craster

The queen eats Craster kippers for breakfast, and if you visit the fishing village where they're made, you'll still see – and smell – them being smoked the traditional way in oak barrels. Lunch on a kipper sandwich with a pint of Northumberland ale, watching boats bob in the sheltered harbour.

Join the St Oswald's Way coastal walking path for the 16km hike to Seahouses.

❹ Farne Islands

Accessed via the coastal town of Seahouses, the Farne Islands are a National Trust-managed cluster of

ABOVE: LINDISFARNE CASTLE

rocky islets inhabited by grey seals, puffins and some 100,000 seabirds. St Cuthbert lived as a hermit on Inner Farne island and died there in 687 CE.

From Seahouses, St Oswald's Way continues north 5km to Bamburgh.

5 Bamburgh

Lording it over a quaint village green and a sprinkle of houses, Bamburgh's castle is just as impressive as Alnwick's but with one added bonus: a glorious swathe of beach below, backed by sand dunes and waist-high coastal grasses.

Spend a full day on St Oswald's Way; it's a 31km walk to Lindisfarne.

6 Lindisfarne

Hauntingly beautiful Lindisfarne, often referred to as 'Holy Island', is one of Britain's most important pilgrimage destinations. A large monastery ruin remains, along with a copy of the *Lindisfarne Gospels* and fairytale Lindisfarne Castle. The sandy island is accessed via a causeway that is cut off twice a day by tides; time your visit carefully.

Take a bus or taxi to Berwick-upon-Tweed, from where regular trains run to London and Newcastle.

ALTERNATIVES

Extension: Kielder Forest

Northumberland International Dark Sky Park is a Gold Tier stargazing haven and Kielder Forest lies within it. Stay at a rural lodge or glamping site; several provide stargazing info packs and telescopes. Kielder Water Park is also on your doorstep for watersports and biking trails, and Kielder Observatory hosts a packed schedule of star-gazing sessions and camps.

A car is the simplest way of getting here.

Extension: Hadrian's Wall

Cleaving Britain in two, stones were first laid for this 117km Roman fortification in 122 CE on the orders of emperor Hadrian, to protect the northern frontier of the empire; its remains are now Unesco-listed. Hexham makes a good base for Housesteads Roman Fort and Museum or Corbridge Roman Town, and you can hike the entire Hadrian's Wall Path in around seven days.

Trains run from Newcastle to Hexham (3-4 hourly; 30-45min).

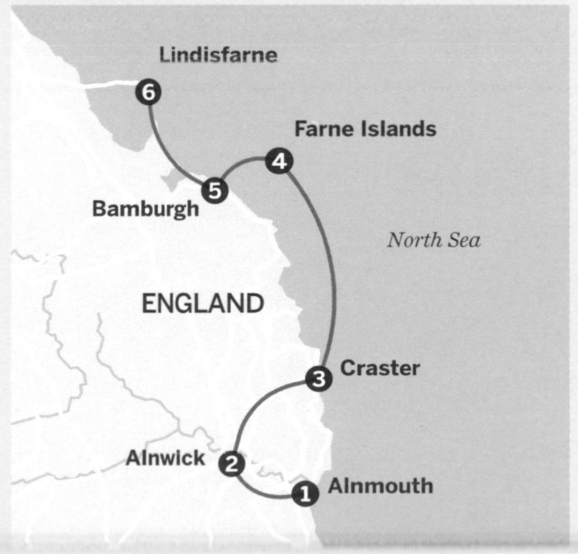

CONNEMARA COASTAL EXPLORER

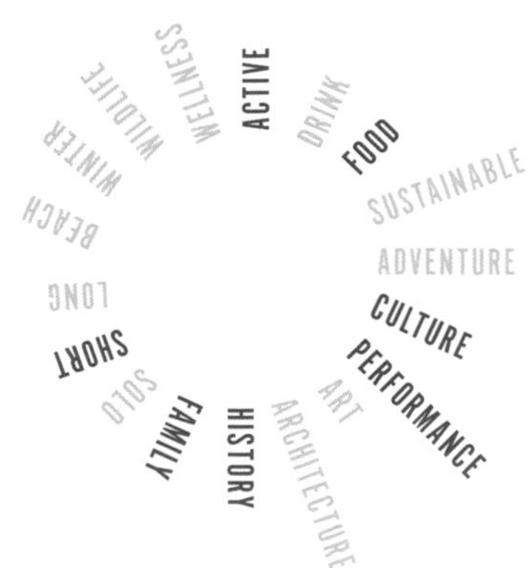

Galway – Westport

Road-trip Ireland's Connemara coast to discover wild mountains, deserted beaches, fresh seafood and live music.

FACT BOX

Carbon (kg per person): 35
Distance (km): 281
Nights: 2-3
Budget: $$
When: May-Oct

❶ Galway

Traditional music, great seafood and a bohemian vibe draw the crowds to Galway, one of Ireland's most vibrant cities. Wander medieval lanes lined with colourful shops, take a walk along the seafront promenade, then follow the music to the nearest available barstool.

Drive 82km west.

❷ Roundstone

Small fishing boats bob in the harbour at Roundstone, a scenic little town renowned for its live music and gorgeous beaches. Pop into Roundstone Musical Instruments to see a traditional *bodhrán* (skin drum) being made, then take a trip to Dog's Bay for powder-soft sands and turquoise waters.

Head 22km north.

❸ Clifden

Connemara's self-declared capital is a Victorian country town that's home to the Connemara Heritage and History Centre. It's also a great place to ride along the beach or cycle the Sky Rd, a 12km loop taking in some of Connemara's most picturesque scenery.

Continue 44km west along the Sky Rd.

❹ Leenane and Killary

Fjord-like Killary Harbour cuts 16km inland to the small village of Leenane and scenic Aasleagh Falls. Mussel rafts dot the inky waters and superb food vans hunker along roadsides offering the freshest of local seafood with dramatic views.

Round Killary Harbour and head 16km north.

ABOVE: ROUNDSTONE HARBOUR

ABOVE: GALWAY BUSKERS

5 Doolough
The steely-grey waters of Doolough are flanked by steep-sided mountains whose splendour is matched by the area's moving history. It was here, in 1849, that one of the most infamous tragedies of the Irish potato famine took place. Look out for the roadside memorial.

Continue 15km north.

6 Louisburgh
A sleepy 18th-century town, Louisburgh offers the chance to learn about Ireland's legendary pirate queen Grace O'Malley at the Granuaile Visitor Centre, and has easy access to a host of Atlantic beaches. You'll find safe sands at family-friendly Carrowmore and sweeping views at Killadoon.

Drive 22km east.

7 Westport
Dynamic and blessed with lively pubs, fine food and superb music, Westport is a handsome Georgian town set on the Carrowbeg River. Wander the streets, visit 18th-century Westport House or cycle the rugged coastline.

Head 80km south back to Galway.

ALTERNATIVES

Diversion: Kylemore Abbey
Now a Benedictine monastery, the grand rooms, neo-Gothic church and poignant mausoleum at Kylemore tell a tale of love and loss, while outdoors, the Victorian walled garden gives way to steep wooded slopes rolling down to scenic Lake Pollacapall.

Kylemore Abbey is about half way between Clifden and Leenane.

Diversion: Croagh Patrick
Forty days and nights of fasting here supposedly gave St Patrick the strength to banish snakes from Ireland, but for the day-tripper, the climb to the conical peak (765m) offers panoramic views of Clew Bay and its hundreds of islands. Come at dawn on the last Sunday in July to join the annual pilgrimage to the summit – shoes optional.

The start of the trail is at Murrisk, about 10km west of Westport.

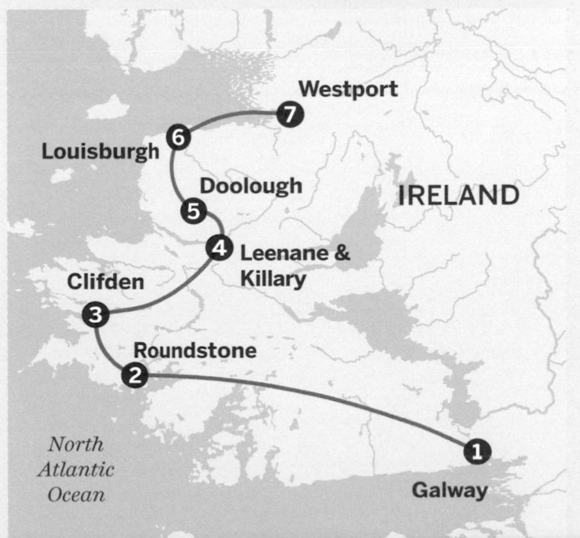

ENGLAND: WHAT THE ROMANS DID FOR US

London – Hadrian's Wall

What did the Romans do for us? Find out on this south-to-north train trip back into England's past.

FACT BOX

Carbon (kg per person): 19
Distance (km): 920
Nights: 7
Budget: $$
When: May–Sep

❶ London

Londinium, largest city in Britannia, was a Roman creation. For an overview, start with the Museum of London's excellent Roman galleries, then see the remains of the 2nd-century amphitheatre at Guildhall Art Gallery, the Temple of Mithras in the Bloomberg Building, and part of the original Roman city wall just outside the exit of Tower Hill Underground station.

ABOVE: BATH'S ROMAN POOLS

Take the train from Victoria Station to Fishbourne (every 30min; 1hr 50min; change at East Croydon).

❷ Fishbourne Palace

Romanised Britons built lavish country villas with all the mod-cons of the age, and one prime example is Fishbourne Palace. Inside, the mosaics, especially the Dolphin Mosaic, are exquisite; outside, the recreated gardens are the oldest in the UK.

Take the train from Fishbourne to Bath (every hour; 3hr; change at Southampton Central).

❸ Bath

The Romans didn't discover the waters at Bath, but they did build a huge bathing complex over them in the city they named Aquae Sulis. Today's 19th-century building contains the original pools and a museum. For modern dunking, Thermae Bath Spa is just around the corner.

Take the train from Bath to Chester (every hour; 4hr; change at Bristol Temple Meads and Birmingham New St).

❹ Chester

The amphitheatre at Deva Victrix (ancient Chester) was the largest in Britannia, built for the entertainment and training of soldiers stationed here; excavated sections

ABOVE: YORK MINSTER

include two entrances used by performers. Nearby, in the middle of a former Roman quarry, the Minerva Shrine has a sculpture of the goddess of war and craftsmanship.

Take the train from Chester to York (every hour; 3hr; various change options).

5 York

When two emperors die and one is proclaimed in your city (Constantine the Great – his statue sits outside York Minster), you've earned your place in Roman history. Eboracum was founded by soldiers in 71 CE; Roman-era remnants include part of the Minster's basilica, and a bath house in the Roman Bath pub.

Take the train to Hexham (every 30min; 1hr 45min; change at Newcastle).

6 Hadrian's Wall

Hadrian's Wall was a full stop in stone on the expansion of the Roman Empire. Begun in 122 CE, it stretches (almost) coast to coast across northern England. Highlights include Housesteads Fort and Vindolanda Museum with handwritten letters from wall residents. Connect the sites along the wall on the AD122 bus, which starts from the abbey town of Hexham.

Trains run from Hexham back to London, changing in Newcastle.

ALTERNATIVES

Extension: Portchester Castle

You get two history hits for the price of one at Portchester Castle. A Roman fort (the best-preserved north of the Alps) was built here in the 3rd century to defend the coast from Saxon attacks. Then, around the early 12th century, a medieval castle was built within the Roman walls, used by royalty for centuries but now a romantic ruin.

Take the train to Portchester from Fishbourne.

Extension: Arbeia Fort, South Shields

To get a sense of what all these ruins originally looked like, head to Arbeia Fort. An impressively reconstructed gate and other buildings show how imposing and intimidating Roman architecture would have been. The commanding officer's house depicts how he lived; the barracks show the less luxurious life of the soldiers.

Take the train from Hexham to Newcastle, then the Metro to South Shields.

THE FRENCH RIVIERA BY RAIL

Cannes – Menton

Few stretches of coastline can match the glamour of the Côte d'Azur – so all aboard for some window-seat railway ritz.

FACT BOX

Carbon cost (kg per person): 3
Distance (km): 66
Nights: 5
Budget: $$-$$$
When: Apr-Jun & Sep-Oct

❶ Cannes

Celebrated for its glitzy film festival, Cannes is a coastal city with swagger. Follow in the footsteps of the stars along La Croisette to the Palais des Festivals, the main festival venue, then spend an afternoon sipping cocktails in one of the beach bars. The city's Old Town is its most attractive asset: follow rue St-Antoine up to Le Suquet, Cannes' atmospheric original village.

Hop on a train along the coast to Antibes (every 30min; 10-15min).

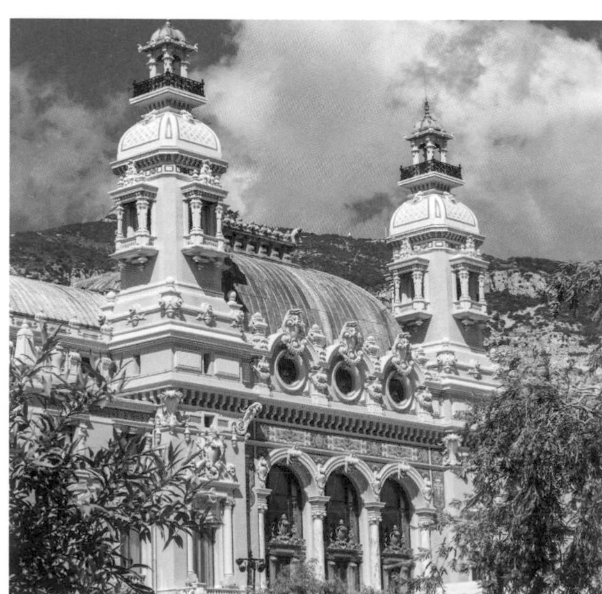

ABOVE: CASINO DE MONTE CARLO, MONACO

❷ Antibes

With its pleasure-boat-filled harbour and narrow streets, Vieux Antibes is the quintessential Mediterranean port. The brilliant Musée Picasso is a highlight, both for its location and collections. Don't miss the views from the 16th-century ramparts, stretching from Nice to the Alps, then hire a bike to ride to Juan-les-Pins, once a favoured haunt for Europe's glitterati.

Continue on the train to Nice (4 hourly; 15-30min).

❸ Nice

With its mix of city grit, old-world opulence and year-round sunshine, Nice is a beguiling stop. The shady alleyways of Vieux Nice make for rewarding wandering, and the legendary Promenade des Anglais guarantees an epic sunset. The city also has excellent art galleries, including the Musée d'Art Moderne, Musée Matisse and Musée National Marc Chagall.

ABOVE: MONTE CARLO HARBOUR, MONACO

Take the train along the coast to Monaco (every 30min; 20-30min).

❹ Monaco

A magnet for high-rollers, hedonists and tax dodgers, the pint-sized principality of Monaco might not be the Riviera's prettiest town, but it's worth a visit for the glam alone. Place a bet at the world-famous casino, visit the excellent Musée Océanographique, and explore the charming Old Town in Le Rocher, home to the principality's royal palace.

Continue east on the train to Menton (every 30min; 10min).

❺ Menton

Last stop on the French Riviera before Italy, seaside Menton offers a glimpse of what the high life here must have been like before the developers moved in. With its sunny climate, shady streets and pastel mansions – not to mention a lovely old port – it's among the most attractive towns on the Côte d'Azur. It also has a fantastic museum dedicated to the artist and film director Jean Cocteau.

Retrace your route back to Cannes by train, or continue over the border into Italy.

ALTERNATIVES

Extension: Mougins

Picasso lived in hilltop Mougins from 1961 until his death. The Old Town is worth visiting for the Musée d'Art Classique de Mougins, which illuminates how ancient civilisations inspired Neoclassical, modern and contemporary art. Equally interesting is the Musée de la Photographie André Villers, with black-and-white photos of Picasso and others from his Riviera milieu.

Buses run to Mougins from Cannes (every 20min; 20min).

Extension: St-Paul-de-Vence

Once, St-Paul de Vence was a small medieval seaside village atop a hill. Then it was discovered by artists: Picasso, Cocteau, Matisse and Chagall are just a few of the illustrious names who lived here. Unsurprisingly, the village is now home to dozens of art galleries as well as the renowned Fondation Maeght, with works by Braque, Kandinsky, Chagall and Miró.

Buses run to St-Paul from Nice (1-2 hourly; 1hr).

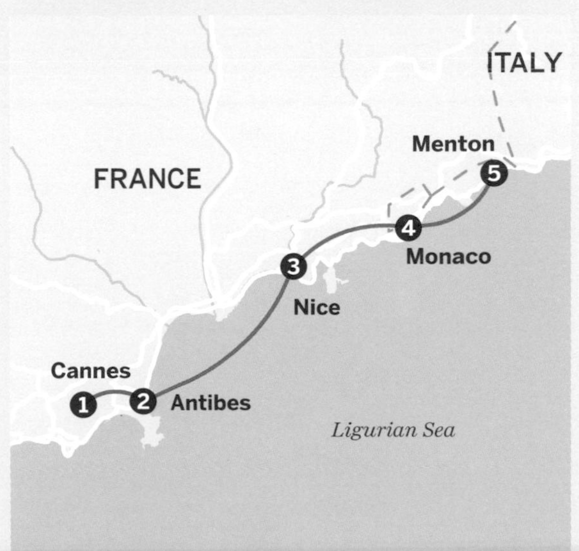

BURGUNDY BY BOAT AND BIKE

Tonnerre – Dijon

Discover the best of Burgundy's wines, cuisine, châteaux and villages on a slow boat- and bike-ride through the French countryside along the Canal de Bourgogne.

FACT BOX

Carbon (kg per person): 19
Distance (km): 228
Nights: 10
Budget: $$
When: May-Sep

❶ Tonnerre

The Canal de Bourgogne officially begins at Migennes, but rent your boat nearby at Tonnerre to see its Baroque and Gothic churches, and take a vineyard bike trip to nearby Chablis whose white wine is enjoyed the world over. Get used to handling the boat with a quiet first day and after five locks moor for the night in Tanlay. If you have time, visit its fairytale château.

Cruise the canal southeast for about eight hours, passing through 14 locks.

❷ Ancy-le-Franc

Ancy-le-Franc's picturesque waterside castle has incomparable Renaissance murals and spectacular gardens. After enjoying the town's charms, continue cruising. Cycling side-trips over the next few days include Fontenay's ancient abbey, idyllic Flavigny-sur-Ozerain (one of France's most beautiful villages) and the fascinating MuséoParc Alésia, recounting battles of the Gauls against Julius Caesar. There's one big adventure before arriving in the next stop, Pouilly-en-Auxois: traversing the *voûte*, a 3km-long tunnel built in 1775, more challenging for skippers than the locks.

Around 30 hours cruising, 5 overnights and 80 locks get you to Pouilly-en-Auxois.

ABOVE: ENJOY BURGUNDY'S WINES ON DECK

ABOVE: SLOW-BOATING THROUGH BURGUNDY

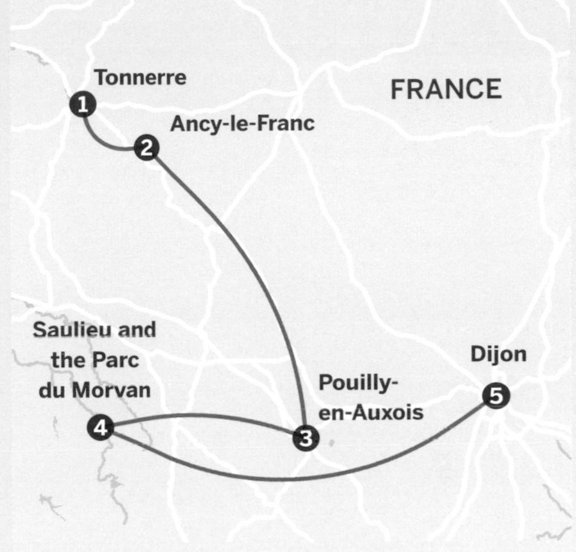

❸ Pouilly-en-Auxois

Pouilly has a fun marina with bars and bistros, but don't miss a trip to the nearby fortress château and medieval hill village of Châteauneuf-en-Auxois, and to Lac de Panthier with a sandy beach that's perfect for swimming.

Cycle west 30km from Pouilly to overnight at Saulieu.

❹ Saulieu and the Parc du Morvan

The gateway to the Morvan's breathtaking natural park is gastronomic Saulieu, perfect for a bistro lunch or gourmet restaurant treat. Then explore the rugged Morvan's splendid forests and lakes.

Some 20 hours cruising, 3 overnights and 54 locks gets you to Dijon.

❺ Dijon

Before Dijon, the canal follows the Ouche river valley through some of Burgundy's most beautiful countryside, including Lac Kir, great for water sports. Despite a host of attractions that can easily fill a weekend, Dijon is never overcrowded with tourists; check out the majestic Palace of the Dukes of Burgundy, fine arts museum and an immense covered food market designed by Eiffel.

Trains run direct to Paris (hourly; 1hr 40min) for international connections.

ALTERNATIVES

Diversion: Beaune

Vineyards stretch right down to Burgundy's wine capital, the walled city of Beaune, its medieval and Renaissance buildings covered with characteristic multi-coloured glazed tiles. This route, perfect by bike, passes through La Côte d'Or, the Gold Coast of some of the world's most famous wine villages – Mersault, Pommard, Vougeot – where visitors are welcomed by friendly *vignerons* for cellar tastings.

Take one of the regular trains from Dijon (30min).

Extension: Besançon

Capital of bucolic Franche-Comté region, imposing Besançon is one of France's forgotten destinations, by-passed by most tourists and guide books. Come and be captivated by Vauban's remarkable hilltop military citadel overlooking the Jura mountains, the bars and bistros along the scenic Doubs river, and the cobbled lanes and leafy squares of the historic Ville Haute.

Hourly trains run from Dijon to Besançon Viotte station (1hr 5min).

GERMANY'S TWIN CAPITALS

Berlin – Cologne

Ride the rails from Germany's capital Berlin to the former West German capital, Bonn, via some of the country's mightiest metropolises.

FACT BOX

Carbon (kg per person): 32
Distance (km): 800
Nights: 10
Budget: $$
When: year-round

❶ Berlin

Berlin is a true cultural heavyweight, whether that means seeing artefacts from the dawn of civilisation in the Museumsinsel or staying out until dawn in the clubs of Kreuzberg. One essential experience is walking through the Tiergarten park to the Brandenburg Gate and Unter den Linden, crossing the threshold from the old West to East Berlin as you go.

Take the train to Leipzig (every 30min; 2hr).

ABOVE: GOETHE-HAUS, FRANKFURT

❷ Leipzig

A rallying point for creative souls escaping the high rents of Berlin, the city of Leipzig has always commanded serious artistic clout. Visit the museums dedicated to two of Germany's most celebrated musical sons: Bach and Mendelssohn.

Continue on the train to Weimar (every 15-30min; 1hr 30min).

❸ Weimar

A studious little spot in the forested state of Thuringia, Weimar is perhaps Germany's brainiest city, a sometime haunt of Goethe, Schiller, Nietzsche and Liszt. It's been something of a design hub too: don't miss the new Bauhaus Museum, a cube-like structure dedicated to the pioneering movement.

Back on the train for Frankfurt (hourly; 4hr), changing at Erfurt.

❹ Frankfurt

Frankfurt gets bad press as Germany's dull financial capital, but look beyond the skyscrapers and you'll

ABOVE: MUSEUM ISLAND, BERLIN

find a spirited town with an international population. Head to taverns in the Old Town to try Apfelwein, a tart, cider-like drink that is the city's favourite tipple.
Take the train to Bonn (every 15min; 2hr).

5 Bonn

The stand-in capital of West Germany for the second half of the 20th century, Bonn retains a rather dignified air, with spires and townhouses lining the banks of the Rhine. Visit Beethoven's house (he was born here) and the Haus der Geschichte, which offers an excellent history lesson on the divided postwar nation and its (mostly) loving reunion.
Continue on the train to Cologne (every 15min; 30min).

6 Cologne

Once a Roman capital, modern Cologne is an irreverent city downstream from Bonn on the River Rhine. The first thing you notice are the gargantuan spires of the cathedral, one of Europe's tallest. Similarly holy is the tradition of the Brauhaus – famously boisterous brewery restaurants where you can try the town's signature Kölsch beer.
Cologne railway station has departures for destinations across Europe.

ALTERNATIVES

Extension: Aachen

Right on the border with Belgium and Holland, Aachen was the one-time capital of the Frankish Empire – its territory extended across Europe, and its legacy is the 9th-century cathedral that rises at the heart of the city. The Romans too, came here to bathe in the hot springs: follow in their sandalsteps with a dip at one of the town's spas.
Direct trains run to Aachen from Cologne (every 30min; 30min).

Extension: Dresden

Dresden is Germany's Renaissance city. Almost totally destroyed by WWII bombs, it was restored brick by brick, and is once again a showpiece of Baroque architecture. A highlight is the Zwinger, a palace-museum in meticulously kept gardens. Cross the Elbe to the Neustadt's lively bars and thriving street art scene.
Dresden has direct trains from Berlin (every 30min; 2hr) and Leipzig (every 30min; 1hr).

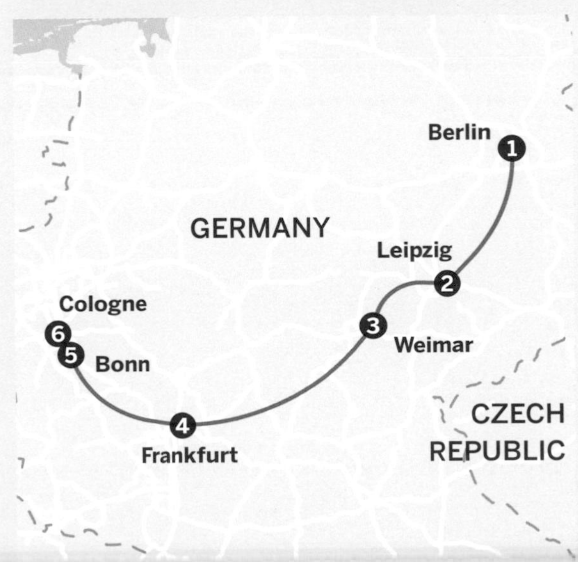

A FEW DAYS ON THE KENNET & AVON CANAL

Bradford-on-Avon – Bath

Cruise through classic English countryside on this canal trip, passing aqueducts, waterside pubs and historic sites on the way.

FACT BOX

Carbon (kg per person): 58
Distance (km): 32
Nights: 4
Budget: ££
When: Apr-Oct

ABOVE: KENNET & AVON CANAL

❶ Bradford-on-Avon

Bradford-on-Avon is a town of historical highlights. It boasts a tithe barn from the 14th century, striking Georgian architecture and a picturesque wharf, which in the 1850s was a major trading point along the Kennet & Avon Canal, a route linking London to the Bristol Channel. So it feels fitting when you visit to slow the pace down, eschew modern transport and experience the setting by narrowboat, with a speed limit of just over 6km an hour. Enjoy waterside coffee shops and pubs, convenient in-town moorings and canal banks so full of nature and plant-life they belie their proximity to towns. Bradford-on-Avon is often described as a mini-Bath, making it the perfect place in which to whet your appetite for that later stop on the trip.

Hire a boat in Bradford-on-Avon (sallynarrowboats.co.uk; book well in advance) and head west and north (2-3hr).

❷ Claverton

Heading west along the Kennet & Avon Canal, you'll pass through wonderful Wiltshire countryside and the impressive limestone aqueducts at Avoncliff and Dundas before reaching Claverton Pumping Station. A 200-year-old environmentally friendly wonder, it

ABOVE: PULTENEY BRIDGE OVER THE AVON, BATH

was built to draw water into the canal system from the River Avon using a large wheel rather than coal. The site was restored in the 1960s and can be visited on pumping days.

Continue in a northwesterly direction towards Bath (2-3hr).

3 Bath

Mooring up just outside of Bath saves you taking on the city's string of narrow locks, but still affords an easy walk into its gorgeous centre. The Scottish engineer John Rennie, who designed the Kennet & Avon Canal, took pains to ensure the bridges and buildings along this stretch reflected Bath's fine honey-hued architecture – as you'll see if you follow the towpath from the Pump House Chimney through Sydney Gardens to Cleveland House. From there, head towards the city centre along the banks of the River Avon and across the majestic Pulteney Bridge. A top range of shops, restaurants and attractions can easily fill a couple of days here, living out your Jane Austen fantasies.

After a day or two in Bath, turn the boat around and wend your way back to Bradford-on-Avon (5-6hr).

ALTERNATIVES

Day trip: Roman Baths

Bath has been welcoming visitors to its central Roman Baths since they were constructed in AD 70 as a bathing and socialising complex; today, they're some of the best-preserved Roman ruins in the world. You can't swim here now, but if you fancy a dip in the city's thermal waters, head to the nearby Thermae Bath Spa.

The Roman Baths and Thermae Bath Spa are both in Bath city centre, a few minutes' walk apart.

Extension: Bristol

Experienced boaters can continue their cruise from Bath to Bristol: be warned, the route is narrow at the start and there are six locks to navigate. After these you join the tidal River Avon, which takes you into Bristol and its historic Floating Harbour, now a hub of art galleries and hip restaurants. Evidence of the city's sea-faring past is everywhere, and nowhere more enjoyably explored than on the SS *Great Britain*.

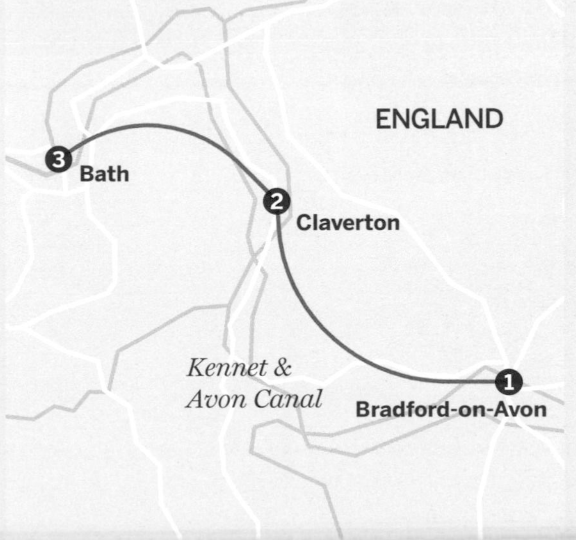

A LITERARY TOUR OF ENGLAND

London – The Lake District

Take a grand tour of English literature to see the locations that inspired some of the world's greatest writers.

FACT BOX

Carbon (kg per person): 96
Distance (km): 710
Nights: 7–10
Budget: $$
When: Apr–Sep

❶ London

Many writers set novels in London, but none has made the city a central character in its own right like Charles Dickens did. His depictions of the 19th-century capital established an image of London that persists even to this day. Explore his works and personal life at an eponymous museum, then dine at Rules, his favourite restaurant.

Drive 90km west.

ABOVE: BRONTË PARSONAGE MUSEUM, HAWORTH

❷ Oxford

Oxford's academic heritage sits alongside a litany of literary giants who called the city home. Raise a glass to CS Lewis (of *Chronicles of Narnia* fame) and JRR Tolkien (of *Lord of the Rings* fame) in the Eagle and Child pub where their writers' group met. Then tour the streets in the footsteps of Lyra Silvertongue from Philip Pullman's *His Dark Materials*.

Continue 130km southwest.

❸ Bath

Glorious architecture, genteel atmosphere and taking of the waters are key attractions of beautiful Bath – and all feature in former resident Jane Austen's works. Visit the Jane Austen Centre, then take a turn past 4 Sydney Place where she lived for three years.

Head 140km northeast.

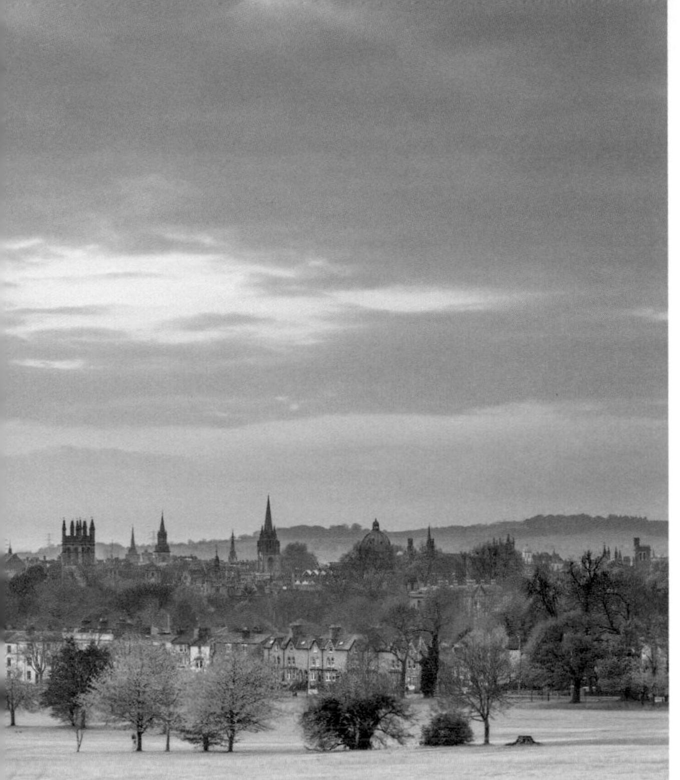
ABOVE: OXFORD

❹ Stratford-upon-Avon

William Shakespeare is arguably the most famous writer ever, and his medieval hometown sees crowds that attest to his continuing popularity. Seek out his birthplace; pay your respects at his tomb in Holy Trinity Church; and enjoy one of his plays at the Royal Shakespeare Theatre.

Drive 250km north.

❺ Haworth

Maybe it was the isolated location or the often-bleak Yorkshire weather, but something in Haworth fired the imaginations of the three Brontë sisters and led to enduring classics of Gothic and Romantic literature such as *Jane Eyre* and *Wuthering Heights*. The parsonage where they lived is now a museum.

Drive 100km northwest.

❻ The Lake District

Naughty rabbits and golden daffodils are two of the scenic Lake District's literary draws. See where Peter Rabbit and friends came to life at Beatrix Potter's Hill Top Farm; then wander Dove Cottage and Rydal Mount, homes to poet William Wordsworth, before heading into the hills to find your own poetic muse.

Drive or take the train back to London.

ALTERNATIVES

Extension: Sedbergh

Bibliophiles should make the detour to Sedbergh on the journey between Haworth and the Lake District. Known as England's 'book town' (similar to Wales' Hay-on-Wye), it has a small but excellent collection of book stores (some doubling up as craft shops, cafes and restaurants) occupying its attractive, stone-built buildings.

Sedbergh is 30km east of Windermere in the Lake District.

Extension: Hardy's Wessex

Follow the Thomas Hardy trail southwest to the Wessex towns and villages that appear (renamed) in Hardy's classic 19th-century novels. Oxford becomes *Jude the Obscure*'s Christminster, Bournemouth is Sandbourne in *Tess of the d'Urbervilles*, Dorchester is Casterbridge in *The Mayor of Casterbridge*.

A car is most convenient, but most places have train and bus services too.

A FOOD TRAIL OF THE BLACK FOREST

Baden-Baden – Triberg

As gorgeous and dark as its namesake gooey gateau, this fabulous forested region is ripe for a food-focused road trip.

FACT BOX

Carbon (kg per person): 29
Distance (km): 235
Nights: 4
Budget: $$
When: year-round

❶ Baden-Baden

This elegant spa town looks to nearby France for its culinary nods. Seek out ivy-swaddled Weinstube im Baldreit for perfectly crisp *flammkuchen* (Alsatian-style pizza), topped with Black Forest ham and paired with local Pinot wines. Or go for coffee and cake at 250-year-old institution Café König, where Liszt and Tolstoy once hung out.

Drive the pretty B3 road 29km south.

❷ Kappelrodeck

Vineyards frame this storybook town of half-timbered houses, much feted for its Pinot Noirs. Try wines and homemade *kirsch* (cherry brandy) at Rebstock Waldulm. Here, in the dark-timber parlour of a 260-year-old farmhouse, chef Karl Hodapp delivers punchy flavours in season-driven dishes like creamy snail soup with wild herbs or quail with plum mustard.

Continue 26km southeast along country lanes towards Baiersbronn.

❸ Baiersbronn

Neatly tucked into the Black Forest's folds, this small town shines with six Michelin stars. Its pride and joy is three-starred Restaurant Bareiss, where head chef Claus-Peter Lumpp adds a pinch of culinary magic to outstandingly sourced ingredients, from own-estate venison to wild mushrooms, truffles and Alsatian pigeon. Book well in advance.

ABOVE: MARKET-FRESH PRODUCE

ABOVE: TRIBERG

Drive 30km south through gentle hills.

❹ Alpirsbach

Presided over by a Benedictine monastery, Alpirsbach was allegedly named after a quaffing cleric who dropped his beer in the Kinzig River and exclaimed 'All Bier ist in den Bach!' ('All the beer is in the stream!'). Sample hoppy, full-bodied Black Forest brews, made with local spring water, at Alpirsbacher Klosterbräu brewery. Guided tours conclude with tastings.

Take a slow, scenic 44km drive south.

❺ Triberg

This town of giant cuckoo clocks (including the world's biggest, according to Guinness World Records) and Germany's highest waterfall, Triberg is also home to Bergseestüble, a woodsy chalet perched beside a little lake on the edge of town. Here you can stop by for no-nonsense Black Forest food with a view. On the menu you'll find the hearty likes of smoked trout, local pork with eggy *spätzle* noodles, and venison goulash with red cabbage and *schupfnudeln* (finger-shaped potato dumplings).

From Triberg, it's a 106km drive back to Baden-Baden.

ALTERNATIVES

Extension: Martinskapelle

A steep road twists precariously up to this 1085m-high peak, past a dark line of fir forest to a medieval chapel. Besides forest walking trails, you'll find wonderful regionally sourced food at family-run Höhengasthaus Kolmenhof. Out front is the spring that is the main source of the Danube, which accounts for the astoundingly fresh trout on the menu.

Martinskapelle is a 13km drive southwest of Triberg.

Diversion: Herzogsweiler

The Black Forest ham at Pfau Schinken in Herzogsweiler is the real deal. Locally reared meat is rubbed with salt and spices, such as coriander, garlic and juniper berries, then dry-cured before being cold-smoked over fir wood and left to mature to retain its intensely smoky, woody flavour. Get an insight into the curing and smoking process on guided tours or stock up in the shop.

Herzogsweiler is an 18km drive east of Baiersbronn.

FRANCE
1 Baden-Baden
2 Kappelrodeck
3 Baiersbronn
GERMANY
4 Alpirsbach
5 Triberg

THE HEART AND SOUL OF YORKSHIRE

Hull – York

Explore traditional villages and historic English cities between the emerald dales and windswept moorlands of 'God's Own County'.

FACT BOX

Carbon (kg per person): 52
Distance (km): 370
Nights: 10
Budget: $$
When: Jun-Sep

❶ Hull

Once a whaling centre, Hull is now home to an important abolitionist museum, plus maritime relics and a cool regenerated dock area. Its well-preserved Georgian old town is a popular filming location – *The Personal History of David Copperfield* used it as a stand-in for Dickensian London.

Drive 55km northeast.

❷ Flamborough Head

Bone-white seacliffs, birds and lonely lighthouses are the highlights of this wild promontory on the Yorkshire coast. Head down the boat ramp at North Landing to paddle in the sandy cove, or visit the RSPB reserve at Bempton Cliffs to look for gannets and puffins.

Continue 60km northwest.

❸ Robin Hood's Bay

Book a fisherman's cottage for an overnight stay in this pretty coastal village. Wedged between two steep cliffs, Robin Hood's Bay was a haven for smugglers in the 18th century. At the bottom of the cobbled high street, its beach is popular for fossil-hunting and rockpooling.

Leave the car and hike 11km along the clifftops to Whitby.

❹ Whitby

Famous for its Captain Cook and *Dracula* connections, Whitby is a lively seaside resort full of Gothic drama, with a ruined medieval abbey crouched on its headland. There's lots to do, including nightly ghost tours, pirate-boat trips and sampling some of the UK's best fish and chips.

Take the X93 or 93 bus (hourly; 20min) back to Robin Hood's Bay; it's then a 105km drive west to Masham.

ABOVE: MASHAM PUB DECOR

ABOVE: WHITBY HARBOUR

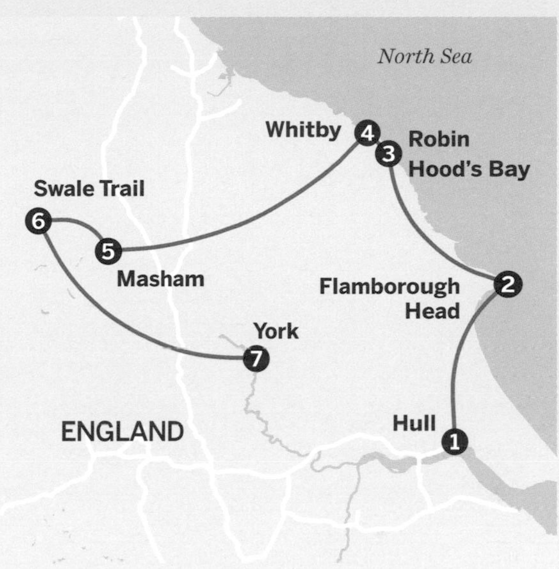

❺ Masham

This handsome market town on the edge of the Dales has beer running through its veins. Make it your base for the brewery at 19th-century Theakstons, then learn about the Theakston family fallout that created the town's second brewery in 1992 at Black Sheep.

Rent a bike from Cycopath Cycles to explore the Dales.

❻ Yorkshire Dales

Professional cyclists train in the Yorkshire Dales, but there's something for every age and level. Young mountain-bikers should try the traffic-free Swale Trail. Find trip planners at cyclethedales.org.uk.

Drive 55km southeast.

❼ York

The ancient capital of the North is a medieval beauty with walkable walls. Learn about its Roman roots in York Minster's Undercroft Museum, then enter a recreation of York's Viking settlement at the Jorvik Centre. Don't miss the Shambles, York's best-preserved medieval street.

York has very regular trains to London.

ALTERNATIVES

Extension: North York Moors National Park

Add some hiking to your trip in the North York Moors National Park. The stark moorlands are etched with footpaths, inflected with heather and home to some pretty villages like Goathland, Hutton-le-Hole and Beck Hole. Steam trains from the North York Moors Railway chug through Goathland, whose station was a filming location for the *Harry Potter* movies.

It's a 14km drive to Goathland from Whitby.

Extension: Harrogate

A less touristy alternative to York, well-to-do Harrogate flourished as a spa town from the early 17th century after mineral-rich springs were discovered in what's now the Valley Gardens. Celebrity visitors included Charles Dickens and Agatha Christie; Harrogate's rise to fame is documented in the Pump House Museum. The original tiled Turkish Baths are still an exquisite, mock-Moorish spa centre.

Harrogate is a 35km drive from Masham in the Dales.

ART AND ARCHITECTURE: AMSTERDAM AND ROTTERDAM

Museumplein – De Rotterdam

Immerse yourself in some Netherlands high culture, admiring art in Amsterdam and taking a cycling tour of Rotterdam's innovative architecture.

FACT BOX

Carbon (kg per person): 0
Distance (km): 103
Nights: 3
Budget: $$
When: year-round

❶ Museumplein, Amsterdam

Amsterdam's blockbuster art museums are conveniently located together in the main square, Museumplein. You'll need at least a couple of days to explore them. On day one, check out the Van Gogh Museum and the likes of Picasso, Klein and Kusama at the Stedelijk. For the second day, spend hours at the city's premier art gallery, the Rijksmuseum.

Take tram 5 from Museumplein to Amstelveen Stadshart (20min), followed by a 15min walk to the Cobra Museum.

❷ Cobra Museum, Amsterdam

Art lovers should definitely make an effort to venture out to this excellent museum. It showcases avant-garde, highly expressionist works by the post-WWII Cobra (Copenhagen, Brussels, Amsterdam) movement artists, including the best-known, Karel Appel.

Take the train from Amsterdam Centraal Station to Rotterdam Centraal Station (every 15min; 40min).

❸ Rotterdam Centraal Station

When you arrive in Rotterdam, don't rush off before getting a good eyefull of Centraal Station and its striking stainless steel angled roof; it was built between 1999 and 2013. When you've had your fill, rent a bicycle for the day's exploration.

It's less than a 10min bike ride southeast from the station to Markthal.

❹ Markthal, Rotterdam

A relatively recent addition to Rotterdam's

ABOVE: RIJKSMUSEUM, AMSTERDAM

ABOVE: OVERBLAAK DEVELOPMENT, ROTTERDAM

architectural achievements, the Markthal building – housing a fabulous food market and apartments – opened in 2014. The 40m-high arched roof has quickly become one of the most eye-catching architectural features in the city.

Overblaak Development is a 1min ride from Markthal.

5 Overblaak Development, Rotterdam

This innovative series of toppled and tilted yellow-and-grey cube houses was designed by architect Piet Blom and constructed between the late 1970s and early '80s. Each home was built to resemble a tree, and together they represent a forest.

Cycle 10min across the pylon Erasmusbrug (a bridge that is an architectural masterpiece in itself).

6 De Rotterdam

A 'vertical city' of three interconnected towers, De Rotterdam is one of the most respected examples of contemporary architecture in the city. It was designed by the firm OMA, whose team included the Pritzker-winning local architect Rem Koolhaas.

Ride back over the bridge and north to Centraal Station (15min).

ALTERNATIVES

Extension: Amsterdam Noord

If you have time in Amsterdam, jump on a ferry across the IJ River to seek out the cutting-edge art scene of the Amsterdam Noord neighbourhood. Former shipbuilding warehouses here have been taken over by artists' studios, and there's plenty of street art spread around to discover too.

Board the free ferry behind Amsterdam Centraal Station to NDSM (5min).

Diversion: Van Nelle Fabriek, Rotterdam

Time your visit to this Unesco World Heritage site for a Saturday and book ahead for a one-hour tour with Urban Guides. The incredible Modernist architectural marvel is constructed mainly of steel and glass and is a former tea, coffee and tobacco factory built between the late 1920s and early 30s.

It's a 10min bike ride west from Rotterdam Centraal Station to Van Nelle Fabriek.

CYCLING LAKE CONSTANCE

Friedrichshafen – Meersburg

One lake, two wheels, three countries. This flat, well-signposted circular bike tour of Lake Constance's German, Austrian and Swiss shores is Central Europe in a nutshell.

FACT BOX

Carbon (kg per person): 0
Distance (km): 246
Nights: 7
Budget: $$
When: Apr–Jun & Sep–Oct

❶ Friedrichshafen, Germany

Zeppelins often hover above Friedrichshafen; the Bauhaus-style museum gives the inside scoop on these mammoth cigar-shaped airships, which first took flight here in 1900. The sculpture-dotted lakefront promenade is appealing for a picnic or stroll.

The cycling trail weaves through orchards, meadows and past harbours to Lindau, 27km east.

ABOVE: STEIN AM RHEIN, SWITZERLAND

❷ Lindau, Germany

Medieval Lindau is postcard-pretty, with its cobbled alleys, gabled houses and lavishly mural-covered centrepiece, Altes Rathaus. The mellow lakefront promenade is atmospheric in the blue dusk, when its lighthouse and Bavarian lion statue glow gold.

The 11km pedal southeast to Bregenz gets steeper and the not-so-distant Alps become clearer.

❸ Bregenz, Austria

Famous for its summer lakefront opera festival, this affluent harbour town's big-hitters include avant-garde Kunsthaus gallery and architecturally striking Vorarlberg Museum, homing in on regional history, art and archaeology. A cable car rises to 1064m Pfänder, affording tremendous lake and mountain views.

Ride for 69km past beaches, through woods and along the lake shore in Switzerland. Continue and cross the border into Germany.

❹ Konstanz, Germany

Roman emperors, medieval traders and the 15th-century Council of Constance bishops have shaped this upbeat university city by the lake. Hugging the banks of the Rhine, Konstanz's crowning glory is a millennium-old cathedral that sits atop Roman ruins.

ABOVE: LAKE CONSTANCE FROM THE PFÄNDER CABLE CAR, BREGENZ

Pedalling 28km west brings you to Stein am Rhein, a brief detour inland from the lake, on the banks of the River Rhine.

5 Stein am Rhein, Switzerland

This Swiss town astride the Rhine enchants first-time visitors with an Old Town that is children's bedtime story stuff. The elongated Rathausplatz is often hailed Switzerland's loveliest square, with half-timbered, frescoed facades lined up in a permanent photo-op.
The scenic long route (best for cyclists) swings 64km along the lake's northern shore to Meersburg. En route, stop off at the Unesco-listed Pfahlbauten (reconstructed prehistoric stilt houses) in Unteruhldingen and the exuberantly Rococo Cistercian abbey church in Birnau.

6 Meersburg, Germany

Framed by vineyards and topped off by a turreted medieval castle, Meersburg is a terrific spot for slipping into lake life, with its tree-lined promenade and wine taverns serving local Pinot Noirs. Check out hospice-turned-wine museum Vineum, and ease aching muscles at lakefront Meersburg Therme spa.
It's an easy 19km pedal back to the starting point in Friedrichshafen.

ALTERNATIVES

Diversion: Rheinfall

Rolling west of Stein am Rhein, the cycling route shadows the river to the hugely impressive Rheinfall, where Europe's largest waterfall thunders, raging at around 700 cubic metres per second as it tumbles 23m into a basin of swirling cascades, billowing spray and foaming white water. Get within a hair's breadth of it on the footpath descending from Schloss Laufen.
The Rheinfall is a 23km ride west of Stein am Rhein.

Day trip: Reichenau and Mainau

Allow an extra day in Konstanz to see a twinset of islands. Dotted with orchards and wineries, Reichenau's Unesco-listed Benedictine monastery gave rise to the Reichenau School of illumination (820 to 1050). Mainau is one giant, Med-inspired botanical garden dreamed up by the Bernadotte family, relatives of the royal house of Sweden.
Reichenau is 7km west of Konstanz; Mainau 6km north.

CORNWALL'S CARRICK ROADS AND ROSELAND PENINSULA BY BOAT

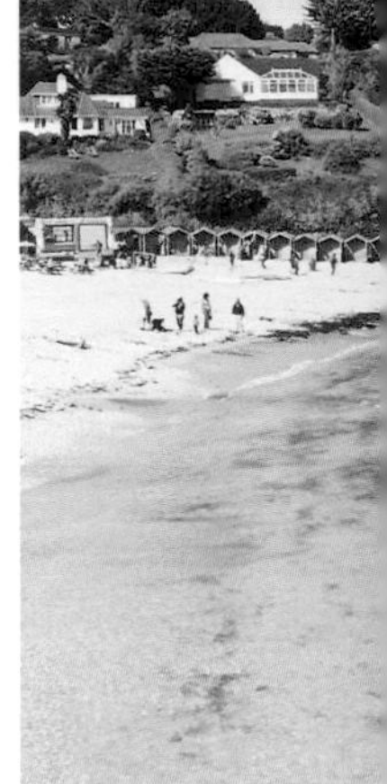

Truro – Falmouth

Ditch the car for a scenic, watery exploration of southwest England in Cornwall's Carrick Roads Estuary, with castles and walks along the way.

FACT BOX

Carbon (kg per person): 3
Distance (km): 41
Nights: 2-3
Budget: $$
When: May-Sep

❶ Truro

A handsome cathedral city, with attractive Georgian streets and a strong food and drink culture, Truro is worth exploring before you board your first ferry. Once on board, the picturesque voyage to Trelissick takes about an hour, passing postcard-perfect countryside, lovely inlets and the occasional ocean-going tanker moored up along the way – a demonstration of how very deep the River Fal is.

From Truro Town Quay, take the ferry to Trelissick (4 daily spring and summer; 1hr).

❷ Trelissick

This National Trust-owned garden and estate is just steps away from the ferry landing – its prime position makes for great views, and walking trails criss-cross the estate. The lush garden has a number of unusual plants as well as an orchard, and it's also a good spot for a cream tea.

Continue on the ferry for the ride to St Mawes (4 daily spring and summer; 45min).

❸ St Mawes

St Mawes sits at the southern end of the Roseland Peninsula, meaning more wonderful views from the village; the region is designated an Area of

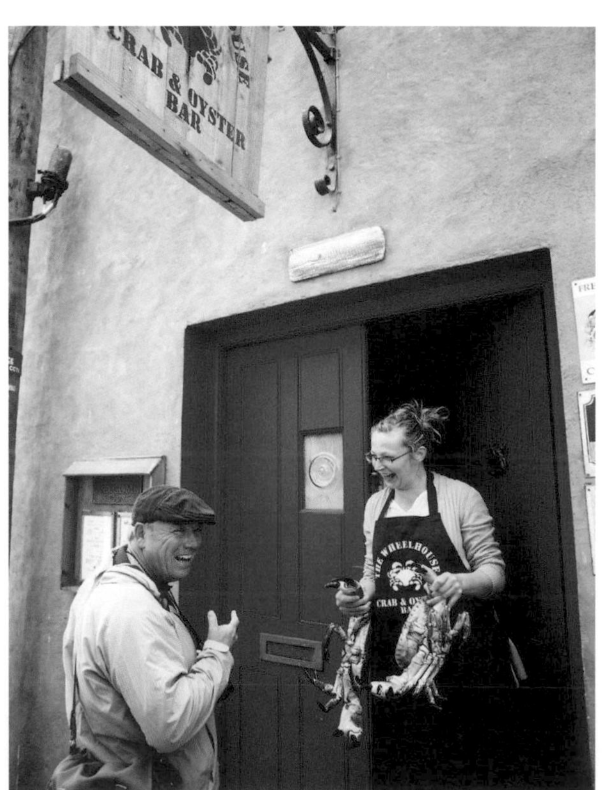

ABOVE: CATCH OF THE DAY, FALMOUTH

ABOVE: MAENPORTH BEACH, FALMOUTH

Outstanding Natural Beauty. The place bustles in high season, so book ahead for hotels and popular restaurants – the most alluring option is the Tresanton hotel and restaurant. While here visit Henry VIII's castle and the sub-tropical Lamorran House Gardens, then strike out on some of the many walking opportunities hereabouts.

Take the ferry to Falmouth's Prince of Wales quay (3 ferries an hour in summer, hourly in winter; 20min).

4 Falmouth

There's lots to keep you occupied in Falmouth, from sandy beaches to hip shops and independent cafes. Falmouth Art Gallery punches above its weight with a collection that includes Pre-Raphaelites and modern works alongside a wealth of contemporary automata. Check out the National Maritime Museum Cornwall (five floors of history, art, science, interactive exhibits, and great views from the Lookout Tower) and Pendennis Castle (a twin to St Mawes' castle across the water). Sunny days can be spent on Gyllyngvase Beach (the closest to the town centre) or watching the boats come in and out of the harbour. Finish with fish and chips at Harbour Lights.

Return to Truro by train (every 30min; 25min).

ALTERNATIVES

Extension: St Anthony's Head

A short trip on the little ferry from St Mawes brings you to Place, access point for some of the loveliest walks in Cornwall. The coastal path is stunning and there are interesting buildings such as St Anthony lighthouse and the church of St Anthony along the way. Bring a picnic.

St Mawes to Place Creek (on the Place Ferry) is a 10-min trip, operating on demand April to October.

Diversion: Trelissick to St Mawes via King Harry ferry

The King Harry chain-ferry acts as a bridge over to the Roseland Peninsula. From the landing point there's a glorious walk that follows the coast most of the way to St Mawes. Take a small diversion to see St Just in Roseland, a beautiful creek-side church with sub-tropical gardens.

Trelissick to Roseland Peninsula via King Harry ferry (10min), then 11km walk to St Mawes.

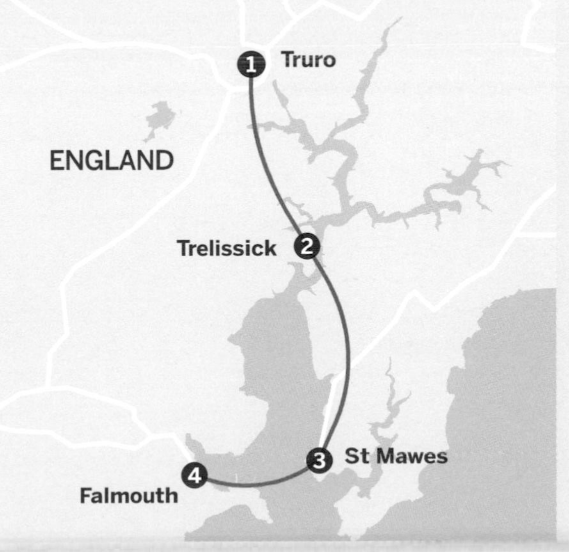

THE HIGHLANDS TO THE ISLANDS

Inverness – Lerwick

Visit the Scotland most people miss: a wild, exposed sprawl of highlands, islands and timeless ruins.

FACT BOX

Carbon (kg per person): 92
Distance (km): 1031
Nights: 14
Budget: $$
When: Apr, May, Aug-Sep

❶ Inverness

Trains, buses and boats will whisk you through the highlands and islands. Start in Inverness, split by the broad River Ness and best viewed from its scenic banks. The hordes go monster-hunting on nearby Loch Ness; for a quieter time, head to the Black Isle (accessible via hourly buses) with its lonely beaches, ecclesiastical ruins and visiting dolphins.

ABOVE: ST MAGNUS CATHEDRAL, KIRKWALL

Take the train to Wick (2-4 daily; 4hr 15min).

❷ Wick

After a scenic train ride tracing the east coast of the Highlands, you'll roll into Wick, once the herring capital of Britain, with its excellent museum, distillery and twin ruined castles.

Jump on the bus to John O'Groats (5 daily except Sun; 30min).

❸ John O'Groats

Don't get hung up on John O'Groats' northerly location; the northernmost point on the Scottish mainland is actually 16km west at Dunnet Head. Instead, stroll up to Duncansby Head for views of the fang-like Duncansby Stacks and the iron-grey North Sea that'll carry you onwards to Orkney.

Take the ferry to Burwick, Orkney (2-3 daily; 40min).

❹ Burwick

Pause when you hit the Orkney shore; within walking distance of Burwick's ferry port is the Tomb of the Eagles, a clifftop Neolithic site whose human occupants were interred with hundreds of eagles.

Take the bus to Kirkwall (2 daily; 40min).

ABOVE: WAULKMILL BAY, MAINLAND ORKNEY

❺ Kirkwall

Devote a couple of days to exploring the tangled wynds (lanes) of the Orcadian capital; poke around the imposing 12th-century St Magnus Cathedral, and drop by the Highland Park and Orkney distilleries for a dram of whisky.

Rent a car or bike in Kirkwall.

❻ Mainland Orkney

Enjoy an 80km loop around Mainland, Orkney's biggest island, stopping at the Iron Age Broch of Gurness, the perfectly preserved Neolithic village of Skara Brae, the moody Ring of Brodgar and the awesome, Viking-graffitied tomb at Maeshowe.

Jump on the ferry to Lerwick (3-4 weekly; 8hr).

❼ Lerwick

Steely-grey but warmly welcoming, the capital of Shetland has an intriguing collection of museums and fortifications. Local buses run to the islands' airport at Sumburgh, just a stroll from the shore-side ruins of Jarlshof, once occupied by Bronze Age, Iron Age, Pictish and Viking Shetlanders.

Finish with a ferry to Aberdeen (daily; 15hr) for onward train and plane connections.

ALTERNATIVES

Diversion: The Eastern Highlands

There's no need to rush north from Inverness. Stop off in Dornoch, with its grand Gothic cathedral; and Golspie, the handsome setting for enormous Dunrobin Castle, former seat of the notorious Duke of Sutherland.

Buses run twice daily from Inverness to Dornoch (1hr) and Golspie (1hr 30min), where you can board the train north to Wick (2-4 daily; 2hr).

Diversion: Sandwick, Shetland

With an extra day in Shetland, head south from Lerwick to Sandwick and board the boat to Mousa, island home to storm petrels, basking seals and an astonishingly preserved broch tower, built over 2000 years ago. Stroll on from to tiny Bigton, where a sand-and-shell causeway crosses to St Ninian's Isle and its ruined 12th-century church.

Local buses connect Lerwick to Sandwick, and Bigton to Lerwick.

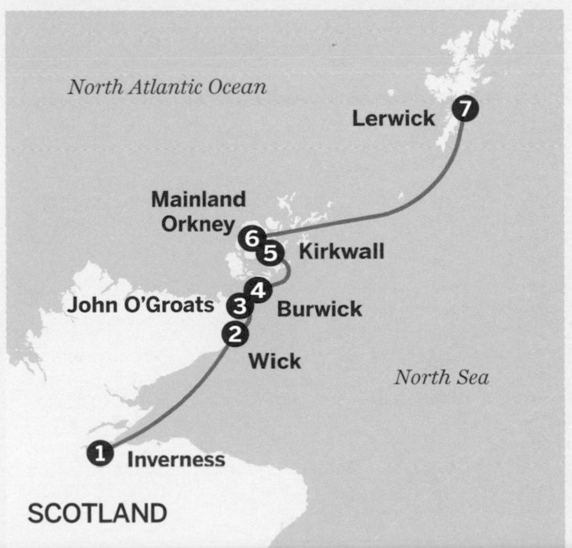

FRANCE'S ROUTE DES GRANDES ALPES

Lake Geneva – Menton

Road-trippers: pack a head for heights and nerves of steel for this wild drive up France's highest mountain passes and down to the azure-blue Med.

FACT BOX

Carbon (kg per person): 83
Distance (km): 662
Nights: 5-10
Budget: €
When: mid-June to mid-Oct

❶ Lake Geneva

The southern shore of Europe's largest alpine lake is the starting grid for this legendary driving route in France. Since 1913, motorists tackling the steep hairpins and remote alpine valleys of the Route des Grandes Alpes have set their odometer to zero in the quaint lakeside town of Thonon-les-Bains, 40km from Geneva Airport in Switzerland.

Drive 155km south through the Chablais region.

❷ Cormet de Roselend

In the local dialect, *cormet* means mountain pass, and this stretch of the trip through the French Alps sports a generous 21 of them. The approach to Cormet de Roselend (1967m) is along the vast, dramatically beautiful turquoise pool of Lac de Roselend. Views of cow-speckled pastures and gentle peaks are eminently pastoral, and hiking trails spaghetti up to the craggy peak of Roc du Vent (2360m). Feasts of hearty potato dishes spiked with tangy Beaufort cheese are memorable at lakeside hotel-restaurant Chalet de Roselend.

Continue 67km southeast to Col de l'Iseran.

❸ Col de l'Iseran

Peaking at 2764m, this unforgettable pass links Savoie's isolated Tarentaise and Maurienne valleys. Gravity-defying tunnels and galleries chiselled in rock dot the dramatic climb. The exceptionally scenic section between chic ski resort Val d'Isère and remote

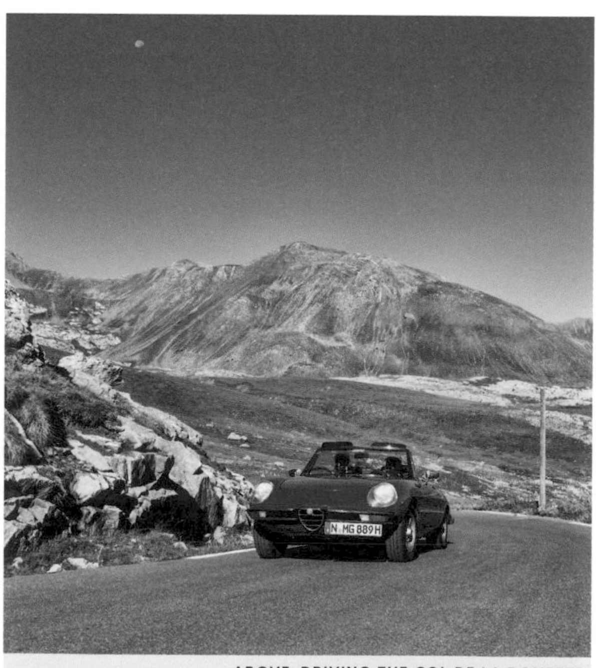

ABOVE: DRIVING THE COL DE LA BONETTE

ABOVE: CORMET DE ROSELEND LAKE VIEW

village Bonneval-sur-Arc – a highly photogenic huddle of stone houses – winds through France's pristine Vanoise National Park.

Drive another 250km south into Alpes-de-Haute-Provence.

4 Col de la Bonette

A holy grail for cyclists, Col de la Bonette (2715m) is the highest paved road in the Alps and one of Europe's highest driveable roads – until 1964 it was a mule track. Go slow to take in the wild panorama of deep valleys, cut-throat gorges, jagged peaks and occasional sightings of golden eagle, mouflon or ibex – all protected by the Mercantour National Park.

Drive 150km through the Tinée, Vésubie and Roya valleys towards the Mediterranean coast.

5 Menton

Last stop on France's Côte d'Azur before Italy, this old-fashioned seaside town evokes the mellow glamour of Riviera highlife decades ago. Pastel-coloured *maisons de village* fill the boutique-rich Old Town, and a museum on the seafront celebrates French poet and filmmaker Jean Cocteau.

Follow the coast 40km west to Nice and its international airport.

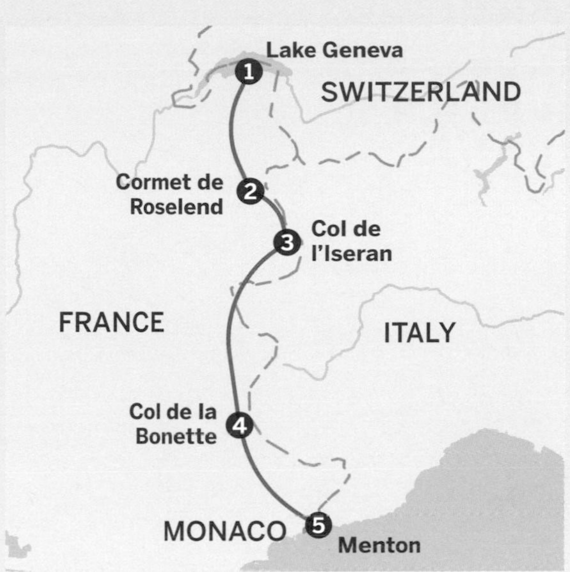

ALTERNATIVES

Diversion: Chamonix

The iconic heart and soul of the French Alps, action-packed Chamonix is dominated by the snow-white dome of Europe's highest peak, Mont Blanc (4807m). Mountaineering started in this region in the 19th century, and the rich and not-so-famous have flocked here ever since to fly down winter ski slopes and hike in summer.

Chamonix is a 40km detour east of the Route des Grandes Alpes from Cluses.

Extension: Monaco

Its world-famous hairpins are not alpine-steep, but they sure are as sharp: Formula One's most iconic race tears through downtown Monaco in May. Squeezed into 200 hectares, the world's second-smallest country seethes with high-rollers and hedonists. Walk the race route, risk a flutter in Monte Carlo Casino, tour the royal palace and aquarium and gawp at super-yachts in the port.

Trains run from Menton to Monaco (every 30min; 10min).

SWITZERLAND'S GLACIER EXPRESS

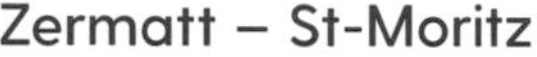

Zermatt – St-Moritz

Bookended by two of Switzerland's oldest, glitziest resorts, this deliciously slow ride aboard the Glacier Express is one of Europe's magical rail journeys.

FACT BOX

Carbon (kg per person): 12
Distance (km): 290
Nights: 1-7
Budget: €€€
When: mid-Dec to mid-Oct

❶ Zermatt

Variety is the spice of Alpine life in the French-German-speaking Valais region, and car-free Zermatt is no exception. Designer fashion boutiques bejewel its opulent main street, while 16th-century pig stalls and timber granaries propped up on stone discs and stilts (to deter pesky rats) infuse its old-world backstreets with rustic charm. Horse-drawn sled, electric taxi and foot are the appealing ways to get around. And then there's the Matterhorn (4478m), rising like a hypnotic shark's fin above town – skiing (winter) and hiking (summer) allow endless ogling.

Jump on the Glacier Express to Andermatt (daily at 08.52, summer only, and 09.52; 3hr), making your obligatory seat reservation for the southern/right-hand side of the train for the best views.

❷ Andermatt

After rattling by train across the awe-inspiring Oberalp Pass – the highest point of this journey at 2044m – you might feel a bit giddy upon arrival in Andermatt. A cocktail of low-key village charm and big wilderness sass, this glamorous mountain resort in central Switzerland is the place to kick back over scenic winter skiing in the SkiArena Andermatt-Sedrun, or summer hiking and cycling on lofty passes around town.

All aboard for the Glacier Express train to Chur (daily at 11.54, summer only, and 12.54; 2hr 30min).

ABOVE: LANDWASSER VIADUCT

ABOVE: CHUGGING INTO CHUR

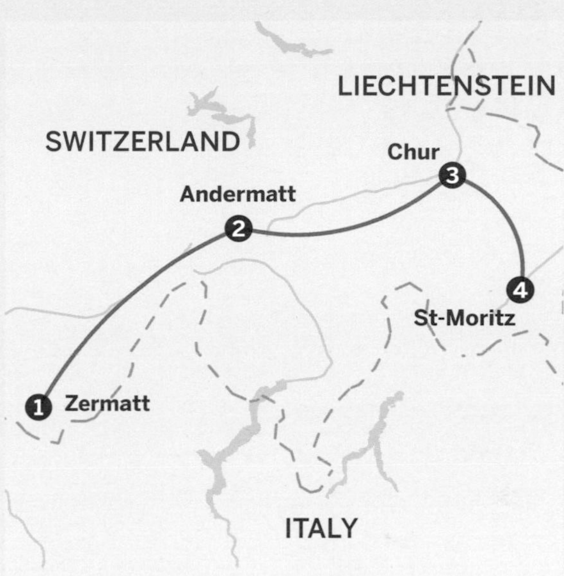

❸ Chur

An overnight stay in Chur provides the opportunity to discover Switzerland's oldest city, inhabited since 3000 BCE. Art galleries, boutiques, restaurants and the church spire of landmark Martinskirche lace narrow lanes and alleys in the Altstadt (Old Town), while in the surrounding Graubünden canton, the stellar Bündner Kunstmuseum near the train station showcases fine arts.

Get back on the Glacier Express to St-Moritz (daily at 14.32, summer only, and 15.32; 2hr). On the first leg to Filisur, look out for the celebrity Landwasser Viaduct: the 65m-high, six-arch bridge features prominently in all Glacier Express adverts.

❹ St-Moritz

Switzerland's slickest Alpine resort, St-Moritz has been luring royals, celebrities and moneyed wannabes since 1864. Its rich natural assets include a shimmering aquamarine lake, emerald forests, mineral springs and magnificent mountains clearly created with world-class skiing and summertime hiking, biking, e-biking and climbing in mind.

Take the train to Zürich (changing in Chur or Samedan and Landquart; hourly; 3hr 30min) for onward connections.

ALTERNATIVES

Diversion: Matterhorn Glacier Paradise

Ride the world's highest-altitude cable car to 3883m to gawp at 14 glaciers and 38 4000m-plus mountain peaks from the panoramic platform; some cabins even come with Swarovski crystals and glass floors. Heart-pumping highlights include an ice palace with glittering sculptures, an ice slide to swoosh down and an exhilarating snow-tubing track.

The cable car runs daily year-round from Zermatt.

Extension: Aletsch Glacier

Appearing like a superhighway of shimmering ice curving around the peak of Aletschhorn (4195m), this 22km-long glacier is Europe's largest and a Unesco World Heritage Site. In the last 20 years, it has shrunk 1km. Hiking to the ice from car-free village of Bettmeralp is an extraordinary experience.

Take the train from Brig, a Glacier Express stop, to Betten (2 hourly; 10min), then the cable car up to Bettmeralp (10min).

AUSTRIAN GRAND TOUR

Vienna – Salzburg

This two-week rail trip presents Austria in a nutshell, waltzing from Vienna's opulent palaces to castle-topped Salzburg.

FACT BOX

Carbon (kg per person): 16
Distance (km): 386
Nights: 14
Budget: $$
When: year-round

❶ Vienna

Start out with a cultural drumroll in Vienna, where Habsburg palaces, world-famous concert halls and a treasure chest of outstanding galleries – from the cutting-edge MuseumsQuartier to the Klimt-filled Upper Belvedere – await. Embrace the coffeehouse scene, gawp at the Gothic cathedral and stroll in canalside gardens.

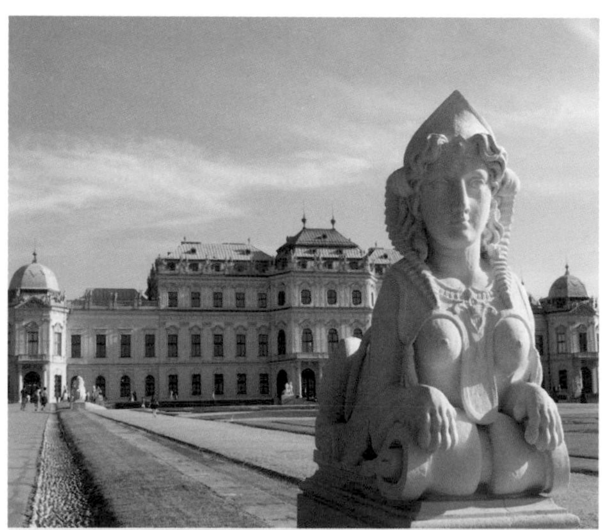

ABOVE: BELVEDERE PALACE, VIENNA

Take the train from Vienna to Melk (every 15min; 1hr).

❷ Melk

With an abbey-fortress perched high above the Danube, Melk is the cherry-on-top of the Unesco World Heritage Wachau Valley. Lording it above the prettily cobbled historic centre, Benedictine Stift Melk's church is a frenzy of Baroque ornament. Other unmissables include the town's richly frescoed library and Marble Hall.

Jump on the train to Linz (every 15-30min; 1hr-1hr 30min).

❸ Linz

Linz is where Austria leaps headfirst into the 21st century. With its gaze fixed firmly on the future, this tech-mad city has an ever-evolving cultural scene too. The Ars Electronica, zooming in on new media, and Lentos, a vast repository of modern art, are the biggest draws – both change colour when illuminated at night.

Take the train to Hallstatt, changing in Attnang-Puchheim (every 30min; 2hr).

❹ Hallstatt

Sheer, wooded peaks and pastel-coloured houses

ABOVE: SALZBURG

reflected in mirror-like waters make Hallstatt the fairest of all the villages in the lake-splashed Salzkammergut region. A funicular hauls you up to Salzwelten, the world's oldest salt mine. While you're up there, free-floating Skywalk platform affords astonishing views of lake and mountain.

Board the train for Bad Ischl (hourly; 20min).

5 Bad Ischl

Once a beloved retreat of the Habsburgs, handsome spa town Bad Ischl is a springboard for dipping into the region's five lakes. Top billing goes to the Italianate Kaiservilla, Emperor Franz Josef's spawling summer residence where he wooed princess-to-be, Elisabeth of Bavaria.

Take the train to Salzburg (hourly; 2hr), with a change at Attnang-Puchheim.

6 Salzburg

Salzburg rounds out the trip with a Baroque bang and an Alpine backdrop. Allow ample time to see the alley-woven, cathedral-topped Old Town, hilltop medieval fortress, and the city's raft of outstanding galleries, museums, cafes and concert halls.

Fast direct trains whisk you back to Vienna (every 30min, 2hr 15min).

ALTERNATIVES

Detour: Wachau Valley

There is something almost poetic about the way orchards and vineyards rib the slopes rising above the Danube in the Wachau Valley. Right in the heart of things is Spitz, a pretty village topped off by the 1000-Eimer-Berg, so-named for its ability to fill 1000 buckets of wine each season. Hiking trails wend their way to medieval castle ruins cresting hillsides and small, family-run wineries.

Buses run from Melk to Spitz (hourly; 50min).

Extension: Werfen

The limestone Tennengebirge range creates a cinematic backdrop to the village of Werfen. Get an early start and take a bus or cable car up to Eisriesenwelt, the world's largest accessible ice caves; wrap up warm to explore the chambers and tunnels of this frozen Narnia. Back in the valley, admire far-reaching views of the Salzach Valley from improbably perched Hohenwerfen fortress.

Trains run from Salzburg to Werfen (every 30min; 50min).

BAVARIA AND SALZBURG DISCOVERER

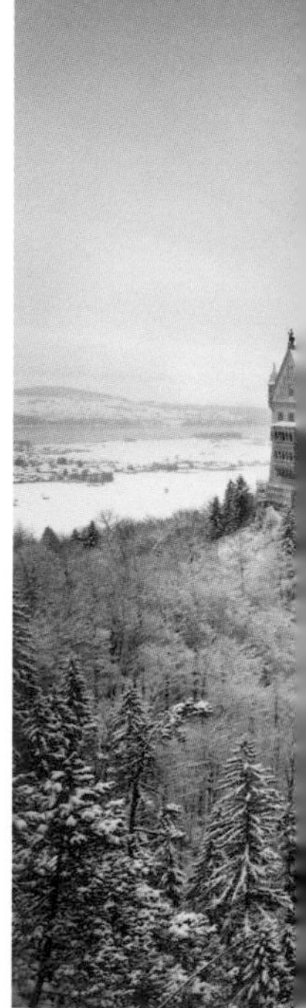

Munich – St Gilgen and Wolfgangsee

Combine the attractions of German Bavaria and northern Austria on a train-and-bus trip linking cities, castles, mountains and lakes.

FACT BOX

Carbon (kg per person): 18
Distance (km): 440
Nights: 4
Budget: $$
When: year-round

❶ Munich, Germany

The historic highlights of Bavaria's capital are packed into the Altstadt (Old Town), easily explored by strolling from Marienplatz, the central square at the heart of the city, admiring the 15th-century Altes Rathaus (Old Town Hall), neo-Gothic Neues Rathaus (New Town Hall) and the opulent Baroque interior of St Peterskirche. Make time for the Nymphemburg Palace, birthplace of King Ludwig II, and the sprawling Residenz, home of Bavaria's rulers from the Middle Ages. Finish your tour sipping a *stein* (glass or mug) of beer in one of the city's atmospheric breweries.

Take the train to Füssen (every 2hr; 2hr).

❷ Füssen and Schwangau, Germany

Of all the castles clustered in far southern Bavaria, best-known is Schloss Neuschwanstein, the dizzyingly romantic confection created by 'Fairytale King' Ludwig II in the late 19th century. But nearby Schloss Hohenschwangau, rebuilt in neo-Gothic style by Ludwig's father Maximilian II, and medieval Hohes Schloss, guarding Füssen's historic Old Town, are quieter and equally fascinating.

Head back to Munich and change for the train to Salzburg (every 2hr; 1hr 45min).

❸ Salzburg, Austria

A riot of Baroque palaces, houses and monasteries,

ABOVE: RAISE A *STEIN* IN MUNICH

ABOVE: SCHLOSS NEUSCHWANSTEIN, FÜSSEN

Salzburg's Altstadt is a timewarp warren of streets lined with elegant houses and shops built during the city's 18th-century heyday. Here you'll find Mozart's birthplace museum, the prince-archbishop's Residenz and the monumental, copper-domed Dom (cathedral). Above looms the 900-year old clifftop Festung Hohensalzburg, one of Europe's largest and best-preserved castles.

Take bus 150 from Salzburg Mirabellplatz to St Gilgen (every 2hr; 45min).

4 St Gilgen and Wolfgangsee, Austria

It'd be criminal to visit Austria without stretching your legs in the Alps; fortunately, the lakes and peaks of the Salzkammergut beckon, just a short bus ride from Salzburg. St Gilgen, an attractive village on the shore of the Wolfgangsee, has its own Mozarthaus (the composer's sister Nannerl lived here). A delightful walking trail leads above the lake's northern shore to St Wolfgang and Strobl; detour to conquer the peak of the Schafberg (1783m) on foot or, during the summer, by rack railway.

Take the bus from Strobl (every 2hr; 1hr) back to Salzburg, which has onward international connections.

ALTERNATIVES

Day trip: Augsburg, Germany

The historic centre of Bavaria's third-largest city is blessed with Renaissance and Baroque merchants' houses, churches and fine museums. The 17th-century Rathaus (Town Hall) houses the gilded, frescoed Goldener Saal ceiling; for more down-to-earth history visit the neat Fuggerei, one of the world's earliest social housing projects, dating from the 16th century.

Frequent trains serve Augsburg from Munich (30min).

Extension: Bad Ischl, Hallstatt and Obertraun, Austria

Bad Ischl's reputation as a spa town saw Emperor Franz Josef build a summer villa here, now open to visitors. To the south is, Hallstatt, sandwiched between mountains and its namesake lake; the ancient salt mine above offers guided tours. Obertraun is the stop-off for the vast Dachstein Caves.

Frequent buses connect Strobl with Bad Ischl (20min), Hallstatt (1hr 30min) and Obertraun (1hr).

SWITZERLAND'S JUNGFRAU BY RAIL

Interlaken – Mürren

This train journey takes you high into the Alps of Switzerland's sensationally beautiful Jungfrau region.

FACT BOX

Carbon (kg per person): 1.9
Distance (km): 47
Nights: 7
Budget: $$$
When: year-round

❶ Interlaken

Interlaken (frequent train connections to Zürich and its international airport) is the springboard for journeys into the Jungfrau region. Straddling piercing blue lakes Thun and Brienz, the town has immediate wow factor, added to by a backdrop of sky-high peaks that includes glacier-capped Eiger (3967m), Mönch (4110m) and Jungfrau (4158m). The country's hottest adventure-sports hub, it offers every pulse-quickening pursuit imaginable, from whitewater rafting to canyoning, glacier bungee-jumping and skydiving.

From Interlaken Ost station, regional trains depart for Grindelwald (every 30min; 33min).

ABOVE: HIKING OUT FROM GRINDELWALD

❷ Grindelwald

Gazing up at the ferocious fang of the Eiger's north face, this chalet-lined resort is a chilled but lively base for summer hiking and winter skiing. A cable car swings up to First, where walking trails and long, sunny, above-the-treeline ski runs await. Kids love the First Flyer zipline, trotti-bike scooters and Europe's longest (15km) toboggan run.

Take the train to Lauterbrunnen (every 30min; 37min).

❸ Lauterbrunnen

Like a scene from a Romantic landscape painting, Lauterbrunnen is framed by wooded cliffs that rise sheer and rugged to snow-capped mountains. The valley's 72 waterfalls include the 297m-high plume of the Staubbach Falls, so beautiful it inspired Goethe's poetic pen. With a sprinkling of laidback cafes and guesthouses, the village is a magnet for BASE-jumpers and ice-climbers.

ABOVE: LAUTERBRUNNEN

Continue on the incredibly scenic ride to Wengen (every 30min; 12min).

4 Wengen

Wengen perches photogenically on a mountain ledge, looking up to the Jungfrau giants and down to the waterfall-laced Lauterbrunnen Valley. This same view has lured visitors, especially Brits, since the early twentieth century. In winter it's a low-key, family-friendly base for skiing (good intermediate terrain plus the World Cup Lauberhorn course); in summer it's all about high-level hiking.

Return to Lauterbrunnen before taking the cable car/train combination up to Mürren via Grütschalp (40min).

5 Mürren

If you thought the Jungfrau region was beautiful so far, Mürren is the icing on the Alpine cake. With its cluster of dark log chalets and grandstand views of Eiger, Mönch and Jungfrau, this is pure wonderland stuff when the flakes fall. In summer, hike the North Face Trail; in winter test your mettle skiing the 16km Inferno Run.

From Mürren, return to Lauterbrunnen via Grütschalp then take the train to Interlaken Ost (55min).

ALTERNATIVES

Day trip: Jungfraujoch

Europe's highest train station (3454m), Jungfraujoch is a once-in-a-lifetime trip. The railway is an engineering feat, burrowing right through Eiger's heart. At the top, views of 4000m peaks and the Aletsch Glacier unfold; escape the crowds on a short hike to Mönchsjochhütte Alpine hut.

From Grindelwald, take the Eiger Express to Eigergletscher (15min) then a train to Jungfraujoch (27min).

Extension: Schilthorn

On clear days at 2970m Schilthorn, the 200-peak, 360-degree panorama reaches beyond the Swiss Alps to Mont Blanc and the Black Forest. The mountain was the backdrop for the 1969 Bond movie *On Her Majesty's Secret Service*. Tackle the Direttissima, a black beast of a piste; or peer into the void on the Thrill Walk at middle station, Birg.

From Mürren, half-hourly cable cars serve Schilthorn (17min).

A SPIN AROUND AUSTRIA'S SALZBURGERLAND

Salzburg – Zell am See

This train journey from castle-topped Salzburg to glacier-capped Zell am See has views to make you yodel out loud.

FACT BOX

Carbon (kg per person): 6.6
Distance (km): 161
Nights: 7
Budget: $$
When: year-round

ABOVE: CHANNELLING SALZBURG'S MUSICAL HERITAGE

❶ Salzburg

Baroque beauty Salzburg wins hearts instantly with its Alpine backdrop, high-on-a-hill medieval fortress, stately Residenz palace and musical legends including Mozart (born here in 1756) and Maria (*The Sound of Music* was filmed here). Creative restaurants, boutique hotels, lavish cafes and concert halls hide in its web of alleys, squares and courtyards.

Jump on the train to Hallein (every 30min; 25min).

❷ Hallein

Snuggling up to the Bavarian border, mountain-rimmed Hallein has an adorable pastel-painted Old Town to wander. Culturally, top billing goes to the Keltenmuseum, zooming in on the town's Celtic past, and the Stille Nacht Museum, paying tribute to Franz Xaver Gruber, composer of the carol *Silent Night*.

Take the train to Werfen (every 30min; 30min).

❸ Werfen

With the jagged limestone peaks of the Tennengebirge rearing above it, Werfen is visually a real drama queen. Arrive early to take a bus or cable car up to Narnia-like Eisriesenwelt, the world's largest accessible ice caves. Afterwards, make for 900-year-old Hohenwerfen fortress for tremendous views of the Salzach Valley.

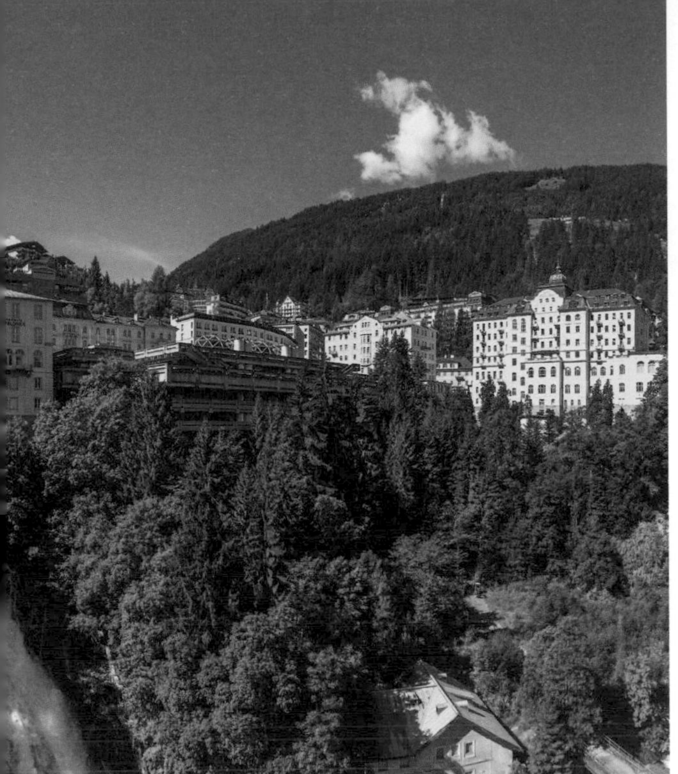

ABOVE: BAD GASTEIN

Take the train to Bad Gastein (hourly; 1hr 30min) changing at Schwarzach-St Veit.

4 Bad Gastein

With thundering waterfalls and Belle Époque villas clinging to its sheer, forest-cloaked cliffs, the genteel spa town of Bad Gastein is a handsome place. Deep in the Gastein Valley, it entices with natural hot springs and radon therapy in mountain caves. A gondola swings up to 2200m Stubnerkogel, where you can hike across a 140m-long suspension bridge with big views in summer, or ski in winter.

Take the fast (IC or railjet) train to Zell am See via Schwarzach-St Veit (every 2hr; 1hr 30min).

5 Zell am See

Perfect for dipping into the Hohe Tauern National Park, the cheerful lakeside resort of Zell am See is surrounded by sky-high peaks, including the glacier-capped 3203m Kitzsteinhorn. Swim, boat and stand-up paddleboard on the piercing blue lake, or hitch a cable car ride up to Schmittenhöhe to hike the five-hour Pinzgauer Spaziergang for phenomenal views of Austria's highest peak, 3798m Grossglockner.

From Zell am See, take the train back to Salzburg (every 30min; 1hr 30min-2hr).

ALTERNATIVES

Detour: Liechtensteinklamm

One of the deepest, longest gorges in the Alps, the Liechtensteinklamm is pure drama, with vertical 300m-high cliffs thrusting above a raging turquoise river. It's named after Johann II, Prince of Liechtenstein, who bankrolled the trail in 1875, with bridges and tunnels hacked into granite-veined cliffs.

It's a 15min train ride from Werfen to St Johann im Pongau where buses run frequently to the Liechtensteinklamm.

Day trip: Bad Dürrnberg

Bad Dürrnberg is home to the trophy sight of the Salzwelten, show mines offering a fascinating insight into the 'white gold' (salt) that once filled Salzburg's royal coffers. It's also a laidback village for Alpine hiking (Bavaria is just over the border), mountain biking, tobogganing and, in winter, skiing and snowshoeing.

Buses to Bad Dürrnberg depart hourly from Hallein train station (11min).

CITY-BREAK IN SECESSIONIST VIENNA

MAK – Secession

Spend a long Austrian weekend taking in the art and architecture of Vienna's Secessionist movement.

FACT BOX

Carbon (kg per person): 0
Distance (km): 6
Nights: 3
Budget: $$$
When: year-round

❶ MAK

Just a short walk from Wien Mitte station, begin this walking tour at MAK, the Museum of Applied Arts. The extensive collection here includes furniture, household objects and ephemera from the Secession era (also known as Art Nouveau), including many stylish items from the Wiener Werkstätte (a cooperative workshop founded in 1903) and a frieze by Gustav Klimt. Across the road is Café Prückel, one of Vienna's most atmospheric coffeehouses – its speciality is apple strudel with cream.

ABOVE: SECESSION BUILDING

Walk 5min north to Georg-Coch-Platz.

❷ Postsparkasse

Otto Wagner's impressive building for the Austrian Post Office is considered a showpiece of early Modernism; he designed both the exterior and interior decoration, right down to the furniture. Inside is a museum dedicated to the architect, complete with a screening room showing film clips.

Walk 30min southwest through the city centre to Museumsplatz.

❸ Leopold Museum

A beautifully designed modern museum with a comprehensive collection of Austrian art, the Leopold holds one of the largest Egon Schiele collections in the world: 42 paintings and many more drawings, watercolours and prints. Paintings

ABOVE: MAK, THE MUSEUM OF APPLIED ARTS

range from exquisite landscapes such as *Setting Sun* to uncompromising canvases like *Cardinal and Nun (Caress)*. There are also works by Gustav Klimt, Oskar Kokoschka and other Austrian Modernists. The Leopold is part of the elegant MuseumsQuartier complex, a great place for a coffee and some people-watching.

Walk 10min southeast to Friedrichstrasse.

4 Secession

Built to serve as the headquarters of the Secessionist art movement, and designed by Josef Olbrich, this gold-topped edifice also stands as a monument to the era. Admire the exquisite exterior, and then head to the basement to marvel at Gustav Klimt's *Beethoven Frieze* – his unsettling interpretation of the composer's ninth symphony. Nearby are more must-see examples of Secession-era buildings, all by architect Otto Wagner: the Otto Wagner Pavillion in Karlsplatz (designed as a metro station), and his apartment blocks at 38 and 40 Linke Wienzeile. Also on Linke Wienzeile is the Naschmarkt which, with over a hundred food and drink stalls, is the perfect spot for a lunch break.

Wien Mitte station is a 25-min walk back across central Vienna.

ALTERNATIVES

Extension: Kirche am Steinhof (Church of St Leopold)

Otto Wagner's only ecclesiastical building stands on a hill in hospital grounds to the west of Vienna. It's a masterpiece: the exterior has a copper dome and four Art Nouveau angels, and the stained-glass is by prominent Secessionist Koloman Moser.

From Wien Mitte it's 12 stops on Metro U3, then a 40-min walk or a 20-min journey on bus 46A or 46B.

Extension: Oberes Belvedere (Upper Belvedere)

A majestic Baroque palace and gardens, just a short tram ride from the city centre, the Oberes Belvedere is home to the Austrian National Gallery. Kokoschka, Schiele and Klimt are all well represented, most famously by the latter's *The Kiss*.

From the Leopold Museum, the Oberes is a 10-min ride on tram D.

A MEANDER ALONG THE MOSELLE

Trier – Cochem

Combine trains, boats and boots to follow the winding path of the Moselle River through Germany's top Riesling-producing region.

FACT BOX

Carbon (kg per person): 3
Distance (km): 122
Nights: 4-6
Budget: $$
When: May-Sep

❶ Trier

Reputedly Germany's oldest city, originally Celtic Trier was conquered by the Romans in 16 BCE. Its imperial heritage is evident in the Römerbrücke (Roman bridge), the blackened stone towers of the monumental Porta Nigra (Black Gate) dating from 160 CE, various baths and the Basilica of Constantine. It's a great place to try Moselle Riesling in the many Weinstuben (wine bars) and, in spring, the region's renowned white asparagus.

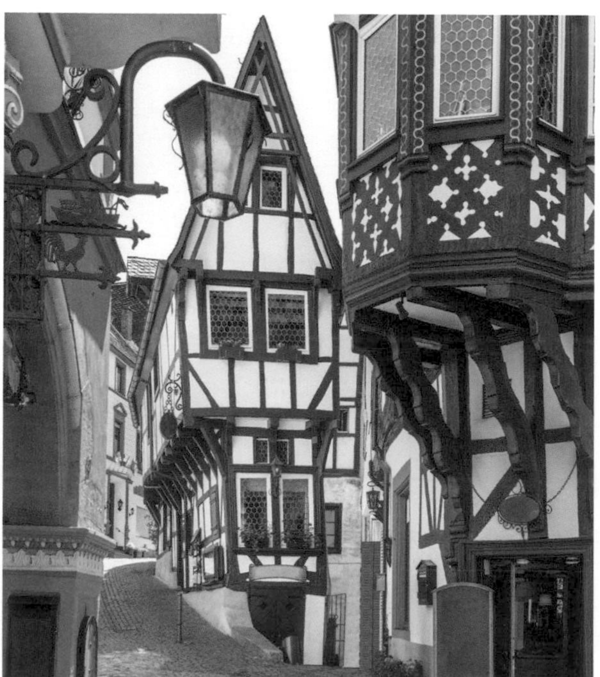

ABOVE: MEDIEVAL BERNKASTEL

Take the train to Ürzig (hourly; 40min), from where it's a 10km walk to Bernkastel-Kues.

❷ Bernkastel-Kues

This attractive twin town straddles the Moselle. On the eastern shore, Bernkastel has an attractive medieval market square lined with timber-framed old shops and houses; and a ruined castle, Burg Landshut, one of the region's many picturesquely battered bastions, wrecked following French assaults in the 17th and 18th centuries.

Scheduled boat tours sail to Traben-Trarbach (May-Oct; 4 daily; 1hr 45min).

❸ Traben-Trarbach

Another Moselle-spanning town, the two sides of Traben-Trarbach are linked by a bridge. Traben, its wealth swollen by the wine trade, is packed with Jugendstil (Art Nouveau) architecture, while the

ABOVE: TRIER

skeletal remains of Grevenburg Castle overlook Trarbach. Sip a glass of *kir moselle* – Sekt (German sparkling wine) tinted with peach liqueur – in the Bellevue Hotel's dinky Jugendstil bar.

Take the Moselweinbahn train from Traben-Trarbach to Ediger-Eller (hourly; 40min; change at Bullay), from where it's a 17km walk to Beilstein.

4 Beilstein

The region's most fairytale village is a cluster of medieval half-timbered houses alongside the river, surrounded by vineyards and overlooked by the ruins of 12th-century Burg Metternich – the courtyard cafe is a scenic spot for a glass of local Riesling.

Walk 14km along the river to Cochem or take bus 716 (2 daily; 20min).

5 Cochem

Top draw in this engaging pastel-hued town is the Reichburg (Imperial Castle) – another medieval fortress ravaged by French attacks in 1689, but restored in lavish neo-Gothic style two centuries later by a wealthy businessman. Explore its pseudo-medieval interior on a guided tour.

Frequent trains from Cochem serve Koblenz (50min), a major rail hub with onward connections.

ALTERNATIVES

Extension: Luxembourg City

Verdant, scenic, sophisticated, the hub of the Grand Duchy is beautifully set in the Alzette and Pétrusse river gorges. Roam the alleys, museums, bars and restaurants of the Unesco-listed Old Town, take in views from the Chemin de la Corniche, and get lost in the 17th-century Bock Casemates, a labyrinth of tunnels and galleries carved into the cliffs.

Frequent trains and buses run from Luxembourg to Trier (50min).

Extension: Koblenz

At the confluence of the Moselle and Rhine rivers, and marked by the Deutsches Eck (German Corner) monument, Koblenz is guarded by crenellated Schloss Stolzenfels and mighty Festung Ehrenbreitstein, looming above the water. It's a leafy, walkable city where three mountain ranges converge, enticing hikers into the hills.

Koblenz is a major rail hub; direct routes include Cologne (1hr) and Frankfurt (1hr 30min).

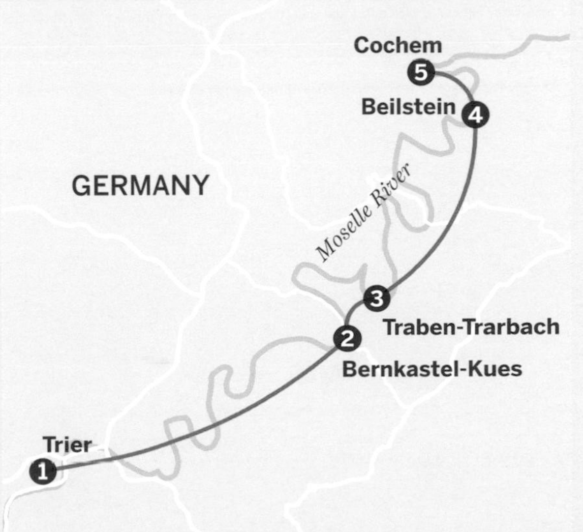

DUTCH CITIES BY TRAIN

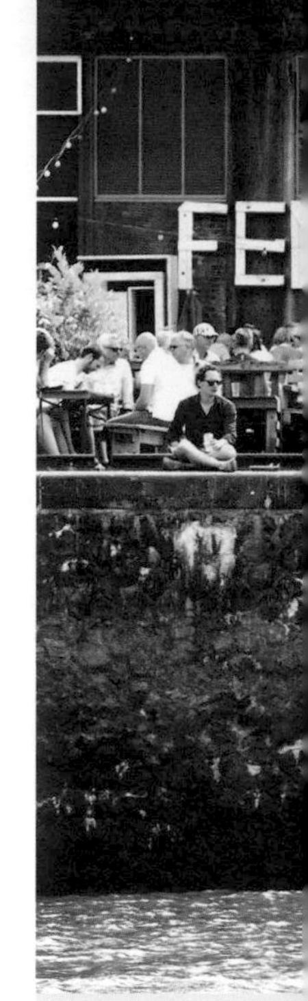

Amsterdam – Utrecht

Take to the rails on a tour of the Netherlands' most interesting cities, taking in world-class galleries, cosy cafe culture and renowned architecture, old and new.

FACT BOX

Carbon (kg per person): 6
Distance (km): 147
Nights: 10
Budget: $$
When: year-round

❶ Amsterdam

Elegant gabled buildings lining Golden Age canals, museums filled with masterpieces, a liberal attitude and a wealth of excellent cafes and restaurants all make Amsterdam one of Europe's great cities. It's best explored by bicycle: stop to admire Dutch masters at the Rijksmuseum and Van Gogh Museum; people-watch in sprawling Vondelpark; cruise through hip hoods such as Jordaan and De Pijp; and take several pitstops in cosy *bruin* cafes (Dutch pubs).

Regular trains link Amsterdam with Leiden (several an hour; 35min).

❷ Leiden

Next stop is another utterly alluring Dutch city: lovely Leiden. The birthplace of Rembrandt is laced with canals and beautiful buildings, and crammed with impressive museums. It has a youthful flair thanks to its student population, and a top-notch food and nightlife scene.

Jump back on the train to Den Haag (several an hour; 15min).

❸ Den Haag

The country's business and bureaucratic hub, grand Den Haag (aka The Hague in English) is home to embassies and stately mansions, and is the seat of the Netherlands' government. It's also renowned for its cultural scene, including the world-class Mauritshuis museum and the Escher in Het Paleis museum, dedicated to the graphic artist MC Escher. The city's credentials are rounded off with an exciting contemporary dining scene.

ABOVE: ST NICOLAASKERK, AMSTERDAM

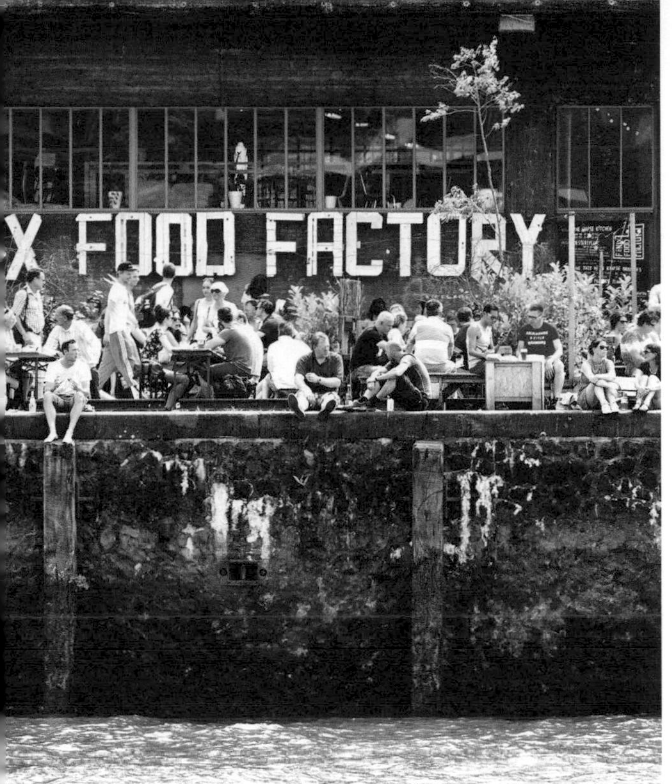

ABOVE: CAFE CULTURE, ROTTERDAM

Take the train to Rotterdam (several an hour; 30min).

4 Rotterdam

Home to Europe's largest port, Rotterdam has a rich maritime history but is not a city stuck in the past – it's a stark contrast to old-world Europe with its innovative architecture, creative spirit, cool art galleries and hip bars and restaurants. Join a guided tour of Rotterdam's architectural achievements, grab a bite to eat inside one of them at the Markthal food market and toast the day on a cafe terrace by the water.

Take to the rails again to get to Utrecht (every 15-30min; 35min).

5 Utrecht

A delightfully scenic town, Utrecht is one of the oldest urban centres in the country and is built around its iconic medieval Domtoren – a 112m-high tower that you can climb for sweeping views over the town's pretty tree-lined canal loop. As a university town, Utrecht is an unsurprisingly vibrant place: there's a fabulous cafe culture along with dynamic arts and live-music scenes.

Head back on the train to Amsterdam (every 15-30min; 30min).

ALTERNATIVES

Diversion: Haarlem

Canals, cobblestoned streets and cosy pubs make Haarlem feel like a mini Amsterdam. Saturdays are a great time to visit, when the Grote Markt brings its namesake square alive with stalls of organic produce, wheels of cheese and Dutch snacks. Admire Old Masters at the Frans Hals Museum, and have a beer at atmospheric Jopenkerk brewery.

Haarlem is easily accessible by train from Amsterdam (several an hour; 15min).

Day trip: Kinderdijk

You can't visit the Netherlands without setting eyes on its windmills, and Unesco-listed Kinderdijk, 16km east of Rotterdam, is the best place to do it: this quintessential Dutch landscape of waterways and marshes has 19 historic windmills. A couple function as museums and there's a bicycle path between the canals (bike rental available).

Fast ferries depart from Rotterdam Erasmusbrug to Kinderdijk (37min).

BELOW: TWILIGHT AT KEIZERSGRACHT CANAL, AMSTERDAM

CONQUER NORMANDY

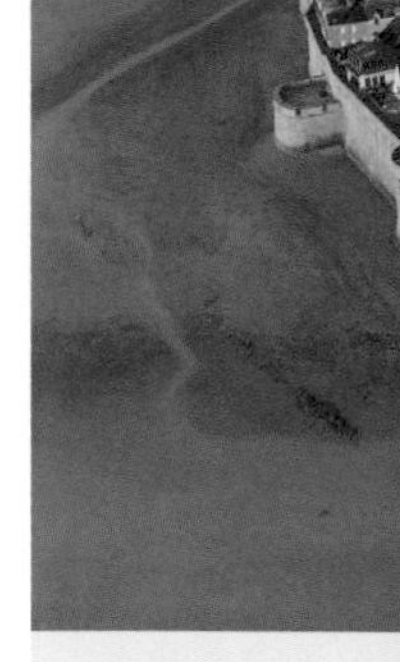

Rouen – Mont St-Michel

Normandy, on France's dramatic northern coast, encompasses medieval cities, clifftop monasteries and haunting D-Day beaches.

FACT BOX

Carbon (kg per person): 14
Distance (km): 350
Nights: 5-7
Budget: $$
When: Apr-Oct

❶ Rouen

Home to soaring Gothic architecture and a beautifully restored medieval quarter, Rouen is packed with historical treasures. Its magnificent cathedral was made famous by Monet, whose 30 paintings captured the changing moods of its dramatic facade. More notorious is the interactive Historial Jeanne d'Arc, which tells the story of Joan of Arc's life and death – the site where she was martyred today contains a postmodern church with breathtaking stained-glass windows.

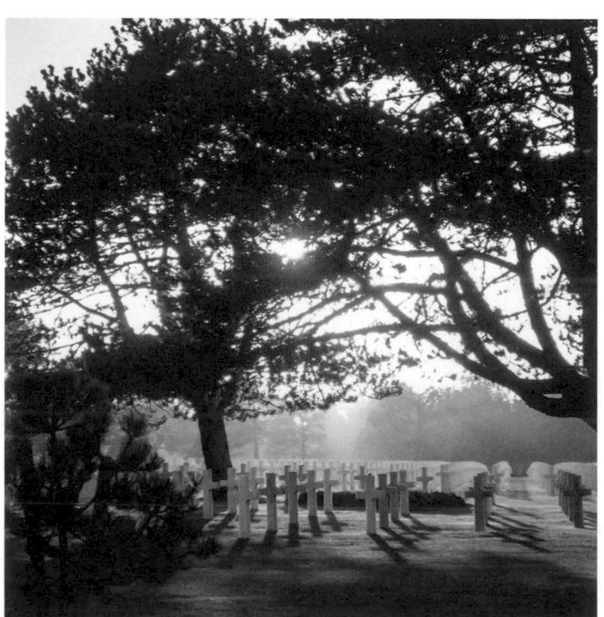

ABOVE: NORMANDY AMERICAN CEMETERY, OMAHA BEACH

Take the train to Caen (every 2hr; 1hr 40min).

❷ Caen

Founded by William the Conqueror in the 11th century, Caen suffered heavily in the 1944 Battle of Normandy. You learn all about the war at the Mémorial de Caen, one of Europe's finest WWII museums. Some vestiges of the past still survive: the restored castle is both home of a small museum complex and a fine lookout over town.

Frequent trains run between Caen and Bayeux (15min).

❸ Bayeux

Tiny, captivating Bayeux makes an excellent base for exploring the D-Day beaches as well as the town's own surprising attractions, namely a jaw-dropping Gothic cathedral that towers over the old centre. Most famous is the Bayeux Tapestry, an atmospherically lit 70m-long piece of needlework created in 1077 to

ABOVE: MONT ST-MICHEL

celebrate the Norman Conquest of England.
Bus 70 goes several times daily (not Sun) to Colleville-sur-Mer (30min).

4 Omaha Beach

The most brutal fighting on D-Day took place on the 7km stretch of coastline near Colleville-sur-Mer. You can take a self-guided historical tour past concrete bunkers along the beach following the Circuit de la Plage d'Omaha. Close by, the Normandy American Cemetery is a powerful testament to the lives lost here, with vast rows of crosses and Stars of David on a bluff overlooking the sea.
Bus back to Bayeux, then catch the twice-daily train to Pontorson (2hr 15min), a 10km taxi ride from Mont-St-Michel.

5 Mont St-Michel

Rising dramatically from the sea, the cliff-topped spire of Mont St-Michel is one of France's most iconic sights. Peel back the centuries as you wander up the narrow, winding passageways that lead to the medieval abbey, where Benedictine monks still hold services and illuminated night-time visits.
Several daily trains go from Pontorson back to Rouen (4hr 30min) and its onward connections.

ALTERNATIVES

Diversion: Honfleur

With a photogenic old harbour overlooking the Seine, Honfleur makes a fine destination for strolling. Leafy Jardin des Personnalités features sculptures from Honfleur history, while nearby Naturospace contains free-flying birds and butterflies in a greenhouse setting. There's also avant-garde art and music at the imaginative Maisons Satie, where composer Erik Satie lived and worked.
Daily express buses link Caen with Honfleur (1hr).

Diversion: Le Havre

The bustling waterfront city of Le Havre is a love-letter to Modernism, evoking France's postwar optimism. All but obliterated during WWII, the centre was completely rebuilt by Le Corbusier acolyte Auguste Perret. Evocative works include the Église St-Joseph with its soaring stained-glass interior, Le Volcan cultural complex and the very 1950s Appartement Témoin.
Hourly trains run between Rouen and Le Havre (1hr).

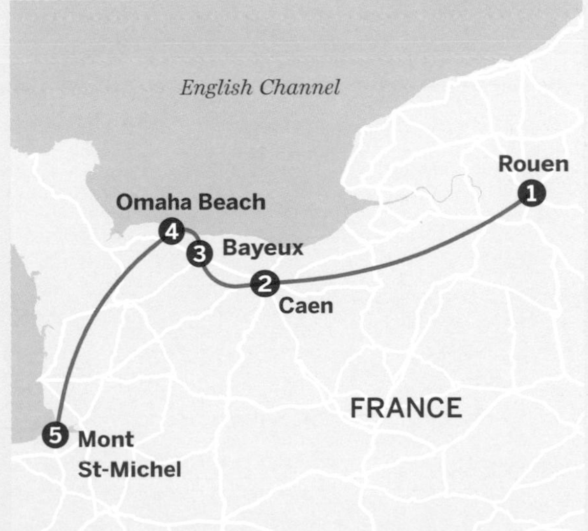

© Guillaume Louyot | Getty Images; MathieuRivrin | Getty Images

GOING OFF-GRID IN THE OUTER HEBRIDES

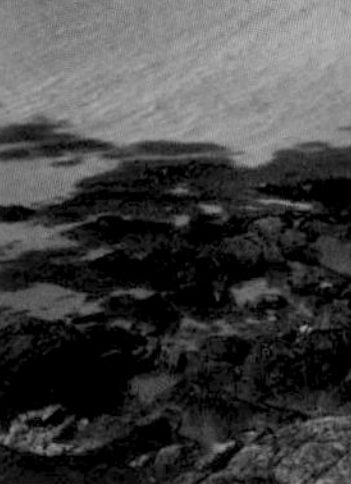

Lewis – The Uists

Remote moors, mountains and machair-fringed beaches astonish on this road trip and ferry skip through Scotland's Outer Hebrides, aka the Western Isles.

FACT BOX

Carbon (kg per person): 56
Distance (km): 308
Nights: 7
Budget: $$
When: Apr-Sep

❶ Lewis

The northern half of an island it shares with Harris, Lewis is a wild, windswept, wave-lashed place to begin your trip. Heading to the far northern tip brings you to the lighthouse-topped Butt of Lewis, which bears the full brunt of the North Atlantic. The island's hinterland is largely desolate, rolling moorland, glittering with lochans (small lakes), while the west has some gorgeous remote white-sand beaches like Mealasta. The island is littered with mysterious prehistoric sites, most famously the late-Neolithic standing stones of Callanish.

Clisham is a 55km drive south of Callanish.

❷ Harris

North Harris is pure drama. Its bare, forbidding mountains top out at 799m Clisham, which – time and energy permitting – can be hiked on a full-day circular walk. Edging south, stop off in the pretty harbour village of Tarbert to try sugar-kelp gin at the Social Distillery or take a boat over to the Shiant Islands for puffin and razorbill sightings. The island's west is necklaced with spectacular dune-flanked beaches, among them Luskentyre, a sweep of frost-white sand and azure sea backed by purple-grey mountains. On the fjord-like east coast, the single-track Golden Rd twists along an otherworldly landscape of moonlike gneiss rock.

Ferries depart from Leverburgh in Harris' south for Berneray (at least 2 daily; 1hr).

ABOVE: CALLANISH STANDING STONES, LEWIS

ABOVE: LUSKENTYRE BEACH, HARRIS

❸ Berneray

Clasped between the islands of Harris and North Uist, beautiful Berneray can be walked in its entirety in a few hours. In many ways this is the Outer Hebrides in microcosm, with machair-frilled beaches, brooding sea views and plenty of peace – as well as lots of wildlife: look out for otters and grey seals. West Beach is the dream, with its 5km ribbon of flour-white sand and insanely turquoise sea.

A causeway links Berneray to North Uist.

❹ The Uists

Silent wilderness, lonely moors and sea lochs enthrall on these twin isles. On North Uist, stop at the Hebridean Smokehouse in Clachan for peat-smoked salmon and scallops, then head south to picnic on white-sand beaches. On South Uist, Loch Druidibeg nature reserve attracts abundant birdlife, with regular sightings of dunlins, ringed plovers, greylag geese and corncrakes on the three-hour circular walk.

From South Uist, island-hop back north to Lewis' main town, Stornoway, or take the daily ferry from Lochboisdale to Mallaig (3hr 30min) in northwest Scotland.

ALTERNATIVES

Extension: Eriskay

Eriskay is home to Prince's Strand, the beach where Bonnie Prince Charlie first set foot on Scottish soil and raised the Stuart standard in 1745, launching the Jacobite Rising. The island's other claim to fame is the SS *Politician*, a cargo ship that ran aground offshore in 1941 with its 250,000 bottles of whisky. The wreck and subsequent looting inspired the film *Whisky Galore!*

Eriskay is connected to South Uist by a causeway.

Extension: Barra

Serene, lovely and almost as far south as you can go in the Outer Hebrides, Barra promises hills, moorland, wildflower-sprouting machair and incredible beaches. Castlebay in the south overlooks Kisimul Castle, on its own wee island. Even the airport is unique: it's the only one in the world where scheduled flights use a tidal beach as the runway.

Barra can be reached by ferry from Eriskay (5 daily Mon-Sat, 2 Sun; 40min).

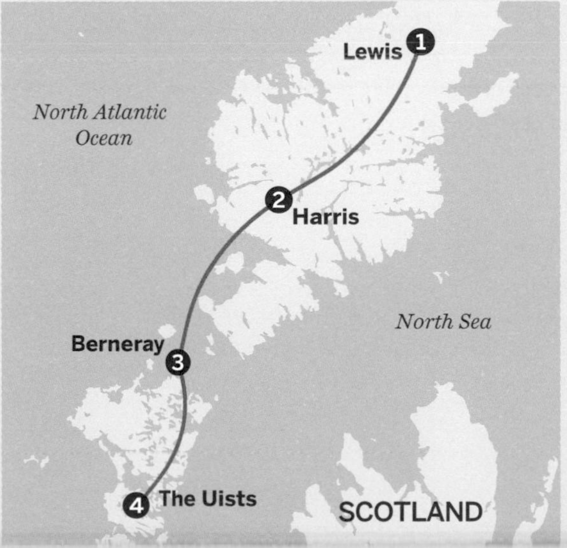

PARIS' BEST BAKERIES AND PATISSERIES ON FOOT

Du Pain et des Idées – Poilâne

Take a tasting tour of the finest cakes, pastries and breads on a walking trail that leads through the heart of the French capital.

FACT BOX

Carbon (kg per person): 0
Distance (km): 10
Nights: 1-2
Budget: $
When: year-round

ABOVE: PARISIAN PATISSERIE

❶ Du Pain et des Idées

A 15-minute walk southeast from Gare du Nord, on Rue Yves Toudic on the western side of the Canal St Martin, your tour begins at one of the prettiest boulangeries in the city. This is the place to buy a croissant, but also popular are the 'snails' – spirals of pastry in different flavours (try the pistachio and chocolate) – and the bread: large hunks of *pain des amis*. Open Mon-Fri only.

Walk 10min south to Rue de Turenne.

❷ Tout Autour du Pain

As the name suggests (it means 'everything about bread'), premium *pain* is the big attraction here: buy the dense, delicious *schwarzbrot* to take home (it will last for days), and an award-winning baguette to eat immediately. Their sweet stuff is great too, with the meltingly good *pain aux raisins* hard to resist.

Walk 15min southwest to Rue des Rosiers.

❸ Yann Couvreur

A fashionable – but friendly – patisserie, serving cakes that look too pretty to eat. Flavours aren't always traditional; the chocolate éclair, for example, is enlivened with Baileys. The *viennoiserie* also goes down a treat – the splendid *kouglof* comes with

ABOVE: CANAL ST-MARTIN

flaked almonds and candied fruit.
Continue 5min southwest to Rue de la Verrerie.

❹ Maison Aleph

Maison Aleph is a modish Syrian-French fusion bakery with unusual cake flavours – orange blossom, rosewater and cardamom – alongside standards such as chocolate and fruit. Fig and cinnamon tart is a typical seasonal offering. A selection of small pastries in one of their elegant boxes makes a fine present (even if just for yourself).
Burn off a few calories with a 30min stroll across the Seine to Rue du Cherche-Midi on the Left Bank.

❺ Poilâne

Internationally famous for its bread, Poilâne sells large loaves of sourdough (*pain au levain*) and rye (*pain de seigle*) which can be bought whole, or by the half or quarter, sliced or unsliced. For a flagship branch, the attractive boulangerie is surprising small, but there's room for croissants, *pains aux raisins* and delicious *punitions* (little biscuits) alongside the loaves.
Poilâne is a short walk from Saint-Sulpice and Sèvres-Babylone Métro stations.

ALTERNATIVES

Extension: Chambelland

An unusual find in Paris, Chambelland not only produces gluten-free bread but also patisserie and cakes, such as Paris-Brest, *tarte aux fruits* and éclairs. There's a nice line in US-style cookies and a vegan banana bread too. Loaves, all certified organic, include five-grain, sourdough and focaccia. Take away or eat in – there's a cafe with outdoor tables.
Chambelland is a 10-min walk east of Tout Autour Du Pain on Rue Ternaux.

Extension: Ten Belles

As well as baking stellar bread (from sourdough to focaccia) and a delightful selection of Anglo-French cakes, Ten Belles also has a laidback cafe with a courtyard terrace. Generously-filled brioche buns, toasted sandwiches and tarts are matched with excellent coffee, making this a great pit-stop. There's always a vegetarian option.
From Yann Couvreur, walk 20min east to Ten Belles, Rue Breguet.

FROM PARIS TO İSTANBUL BY TRAIN

Paris – İstanbul

Follow on the footplate of the Orient Express, the original luxury train across Europe, visiting multi-country Art Nouveau highlights along the way.

FACT BOX

Carbon (kg per person): 20
Distance (km): 3233
Nights: 4-7
Budget: $$
When: Jun-Sep

❶ Paris, France

There's no shortage of Art Nouveau and fin-de-siècle glamour in the French capital: enjoy a cabaret at the Moulin Rouge; absorb the masterpieces of the Musée d'Orsay; browse beautiful department stores (particularly Galeries Lafayette and La Samaritaine); and ride the Métro. The extravagance of the epoch is epitomised in Le Train Bleu, the Gare de Lyon's opulent restaurant.

Direct trains to Munich depart Paris Gare de l'Est each afternoon except Saturday (5hr 40min).

❷ Munich, Germany

Bavaria's capital is an exuberant delight, from the Baroque excesses of Nymphenburg Palace to busy breweries and beerhalls. Sticking with the theme of 19th and early 20th-century style, browse exceptional art in the Neue Pinakothek and Pinakothek der Moderne galleries.

Sleeper trains to Budapest depart Munich Hauptbahnhof nightly (10hr); direct day trains are quicker (7hr).

❸ Budapest, Hungary

Buda, on the Danube's west bank, is dominated by the medieval monuments of Castle Hill, while Pest, across the Chain Bridge, is studded with Secessionist (Art Nouveau) architecture. Admire the 1902 Parliament building, enjoy a performance at the elegant State Opera House or Liszt Music Academy then, back in Buda, soak amid sumptuous Belle Époque tilework in the Gellért Thermal Bath.

The sleeper to Bucharest departs Budapest Keleti Station nightly (16hr).

ABOVE: ENGLISCHER GARTEN, MUNICH

ABOVE: PARLIAMENT FROM THE FISHERMEN'S BASTION, BUDAPEST

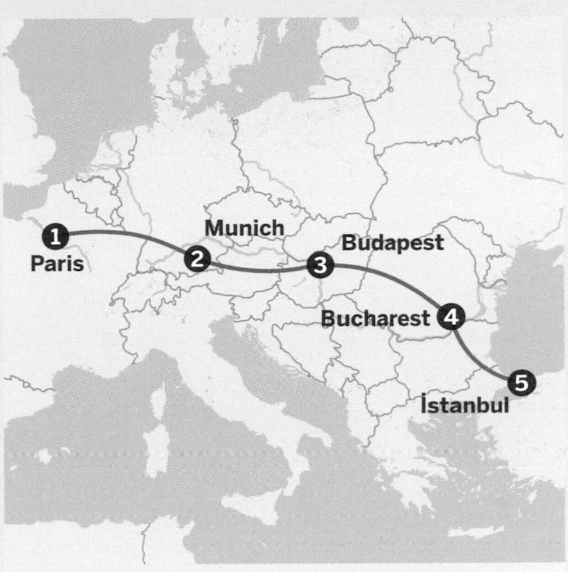

❹ Bucharest, Romania

Romania's capital can seem overwhelmed by Communist-era relics such as the Stalinist Palace of Parliament, but less austere treasures to unearth include the grandiose Athenaeum with its fabulous fin-de-siècle dome and dazzling mosaics. And another far cry from the sober Cold War days can be found in the city's streets, lined with buzzy bars and cafes hosting hopping nightlife.

Sleeper trains depart Bucharest Gara de Nord for İstanbul Halkalı Station nightly (Jun-Sep; 17hr); otherwise change in Bulgaria. Frequent suburban Marmaray trains run from Halkalı to central Sirkeci Station (35min).

❺ İstanbul, Turkey

The historic Sultanahmet district spreads south of Sirkeci Station, erstwhile eastern terminus of the Orient Express, encompassing most of the ancient city's historic highlights. Here you'll find the astonishing 6th-century basilica known as Hagia Sophia, the gorgeously tiled Blue Mosque, and opulent Ottoman Topkapı Palace.

İstanbul Airport has flights to most corners of the globe.

ALTERNATIVES

Day trip: Vienna

An easy stop-off between Munich and Budapest, the Austrian capital is awash with Art Nouveau (aka Jugendstil), showcased at the Secession building. Or head to the Baroque Belvedere for a wealth of paintings by Gustav Klimt, absorb Gothic grandeur at the Stephansdom (St Stephen's Cathedral), and gawp at imperial swagger at the Hofburg and Schönbrunn palaces.

Several trains daily between Munich and Budapest call at Vienna (4hr).

Diversion: Zagreb, Belgrade and Sofia

For an alternative routing, catch the sleeper from Munich to Croatia's Zagreb for historic streets and quirky museums. Continue to Belgrade in Serbia for nightlife, and to Bulgaria's Sofia for Roman, Ottoman and Communist-era monuments.

Nightly sleepers run Munich-Zagreb (9hr). Trains run daily Zagreb-Belgrade (7hr) and Belgrade-Sofia (10hr; direct summer only). Sofia-İstanbul (8hr) sleepers run Jun-Sep.

A RAIL JOURNEY ALONG THE RHONE VALLEY

Lyon – Orange

Head to southeast France and soak up the wide-ranging appeal of the Rhône Valley, packed with Roman ruins, grand markets and stunning natural scenery.

FACT BOX

Carbon (kg per person): 11
Distance (km): 305
Nights: 5-7
Budget: $$
When: Apr-Dec

❶ Lyon

Sitting pretty between the Rhône and Saône rivers, Lyon has lured visitors since its founding in 43 BCE. Riverside strolls make a fine introduction to the city, followed by a ramble through Vieux Lyon, with its brick streets, Renaissance architecture and historical museums. Nearby, lofty Fourvière offers sweeping views over France's third largest city. Other highlights include grand food markets, dining in a *bouchon* (a traditional Lyonnais bistro) and nightlife in bohemian-loving Croix Rouge.

ABOVE: *BOUCHON* DINING, LYON

Vienne-bound trains run from Lyon's three main train stations (hourly; 30min).

❷ Vienne

France's Gallo-Roman heritage is alive and well in this relaxed riverfront city whose old quarter hides spectacular Roman ruins. Get an overview of ancient Vienne amid fine mosaics and sculptures at the Museé Gallo-Romain. Then take a peek at a remarkably well-preserved Roman temple and the Théâtre Antique, an 11,000-seat amphitheatre that still hosts big events (like the summertime Jazz Festival).

Take the train to Montélimar (hourly; 1hr 15min).

❸ Montélimar

Montélimar boasts an atmospheric Old Town and a grassy, tree-shaded, cafe-lined promenade that meanders through its centre. There are fine views from the 12th-century Château des Adhémar's hilltop

ABOVE: THÉÂTRE ANTIQUE, ORANGE

perch, but the real reason to come to Montélimar is to indulge in its famous nougat. You can take a factory tour and load up on delectable souvenirs at Arnaud-Soubeyran, going strong since 1837.

Take bus 76 to Vallon-Pont-d'Arc (4 daily; 1hr).

4 Gorges de l'Ardèche

The tiny town of Vallon-Pont-d'Arc is gateway to a dramatic region of river gorges and caverns. You can spend the day pondering Paleolithic art at the Caverne du Pont-d'Arc, followed by a leisurely paddle around horseshoe bends and beneath natural stone arches on the scenic Ardèche River.

Take the bus back to Montélimar and catch the train to Orange (hourly; 40min).

5 Orange

Once the largest city in Gaul, 2000-year-old Orange contains a dazzling collection of Roman works. Take in the view of city and the distant mountains from the Colline St-Eutrope, then descend into the Unesco World Heritage centre for a close-up look at the ancient theatre, triumphal arch and amphora-filled art museum.

Frequent trains head back to Lyon (2hr 20min) for onward connections.

ALTERNATIVES

Extension: Beaujolais

Some 50km north of Lyon lie the rolling vineyards and sleepy villages of the Beaujolais region, synonymous with luscious red wines. Book tastings and tours at the 16th-century Château de Juliénas and the 17th-century cellar at Caveau de Cru Morgon, followed by a meal at the celebrated Franco-Japanese restaurant Au 14 Février.

Rent a car in Lyon for the one-hour drive up to Beaujolais country.

Extension: Avignon

One of Provence's most captivating towns, Avignon has a rampart-ringed old town, leafy squares, brilliant restaurants and one very famous medieval bridge. The town's showpiece is the Palais des Papes, the largest Gothic palace ever built. Equally famous (among foodies), the sprawling food market Les Halles showcases Provence's finest ingredients.

Frequent trains link Orange with Avignon (22min).

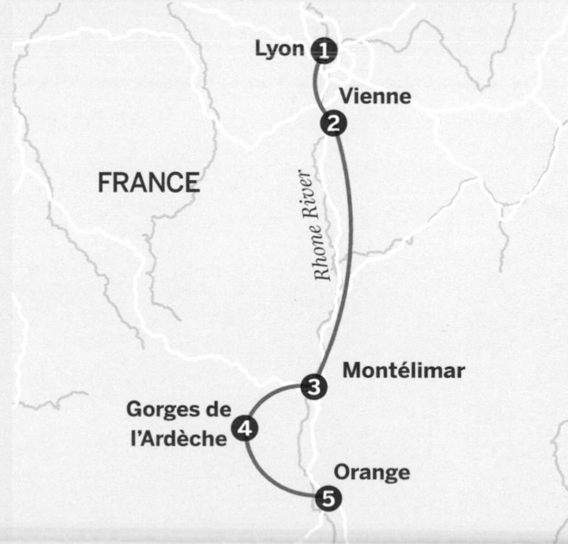

FAMILY FUN IN PEMBROKESHIRE

Saundersfoot – Freshwater West
Take a short road trip along Pembrokeshire's southern coastline for dramatic Welsh scenery, pretty fishing villages and family-friendly beaches.

FACT BOX

Carbon (kg per person): 4.5
Distance (km): 36
Nights: 3-4
Budget: $$
When: year-round

❶ Saundersfoot

Starfish, limpets and crabs scuttling amongst the seaweed are just some of the delights you might see on the shoreline while rockpooling at Saundersfoot. This small seaside town is a perfect starting point for families looking to explore Pembrokeshire's southern shores thanks to its manageable size, convenient parking and wide flat beach, which starts off sandy but becomes rockier as you head east. The coast path, which follows an old railway line, is suitable for buggies on this stretch.

Drive 5km south.

❷ Tenby

Tenby may be the biggest resort town in the region, but it still exudes a genuine quaintness and charm thanks to its medieval walls, picturesque fishing harbour and sloping streets of pastel-coloured townhouses. Family-friendly fun comes in the form of bodyboarding, swimming and sandcastle-building at one of the town's four fine beaches. Younger children will love the lifeboat launches at the RNLI station, while older kids can try coasteering: exploring the coastline by climbing, jumping and swimming in sea caves.

Continue 10km west.

ABOVE: TENBY HARBOUR

ABOVE: MANORBIER BEACH

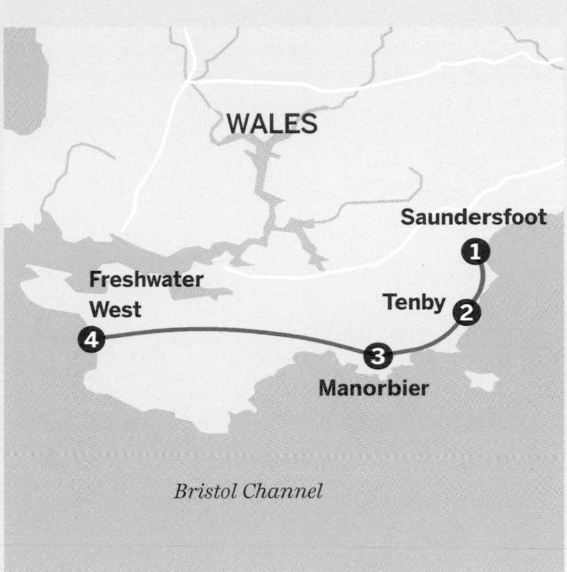

❸ Manorbier

Manorbier is a small seaside village with a sandy cove that sits amid one of the most scenic sections of the Pembrokeshire Coast Path. Hikes both east and west are easily accessed from the beach, which is popular with surfers. Though the paths are steep in parts, the views are wonderful, and there are plenty of lovely picnic spots along the way from which to enjoy them. Kids will love to wander around Manorbier Castle, built in Norman times, to finish off the day.

Drive 21km west.

❹ Freshwater West

Freshwater West is an essential stop for adventurous families. There's a sweeping sandy beach backed by high dunes that are perfect for games of hide and seek, plus rocks and boulders just begging to be climbed upon, and one of the best surf breaks in Wales. Children can also seek out the shrine to house elf Dobby from *Harry Potter* on the beach (a key movie scene was filmed here). And be sure to stop for a lobster roll or ice cream at Café Môr, a solar-powered boat-cum-beach-shack in the car park.

Drive 170km east to Cardiff for onward connections.

ALTERNATIVES

Day trip: Wildlife-watching off St Davids

Boats from St Davids (Britain's smallest city) delve into the nutrient-rich waters of the Celtic Deep, home to a range of marine wonders including grey seals, porpoises, puffins, guillemots and razorbills. Offshore sailings further afield can even deliver dolphin and whale sightings.

St Davids is 55km west from Freshwater West.

Diversion: Barafundle Bay

The fact you have to walk for half an hour from the car park at Stackpole to reach the golden sands and wonderfully clear water of Barafundle Bay is a helpful limit on visitor numbers. The walk itself, along cliff tops, dunes and through fragrant pine forest, is part of the Pembrokeshire Coast Path and a joy in itself.

Barafundle Bay is 10km southwest of Manorbier.

SCOTLAND THROUGH ITS WHISKY

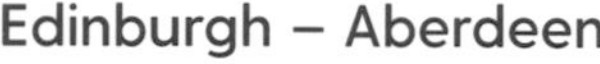

Edinburgh – Aberdeen

Summon your adventurous spirit for a singular malty journey north through Scotland, all in celebration of the nation-defining drink, whisky.

FACT BOX

Carbon (kg per person): 23
Distance (km): 360
Nights: 8-10
Budget: $$
When: year-round

MACALLAN DISTILLERY, CRAIGELLACHIE

❶ Edinburgh

Scotland's capital is the perfect place to pick up the whisky trail at the Scotch Whisky Experience on the Royal Mile, just down from the castle. Discover why whisky means so much to the Scots at this full-on sensory experience: it includes a whisky-barrel ride through the production process, a lesson on the drink's mighty importance to the country's history and culture, tasting tips and much more.

Take the train to Pitlochry (every 1-2hr; 2hr).

❷ Pitlochry

Pastoral lowland Perthshire opens into the county's wilder highlands around pretty Pitlochry, where there are two especially handsome whisky distilleries, Blair Athol and Edradour. The smallest of Scotland's traditional distilleries, Edradour is among the nation's most boutique whisky experiences with every part of the distilling process done by hand.

Get back on the train and head to Dalwhinnie (every 1-2hr; 50min).

❸ Dalwhinnie

Heading up into splendid mountain scenery, the train brings you to Dalwhinnie on the edge of Cairngorms National Park. Dalwhinnie Distillery is Scotland's

ABOVE: EDRADOUR DISTILLERY, PITLOCHRY

highest – they say it's often snowmelt that flavours the single malt made here.

Take the train to Aviemore (every 1-2hr; 50min) and rent a car to drive 67km to Dufftown.

❹ Dufftown

You'll smell the waft of malt before you arrive in the deservedly dubbed 'whisky capital of the world', Dufftown, in the heart of the famous Speyside whisky region. Just as Rome was founded on seven hills, locals say Dufftown was founded on seven stills. Six active distilleries await, most famously Glenfiddich, producing the planet's best-selling single malt.

Drive 85km to Aberdeen

❺ Aberdeen

The handsome granite city of Aberdeen is where you'll encounter some of the nation's classiest whisky bars. With 600 eclectic malts, the Grill is still considered among the finest addresses for sampling Scotland's favourite spirit, 150 years after its doors first opened. Follow up at the Tippling House, where you can indulge in a 'whisky flight' paired with cheese, chocolate and the like.

You can get the train back to Edinburgh from Aberdeen (every 30min; 3hr).

ALTERNATIVES

Day trip: Aberfeldy

Smart, stone-built Perthshire market town Aberfeldy sits at a whisky crossroads. Dewar's Distillery here adroitly demonstrates whisky's two principal forms (blended and single malt) at their distillery, interactive museum and whisky bar/lounge. The Scots call whisky 'water of life': a double-meaning in Aberfeldy, where some of Scotland's most dramatic white-water rafting beckons on the Tay.

Buses run to Aberfeldy from Pitlochry (40min).

Day trip: Craigellachie

From Dufftown, take a jaunt to Craigellachie village on the River Spey for a whisky-themed wander on the Speyside Way. Sip a dram at one of Scotland's greatest whisky bars, the Highlander Inn, which offers the best Japanese selection outside Japan. Or drop into the Macallan distillery. Then divert to the Speyside Cooperage to observe the art of whisky-barrel manufacture.

Walk 8km along the River Fiddich or drive via the A94.

SAILING THE ISLES OF SCILLY

St Mary's – St Martin's

Island-hop across England's westernmost archipelago, savouring spectacular beaches, secret coves, cute fishing villages and prehistoric sites.

FACT BOX

Carbon (kg per person): 0.5
Distance (km): 26
Nights: 4-7
Budget: $$
When: Apr-Oct

❶ St Mary's

A walk around St Mary's (the largest Scilly at a whopping 5km long) reveals 4000-year-old hut remains and burial mounds, gorgeous sandy beaches and 18th-century ramparts built around Star Castle. 'Capital' Hugh Town is an engaging village with the archipelago's tastiest seafood and the Isles of Scilly Museum, where you can brush up on the local history.

Scheduled trips to St Agnes sail most days during the season, weather permitting (15min); boat taxis are available on other days.

❷ St Agnes

Rocky, edge-of-the-world St Agnes is home to the UK's southwesternmost pub, the Turk's Head, serving fine fish and chips washed down with a pint of proper ale. Visit Bronze Age remains on Gugh, swim from pristine Covean Beach or SUP from the activity centre.

Return to St Mary's to catch the twice-daily boat to Bryher from Hugh Town (15min).

❸ Bryher

The smallest and wildest of the populated islands is a heather-clad, wildflower-spangled beauty. Samson Hill, studded with ancient burial chambers, provides panoramic views across the whole archipelago, while waves charge across windswept Hell Bay.

ABOVE: BRYHER ISLAND RESIDENT

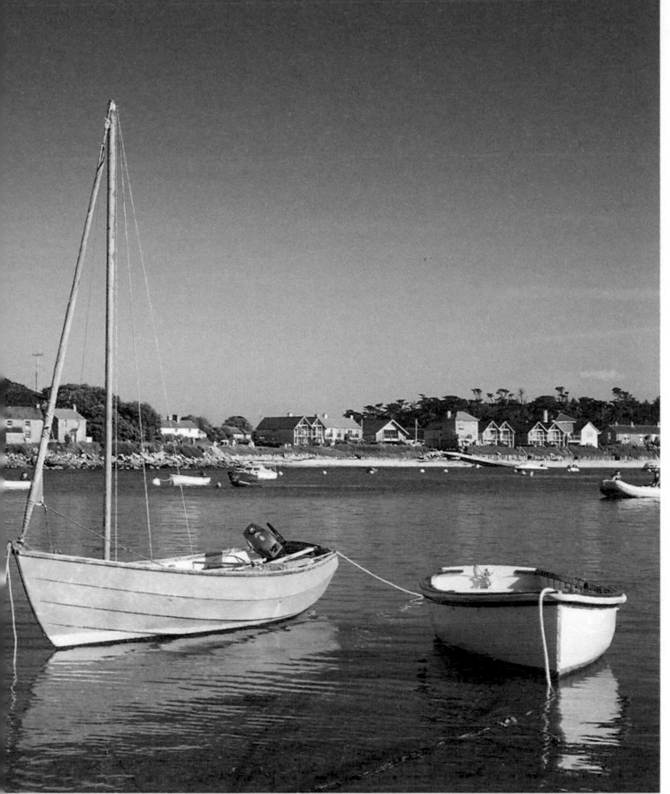

ABOVE: NEW GRIMSBY HARBOUR, TRESCO

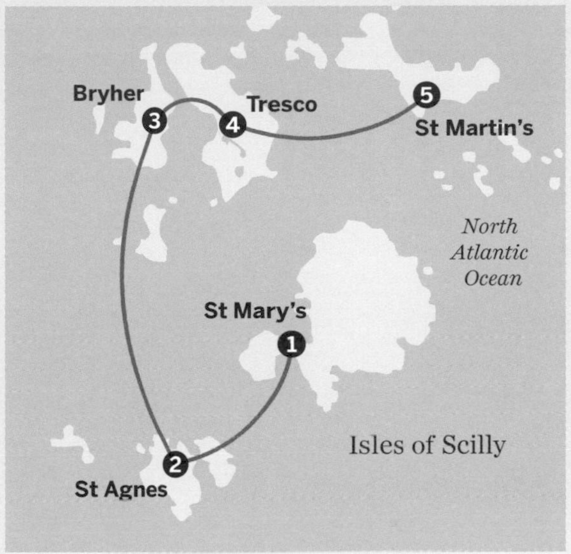

Boats to Tresco, just 250m across the water, run most days (5min). At the lowest spring tides it's possible to walk across – seek local advice.

❹ Tresco

The lush subtropical Tresco Abbey Garden, planted from 1834 amid the ruins of a medieval Benedictine priory, represent the main draw to this island. But along with the famous plants and flowers, there's earlier history to be found on Tresco at King Charles's Castle and Cromwell's Castle, built in the 16th and 17th centuries, along with dazzling white-sand beaches to sun on such as Appletree Bay. Kayaks and sailing dinghies can be hired here for getting out on the water.

Sail to St Martin's via St Mary's on a scheduled service (15min each leg) or hire a boat taxi.

❺ St Martin's

For more fun in the water, book an aquatic activity in Higher Town. Choose from scuba-diving excursions to explore the many wrecks around the islands and snorkelling with Atlantic grey seals. Par Beach, arguably Scilly's finest sweep of sand, flanks St Martin's southern shore.

Take the boat back to Hugh Town on St Mary's.

ALTERNATIVES

Day trip: Annet, Bishop Rock and Western Rocks

The Isles of Scilly draw wildlife-lovers and birders with their internationally important breeding population of Atlantic grey seals, plus numerous seabird colonies. Puffins breed late April-August; spot them – plus guillemots, razorbills, shags and many others – on a boat trip to the Western Rocks via Annet and Bishop Rock.

Wildlife-watching trips run most days in summer from St Mary's.

Extension: Penzance and Marazion

The ferry linking the Isles to mainland Cornwall sails from lively Penzance, with grand Georgian and Victorian architecture along Chapel St, notably the kitsch Egyptian House, and the Penlee House Gallery, displaying works by Newlyn School and Lamorna artists. Just 5km to the east, St Michael's Mount rises from the waves off Marazion.

The *Scillonian III* sails to/from St Mary's at least daily late Mar-early Nov (2hr 45min).

CYCLE THE CHÂTEAUX OF THE LOIRE

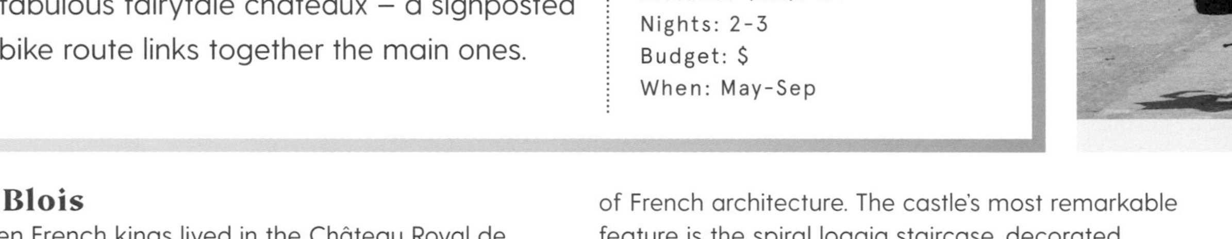

Blois – Amboise

The Loire Valley is home to France's most fabulous fairytale châteaux – a signposted bike route links together the main ones.

FACT BOX

Carbon (kg per person): 0
Distance (km): 114
Nights: 2-3
Budget: $
When: May-Sep

❶ Blois

Seven French kings lived in the Château Royal de Blois, whose four grand wings span several periods of French architecture. The castle's most remarkable feature is the spiral loggia staircase, decorated with salamanders and curly Fs, heraldic symbols of François I. The richly painted State Room and the King's and Queen's Chambers are redolent with historical intrigue.

ABOVE: CHÂTEAU ROYAL DE BLOIS

Follow the signposted bike route east from Blois to Chambord (20km).

❷ Chambord

The grandest and largest of the Loire châteaux, Chambord boasts an astonishing 440 rooms, 365 fireplaces and 84 staircases. Begun in 1519 by François I as a weekend hunting lodge, it grew into one of the most ambitious – and expensive – architectural projects ever attempted by a French monarch. The world-famous double-helix staircase – reputedly designed by the king's chum Leonardo da Vinci – ascends to the great lantern tower and rooftop.

Cycle 16km south to Cheverny along the bike path.

❸ Cheverny

For sheer elegance, Cheverny is hard to top. The château's sumptuous rooms include the formal Dining

ABOVE: CYCLING TO CHAMBORD

Room, with panels depicting the story of Don Quixote; the King's Bedchamber, with ceiling murals and tapestries illustrating stories from Greek mythology; and a children's playroom complete with toys from the time of Napoléon III. Tintin fans may recognise the château's facade as the model for Captain Haddock's ancestral home, Marlinspike Hall.

Cycle 42km west to Amboise, through countryside and along the Loire.

❹ Amboise

Towering above town and the Loire, the Château Royal d'Amboise was a favoured retreat for France's Valois and Bourbon kings. The ramparts afford thrilling views of the town and river, and you can visit the furnished Logis (Lodge) and the Flamboyant Gothic Chapelle St-Hubert (1493). Leonardo da Vinci spent the last five years of his life at nearby house of Clos Lucé (and is supposedly buried in the chapel). Models of his inventions are on display both inside and around the 7-hectare gardens.

Cycle 36km back to Blois for onward train connections.

ALTERNATIVES

Extension: Château de Chenonceau

Spanning the Cher River atop a graceful arched bridge, Chenonceau looks like it's dropped straight out of a story book. The pièce de résistance is the 60m-long, chequerboard-floored Grande Gallerie over the Cher, but it's also known for its yew-tree maze and formal gardens.

Follow the bike path from Amboise (14km) to get here.

Extension: Château d'Azay-le-Rideau

Romantic, moat-ringed Azay-le-Rideau, built in the early 1500s on an island in the middle of the Indre River, is adorned with elegant turrets, perfectly proportioned windows, delicate stonework and steep slate roofs. The writer Honoré de Balzac called it a 'multifaceted diamond'.

Azay-le-Rideau is 55km west of Amboise (or take the train to Tours, just 26km from Azay).

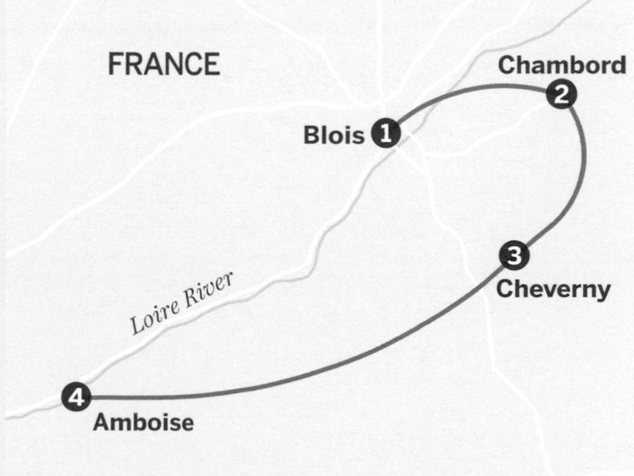

A ROMP THROUGH ROMAN PROVENCE

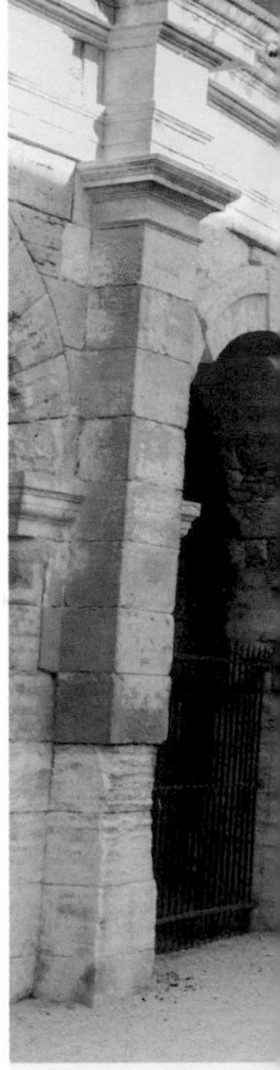

Arles – Vaison-la-Romaine

Seek out France's finest Roman remains in Provence, along with famously fabulous food and hours of southern sunshine.

FACT BOX

Carbon (kg per person): 9
Distance (km): 236
Nights: 5
Budget: $
When: Apr-Oct

❶ Arles

Arles, formerly known as Arelate, was part of the Roman Empire from the 2nd century BCE, and the city is home to an impressive amphitheatre known as Les Arènes. Though not as large as the one in nearby Nîmes, it still sometimes sees blood spilled, just like in the old gladiatorial days (it hosts gory bullfights and Courses Camarguaises, the local variation). Likewise, the 1st-century Théâtre Antique is still used for open-air performances. Elsewhere around town, you can visit subterranean Roman foundations and buried arcades under Place du Forum. Emperor Constantine's partly preserved 4th-century private baths, the Thermes de Constantin, are a stroll away.

Take the train to Nîmes (hourly; 20min).

❷ Nîmes

Nîmes's monumental amphitheatre is the best preserved in France. Built around 100 CE, the arena once seated 24,000 spectators and staged gladiatorial contests and public executions. Directly opposite, the Musée de la Romanité houses more than 5000 archaeological finds, including mosaic floors, statuary and coins. Also worth visiting are the Maison Carrée, a soaring limestone temple dating from 5 CE, and the Jardins de la Fontaine, built on top of a Roman temple site. The 30m-high Tour Magne offers sweeping views over the city.

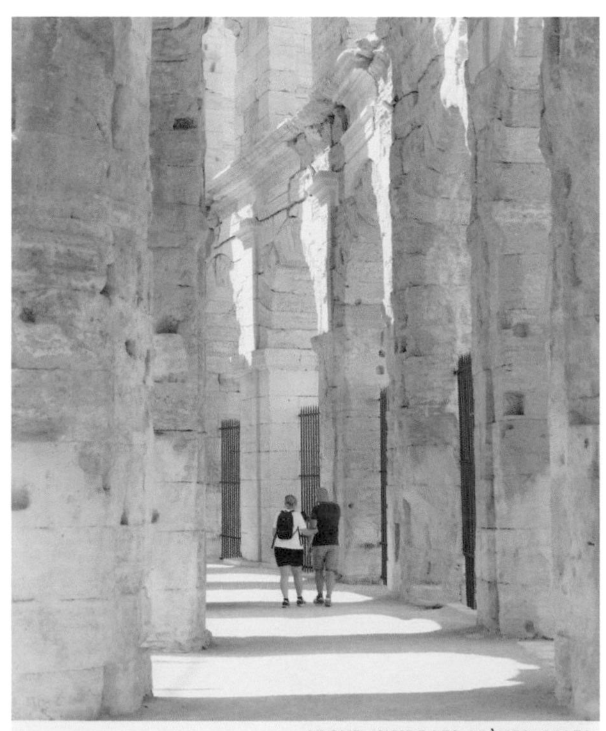

ABOVE: INSIDE LES ARÈNES, ARLES

ABOVE: BOULES BY THE ARENA, ARLES

Catch a local train to Avignon, then another to Orange (hourly; 35min to Avignon, 22min to Orange).

❸ Orange

A different side of Rome is revealed in Orange – more cultural than gladiatorial. At 103m wide and 37m high, the Théâtre Antique is one of only three Roman theatres that have survived intact (minus a few mosaics and the original roof); it originally seated 10,000 spectators. For bird's-eye views of the theatre and town in general, follow montée Philbert de Chalons or montée Lambert up to Colline St-Eutrope, the ever-vigilant Romans' lookout point.

Catch ZOU! Line 4 bus to Vaison-la-Romaine (several times daily Mon-Sat, 2 on Sun; 1hr).

❹ Vaison-la-Romaine

The ruined remains of Vasio Vocontiorum, the Roman city that flourished here between around 100 BCE and 450 CE, are spread over two sites in the heart of this small hill town. Also noteworthy is the pretty Pont Romain, a bridge which has stood both the test of time and severe floods.

To get back to Arles, catch the bus back to Orange, then a direct train from there to Arles.

ALTERNATIVES

Extension: Pont du Gard

The extraordinary three-tiered aqueduct known as the Pont du Gard was once part of a 50km-long system of channels built around 19 BCE to transport water from Uzès to Nîmes. At 48.8m high, and graced with 52 precision-built arches, it was the highest such structure in the entire Roman Empire, and remains a splendid sight to this day.

Several buses a day from Nîmes stop in Collias and Remoulins, near Pont du Gard (40min).

Extension: Glanum

At the amazingly well-preserved town of Glanum, 2km south of St-Rémy-de-Provence, the everyday lives of ordinary Gallo-Romans feel extraordinarily close. Founded around 27 CE, excavations have revealed houses, temples, columns, baths and the ancient forum – as well as France's oldest triumphal arch.

Buses travel from Arles to St-Rémy-de-Provence (3 daily Mon-Sat; 1hr).

WALES' WILD COASTAL WAY

St Davids – Aberdaron

This coastal road trip runs along Wales' western shores, taking Cardigan Bay, legends, mountains, hidden bays and castles all in its stride.

FACT BOX

Carbon (kg per person): 39
Distance (km): 311
Nights: 7-10
Budget: $$
When: Mar-Oct

❶ St Davids

Start in dinky St Davids, Britain's smallest city (population 1600), dominated by Wales' most impressive Norman cathedral. Named after the country's patron saint, born here around 500 CE, it's a pretty coastal honeypot, with a relaxing vibe and terrific beaches, including the vast sandy sweep of surf-magnet Whitesands.

Drive 24km northeast.

❷ Fishguard

Culturally vibrant Fishguard is shoehorned between its port (with ferries to Ireland) and fishing village. Come for nautical-flavoured pubs and breezy clifftop hikes along the coast path – lighthouse-topped Strumble Head (west) and Dinas Island (east) are standouts. As you walk, keep an eye out for seals, dolphins and, in summer, puffins.

Continue 29km north.

❸ Cardigan

Cardigan is an attractive springboard for exploring the Ceredigion coast. Its cake-topper of a 900-year-old castle holds an important place in Welsh hearts as the host of the first National Eisteddfod (competitive poetry and music festival) in 1176. An 8km drive down

ST DAVIDS CATHEDRAL

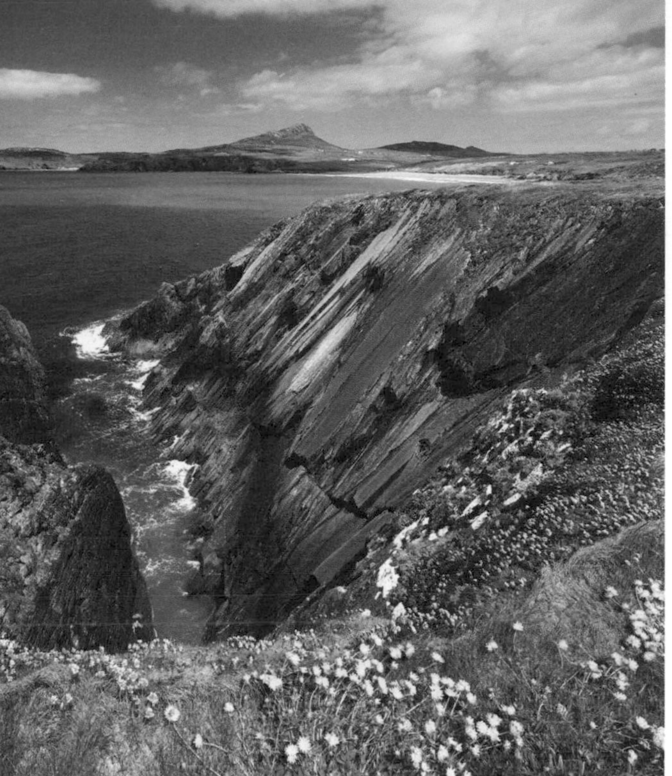

ABOVE: PEMBROKESHIRE COASTLINE

single-track lanes brings you to the delightfully secluded, cliff-wrapped bay of Mwnt.

A 62km drive takes you north to Aberystwyth.

4 Aberystwyth

A nicely laidback university town by the sea, Aberystwyth has a lively arts, food and cultural scene. A line of grand Georgian houses backs the promenade, Victorian Royal Pier and the long pebble beach. The clincher is the hilltop National Library, holding millions of rare books, including the 13th-century *Black Book of Carmarthen* (the oldest existing Welsh text).

Drive a highly scenic 196km north, with Snowdonia's peaks on the horizon.

5 Aberdaron

The ravishing Llŷn Peninsula flicks out like a dragon's tail into the Irish Sea. At its southernmost tip is prettily whitewashed Aberdaron, whose gorgeous 1.6km-long beach attracts kayakers, surfers and rock-poolers alongside the bathers. Boats from here chug across to wildlife-rich Bardsey Island, where 20,000 saints supposedly lie buried.

From Aberdaron, it's a 256km drive back to St Davids or a 300km drive to Cardiff.

ALTERNATIVES

Day trip: Snowdon

You can't drive this far north in Wales without feeling the tug of Snowdonia's dark, brooding mountains, so swap the coast for a detour inland to Snowdonia National Park. Wales' highest of the high is 1085m Snowdon. On cloud-free days you'll be flabbergasted by the peak's out-of-this-world views, which reach over ridge and shimmering sea as far as Ireland.

Snowdonia National Park is 85km northeast of Aberdaron.

Diversion: Ynyslas

Shaped by shifting tides and the elements, Ynyslas' magnificent 5km ripple of sand dunes form part of Dyfi Unesco Biosphere Reserve. Take the boardwalk through the dunes, where marram grass and wildflowers attract rare butterflies and birds, and you'll emerge on the broad arc of beach to the roar of the pounding surf.

Ynyslas is 14km north of Aberystwyth along the Coastal Way.

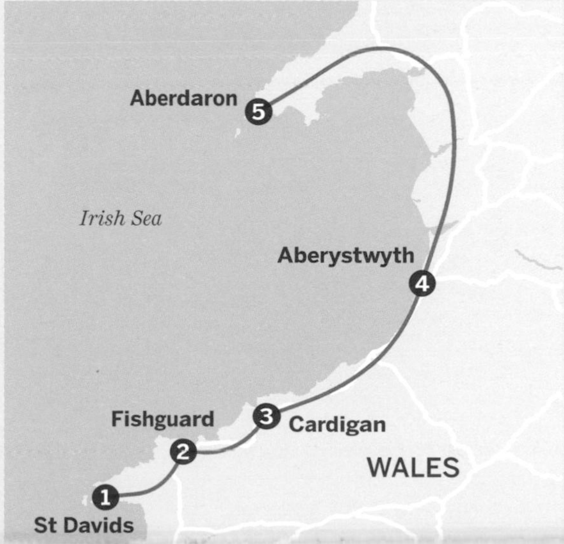

MEDITERRANEAN FERRY ODYSSEY

Marseille, France – Sicily, Italy

Island-hop from mainland France to rugged Corsica and on to two Italian outcrops full of gorgeous beaches and ancient history.

FACT BOX

Carbon (kg per person): 30
Distance (km): 1825
Nights: 14
Budget: $
When: Apr-Sep

❶ Marseille, France

Once gritty, Marseille has undergone one of Europe's biggest regeneration schemes, making the revamped city a cool destination to rival its Med-coast siblings along the French Riviera. Explore the spruced-up port – the shops and cafes of Les Docks, the striking Museum of Civilisations of Europe and the Mediterranean, the Foster+Partners mirrored pavilion – as well as hipster Cours Julien and the views from Basilique Notre-Dame.

ABOVE: PALERMO'S PIAZZA BELLINI, SICILY

Around 13 sailings a week depart from Marseille for Corcica's Ajaccio (10hr 30min).

❷ Corsica, France

The Marseille ferry pulls into Ajaccio, capital of Corsica, birthplace of Napoleon and gateway to the wild west of this ravishingly mountain-rucked isle. Sip a welcome drink in one of Ajaccio's cafes then head, via spectacular sands and the quintessentially Corsican hill town of Sartène, to the island's southernmost tip. Here, the labyrinthine town of Bonifacio teeters atop limestone cliffs and, on clear days, next stop Sardinia can be seen.

Buses connect Ajaccio and Bonifacio, via Scopetto (2 daily; 4hr). Several sailings a day link Bonifacio and Sardinia's Santa Teresa di Gallura (1hr).

❸ Sardinia, Italy

After the quick ferry ride, you've swapped French Med for Italian – though Sardinia has its own unique

ABOVE: BONIFACIO, CORSICA

flavour, not least roast suckling pig, an island fave. Trains run from Olbia (a two-hour bus ride from Santa Teresa) to southerly capital Cagliari. Hop on and off to enjoy elegant Oristano and its nearby pristine beaches, or Macomer and the branch-line to hilltop Nuoro, known as the 'Athens of Sardinia'. Or rent a car to make forays into the rugged interior, seeking out *nuraghe* (enigmatic prehistoric edifices). Cagliari itself is worth a day or two, thanks to elegant *palazzi*, seafront cafes and Roman remains.

Take the train from Olbia to Cagliari (3 daily; 4hr). Ferries run from Cagliari to Palermo up to twice a week (12hr).

4 Sicily, Italy

A slow boat swaps chic Cagliari for grungier Palermo. Like Marseille, though, this is a city on the rise: seized mafia money has helped clean up the palaces and piazzas. Get lost down the side-alleys of pedestrianised Via Maqueda and eat, eat, eat – the street food here is world-beating. Hit the three main markets (Capo, Vucciria and Ballarò) and the Cala quarter, slow-food central.

Palermo's airport connects to major European hubs. Ferries from Messina link to the Italian mainland.

ALTERNATIVES

Extension: Naples

Not done with Italy? Continue by rail from Sicily to the Italian mainland: trains are loaded aboard ferries to cross the Straits of Messina. Head up the Cantabrian coast to roguish, resurgent Naples for world-class modern art (visit the Madre gallery) and to decide which slice is best on Via Tribunali, aka 'Pizza Alley'.

It's possible to ride the train direct from Palermo to Naples (1 daily; 9hr 30min).

Day trip: Îles Lavezzi, Corsica

Tick off more islands by adding a day trip to this idyllic archipelago – now a nature reserve – off the Corsican coast. The uninhabited Lavezzi islets are dotted with granite boulders and lapped by crystal-clear waters, excellent for snorkelling. As the boat passes, ogle uber-exclusive Île Cavallo, home to millionaires' retreats.

Half- and full-day trips to the Îles Lavezzi run from Bonifacio.

SUSSEX GARDENS ROAD TRIP

Leonardslee – Great Dixter

Buy a National Trust pass and go at your own pace on this driving tour of some of the most beautiful and varied gardens in England.

FACT BOX

Carbon (kg per person): 24
Distance (km): 194
Nights: 2-3
Budget: $
When: year-round

❶ Leonardslee

A full-scale restoration project has returned these gardens to their former magnificence. Visit in spring to see azaleas, camellias and rhododendrons; in autumn for the eye-popping foliage; or at any time to clamber about the rock garden and see wallabies and deer.

Drive 7km northeast.

❷ Nymans

These National Trust-run romantic gardens are set around a partially ruined manor house and combine a mixture of formal planting and wilder woodland with masses of rare plants. Don't miss the charming rose garden full of old-fashioned varieties. The views over the High Weald are lovely, too.

Continue 15km east.

❸ Wakehurst

One of the big-hitters of the gardening world, Wakehurst has over 500 acres of woodland, plants and flowers. An intriguing range of habitats includes English meadows, a North American prairie, a silver birch wood and a Himalayan glade. Owned by the National Trust but run by the Royal Botanic Gardens Kew, it's the home of the Millennium Seed Bank, a major conservation project storing more than 39,000 plant species.

Drive 12km northwest.

ABOVE: SHEFFIELD PARK

ABOVE: GREAT DIXTER

❹ Sheffield Park

Another National Trust site, Sheffield Park is a mix of parkland and garden, and includes not one but four lakes. The garden here was designed with autumn in mind and is a real stunner then, though the estate is a splendid sight in any season; daffodils and bluebells put on a notably good show in spring.

Head 33km east.

❺ Bateman's

The home of Rudyard Kipling, this National Trust-owned Jacobean house and gardens has maintained the sense of a family garden, albeit a grand one. The combined elements of walled kitchen garden, lily pond, rose garden, orchard and meadow are an evocative blend.

Drive 22km east.

❻ Great Dixter

Established by the garden writer Christopher Lloyd, and now run by a trust, Great Dixter is many people's idea of the perfect garden – the flower borders are much copied. It's a special place, with every aspect – pond, orchard, topiary and vegetable garden – flowing seamlessly into one another.

To return to London, drive 105km northwest.

ALTERNATIVES

Extension: Bluebell Railway

From Sheffield Park, take a return trip to East Grinstead (40min each way) on the Bluebell steam railway. The 18km trip goes through glorious countryside, with stops at Horsted Keynes and Kingscote – buy a Rover ticket and you can hop off and on as much as you like, pottering around restored stations or walking on local trails.

Sheffield Park station is a 10-minute walk on a designated footpath from Sheffield Park gardens.

Extension: Pashley Manor Gardens

A family owned English country garden, known for its tulip, rose and dahlia displays, Pashley Manor encompasses a bluebell walk, a kitchen garden, a good number of old oak and beech trees, and views across the Sussex countryside. A licensed cafe with a terrace looks on to fountains and water features.

Pashley Manor is a 13km drive northeast of Bateman's.

TRACKING SWITZERLAND'S FIRST TOURISTS

Leukerbad – Grindelwald

Travel through Switzerland's scenic Bernese Oberland and recreate the world's first package holiday, recorded in 1863 by British traveller Jemima Morrell.

FACT BOX

Carbon (kg per person): 5
Distance (km): 140
Nights: 7-10
Budget: $$
When: Jun-Sep

❶ Leukerbad

Take the train from Geneva to Leuk and then bus 471 (3hr 30min total) to get to Leukerbad, start of this history-following trip. This spa town has been recognised for its healing hot springs since Roman times – there are numerous public baths to dip in here, including Europe's largest Alpine thermal spa. The town sits below the 2350m Gemmi Pass, a formidable barrier to the Bernese Oberland. Formidable, but not insurmountable: it's a stiff, zigzag climb – or easy cable car – up the rock wall to a stark, high-altitude plateau. Hike across this – via the 18th-century Schwarenbach Inn – to descend by foot or cable car into the Kander Valley.

Leukerbad to Kandersteg is 19km; cable cars run almost year-round.

❷ Kandersteg

Kandersteg is a picture-book base for exploring the well-waymarked paths around the valley and up into the surrounding mountains. When Miss Jemima moved on from here, she was up at 4.30am to catch a carriage to Spiez, on Lake Thun. No such early wake-up is required now: a smooth train runs to the deep-blue water's edge, where paddle steamers depart for Interlaken. From this between-lakes town, trains glide into the Middle Earth-like Lauterbrunnen valley, and the rack-and-pinion Wengernalpbahn railway climbs to winsome Wengen.

Take the train to Spiez (hourly; 30min), the boat to Interlaken (at least 3 daily; 1hr 20min), the

ABOVE: LAUTERBRUNNEN VALLEY FROM WENGEN

ABOVE: LAKE THUN, NEAR KANDERSTEG

train to Lauterbrunnen (hourly; 30min), and the Wengernalpbahn to Wengen (every 30min; 15min).

❸ Wengen

Wengen is less a village, more a balcony overlooking the Alps: mighty snow peaks – including Eiger, Mönch and Jungfrau – loom beyond its geranium-bright windowboxes. Hikes will take you within almost-touching distance of these spectacular mountains, or the rack railway will whisk you up to the Kleine Scheidegg pass – jump off at the pass for the Jungfraubahn, a breathless trip up to 3454m Jungfraujoch, Europe's highest railway station.

If not hiking, continue on the Wengernalpbahn from Wengen to Kleine Scheidegg (hourly; 25min) and onward from Kleine Scheidegg to Grindelwald (hourly; 40min).

❹ Grindelwald

This classic Alpine resort is surrounded by natural superstars: the crown-like Wetterhorn, the glistening Oberer and Unterer glaciers, the Eiger's redoubtable north face. Outdoor adventures aplenty are possible, with cable cars linking to higher elevations.

Take the train back to Interlaken (every 30min; 35min) for onward, international connections.

ALTERNATIVES

Day trip: Oeschinensee

Perched 1578m up above Kandersteg, Lake Oeschinen is a stunner. Only accessible by cable car or on foot, this dazzle of turquoise sits amid peaks, pine forest and wildflowers. There are numerous hiking trails. In summer, brave a swim or hire a rowing boat; in winter, try skiing, sledging or ice-fishing.

Cable-cars run from Kandersteg (or it's a 5km walk); the lake is a 15min walk from the upper station.

Day trip: Faulhornweg

Perhaps Europe's best day-hike, this 16km route runs from Schynige Platte to First, combining a cog railway, flower-filled pastures, blinding-blue lakes, a high mountain hut and views of immense snowy summits. A microcosm of the Swiss Alps.

Trains run from Grindelwald to Wilderswil (every 30min; 30min) from where there's a rack railway to Schynige Platte (hourly; 50min). Gondolas run from First to Grindelwald (8am-6pm; 25min).

RIDING THE HEART OF WALES RAILWAY

Llanelli – Knighton

Ride the scenic Heart of Wales Line (HoWL) to take in spa towns, sheep-filled fields, ancient history and some rather odd sports events.

FACT BOX

Carbon (kg per person): 5
Distance (km): 127
Nights: 2-4
Budget: $
When: year-round

❶ Llanelli

The Carmarthenshire market town of Llanelli sits on the Loughor Estuary. The tinplate industry saw it prosper from the late 18th century and a few grand mansions from this period remain – browse Parc Howard Museum, a fine Italianate pile, and nip into elegant Llanelly House for afternoon tea in the old drawing room. You can board the northbound Heart of Wales Line from the town, but better is to walk the easy 10km to Bynea, the next stop, via the Millennium Coast Path, with views to the Gower Peninsula. The flat trail also passes Llanelli Wetland Centre, an expanse of lagoons, ponds, nesting areas and streams bursting with birds.

Take the train from Bynea to Llandeilo (4 daily Mon-Sat, 2 Sun; 31min) – Bynea is a request stop so stick out your arm.

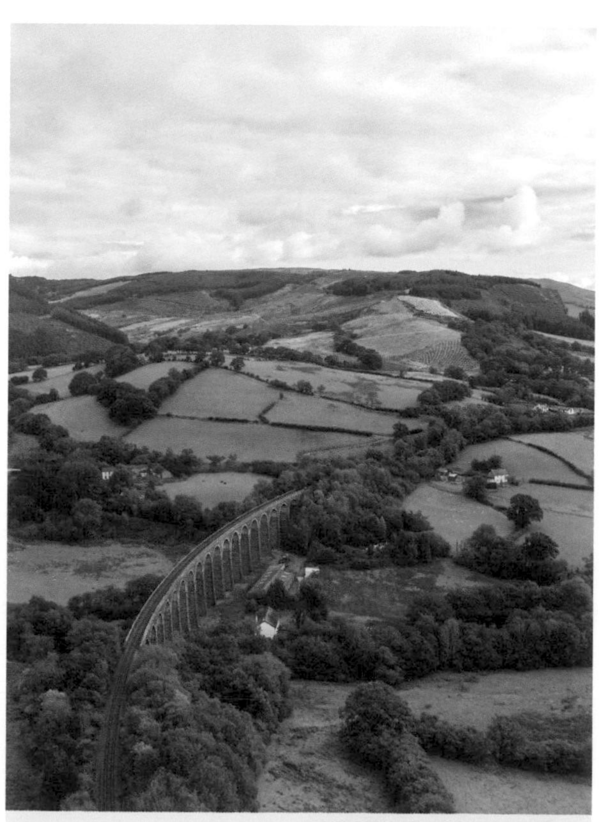

ABOVE: CYNGHORDY VIADUCT

❷ Llandeilo

Lovely Llandeilo, in Carmarthenshire's Tywi Valley, has a colourful main street lined with independent shops and cafes, plus some heavyweight history nearby: to the west, the ancient woodland and ruined castle of Dinefwr Park; to the southeast, mighty Carreg Cennen, a 12th-century turreted fortress on the edge of the Brecon Beacons mountains. Walk to one or

ABOVE: CARREG CENNEN, NEAR LLANDEILO

both. When you leave, bag a window seat for the northbound train as it enters the line's dramatic central section.

Continue on the train to Llandrindod (1hr 30min).

3 Llandrindod Wells

To reach this town at the foot of the Cambrian Mountains, the HoWL makes quite the journey, running via 18-arch Cynghordy Viaduct and the long tunnel to middle-of-nowhere Sugar Loaf station, named for the nearby peak. Disembark at Llandrindod to visit the National Cycle Museum and the natural springs that attracted Victorian wellness-seekers. There are also walks aplenty nearby.

Jump back on the train for Knighton (35min).

4 Knighton

The narrow lanes of Knighton sit on the River Teme – and on the border between England and Wales. The town is at the heart of excellent walking country: two National Trails (Offa's Dyke and Glyndŵr's Way) can be accessed here. If you don't want to get straight back on the train, hike to Bucknell (8km), the train's next stop, along the Heart of Wales Line Trail.

End the trip with a train to Shrewsbury (40min) and its onward connections.

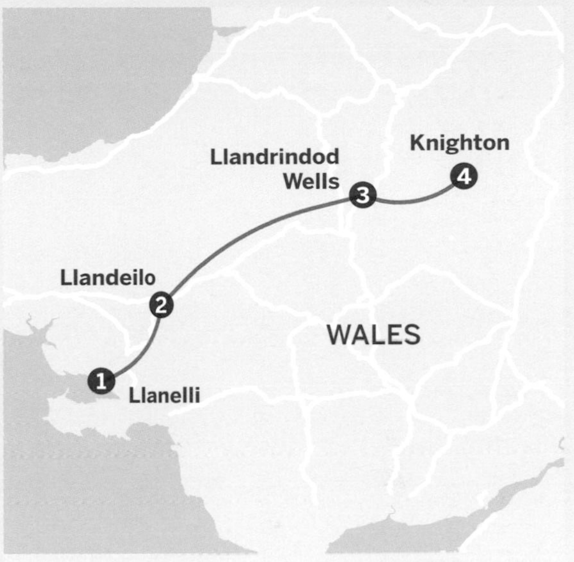

ALTERNATIVES

Diversion: Llanwrtyd Wells

No longer drawing tourists to its healing waters, Llanwrtyd has reinvented itself as a hub of strange sports: this is the home of the annual World Bog Snorkelling Championships, the biannual Alternative Games and all manner of other active oddities. While here, head to the ramshackle Neuadd Arms to raise a pint of chestnutty Heart of Wales Bitter, brewed on site.

Llanwrtyd Wells is six stops north of Llandeilo (1hr).

Extension: Shrewsbury

Linger at the line's end. The county town of Shropshire, and birthplace of Charles Darwin, Shrewsbury has a handsome half-timbered Tudor centre, with a number of tasty indie cafes and delis along its cobbled lanes. There's a castle, abbey and medieval church to admire, and boat trips to take on the River Severn.

Direct trains connect Shrewsbury to hubs such as Birmingham (every 30min; 1hr 20min) and Manchester (hourly; 1hr 20min).

THE DANUBE CYCLE PATH

Passau, Germany – Vienna, Austria

Take the Danube Cycle Path for a joyous pedal between vineyards, orchards and castle-topped villages, tracing the course of one of Europe's mightiest rivers.

FACT BOX

Carbon (kg per person): 0.1
Distance (km): 310
Nights: 7-9
Budget: $$
When: Apr-Oct

❶ Passau, Germany

Nudging the Austrian and Czech borders, Passau was once the largest bishopric in the Holy Roman Empire. The proudly Baroque Old Town perches on a peninsula above the confluence of three rivers: the Danube, the Inn and the Ilz. Its web of lanes, tunnels and archways wrap around an opulent, Italianate cathedral.

Cycle 45km southeast into Austria.

❷ Schlögener Loop, Austria

The icing on the scenic cake in this uplifting stretch of the Danube Valley is the Schlögener Loop, where the river makes a complete 180-degree turn through the wooded granite hills. For sensational views of the Loop, hike for thirty minutes above the village of Schlögen to reach Schlögener Blick.

Take the passenger ferry across the river from Schlögen and continue riding southeast for 46km.

ABOVE: CYCLING THE DANUBE PATH IN AUSTRIA

❸ Linz, Austria

The Danube Valley broadens as you roll on through fertile plains towards culture-crammed Linz. Check out Lentos Kunstmuseum for modern art, and hands-on Ars Electronica Center for new technology, science and digital media. Afterwards, stroll the Baroque and Renaissance Old Town, centred on cafe-rimmed Hauptplatz.

Continue another 60km east.

❹ Grein, Austria

Many cyclists rave about Grein being their favourite part of the ride and it's a beauty for sure. Jutting above the village, its whimsically turreted Baroque castle hides an impressive arcaded inner courtyard, while the 18th-century theatre in a former granary is the oldest still in operation in Austria.

Cycle 44km east.

ABOVE: PASSAU, GERMANY

❺ Melk, Austria
Few sights lift spirits like the Benedictine abbey-fortress of Melk, high on a hill above the Danube, where you can tour the Baroque-gone-mad church, library and Marmorsaal (Marble Hall). The next leg of the ride takes you deep into the Wachau region, with terraced vineyards, apricot orchards, forested slopes and castles on almost every bend.

Take your time on the 35km ride to Krems.

❻ Krems an der Donau, Austria
Krems entices with a laidback historic centre, Grüner Veltliner and Riesling white wines from local vineyards, and plenty of atmospheric places to eat and stay. The gallery-heavy Kunstmeile area and lavish Stift Göttweig abbey are well worth visiting.

The final leg is 80km east to Vienna.

❼ Vienna, Austria
A fittingly grand end to this grand bike ride, the Austrian capital reels you in with its magnificent ensemble of palaces, galleries, parks, concert halls and coffee houses. Spend at least a couple of days resting and exploring before the journey home.

Vienna has dozens of international flights and train connections.

ALTERNATIVES

Extension: Spitz and Dürnstein
Detour to vine-swathed Spitz to taste local wines in *heurigen* (wine taverns), and visit Dürnstein for its romantic, eyrie-like castle where England's King Richard the Lionheart was once imprisoned.

Spitz is 18km from Melk; Dürnstein is at the 29km marker on your ride from Melk to Krems.

Extension: Tulln
Austrian Expressionist artist Egon Schiele (1890-1918) was born in Tulln and his hometown has two museums devoted to his life and work (one of which was his birthplace). The town is also one of Austria's oldest, and it has some wonderful parks and gardens in which to take a break; Tulln's nickname is Blumenstadt (City of Flowers).

Tulln is about 45km into the ride from Krems to Vienna.

EASTERN EUROPE

You'll be spoilt for choice in this vast and varied corner of Europe. Dip into Black Sea spas or Art Deco thermal pools, seek out Gothic castles and Dalmatian beaches, tick off Brutalist architecture in Bratislava, explore historic Baltic capitals or go off the beaten track to discover Albania's Ottoman-era towns and Accursed Mountains. And throughout the region, lace up your boots or don a helmet to hike or bike epic trails through some truly wild national parks.

EASTERN EUROPE
BALTIC SEA
ESTONIA
LATVIA
LITHUANIA
RUSSIA
BELARUS
POLAND
CZECH REPUBLIC
SLOVAKIA
MOLDOVA
UKRAINE
HUNGARY
ROMANIA
SLOVENIA
CROATIA
BOSNIA
SERBIA
BLACK SEA
MONTENEGRO
BULGARIA
ALBANIA
TURKEY
ANEAN SEA
AEGEAN SEA
1
2
3
4
5
6
7
8
9
10
11
12

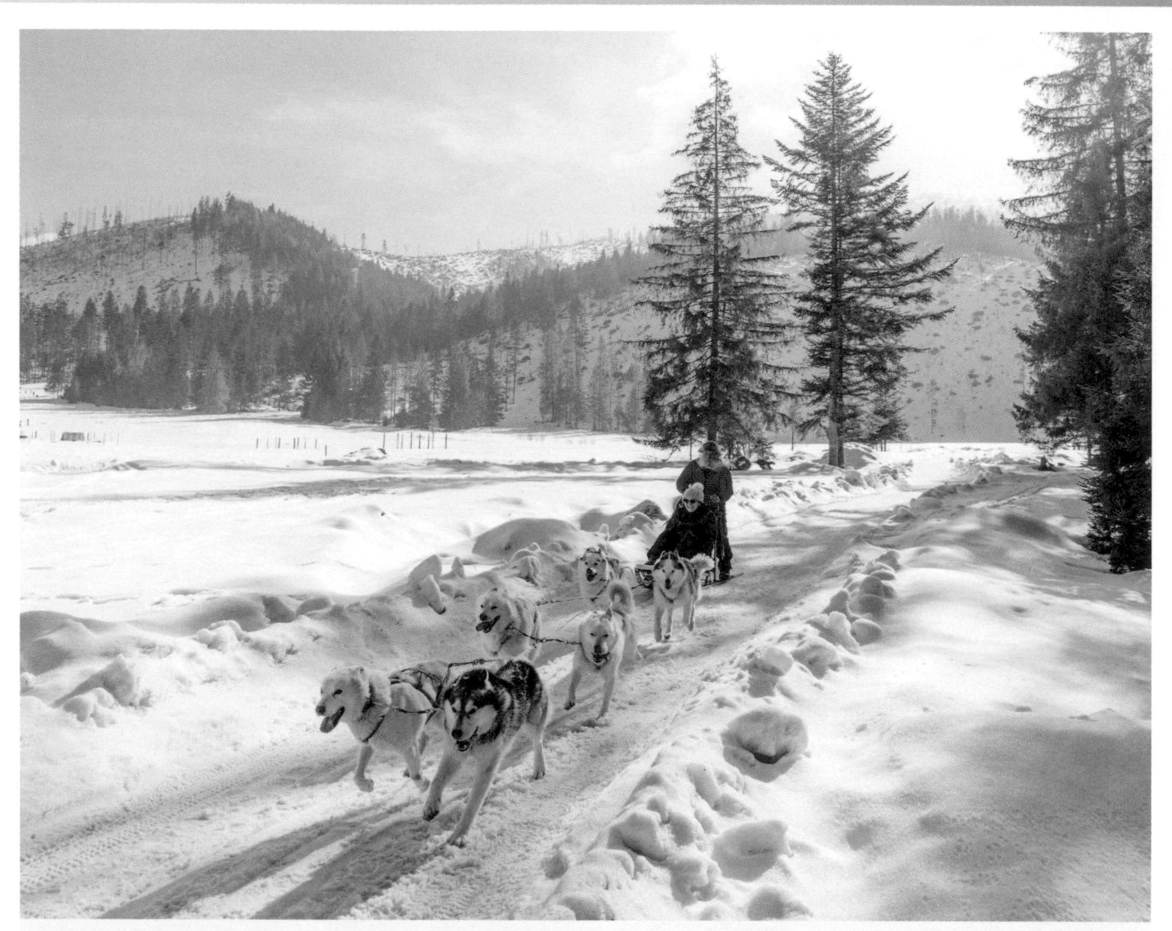

ABOVE: ZAKOPANE, TATRA NATIONAL PARK, POLAND

CONTENTS

GOING GOTHIC IN ROMANIA'S TRANSYLVANIA AND WALLACHIA

Brașov – Sighișoara

Mountain passes, vampire sites, Gothic architecture and medieval cities come together on this unforgettable cycling challenge.

FACT BOX

Carbon (kg per person): 0
Distance (km): 487
Nights: 9-10
Budget: $
When: Apr-May

❶ Brașov

How could touring Transylvania's most Gothic locales not begin in mesmeric Brașov, rimmed by mountains, stippled with brooding church spires and wrapped in watchtower-studded walls? The highlight? Imposing, 15th-century Biserica Neagră (Black Church). Legends are plentiful in a settlement that was one-time haunt of Vlad the Impaler, the bloodthirsty ruler on whom the Dracula legend was based.

Cycle 173km southwest to Curtea de Argeș.

ABOVE: TRANSFĂGĂRĂȘAN ROAD

❷ Curtea de Argeș

The gruelling onward pedal to the ancient capital of Wallachia (the foundation of modern Romania) can be broken at Bran Castle, 27km along this route. This fairytale structure was Dracula's supposed lair – untrue, but it's a breathtaking place nevertheless. Your turning point northwards through the mountains into Transylvania, Curtea de Argeș is the burial place of several Wallachian princes and another very legend-steeped place: locals tell of the architect's wife being buried alive under the foundations of the impressive cathedral.

Ride 27km north to Arefu.

❸ Arefu

Forge north towards the foreboding Făgărăș Mountains. Landscapes become dramatic as you approach Arefu, surrounded by steep, green, forest-carpeted valley. Nearby is Poenari Castle, a crag-top ruin once home to Vlad the Impaler.

Cycle 124km north along the 7C Transfăgărășan to Sibiu.

ABOVE: BRAN CASTLE

❹ Sibiu

It's another tough leg as you ascend the twisting Transfăgărășan road across the Făgărăş Mountains into Transylvania, passing close to Romania's highest peak, Moldoveanu (2544m). As well as the difficulty of terrain, the sheer beauty of this upland area – with handsome vistas provided by both the peaks and by snaking Lake Vidaru – means making a mountainside stopover might be agreeable: you could try Hotel Posada Vidraru on the southern edge of the lake, with a spa for soothing those aching limbs. Then descend to cosmopolitan Sibiu. This architectural treasure-trove demands a day of exploring; start with the Gothic masterpiece that is its landmark Lutheran cathedral.

Ride 96km northeast to Sighişoara.

❺ Sighişoara

Approach spellbinding Sighişoara via route 106: a full day's biking through bucolic, time-lost little villages. Vlad the Impaler's birthplace (Vlad Dracul's House) is in Sighişoara, but even if your verve for vampires is waning, the city's Unesco-listed walled Old Town can't fail to fascinate.

Cycle 132km back to Braşov for onward rail connections.

ALTERNATIVES

Diversion: Făgărăş

It's worth detouring to see one of Romania's best-preserved castles, Făgărăş Citadel, a handsome and impressively intact stronghold once the base of Vlad the Impaler. Its 3m-thick walls went up during the 14th century and are strikingly encircled by a moat. Inside, the Val Literat Museum, is dedicated to religious art, ceramics, medieval armour and craftsmen's guilds.

Făgărăş lies on the return cycle route between Sighişoara and Braşov.

Extension: Snagov

If you're returning to capital Bucharest post-pedal, set aside time for this special and justifiably popular lakeside village. The 14th-century Snagov Monastery, photogenically gracing an island in the lake, is thought to be the final resting place of Vlad the Impaler, who also built a prison (still visible) and torture chamber here.

Snagov is 39km south of Ploieşti and 40km north of Bucharest

BRUTALIST TOUR OF EASTERN EUROPE

Berlin, Germany – Budapest, Hungary
Take a four-country train trip behind the former Iron Curtain and discover poignant museums and era-defining Brutalist architecture.

FACT BOX
Carbon Cost: 36kg
Distance: 878
Number of Nights: 7
Budget: $$
When: Apr-Oct

❶ Berlin, Germany

Cosmopolitan Berlin is both cutting-edge and uncompromising in confronting its turbulent past. The Berlin Wall, raised in August 1961, instantly became a tangible symbol of the divide between the capitalist west and Communist east, and some sections still remain. At the most intact stretch, the East Side Gallery, the austere concrete construction comes harrowingly alive with vivid, overtly political artworks. Nearby, bombastic Karl-Marx-Allee was the East's erstwhile show space for the best of Socialist Realism architecture. For an unsettling peep into Communist East Germany's machinations, visit the Stasi Museum, where displays cover the activities of the state's secret police.

Take the train to Prague (every 2hr; 4hr 15min).

❷ Prague, Czech Republic

Entrancing Prague is better-known for its beautiful Baroque and Gothic buildings, but the Communist era's influence is visible too. Today's Crowne Plaza Hotel once numbered among the finest Stalinist structures outside Russia – the formerly red edifice-topping star was repainted green by the new owners to represent the colour of money rather than that of communism. A stroll in lovely Letná Park, scene of key 1989 Velvet Revolution protests, takes you to the Expo 58 Restaurant, winner of best pavilion at the 1958 Brussels World Fair and the period's most striking example of Czechoslovak architecture.

Continue on the train to Bratislava (hourly; 4hr).

ABOVE: MUSEUM ISLAND, BERLIN

ABOVE: LETNÁ PARK, PRAGUE

❸ Bratislava, Slovakia

Sitting close to the Austrian border and the Communist Bloc's frontier with the West, the Slovak capital was formerly used as an advertisment for Soviet Brutalist architecture. Along Bratislava's Danube waterfront you can scarcely miss Most SNP, the bizarre bridge with a spaceship-like capsule suspended above, nicknamed the UFO and hailing from 1972. This links Bratislava's Old Town with the jarringly contrasting 1970s-built suburb of Petržalka, the biggest Soviet housing project outside the USSR. And Brutalism gets odder still in the stark upside-down pyramid shape of the Slovak Radio Building north of the Old Town.

Take the train to Budapest (every 1-2hr; 2hr 30min).

❹ Budapest, Hungary

Once you've enjoyed central Budapest's spectacular sights, it's time to spurn the bourgeois city centre and head to the suburbs and Memento Park. Here, the city authorities have amassed a huge, hideous and fascinating collection of Communist-era monuments: comical, but simultaneously chilling.

Budapest has international train and air connections.

ALTERNATIVES

Diversion: Dresden

Dresden presents a radically different version of the post-WWII East. All but razed by Allied bombing, this former bastion of outstanding German architecture was faithfully rebuilt in the postwar decades. It's hard not to be impressed by the scale of the sublime reconstruction at the iconic collection of palatial buildings, sculptures and gardens that is the Zwinger, a Baroque masterpiece housing three museums.

Trains run from Berlin (hourly; 2hr).

Diversion: Banská Bystrica

Banská Bystrica is home to Slovakia's best museum, the Museum of the Slovak National Uprising (SNP), within a giant flying saucer-shaped wedge riven in two. Diplays evocatively chart the uprising. Aftewards, head into the Veľká Fatra mountains to hike a section of the spectacular Cesta Hrdinov SNP (Path of the Heroes of the Slovak National Uprising).

Trains run from Bratislava (every 2 hours; 3hr 15min).

A TRAIN TRIP DOWN THE VISTULA

Kraków – Gdańsk

Encounter historic and contemporary Poland by following the country's longest river, the Vistula, starting in former royal capital of Kraków.

FACT BOX

Carbon (kg per person): 34
Distance (km): 852
Nights: 21
Budget: $$
When: year-round

❶ Kraków

Kraków is a visual delight. The Main Market Square is dominated by the ornate 16th century Cloth Hall and overlooked by the striking brick-built St Mary's Basilica. Other top sights include Wawel Royal Castle and the old Jewish district of Kazimierz. Day trips can be made to the Wieliczka Salt Mine, with its incredible underground salt sculptures, and the Schindler's Factory museum.

Jump on the train to Warsaw (at least hourly; 2hr 30min)

❷ Warsaw

Take your time exploring Poland's indefatigable capital. Begin by surveying the city from the observation platform of the Communist-era Palace of Culture & Science, then dive into the remarkably restored Old Town. Take time, too, to ponder wartime Warsaw at the Warsaw Rising Museum and the parts of the Jewish city which still survive, along with the beautiful palaces and parks of Wilanów and Łazienki.

Take the train to Płock, changing in Kutno (hourly; 3hr).

❸ Płock

Perched dramatically on a cliff high above the Vistula, Płock has the remnants of a Gothic castle and a glorious cathedral dating back to the 12th century. The Mazovian Museum houses Poland's finest collection of Art Nouveau pieces, along with regional history displays.

ABOVE: A SCENIC RIDE IN KRAKÓW

ABOVE: WARSAW'S OLD TOWN

Return by train to Kutno for connections to Toruń (every 2hr; 2hr 45min-3hr 30min).

❹ Toruń

Unesco World Heritage-listed Toruń is a magnificent walled 13th-century town on the banks of the Vistula. Its Old Town is stacked with ancient buildings, churches and museums – including one dedicated to local sweet speciality, gingerbread (you can even make your own). It's fun (and free) to stroll around the remnants of the town's medieval fortifications.

To continue to Gdańsk, head to either Bydgoszcz or Iława for train connections (5 daily; 3hr 15min).

❺ Gdańsk

This Baltic port's crown jewel is its historic Main Town. Zone in on the touristy Długi Targ (Long Market), and Gothic St Mary's, the world's largest brick-built church. The superb Museum of the Second World War is a bold addition to Gdańsk's waterfront and well worth a visit. Near the city, you can be whisked back to medieval times at the marvellous Gothic castle at Malbork and then head for some relaxation on the beach at Sopot.

Gdańsk Lech Wałęsa Airport has onward international connections.

ALTERNATIVES

Extension: Kazimierz Dolny

Founded in the 14th century, this quaint river port has long attracted artists and free thinkers. Fine buildings surround its Old Town square. Enjoy a day trip to the early 16th century castle at Janowiec, a five-minute ferry ride across the Vistula followed by an easy 2km walk.

From either Kraków or Warsaw take a train to Puławy, from where there are frequent buses to Kazimierz Dolny (45min).

Extension: Zakopane

Depending on the season, this appealing mountain resort offers hiking in the Tatra National Park and skiing on several nearby mountains. The town's decorative, early 20th-century timber architecture gives the place a unique look, and there are collections of local arts and crafts to discover in institutions such as the Museum of Zakopane Style.

Frequent trains connect Zakopane with Kraków (2hr 30min).

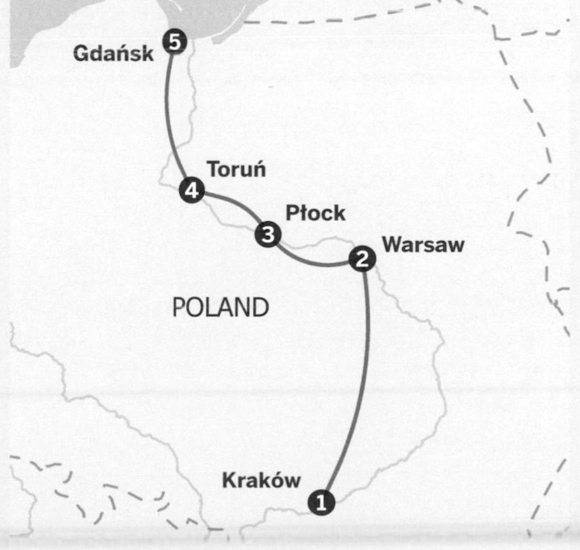

SPAS AND MORE IN BUDAPEST

Széchenyi – Király and Rudas

A few days in Hungary's capital are enough for a dip into the city's world-class spas and to take in some of its culture and nature too.

FACT BOX

Carbon (kg per person): 0
Distance (km): 17
Nights: 3-4
Budget: $$$
When: year-round

ABOVE: PEDAL POWER, MARGARET ISLAND

❶ Széchenyi Thermal Bath

Of the many spas, baths and public pools in the Hungarian capital, the one unmissable option is the vast neo-Baroque Széchenyi Thermal Bath. Set in the city's main park (worth a visit in itself), this sprawling complex houses an impressive 18 pools, plus ten saunas and steam chambers. This is also where you'll see the famous water-wallowing chess players testing their skills. From here, it's a pleasant 3km walk through the streets of Pest to the country's spectacular Parliament building, inspired by the UK's Houses of Parliament: daily English-language tours are available.

Next stop Margaret Island sits between Buda and Pest in the middle of the Danube.

❷ Margaret Island

Perfect on a summer's day (but enjoyable year-round), this expansive River Danube island-park houses the Danubius Health Spa Resort Margitsziget, the ruins of a 13th-century nunnery, and the Palatinus Waterpark, popular with families. The island offers lots of lovely walks or cycling too, as well as a 5km running track.

To get to the Gellért, cross the river to the Buda side of the city.

ABOVE: GELLÉRT THERMAL BATH

❸ Gellért Thermal Bath

Highlight of the Art Deco Gellért Thermal Bath, opened in 1918, is its 33m-indoor pool, decorated with imposing marble columns and water fountains under a vaulted ceiling that opens in the summer. Don't miss the outdoor sun terrace and pool, too. Behind the baths, the Buda Castle area is worthy of at least a day's exploration, including quaint fishermen's houses, the 13th-century Matthias Church and, atop the hill, the Hungarian National Gallery in the former Royal Palace.

From the Gellért, the next two baths lie a stroll north along the Danube.

❹ Király and Rudas Turkish baths

Featuring Ottoman architecture from the 16th and 17th centuries, this delightful duo offers the closest you'll get to the original Turkish *hammam* experience. For some post-soak culture, cross the Danube back to Pest and the impressive Art Nouveau interiors of the Zeneakadémia, a great place to hear the music of Hungarian composer Franz Liszt. Nearby, the diminutive Hungarian State Opera House, Baroque St Michael's Church and grand St Stephen's Basilica are all must-sees.

Budapest's Ferenc Liszt Airport, 24km southeast of the city, has international connections.

ALTERNATIVES

Extension: Kerepesi Cemetery

In Kerepesi Cemetery, Hungary's turbulent 20th-century history is captured through arresting statuary and sculpture, Art Nouveau mausoleums, Soviet army graves, the monumental Brutalist Workers' Movement Pantheon, and lovely tombs of Hungary's world-class chess players, musicians, politicians, poets, writers, artists and actors.

The cemetery is on the Pest side of the city, a 3km walk from the river.

Extension: Trams, trains and railways

Inventive transport options for day trips include the Cogwheel Railway to the forested Buda hills, and the Children's Railway, whose child-staffed trains run to Hűvösvölgy's forest and past 19th-century villas. In the city, a dusk riverside trip on tram line 2 is a treat, while the four metro lines bear their own distinctive architecture – line 1 has impressive original Art Nouveau motifs, while line 3 has Soviet throwbacks to Social Realism.

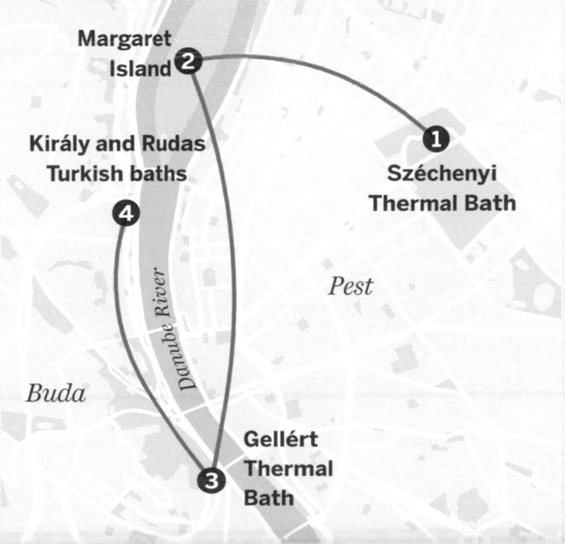

SOVIET SPA-STAYS ON THE BLACK SEA

Odessa – Sochi

The magnificent Soviet-era Black Sea sanatoriums make for an unforgettable road trip from Ukraine's Odessa to Russia's Sochi, via the Crimean resort of Yalta.

FACT BOX

Carbon (kg per person): 163
Distance (km): 1306
Nights: 14
Budget: $$$
When: year-round

❶ Odessa, Ukraine

With its iconic Potemkin Steps, Neoclassical architecture, 19th-century market and diminutive but moving Jewish Museum, Odessa, the 'Pearl of the Black Sea', makes a great starting point for this trip. Not only can you rent an inter-country one-way car here, but the city also has a clutch of great sanatoriums to visit. Don't miss the eponymous Sanatorium Odessa, a faded Constructivist beauty located in the beach resort of Arcadia and, 17km away, the monumental Kuyalnik, set on a sulphur-rich black-mud estuary; the star of its sprawling show is a pyramid-roofed swimming pool – though the concert hall and kitsch restaurants are pretty special too.

Drive 576km east and south along the coast.

❷ Yalta, Crimea

If you know your architecture, you may be familiar with the saw-toothed concrete cogs that make up the Druzhba Sanatorium on the outskirts of resort town Yalta. Designed in 1985, the 'friendship sanatorium', a joint project between the then Czechoslovakia and the USSR, offers lots of opportunities for R&R, including dance halls, a cinema, a concert venue and a gorgeous pool. If you only stay in one sanatorium on this trip, make it this one – though the Miskhor, 12km

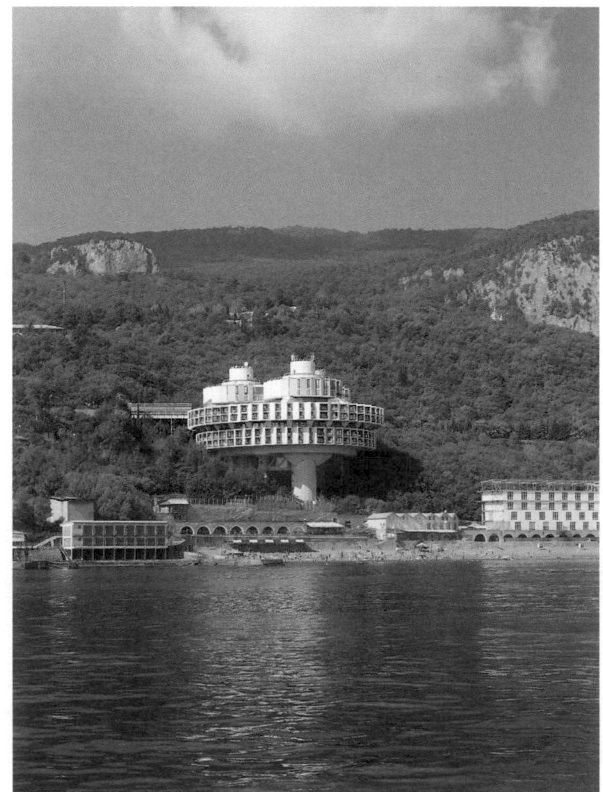

ABOVE: YALTA'S DRUZHBA SANATORIUM, CRIMEA

ABOVE: ODESSA'S KUYALNIK, UKRAINE

south, offers serious competition. Built in 1974, this huge Brutalist complex is set in a botanical garden a short walk from the beach.

Continue 730km east along the coast, crossing the 19km-long Kerch Bridge, Europe's longest.

3 Sochi, Russia

The Russian Riviera location of the 2014 Winter Olympics has plenty of must-see attractions – including Stalin's holiday *dacha* – and the landscape is a walker's heaven of waterfalls and caves, but it's the area's 65 sanatoriums that are the biggest draw. Bathing in stinky sulphurous mud is an option, and treatment centres have thankfully replaced leeches with chocolate body wraps and pedicures. Like Neoclassical grandeur? Check out the Metallurg (perhaps giving its magnet therapy and mud tampons a miss). Prefer a grand Modernist behemoth? Take your pick from the Beliye Nochi (White Nights) Health Resort, with its jazzy murals, lush gardens and private beach; the luminous indoor pool of the eight-floored Yuzhnoe Vzmorye on Southern Beach; or the Sanatorium Oktyabrskiy, whose amenities take in an extensive health centre, cinema and even a mini zoo.

Drop off the car at Sochi International airport, which has onward connections.

ALTERNATIVES

Day trip: Solokhaul

In the forested foothills of Russia's Caucasus Mountains lies the village of Solokhaul, home to mainland Europe's most northerly tea plantation and the cute Koshman's House Museum, named for the family who first planted tea here back in 1901. The journey to it, along picturesque switch-back roads offering great views back to the Black Sea, is lovely.

Solokhaul is 42km north of Sochi.

Sidetrip: Alushta

Unless you're best friends with Elon Musk, you're not likely to be heading into space anytime soon. But in the Crimean resort of Alushta you can act out your cosmonaut fantasies at the Guardian of the Empire complex. Stay in a space cabin, eat 'space food', play with space-flight simulators, wander a space garden and explore the history of space exploration in the museum.

Alushta is 37km northeast of Yalta.

A GRAND TOUR OF ALBANIA

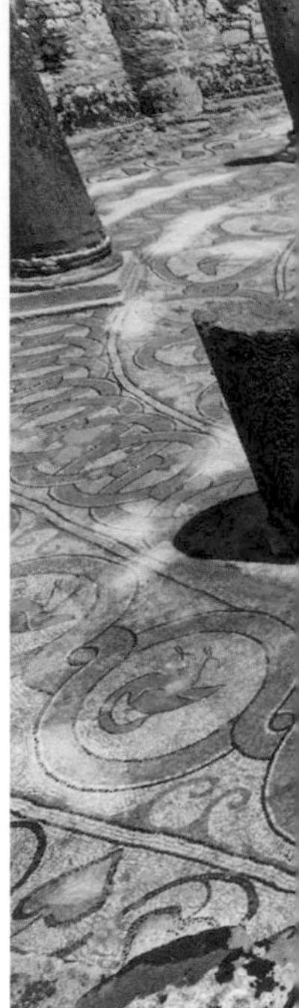

Tirana – Butrint

Take a rollicking road trip among karst peaks, lakes, Ottoman-era towns and Roman ruins in this underrated Balkan country.

FACT BOX

Carbon (kg per person): 150
Distance (km): 515
Nights: 14
Budget: $
When: Apr-Oct

❶ Tirana

Albania's rough-and-ready capital won't win many beauty awards, but it's strewn with fascinating remnants of the nation's Communist past, from the space-age Pyramid of Tirana complex to the triumphant murals of Skanderbeg Square.

Drive 100km north.

ABOVE: PYRAMID OF TIRANA

❷ Lake Shkodra

The reed-strewn leagues of Lake Shkodra mark Albania's border with Montenegro. While away lazy days paddling in the shallows, or hire bikes to cycle across humpbacked Ottoman bridges and through bucolic lakeside hamlets.

Access the Accursed Mountains as part of a mini-package tour, such as those offered at Lake Shkodra Resort.

❸ Accursed Mountains

The Accursed Mountains look like a mini-Yosemite in Albania. The classic two or three day-circuit out of Shkodra town sees you boarding the Lake Koman Ferry – a little boat puttering among leviathan mountains – before transferring by road to the mountain village of Valbona. The following day, set out on a spectacular 17km hike over the windy pass to Theth.

Tours transfer back to Shkodra town. From Shkodra, drive 190km south.

❹ Berat

Berat presents a vision of bygone Balkan life: whitewashed Ottoman houses cascading down to a rushing river, and a crumbly castle watching over

ABOVE: BAPTISTERY OF BUTRINT

the rooftops. Head to the atmospheric Mangalem Quarter to see a cluster of beautiful mosques – among them, the 16th-century Sultan's Mosque, with intricate painted ceilings.

Back in the car, drive 115km south.

5 Gjirokastra

Gjirokastra's crowning glory is its 12th-century castle, a hulking fortress full of shadowy corridors and – rather unexpectedly – a US spy plane, which was (allegedly) shot down during the Cold War. The town beneath its ramparts sees Ottoman-era mansions straddling steep cobbled lanes.

Drive 80km south.

6 Butrint

The ancient ruins of Butrint sit beside a coastal lagoon at the southernmost edge of Albania: wander thick woods in search of a flooded theatre, Roman temples and a Byzantine church. The neighbouring beach resort of Ksamil is a fine place to wind up your odyssey on Ionian sands.

From Butrint, drive 30km north to Saranda to take the ferry to Corfu, whose airport has European connections. Alternatively it's 300km back to Tirana.

ALTERNATIVES

Extension: The Albanian Riviera

The Llogara Pass is a nerve-shredding mountain road, performing switchbacks as it teeters high over the Ionian Sea. The reward for crossing it is arriving on the Albanian Riviera: an idyllic coastline of sandy bays, seaside castles and even a Cold War-era submarine silo. The ethnically Greek town of Himara is a fine base in which to submerge yourself.

Himara is 70km north of Butrint.

Extension: The Peaks of the Balkans Trail

Take at least 10 days to hike this epic trail, zigzaging between Albania, Kosovo and Montenegro in a 190km lap of the Accursed Mountains. In Albania, you pass under Maja Jezercë, highest peak in the Dinaric Alps; Kosovo claims the steeply-sided Rugova Gorge; while the Montenegrin section winds among lakes and forests.

Organise a hike through a local operator: start in Valbona or Theth.

BOSNIA & HERCEGOVINA'S BEST BITS

Sarajevo – Sutjeska National Park

Road-trip through the centre of Bosnia & Hercegovina, exploring an area richly endowed with Ottoman architecture and natural beauty.

FACT BOX

Carbon (kg per person): 109
Distance (km): 546
Nights: 7
Budget: $
When: Apr-Sep

ABOVE: SARAJEVO SKYLINE

❶ Sarajevo

Set in a steep-sided bowl surrounded by mountains, the Bosnian capital has a vibrancy and spirit all its own. You'll feel it most strongly in Baščaršija, the historic market area crammed with shops, cafes and restaurants, and peppered with mosques, churches and synagogues. Sarajevo's sad history is laid bare in museums scattered across the city.

Drive 92km north, taking the bridge across the Bosna River.

❷ Travnik

Once the seat of Bosnia's Ottoman governors, Travnik is now a quiet backwater – but it's well worth stopping for a few hours to explore the Many-Coloured Mosque and 15th-century citadel, and to dine at one of the restaurants lining the gurgling Plava Voda stream.

Continue 67km, turning sharply right at Donji Vakuf and following the Vrbas River.

❸ Jajce

Gorgeous little Jajce sits on a crag above the confluence of two rivers, one of which tumbles into the other by way of a 21m-high waterfall. The walled Old Town above it is capped by a fortress dating

ABOVE: STARI MOST, MOSTAR

from the medieval period, when Jajce was the capital of the short-lived Kingdom of Bosnia.

Double back to Donji Vakuf, cross the river, and follow the signs to Mostar (164km).

4 Mostar

Mostar's famous 16th-century bridge, elegantly spanning the green waters of the Neretva River, is one of the most recognisable sights in the Balkans. It was notoriously destroyed during the 1990s war and painstakingly rebuilt, but Mostar overall remains a deeply divided city. It's still surprisingly beautiful, though, and a wonderful place to while away a day or two.

Head 121km east.

5 Sutjeska National Park

Whether or not you're in the mood for hiking or mountain-biking (both on offer here), the drive through this mountainous and heavily forested national park is well worth the detour on the way back to Sarajevo. Drop into the park information centre in Tjentište for advice on trails, and stop to check out the vast, angular Partisan Memorial, commemorating the major WWII battle that took place here.

Drive 102km northwest back to Sarajevo.

ALTERNATIVES

Diversion: Blagaj

Call into peaceful Blagaj to visit its famous *tekke*, a meeting house for the mystical Sufi branch of Islam. The nature-loving Sufi dervishes who built it chose the site for its awe-inspiring beauty – it's right by a cavern where the turquoise Buna River emerges from a sheer cliff.

Blagaj is only 12km southwest of Mostar and barely a 5min diversion from the road to Tjentište.

Diversion: Trebinje

In the southern corner of the country, wedged between the Croatian and Montenegrin borders, Trebinje is an attractive city where life moves as slowly as the Trebišnjica River which passes through it. There's a leafy central square, a tiny walled Old Town and an interesting Orthodox church perched on the hill above it.

Slot this 100km detour in between Mostar/Blagaj and Sutjeska National Park.

DALMATIAN COAST DISCOVERER

Split – Dubrovnik

Ferry-hop between the wineries, oyster farms and Roman palaces of Croatia's delicious Dalmatian Coast.

FACT BOX

Carbon (kg per person): 14
Distance (km): 211
Nights: 7-10
Budget: $$
When: Apr-Oct

❶ Split

The remarkable Diocletian's Palace remains the hub of urban life in Split, 1700 years after it was built. The imperial apartments and temples of this Roman palace are now home to shops, bars and restaurants, beyond which lie Renaissance squares, a vibrant promenade and the turquoise waters of the Adriatic.

There are 14-16 ferries a day to Hvar (55min).

❷ Hvar

The most fashionable island on the Dalmatian Coast, Hvar offers historic architecture, some of Croatia's best wine and countless quiet coves and beaches. Hvar town has a well-preserved core, but for quiet lanes and real character head north to Stari Grad. Around it lie terraced fields of olives, grapes, figs and lavender first cleared by the Ancient Greeks, while craggy peaks and lush forests beg to be explored.

Ferry services run twice a day to Korčula (1hr 10min).

❸ Korčula

Historic towns sit between small-scale olive groves and vineyards on Korčula, a quietly traditional island inhabited since Mesolithic times. Its eponymous main town is often referred to as 'Little Dubrovnik' thanks to its Gothic, Renaissance and Baroque palaces, ornate churches and grand squares. Elsewhere, you'll find traditional folk music, sandy beaches and some of Croatia's best white wines.

ABOVE: DUBROVNIK

ABOVE: DIOCLETIAN'S PALACE, SPLIT

Frequent ferries run between Korčula Domince and Orebić (20min) from where you can catch a bus to Ston (1hr 20min).

4 Ston

Once an important salt-producing town, Ston was part of the Republic of Dubrovnik and still lies behind a protective ring of 14th-century walls. Its car-free streets are lined with medieval buildings, but the true allure is its food and drink. Nearby at Mali Ston are oyster beds known to produce some of the best oysters in the world, while the Pelješac Peninsula's red wines are highly prized.

Buses run five times a day to Dubrovnik (1hr 10min).

5 Dubrovnik

Dubrovnik remains one of the best-preserved walled towns in Europe, with steep lanes lined with elegant townhouses and Baroque churches overlooking grand squares. The Old Town is surrounded on three sides by the waters of the Adriatic, and a walk along its 9th-century city walls offers the chance to visit fortresses, wander museums, and relax on tiny beaches and wooded peninsulas with timeless views.

Dubrovnik has regular flights to cities across Europe.

ALTERNATIVES

Extension: Vis

Inhabited since the 6th century BCE and home to Greek cemeteries, Roman baths and an English fortress, Vis' more recent history as a military base has left it with an unspoiled beauty. Stroll the waterfront in Vis town, wander through the jumble of 17th- and 18th-century townhouses in Komiza, paddle the dramatic coastline or tour local vineyards.

Ferries run once a week between Hvar and Vis, but private transfers can easily be arranged.

Extension: Pakleni Islands

These 16 densely forested islands, just off the southern coast of Hvar, are best explored by gliding through the translucent waters in a kayak. Discover hidden coves and bays, quiet lagoons and pebble beaches, or stop to snorkel off Mlini Beach and hike in the pine forests.

Hire a kayak in Hvar town or take a guided tour.

MAGNIFICENT MONTENEGRO

Podgorica – Lake Skadar National Park
Take a short driving circuit through the mountainous heartland of Old Montenegro and journey along the country's glorious coastline.

FACT BOX
Carbon (kg per person): 37
Distance (km): 186
Nights: 5
Budget: $$
When: May-Oct

❶ Podgorica
Half a day is enough to get a feel for the unassuming Montenegrin capital. Take a stroll through the old Ottoman core of Stara Varoš then cross the Morača River to the shops and bars of Nova Varoš.
Drive west and follow the signs for 36km.

ABOVE: SVATI STEFAN

❷ Cetinje
Set within the craggy limestone mountains that for centuries formed the Montenegrin heartland, the old royal capital is a fascinating place: a small town jam-packed with palaces, churches, parks, museums and art galleries.
Continue 20km to Lovćen National Park.

❸ Lovćen National Park
Crisscrossed by hiking and mountain-biking trails, Lovćen is the Black Mountain which gave Montenegro its name. Be sure to visit the dramatic mountaintop Njegoš Mausoleum, featuring gigantic sculptures by acclaimed 20th-century artist Ivan Meštrović.
The narrow, serpentine 35km-long road heading down to the fjord-like Bay of Kotor is a white-knuckle ride, but breathtakingly beautiful.

❹ Kotor
One of the Adriatic's most beguiling walled towns, Kotor is encircled by fortifications which climb steeply up the mountain behind. It's set on the steely waters of the bay, where boats await to zip you out to beaches and a secret Yugoslav-era submarine dock. Spend a day wandering the tight medieval lanes and exploring the many churches and the quirky Cat Museum.

ABOVE: OUR LADY OF THE ROCKS CHURCH, PERAST

Head south, through the tunnel, for 23km.

5 Budva

Montenegro's premier summertime destination may be overdeveloped, but the long sandy beach is still a stunner and the medieval walled town oozes charm.

Continue east along the Adriatic coast for 10km.

6 Sveti Stefan

This photogenic walled island village, connected to the coast by a causeway, is best viewed from the highway above. The whole thing is now a luxury private resort but it's surrounded by spectacular beaches, some of which are open to the public.

Drive 33km along the coast, taking the Sozina tunnel through the mountains.

7 Lake Skadar National Park

Lake Skadar is one of Europe's most important habitats for birds, and there are plenty of boats waiting to get you out spotting them. The lake itself is dotted with island monasteries, while the surrounding mountains enclose lost-in-time winemaking villages and ruined castles.

Head 29km north back to Podgorica, passing the airport on the way.

ALTERNATIVES

Diversion: Perast

Along the bay from Kotor, Perast is a village comprised almost entirely of grand Venetian-era palaces and churches. Be sure to take the quick hop across the bay to Gospa od Škrpjela (Our Lady of the Rocks), a man-made island topped with a blue-domed church. When you return, secure a seat at one of the excellent waterfront restaurants.

To get to Perast, head north from Kotor and follow the bay for 12km.

Diversion: Ulcinj

At the southern end of Montenegro's Adriatic Coast, Ulcinj is one of its most appealing beach towns. The atmospheric walled town occupies a crag high above the main beach, while the newer part of Ulcinj is punctuated by the minarets of numerous mosques. Wander along the coast to rocky swimming spots or take a drive out to sandy Velika Plaža.

To reach Ulcinj from Sveti Stefan, continue south along the coastal highway for 55km.

SOAKING UP THE BEST OF SERBIA

Novi Sad – Niš

Criminally under-visited, Serbia is one of Europe's best-kept secrets. Uncover the off-the-radar gems of this wildly diverse country – before everyone else does.

FACT BOX

Carbon (kg per person): 18
Distance (km): 600
Nights: 10
Budget: $$
When: year-round

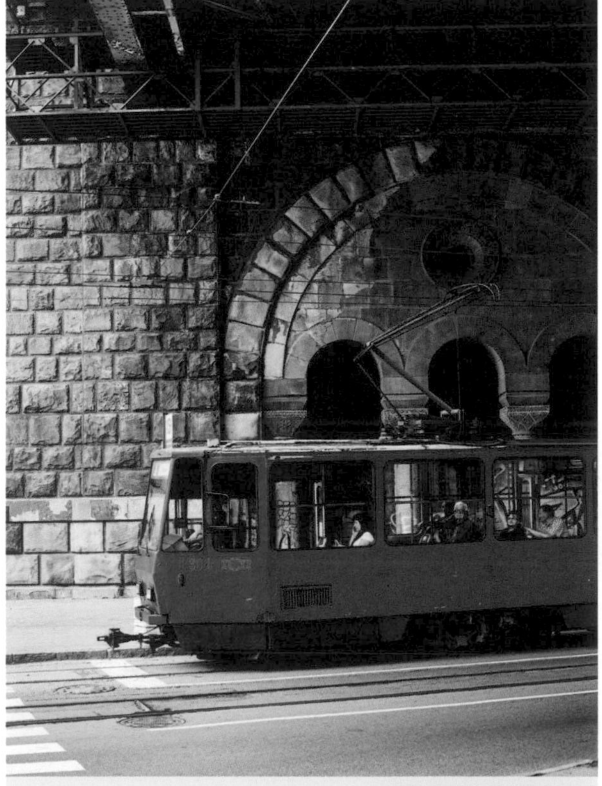

ABOVE: BELGRADE TRAM

❶ Novi Sad

Nicknamed the 'Athens of Serbia', Novi Sad is a freewheeling, artsy student city that hosts Southeast Europe's largest – and most frenetic – music festival (EXIT) each July. Creep through tunnels beneath the mighty 18th-century Petrovaradin Fortress, inspect the city's many galleries and museums, dip your toes in the Danube River from Štrand beach and cavort in any of the many bars lining boozy laneway Laze Telečkog.

Take the train to Belgrade (every 2hr; 1hr 30min). Note transport information is often written in Cyrillic, but many Serbs speak English and can usually translate for you.

❷ Belgrade

Bacchanalian Belgrade is an underrated hedonistic hotspot. During summer, hit a *splav* (floating nightclub) for till-dawn partying; the rest of the year, park yourself at any of the clubs or *kafane* (taverns) dotting the mega-cool, inner-city Dorćol neighbourhood. Tear yourself away from the *rakija* (strong Serbian spirit) for a culture and history fix at Kalemegdan Citadel (built 279 BCE), the Nikola Tesla Museum and the boho Skadarlija quarter.

Take the bus from BAS, Belgrade's main bus station, to Zlatibor (10 daily; 4hr).

ABOVE: EXIT FESTIVAL, NOVI SAD

❸ Zlatibor

This bucolic, mountainous region is a winner any time of year. Come for hiking, herb-picking and horse riding in warmer weather, and superlative skiing and hearty local cuisine beside a roaring fireplace in winter. Take in the verdant views year-round from the 8.95km-long Gold Gondola, the world's longest panoramic gondola ride. Zlatibor (aka 'golden pine') is also a renowned wellness destination: spas and health centres abound.

Take the bus to Niš (2-3 daily; 6hr).

❹ Niš

Encompassing everything from Ancient Roman sites to modern-day Roma culture, Niš is a marvellous mishmash of a city. Roman emperor Constantine the Great was born here in 272 CE; check out the mosaics at his luxurious restored palace, Mediana. The early 19th century Ćele Kula (Skull Tower) and the well-preserved Red Cross Concentration Camp offer some sobering glimpses into the city's often-tragic past; the hip bars of Tinker's Alley typify the buzzy Niš of now.

Frequent buses travel north back to Belgrade and Novi Sad; international buses run daily to nearby countries including Bulgaria, North Macedonia and Greece.

ALTERNATIVES

Extension: Mokra Gora

Quaint, quirky Mokra Gora is home to Drvengrad – a surreal mini-village built by Serbian art-house director Emir Kusturica for his film *Life is a Miracle* – and the old-fashioned Šargan Eight narrow-gauge railway, which chugs and twists up a mountain peak and through 22 tunnels.

Hop on a Mokra Gora-bound bus for the scenic 43km journey from Zlatibor.

Extension: Sremski Karlovci

This tiny, picturesque village preens beside the Danube, jam-packed with centuries-old gems including ornate palaces, cathedrals and a working theological seminary. It's also locally famous for its wines; don't miss Bermet, a dessert wine made to a secret recipe passed down through generations.

Public buses 60, 61 and 62 run all day between Sremski Karlovci and Novi Sad (30min).

BIKE SLOVENIA GREEN

Kranjska Gora – Koper

The Bike Slovenia Green cycling route, launched in 2020, takes riders to green-certified destinations through valleys and vineyards from the Alps to the Adriatic.

FACT BOX

Carbon (kg per person): 1.5
Distance (km): 252
Nights: 6
Budget: $$
When: mid-May to mid-Oct

❶ Kranjska Gora

Begin in this World Cup ski resort (reached by regular trains from capital Ljubljana), lodged in Slovenia's densely forested, Alpine northwest corner. A former Austro-Hungarian railway-cum-bike path ushers cyclists along the route's first stage, from the gateway of the Julian Alps to the natural riches found in the south.

Cycle 37km.

❷ Bled

This handsome town sits on Slovenia's most recognizable icon, tectonic and glacial Lake Bled. Celebrating the end of this stage means taking in a sunset at the water's edge while snapping shots of Bled Castle and the famous Church of the Assumption, perched on an island in the lake's centre.

Cycle 44km.

ABOVE: CHURCH OF THE ASSUMPTION, LAKE BLED

❸ Lake Bohinj

Riding from Lake Bled to Lake Bohinj through Triglav National Park is a route highlight. Once in Bohinjska Bistrica, Bohinj's main town, enjoy local cheeses, sausages, fruits and vegetables, and craft beer. The next stage starts with a panoramic train ride before getting back on the saddle.

Take the train (10 daily; 35min) from Bohinjska Bistrica to Most na Soči (35min), then cycle 32km.

❹ Goriška Brda

The Bohinj Railway, built during the time of the Austro-Hungarian Empire to connect Prague to Trieste, delivers riders from the mountains to the valley of the electric-emerald Soča River. From there, cyclists climb to Slovenia's most famous wine region, Goriška Brda, a panorama of terraced vineyards on the Italian border.

Cycle 52km.

ABOVE: PEDALLING OUT OF KRANJSKA GORA

❺ Komen

The route hurtles south towards Italy, then hugs the border as it heads toward the perched medieval village of Komen in the rocky Karst Plateau region. Here spelunkers and wine lovers are equally at home in a cave-riddled landscape where vineyards abound.

Cycle 45km.

❻ Lokev

Continuing through the Karst region toward the coast, riders come to Lokev, just above the Istrian Peninsula, shared with Croatia. This village, home to less than 1000 people, sits between two Slovenian highlights: the Lipica Stud Farm, home of the Unesco-nominated Lipizzaner horses; and the Unesco-inscribed Škocjan Caves.

Cycle 42km.

❼ Koper

On the Adriatic Sea, in a bay with Trieste just north and Croatia to the south, Koper is Slovenia's largest port and an overlooked coastal getaway with a history dating back millennia. Riders can celebrate the end of the route with fresh seafood, and birdwatching at the Škocjanski Zatok Nature Reserve.

There are three direct trains per day from Koper to Ljubljana (2hr to 2hr 30min), for onward connections.

ALTERNATIVES

Extension: Lipica

Since 1576, Lipizzaner horses have been bred in the tiny village of Lipica. Hugging the Italian border just northeast of Trieste, the Lipica Stud Farm claims a handful of superlatives: it's the original farm for the royal Lippizaners, among the oldest breeds on the continent; and is Europe's oldest stud farm dedicated to one horse breed.

Lipica is a 4.5km detour from Lokev.

Extension: Walk of Peace

A brief detour in the Soča Valley, the 230km Walk of Peace hiking trail follows WWI's Isonzo Front, which pitted Austro-Hungarians against Italians. Museums, cemeteries and renovated trenches (within today's Slovenia) provide an important perspective, allowing active travellers to embrace beautiful landscapes and culture.

Cycle 5.5km from Most na Soči to Tolmin.

BEST OF THE BALTICS

Tallinn – Vilnius

Take a Baltic-spanning journey between the historic capitals of Estonia, Latvia and Lithuania, stopping off at the beach along the way.

FACT BOX

Carbon (kg per person): 17
Distance (km): 602
Nights: 5-7
Budget: $$
When: Jun-Aug

❶ Tallinn, Estonia

Turrets and church towers lend a fairytale ambience to Estonia's capital city. At its core is one of the most charming and best-preserved medieval walled towns in Europe. Soviet structures add grit to the surrounding neighbourhoods, some transformed into hip shopping and entertainment precincts, but there's nothing Soviet about the cutting-edge dining scene.

ABOVE: FIND FUNKY CAFES IN RĪGA

Take the bus from Tallinn's Central Bus Station to Pärnu (hourly; 2hr).

❷ Pärnu, Estonia

Estonia's prime summertime destination sits alongside a gorgeous sandy beach and the chilly Baltic Sea. A few traces of medieval Pärnu remain, but mainly it looks and feels like the fashionable seaside resort that it became at the tail end of the Tsarist era. It's still largely a place of parks, leafy promenades, Orthodox churches and grand Art Nouveau homes.

Jump on one of the frequent buses to Rīga (2hr 30min).

❸ Rīga, Latvia

Latvia's capital also has a medieval heart, but Rīga is much more famous for the Art Nouveau houses of its 'Quiet Centre'. Elaborate sculptural reliefs and quirky designs grace the facades of more than 750 buildings, making the city the biggest showcase for this flamboyant architectural style in the world.

ABOVE: TALLINN

In summer Rīga's parks fill with people, and beer gardens pop up in the city squares.

Take the bus from Rīga's bus station to Bauska (every 20-30min; 2hr 45min).

4 Bauska, Latvia

The small Latvian town of Bauska sits at the confluence of two rivers, which a pretty castle (now a local history museum) still guards. However, the main reason to stop here is to visit nearby Rundāle Palace, the extravagant 18th-century seat of the Duke of Courland. The surrounding gardens are equally lavish, inspired by those at Versailles. Buses head between the town and palace hourly.

Direct buses head between Bauska and Vilnius (a few daily; 3-4hr).

5 Vilnius, Lithuania

The Lithuanian capital is known for yet another architectural style: Baroque. Its Old Town – the biggest in Eastern Europe – is a Unesco World Heritage Site, with plenty of cobbled lanes and gorgeous buildings to explore. Catch the funicular up to the 15th-century castle which overlooks it all.

Vilnius International Airport has onward connections.

ALTERNATIVES

Diversion: Tartu, Estonia

Situated deep in the Estonian heartland, the pretty little riverside city of Tartu gravitates around its historic university. Interesting museums will keep you busy during the day (including the strikingly modern Estonian National Museum), while the student-filled bars buzz into the night.

Buses (every 30min) and less-regular trains (8 daily) head from Tallinn to Tartu (2hr 30min). There are also frequent buses to/from Pärnu.

Day trip: Gauja National Park, Latvia

This large expanse of pine forest shrouds the Gauja River as it flows past a series of castles and small towns. Latvians head here to hike, mountain-bike, canoe, cross-country ski and generally escape into nature. If you're more adventurous, there's bungee jumping, bobsledding, tobogganing and a high-ropes course.

Regular trains and buses head from Rīga to the park's pretty towns of Sigulda and Cēsis.

FAROE ISLANDS

6

NORTH ATLANTIC OCEAN

THE NORDICS

Go north to find family-friendly frolics and a feast of Danish design, sweat it out in a trad Finnish sauna, chug through the fjords in Norway or discover the geysers, glaciers, geothermal pools and awesome natural landscape of otherwordly Iceland. And bring your binoculars: the wild Nordics promise brilliant birdwatching and the chance of spotting seals, whales and Arctic foxes.

SWEDEN
FINLAND
NORWAY
BALTIC SEA
NORTH SEA
DENMARK
1
2
3
7
8
9
11
12

ABOVE: TÓRSHAVN, FAROE ISLANDS

CONTENTS

THE STOCKHOLM-HELSINKI FERRY

Stockholm – Helsinki

This legendary ferry service links two enticing Nordic capitals, with a stop to explore the little-visited Åland Islands on the way.

FACT BOX

Carbon (kg per person): 66
Distance (km): 600
Nights: 8
Budget: $$$
When: Apr-Oct

❶ Stockholm, Sweden

Stockholm sprawls over a series of islands, each with its own distinct character. Start on Gamla Stan, the historic core of the city, with gabled townhouses and cobbled lanes winding beside the Royal Palace. Djurgården is known as museum island – the extraordinary Vasa Museum here contains a perfectly preserved 17th-century warship – while Södermalm has a reputation as something of a hipster colony, with artisan coffee shops and boutiques. One of the defining memories of Stockholm, however, is sailing out of its harbour: board the ferry for Helsinki and you'll pass rocky skerries, lighthouses and the summerhouses of the Stockholm Archipelago, before you reach the open water of the Baltic.

It's roughly 7hr by ferry from Stockholm to Mariehamn – the boat can be an adventure in itself, with many passengers making the most of the duty-free alcohol on board.

❷ Mariehamn, Åland islands

Åland is an archipelago of some 6500 islands, adrift in the cold waters between Sweden and Finland. They're somewhat politically adrift too – a demilitarised region that's Swedish-speaking but which technically belongs to Finland. Disembark in the

ABOVE: TEMPPELIAUKIO, HELSINKI

ABOVE: KASTELHOM CASTLE, ÅLAND

capital, Mariehamn, a low-key port of some 11,000 souls, and set a course for the excellent Maritime Museum. Mariehamn is also a departure point for cycling excursions along Åland's flat country roads – for a full-day adventure pedal north through leafy countryside to the battlements of Kastelholm Castle. For something a little easier, cycle the short distance south from the city to Järsö for a panoramic prospect over myriad islets and sounds.

Hop on a ferry to make the 10hr crossing from Mariehamn to Helsinki.

3 Helsinki, Finland

The approach into Helsinki harbour is every bit as dramatic as the departure from Stockholm: the ramparts of the Suomenlinna Fortress guarding the city and the tower of the cathedral vying with the masts of docked ships. Russian and Nordic influences converge in Helsinki's architecture – where the Swedish capital is a bastion of tradition, its Finnish counterpart goes big on pioneering modern structures. Step into the metallic wave of the Kiasma Museum of Contemporary Art, or visit the intergalactic Temppeliaukio, an extraordinary church built into the bedrock of the city.

Helsinki Airport has international connections.

ALTERNATIVES

Extension: Uppsala

With leafy squares and ancient spires, the university town of Uppsala offers relative serenity close to Sweden's busy capital. Head to the 13th-century Domkyrka, Scandinavia's biggest cathedral, resting place of revered Swedish kings; or the Botanical Garden, where greenhouses conserve tropical species in the chilly reaches of Northern Europe.

Take the train from Stockholm to Uppsala (72 daily; 30min).

Extension: Tallinn

If you're craving more ferry journeys, sail from Helsinki to the Estonian capital Tallinn. Perhaps the most picturesque city on the Baltic, its jumble of watchtowers, spires and red rooftops extend down a hillside: lose yourself in the narrow alleyways seeking out traditional bars and restaurants.

The ferry from Helsinki to Tallinn takes 2hr 30min.

FAR-NORTH NORWAY BY BUS AND TRAIN

Oslo – Nordkapp

Journey north for summer Midnight Sun, winter Northern Lights, majestic scenery and encounters with Norway's indigenous Sami people.

FACT BOX

Carbon (kg per person): 564
Distance (km): 4460
Nights: 7
Budget: $$$
When: May-Sep for 24hr daylight, Oct-Feb for Northern Lights

❶ Oslo

Alongside its urban attractions, Oslo is a city with a surprisingly wild side. The wonderful woodland of Nordmarka offers great hiking and cycling, and further afield in Ekebergparken, sculptures by world-famous artists are dotted amongst the trees and meadows.

ABOVE: VIGELAND PARK, OSLO

Catch the Flytoget train to Gardermoen International Airport for a 2hr flight to Tromsø.

❷ Tromsø

Some 400km north of the Arctic Circle at 69°N, Tromsø bills itself as Norway's gateway to the Arctic. In previous centuries the town was a centre for seal hunting, trapping and fishing, and later a launch pad for several notable Arctic expeditions, including some led by Roald Amundsen. These days it's a great base to spot the Northern Lights: organised aurora-spotting expeditions run nightly in season (roughly Oct-Feb).
Take the bus to Alta (daily; 6hr 30min).

❸ Alta

Alta might not be Norway's prettiest town, but it does have attractions – including the Northern Lights Cathedral, a swirling pyramid structure built in 2013 and clad in rippling titanium sheets; and the Alta Museum, where you can see 6000 late Stone Age petroglyphs carved into the surrounding cliffs.
An express bus connects Alta with Honningsvåg (daily; 4hr).

❹ Honningsvåg

Deep in Norway's Arctic North, tiny Honningsvåg is the

ABOVE: NORTHERN LIGHTS, TROMSØ

next stop en route to Nordkapp, the 307m cliff that marks Norway's northern tip. A port on the Hurtigruten ferry route, it's also a good place to try northern pastimes such as fishing for king crabs, quad-biking and snowmobiling.

The 36km-long E69 road connects Honningsvåg with Nordkapp. Buses connect the two – ask Honningsvåg's tourist office for advice, especially in winter, when the road is sometimes snowbound and must be visited in a vehicle convoy. Alternatively, the Hurtigruten ferry makes the 3hr 30min journey to Nordkapp.

5 Nordkapp

Touristy it may be, but Nordkapp is an essential sight: it's the northernmost point in Europe accessible by road, nearer to the North Pole than to Oslo. Nordkapp sits at latitude 71°N, where the sun never drops below the horizon from mid-May to the end of July. It's hauntingly beautiful, despite the busloads of visitors who come to see it. In good weather you can gaze down at the wild surf more than 300m below.

Catch the bus back to Honningsvåg, then climb aboard the Hurtigruten for a ferry ride along the coast to Tromsø, followed by a return flight to Oslo.

ALTERNATIVES

Extension: Karasjok

Karasjok (Kárásjohka in Sami) is the capital of Sami Norway. It's home to the Sami Parliament and library, NRK Sami Radio, a Sami theme park and the wonderful Sami Museum, which has displays of traditional clothing, tools and artefacts, and works by contemporary Sami artists. In winter, Karasjok is also one of the best places in Norway to go dog-sledding.

Twice-daily buses connect Karasjok with Alta.

Extension: Senja

Senja, Norway's second-largest island, rivals Lofoten and Vesterålen for natural beauty, yet attracts a fraction of the visitors. It warrants patient exploration, and at least two full days to fully appreciate its quiet beauty. Whale-watching, fjord trips and midnight-sun kayaking are just a few of the activities on offer.

There's barely any public transport, so a car is the only practical way to explore.

NORWAY'S COASTAL ROAD

Trondheim – Bodø

Set off on an epic coastal road trip (with several ferries) taking in Norway's natural wonders, from glaciers and fjords to isolated islands.

FACT BOX

Carbon (kg per person): 160
Distance (km): 905
Nights: 8-12
Budget: $$$
When: Apr-Oct

❶ Trondheim

Though the Coastal Rd (locally known as the Kystriksveien) starts in the little town of Steinkjer, Trondheim – Norway's handsome third city – is the more obvious gateway. Walk to the bridge of Gamle Bybro for views of the colourful fishing warehouses that straddle the Nidelva River.

From Trondheim, it's a 320km drive north, including a 30min ferry ride, to Leka. It's not necessary to book ferries in advance on the Kystriksveien.

ABOVE: GLOMFJORD

❷ Leka

The island of Leka is a geological wonder, with rocky landscapes that look like lost chunks of the American West. On the west coast you'll find prehistoric cave paintings, while on the east are Viking burial mounds.

Continue 115km north (with 2 ferries).

❸ Torghatten

Torghatten is famous as the mountain with the hole in it, and it's certainly an arresting sight – a natural sea arch marooned high among coastal cliffs. It's an easy twenty-minute walk up to the cavernous hole from the car park.

It's 50km (including a 1hr ferry trip) to Vega.

❹ Vega

Vega is a little archipelago, defiantly scattered in stormy northern seas. It's home to circling sea eagles,

ABOVE: NIDELVA RIVER, TRONDHEIM

wind-blasted bays and a rich eider down cultivating tradition. The slick visitor centre on the north shore provides an overview of how islanders create the world's most expensive duvets.

The most spectacular stretch of the Kystriksveien stretches 290km northward to Glomfjord: expect multiple ferry crossings over the fjords.

5 Glomfjord

The scenery ratchets up a notch as you approach Glomfjord: the fjords get steeper, the mountains taller and the icy mass of the Svartisen glacier looms. In Glomfjord it's easy to organise ice-climbing excursions on the glacier's Engabreen outlet – sinking axes and crampons into veins of blue ice, high above a turquoise lake.

Drive 130km drive north.

6 Bodø

Bodø marks the end of the Kystriksveien – and what a finale. Just before you get to town you'll pass a bridge over the Saltstraumen – said to be the world's most powerful tidal currents, with fast-flowing waters surging under the span.

Drop the car at Bodø airport, which has flights back to Oslo for international connections.

ALTERNATIVES

Extension: Lovund

Even more remote than Vega, Lovund consists of just a few hundred souls and a cluster of red houses, all huddled on a tiny speck of land far out at sea. The highlight of the year is the arrival of puffins, with hundreds of thousands of them coming to nest on Lovund's slopes from April until August.

Sail to Lovund from the port of Stokkvågen on the Kystriksveien (year-round; 3 daily; 1hr).

Extension: Lofoten

Lofoten needs no introduction: this archipelago is one of Norway's most photographed corners, with vertical cliffs dropping into chilly Arctic waters. From the ferry port, drive the meandering E10 road northeast, stopping at little galleries and remote fishing villages, Vestvågøy's Viking Museum and Vågan's imposing church.

It's a 3hr ferry ride from Bodø to the southern tip of the Lofoten and the start of the E10.

ROAD-TRIP ICELAND'S WESTFJORDS

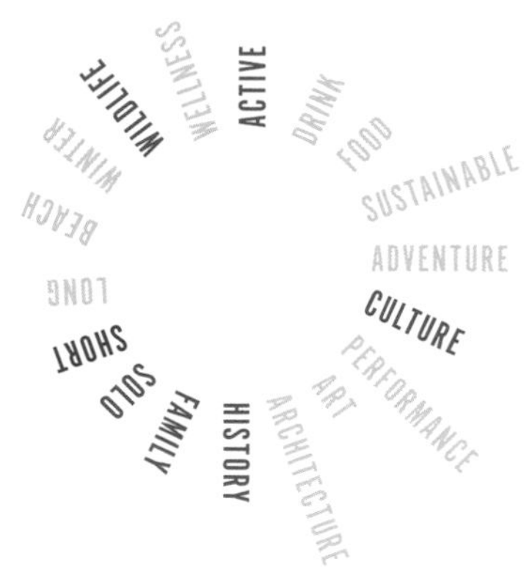

Ísafjörður – Bíldudalur

The remote, beautiful Westfjords offer a spectacular procession of waterfalls, geothermal pools, birds, beaches, sagas and, of course, fjords.

FACT BOX

Carbon (kg per person): 24
Distance (km): 188
Nights: 5–7
Budget: $$$
When: May–Sep

❶ Ísafjörður

The Westfjords 'capital' has a cluster of interesting museums amid its pretty timber buildings, and plenty of activities, from kayaking in the fjord to whale- and seal-watching boat trips. Book dinner in atmospheric, 18th-century Tjöruhúsid.

Ísafjörður's West Tours run day trips to Vigur.

❷ Vigur

A perfect day out from Ísafjörður, Vigur is a tiny island with a lot of charm, where Iceland's oldest lighthouse (1837) watches over frolicking seals. Take a hike for stunning fjord views and pop into the cafe for excellent cakes.

Back in Ísafjörður, it's a 40km drive on to Súđavík.

❸ Súđavík

Your best chance of spotting Iceland's only native mammal is at Súđavík's Arctic Fox Centre. Exhibits cover the animals' habitat and habits while outside a live, orphaned fox is as cute as you'd expect.

From Ísafjörđur drive 50km south.

❹ Þingeyri Peninsula

Ride a horse or a fat bike, or get those hiking boots on to explore the landscapes (including the Westfjords' highest peak, Kaldbakur) and history (Gísli, protagonist of an eponymous saga, lived here in Haukadalur valley in the 10th century) along Þingeyri (pronounced 'thingay-ree') Peninsula.

Drive 33km southeast.

❺ Dynjandi

The most impressive of the Westfjords' many

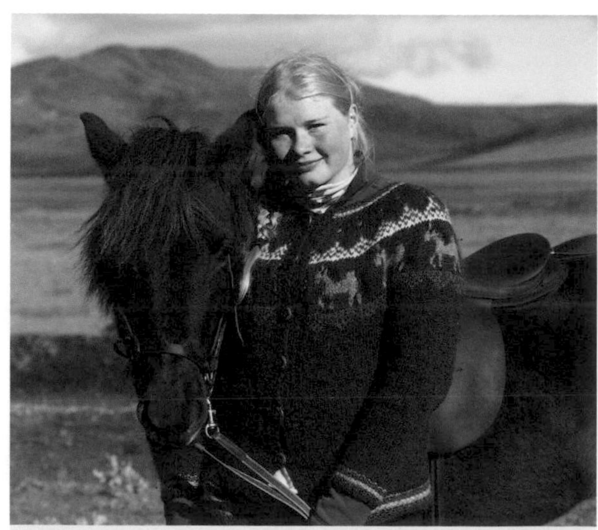

ABOVE: SADDLE UP IN THE WESTFJORDS

ABOVE: ÍSAFJÖRÐUR

impressive waterfalls, Dynjandi has a series of smaller cascades culminating in the 100m drop of the main fall, theatrically spread out across a wide cliff. Climb the path to the side of the tumbling water to appreciate the full effect.

Drive 45km southwest.

6 Reykjafjarðarlaug

However chilly the weather, you're always near a warming hot-pot – an outdoor, geothermally heated pool – in the Westfjords. One of the most scenic is at Reykjafjarðarlaug, which combines a large fjord-facing swimming pool with a cosy, turf-fringed, 45°C 'tub' behind.

Drive 20km west.

7 Bíldudalur

A fine fjord setting, a surrounding curtain of massive cliffs and connections to both the *Saga of Gísli* and legendary sea monsters (four different types are said to have been seen in the waters here – discover more in the local museum) make Bíldudalur a fun, photogenic final stop.

Bíldudalur Airport has rental car drop-off and daily (except Sat) flights to Reykjavík

ALTERNATIVES

Extension: Hornstrandir Nature Reserve

It's hard not to indulge in hyperbole when describing the magnificent scenery of Hornstrandir, the most remote part of the Westfjords. No facilities and unpredictable weather mean joining an organised hiking tour is the best way to see the monumental cliffs, wild Arctic foxes, plunging waterfalls, masses of sea birds and colourful wildflowers.

Day and overnight tours depart from Ísafjörður (summer only).

Extension: Rauđisandur and Látrabjarg

Depending on the weather, Raudisandur (Red Sand Beach) can be anything from muddy brown to yellow to vivid orange. No matter the colour, the 10km stretch of sand is an impressive sight. A short distance but a bumpy drive along the coast, Látrabjarg is a birdwatcher's heaven, where millions of puffins, guillemots and razorbills breed.

The roads to these sites, the 614 and 612, are 4WD territory so come prepared and drive carefully.

ICELAND ON AND OFF THE BEATEN TRACK

Reykjavík – Blue Lagoon

Waterfalls, volcanoes, icebergs, glaciers, beaches: attractions aplenty sit on, and just off, Iceland's famous, country-circling Ring Road.

FACT BOX

Carbon (kg per person): 165
Distance (km): 1380
Nights: 7-10
Budget: $$$
When: May-Sep

❶ Reykjavík

The world's most northerly capital, Reykjavík buzzes with museums, restaurants and bars. Climb Hallgrímskirkja's spaceship-like church tower for city views, eat out in port-turned-culinary-hotspot Grandi, shop in Laugavegur's many stores and generally enjoy civilisation before heading out into the wilds.

Drive 80km north along the Ring Road (aka Route 1).

ABOVE: AKUREYRARKIRKJA, AKUREYRI

❷ Borgarnes

First stop should be harbourside Borgarnes to stretch the legs and visit the Settlement Centre, where exhibitions detail Iceland's history and nature and tell the much-loved tale of the legendary Egil.

Drive 320km northeast.

❸ Akureyri

Iceland's second city is the place to fine-dine and drink local beer, visit the world's highest latitude botanical gardens, and take easy sidetrips to Húsavík (for whale watching) and the mighty Goðafoss (Waterfall of the Gods).

Drive 280km southeast.

❹ Seyðisfjörður

Multicoloured streets filled with the homes of artists, musicians and craftspeople and surrounded by snowcapped mountains greet you in this tiny east-coast town. The summer ferry from Denmark adds a lively feel once a week.

Drive 280km southwest.

❺ Jökulsárlón

Neon-blue icebergs in fantastical shapes drift through this lagoon, formed by a retreating glacier and sitting

ABOVE: JÖKULSÁRLÓN

right next to the Ring Road. Join a cruise among the frozen formations for a closer look.

Drive 200km west.

❻ Reynisfjara

Black sand, looming basalt columns, spiky offshore islands and crashing waves make Reynisfjara feel like a beach designed by Tim Burton. During the summer, puffins add a splash of colour.

Drive 80km west.

❼ LAVA Centre

Iceland exists because of volcanic activity and nowhere does a better job of explaining this than the LAVA Centre, with displays on the country's unique geology, a live earthquake map (there are dozens every day), and views of the nearby active Hekla volcano.

Drive 140km west.

❽ The Blue Lagoon

A soothing dip in the country's most famous pool is a just reward after the adventures of the Ring Road, as is a meal and glass of something bubbly in the excellent on-site restaurant. Book in advance.

Reykjavík and the country's main airport, Keflavík, are a short drive from the Blue Lagoon.

ALTERNATIVES

Extension: The Westfjords

If you thought parts of the Ring Road were isolated and spectacular, the Westfjords raise the bar higher. Looking like fingers jutting into the ocean from Iceland's northwest, the region is a people-light, jawdropping-scenery-heavy mix of waterfalls, fjords, cliffs and locations connected to the ghosts, heroes and battles recounted in the country's sagas.

Hike in Hornstrandir or go riding on a sturdy Icelandic horse.

Extension: Grímsey

The whole of Iceland is a long way north, but only tiny Grímsey island actually lies on the Arctic Circle – where the summer sun never sets and the winter nights are endless. You can dive or snorkel, hike to see birds (who outnumber humans 10,000 to one), or just have your photo taken on the Circle – be quick as the earth's tilt means it's moving further north all the time.

A ferry runs here from Dalvík on the mainland.

FAROE ISLANDS EXPLORER

Sørvágur – Viðareiði

Explore the 'Land of Maybe' by bus, visiting soaring sea cliffs, shimmering waterfalls and remote turf-roofed villages on a trans-archipelago adventure.

FACT BOX

Carbon (kg per person): 5
Distance (km): 190
Nights: 5-7
Budget: $$-$$$
When: May-Aug

❶ Sørvágur

The village alongside Vágar international airport is the base for visits (cycle or hike) to two of the archipelago's most alluring settlements. Tiny Bøur, a 4km stroll along the northern shore of Sørvágsfjørður, provides spectacular views across to the stegosaurus-esque island of Tindhólmur and to neighbouring Gáshólmur. It's a longer leg-stretch (10km) to magical Gásadalur, overlooking a waterfall plunging over a cliff into the Atlantic.

Take bus 300 to Tórshavn (5 daily; 1hr).

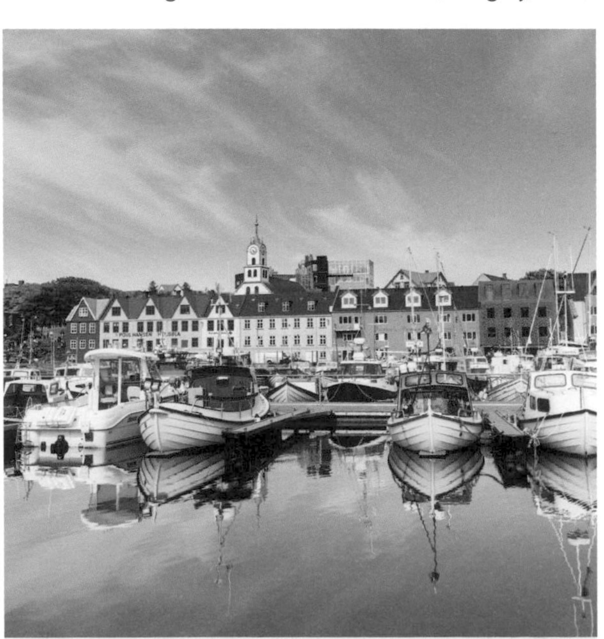
ABOVE: TÓRSHAVN

❷ Tórshavn

The Faroes' vivacious capital is the place for fine seafood landed at its working harbour, lined with colourful warehouses, cafes and shops. The historic old quarter, Tinganes, is a warren of tar-blackened timber houses huddled on a rocky spit. From Tórshavn, bus-based excursions visit the muscular, 700-year-old cathedral and the even older wooden Roykstovan farmhouse at Kirkjubøur, and – changing at Oyrarbakki – the Viking graveyard at Tjørnuvík, from where a testing hike visits Saksun, arguably the most picturesque hamlet in the whole of the Faroe Islands.

Continue the trip on bus 400 from Tórshavn to Oyrarbakki (several daily; 45min), for connections on bus 201 to Gjógv (45min).

❸ Gjógv

Wedged into a cleft between mountains and sea, this photogenic fishing village is a great place to

ABOVE: GJÓGV

learn about Faroese traditions on regular cultural evenings. It's also a hub for hikes along the coast and inland towards the archipelago's highest peak, Slættaratindur (882m).

Take bus 201 back to Oyrarbakki (45min) then bus 400 to Klaksvík (several daily; 1hr).

4 Klaksvík

The islands' second-largest settlement boasts an extraordinary setting between two inlets surrounded by sheer, hulking mountains. Hikes up these hills reward with far-reaching views: bus 504 runs to the island of Kunoy daily, another option for fine fjord walks.

Bus 500 runs to Viðareiði (1 daily; 40min).

5 Viðareiði

The northernmost village on the northernmost island of the archipelago, Viðareiði affords wonderful vistas towards the easternmost island, Fugloy, and the chance to hike to the edge of the Enniberg sea cliff – at 754m, Europe's second-highest.

Return to Klaksvík for frequent buses to Tórshavn.

ALTERNATIVES

Extension: Mykines

Rugged and heart-stoppingly wild, the Faroes' westernmost island is a short boat journey from Sørvágur – when capricious weather doesn't stymie sailings. Rewards for visitors include timber-and-turf Mykines village; fabulous hiking to Mykineshólmur or Knúkur peak (560m); and in summer, millions of breeding seabirds, from puffins to Arctic terns.

Ferries sail from Sørvágur to Mykines twice daily May-August (45min).

Day trip: Vestmanna

Sea cliffs soaring 600m from the Atlantic just north of Vestmanna town host raucous hordes of nesting fulmars, razorbills and guillemots; boat tours from the nearby harbour also typically chug past rafts of puffins bobbing on the waves. It's an almost overwhelming experience.

Buses 100/300 run several times daily from Tórshavn to Vestmanna (55min); some require a change at Kollafjørdur.

BELOW: THERE ARE 18 FAROE ISLANDS

DENMARK FOR FAMILIES

Copenhagen – Billund

Denmark is so good for kids because it's designed that way, with urban architecture integrating playgrounds, harbour baths and plenty of entertaining museums.

FACT BOX

Carbon (kg per person): 8
Distance (km): 270
Nights: 7
Budget: $$
When: May-Sep

❶ Copenhagen

Kids love Copenhagen: it's calm, bike-friendly and low in traffic. First stop should be Tivoli, an amusement park with gardens, lakes, rides and even live ballet, the sets for which are often designed by the Queen of Denmark. During the summer there are fireworks every weekend. Fantastic hands-on science museum Experimentarium could keep a family entertained for days, and the beautifully designed National Aquarium has everything from piranhas to otters. The National Museum of Denmark has an entire children's wing, where kids can dress up as Vikings or pretend to be in 19th-century school. In summer you can swim in the city canals, and the beach is only a few Metro stops away.

Take the train from Copenhagen Central to Odense (every 30min; 1hr 10min).

ABOVE: TIVOLI GARDENS, COPENHAGEN

❷ Odense

Odense is the birthplace of children's author Hans Christian Andersen, Denmark's favourite fablist. Plenty of sights in the picture-book-pretty town celebrate the author and his stories, such as *The Little Mermaid* and *The Steadfast Tin Soldier*. At Odense's children's centre, kids can immerse themselves in fairytales via dressing up and music, while at Egeskov Castle, they

ABOVE: LEGOLAND, BILLUND

can clamber along treetop walkways in the grounds. Other family-friendly pursuits include boating along the river; wandering Odense's outdoor museums, with reconstructed homes through the ages; or visiting interactive museum Tidens Samling, where you can explore rooms styled from the 1900s to the 1980s.

Take the train from Odense to Vejle (every 30min; 1hr), then a 30min bus to Billund.

3 Billund

Billund is the Lego heartland, where everybody's favourite little plastic blocks are designed. Danes take creativity through play seriously, and the town has several attractions devoted to playful construction. Danish Legoland has Duplo-themed rides for younger kids and rollercoasters for older ones, as well as a Lego miniature city and copious elaborate Lego sculptures. You can stay in the Lego hotel, a block-lover's dream, with themed rooms and Lego zones. Within walking distance is the Bauhaus-style Lego House, filled with zones of creativity and with Lego robots serving food in the cafe. It's also an easy trip to a holiday village attached to Scandinavia's biggest waterpark, Lalandia, where there are slides, a lazy river and waves.

Billund has trains back to Copenhagen and international flights from its airport.

ALTERNATIVES

Extension: Helsingør

The Unesco-listed, Renaissance Kronborg is Hamlet's Elsinore, an imposing castle at the water's edge. During the summer Shakespeare festival, actors play the Bard's scenes in a way that's accessible and wildly entertaining for kids – as are the hands-on medieval games. Characters from the plays roam around the castle, and the finale is a sword fight in the upper hall.

Take the train from Copenhagen to Helsingør (every 15-20min; 45min).

Extension: Ribe

Ribe, Denmark's oldest town, is close to a remarkable reconstructed Viking village, an amazing place for families to visit. You can wander through the Viking houses, learn to sword-fight, make bread over a fire, and see reenactments of battles. International (but mainly Scandinavian) Viking enthusiasts even come here to live as Vikings for days at a time.

It's easiest to drive to Ribe from Billund (70km).

DISCOVER DANISH DESIGN

Copenhagen – Aalborg

Travel around eastern Denmark to indulge in person-centred, pared-down Danish style at its source.

FACT BOX

Carbon (kg per person): 11
Distance (km): 724
Nights: 7
Budget: $$
When: May-Sep

❶ Copenhagen

The Danish capital is also its design headquarters, a place where style-hunters can lead their best lives. Stay at one of its many design hotels (such as the Radisson, where every detail was designed by Arne Jacobsen), and enjoy strolling around a city where even the street lamps are iconic pieces of mid-century design. Must-sees include the Design Museum, which will make lovers of the perfect chair giddy with joy, and BLOX, which houses the Danish Architecture Centre. Browse design shops in the city centre, and visit Copenhagen's extraordinary CopenHill, an innovative piece of architecture from the Bjarke Ingels Group in which aesthetic ambition combines with a sense of civic duty: a dry ski slope on the rooftop of a waste-burning power station.

Take the train to Aarhus from Copenhagen Central (every 10min; 3hr).

ABOVE: COPENHAGEN OPERA HOUSE

❷ Aarhus

On the east coast of Jutland, the waterside university city of Aarhus has a City Hall designed by Arne Jacobsen, and *Your Rainbow Panorama* at the ARoS contemporary art museum, a work of art by Olafur Eliasson that provides colourful, sweeping views of the city. See Aarhus from another angle from the rooftop of the city's Sailing department store, with a glass-floored and glass-walled walkway over the street. Aarhus' harbour is another architectural and design highlight, with the harbour bath, circular diving pool, panoramic walkway and saunas, also designed by the Bjarke Ingels Group. In summer there's the Infinite Bridge walkway, a circular pier by architects Gjøde & Povlsgaard.

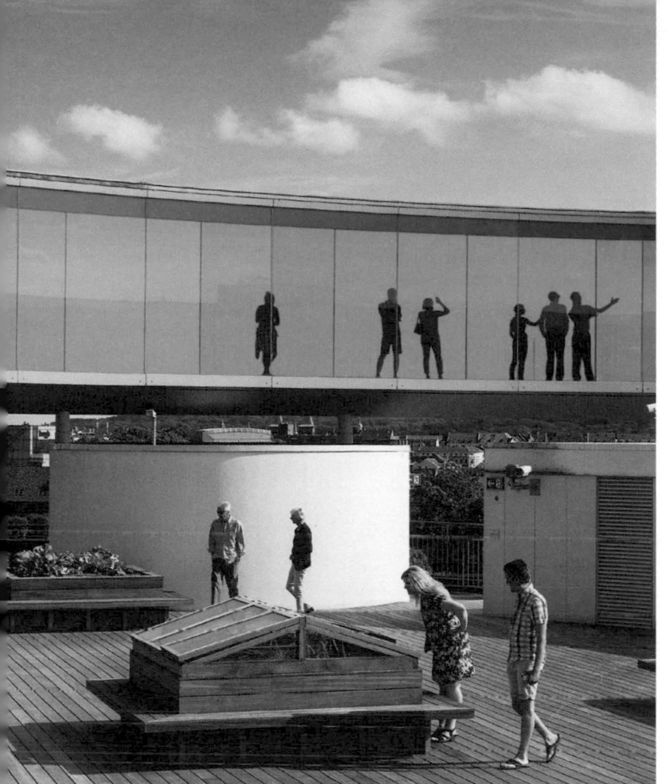

ABOVE: ARoS ART MUSEUM, AARHUS

Take the train to Aalborg (every 15-45min; 1hr 35min)

❸ Aalborg

Aalborg's Utzon Centre is named after Jørn Utzon, the city's most famous local and architect of Australia's Sydney Opera House. The centre was his last building: inspired by the city's industry and shipyards, it has dramatically high curved ceilings. As well as exploring Utzon's works and other great Danish architecture and design, the centre offers great views over the Limfjord. CF Møller transformed the Aalborg waterfront into a promenade influenced by the rise and fall of dunes, and ADEPT architects designed Vestre Fjordpark, with harbour baths, walkways and places to boat and kayak. Aalborg's Kunsten Museum of Modern Art is a mid-century masterpiece, designed by Finnish architect Alvar Aalto.

Take the train back to Copenhagen (hourly; 4hr 15min)

ALTERNATIVES

Extension: Louisiana Museum of Modern Art

Denmark's foremost collection of contemporary art is set in a garden on the banks of the Oresund. In 1958, the original 19th century house was expanded into an art museum: architects Bo & Wohlert created a subtle and understated building, designed to allow interplay with its surroundings.

Take the train from Copenhagen Central to Humlebaek station (every 20min; 40min)

Extension: Charlottenlund

An essential day trip from Copenhagen for design lovers, this small town to the north contains not only the Zaha Hadid-designed Ordrupgaard (a museum with works by Hammershøi, Monet and Gauguin), but also the 1942 home of one of Denmark's most venerated mid-century designers, Finn Juhl.

Take bus 23 from close to Copenhagen Central to Charlottenlund (every 20min; 55min).

A FINNISH SAUNA ODYSSEY

Helsinki – Rovaniemi

Never were cold climates such hot stuff as on this steamy pilgrimage through Finland's sauna experiences.

FACT BOX

Carbon (kg per person): 101
Distance (km): 2200
Nights: 6-7
Budget: $$
When: Dec-Mar

❶ Helsinki

To discover why it is so important (and it really, really is) to Finns to take off their clothes and sweat it out in front of strangers in temperatures of 80°C, capital Helsinki is a good starting point. Here, they even have sauna guides to talk you through sauna culture and etiquette. Start in style at Löyly Design Sauna in Hernesaari, a large, slatted wooden construction with a terrace opening onto Helsinki waterfront, as well as a restaurant and viewing platform. The surrounding parkland is very pretty, and you can take a boat out to Pihlajasaari, an island nature reserve with lovely beaches.

Train the train to Turku (hourly; 2hr); take a guided tour through Visit Turku (Monitori Market Square, Aurakatu 8) to Herrankukkaro.

❷ Herrankukkaro

Swap the train for a guided tour through forested scenery on some of the lonely islands in the extensive archipelago to the south of Turku. Your goal? The hidden-away resort and former fishing village of Herrankukkaro, sporting the world's biggest underground smoke sauna – the best kind of sauna for Finns – where aromatic birch twigs create an atmospheric, smoky sweat-athon.

Return to Turku with the guided tour and take the train to Tampere (6 daily; 2hr).

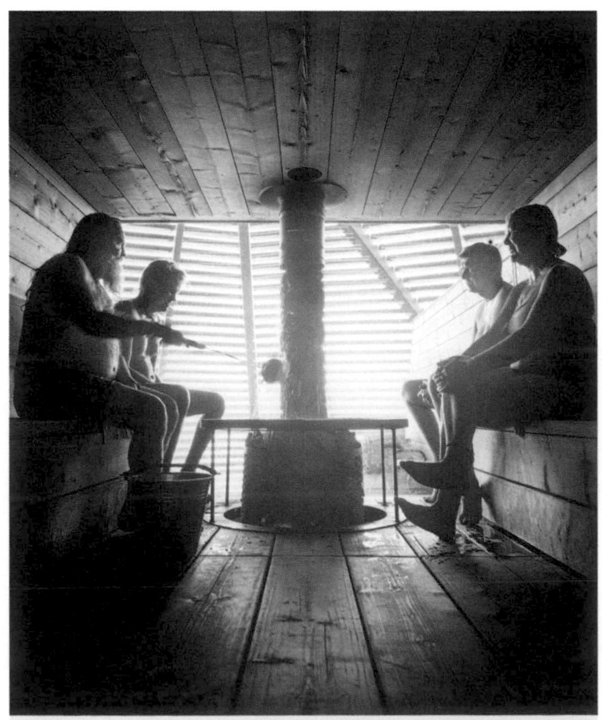

ABOVE: LÖYLY DESIGN SAUNA, HELSINKI

ABOVE: FINLAND'S SAUNAS COME WITH A VIEW

❸ Tampere

Tampere boasts the oldest still-used sauna, Rajaportti, an edifice built in 1906 and retaining both its original wood-smoke heating system and a cast of local characters you just do not see in more touristy sauna-going spots. It's Finland's most traditional sauna.

Back on the train, head for Kuopio (every 3hr; 3hr 15min).

❹ Kuopio

The characterful sauna experience reaches new heights in this Finland Lake District city. On a beautiful lakeshore sits Jätkänkämppä Lumberjack Lodge, a log-cabin complex built in the 1950s for local lumberjacks. Its traditional wood-smoke sauna is often available, alongside traditional 'lumberjack nights' with costumed characters and platters of local specialities.

Take the train north to Rovaniemi (3 daily; 7hr).

❺ Rovaniemi

Lapland's capital is more than just Santa Claus' home – eclectic sauna activities beckon too. Come in winter and, at the Arctic Snow Hotel, you can burrow into a sauna constructed entirely from snow and ice.

Take the train back to Helsinki (1 daily; 8hr).

ALTERNATIVES

Diversion: Juokslahti

On a Lake District shore between Tampere and Kuopio, Sauna Village claims to be the 'sauna region of the world' – there are over 20 here, including 18th- and early 20th-century examples, some still in use. Guides give an overview of Finnish sauna culture.

Sauna Village is 13.5km northeast of Jämsä. From Tampere, take a train to Jämsä, then a bus Juokslahti, asking the driver to drop you at the Sauna Village turn-off, from where it's a 200m walk.

Diversion: Ylläs

The biggest ski resort in Finland might waylay you with its winter sports, but you should certainly veer off-piste too for the world's only sauna-in-a-gondola ride. From the resort complex you can get a sweat on whilst drifting up to or down from the snowy slopes.

Ylläs is 175km northwest of Rovaniemi. Rent a car in Rovaniemi, or take buses from Rovaniemi to Kitillä and then Kitillä to Ylläs.

AROUND THE ISLAND OF ICELAND

Reykjavík – Þingvellir National Park
Buckle up for a wild drive around Iceland's Ring Road taking in volcanoes, glaciers and thundering falls.

FACT BOX

Carbon (kg per person): 181
Distance (km): 1451
Nights: 14
Budget: $$
When: May-Sep

❶ Reykjavík

Renowned for its quirky character, vivid street art, live music and wild nightlife, Reykjavík is also the place to delve into the country's history in state-of-the-art museums and galleries, as well as pausing over coffee in hip cafes and feasting on seafood.

Drive 327km east.

ABOVE: ÞINGVELLIR NATIONAL PARK

❷ Skaftafell

Skaftafell, the southern end of Vatnajökull National Park, offers thundering waterfalls, massive glaciers, tangled birch forests and ice-choked lagoons, but you'll need to tackle the more remote trails to get a taste of real wilderness. Nearby, blue-green icebergs litter Jökulsárlón, while brooding Vatnajökull, Iceland's largest ice cap, looms above all.

Continue east for 339km.

❸ Seyðisfjörður

Brightly coloured wooden houses sit below snowcapped peaks and tumbling waterfalls in Seyðisfjörður, the liveliest and most bohemian Eastfjords town. Visit the museum and cultural centre, hike deserted valleys, and discover tiny fishing villages, diverse birdlife and playful dolphin pods.

Drive north for 186km.

❹ Dettifoss

In a country of impressive waterfalls, Dettifoss rules supreme, its thundering chutes sending up a spray that can be seen up to 1km away. On a sunny day vivid double rainbows appear above the churning glacial waters.

Continue 69km west.

ABOVE: DETTIFOSS WATERFALL

5 Reykjahlíð

The gateway to the Mývatn wonderland, Reykjahlíð is surrounded by an otherworldly landscape of mudpots, fumaroles and volcanic craters. Explore giant lava fields, climb those craters, discover bubbling mud or simply wallow in the milky local nature baths.

Drive 83km west.

6 Akureyri

Along with a handful of churches, museums and galleries, relaxed Akureyri, Iceland's second-largest city, has surprising botanical gardens and a fantastic summer arts festival. Head north to go whale watching at Hauganes or catch the ferry to Grímsey to cross the Arctic Circle.

Drive 400km west and south.

7 Þingvellir National Park

Starkly beautiful Þingvellir is Iceland's most important historical site. This giant rift valley held the world's first democratic parliament, and along with scattered historical remains you can see two of the earth's tectonic plates cleaving apart among glacial lakes, waterfalls and rugged rock formations.

Drive 47km back to Reykjavík.

ALTERNATIVES

Extension: Snæfellsnes Peninsula

Parading everything from giant ice caps and volcanic peaks to ghostly lava fields and golden beaches, the Snæfellsnes Peninsula is a microcosm of Iceland. Snæfellsjökull peak dominates the landscape but if hiking isn't your thing, you can go whale watching, lounge in geothermal pools, walk windswept beaches or follow legends of Iceland's 'little people'.

It's a 300km drive around the peninsula.

Extension: Strokkur and Gulfoss

The original hot-water spout, Geysir is considered inactive these days, but its neighbour Strokkur reliably shoots hot water 15m to 30m into the sky roughly every five minutes. Nearby are steam vents, bubbling mud pools and colourful springs. At Gulfoss, 10km on, a spectacular double waterfall drops 32m down a narrow ravine.

Plenty of guided tours get to Strokkur and Gulfoss from Reykjavík.

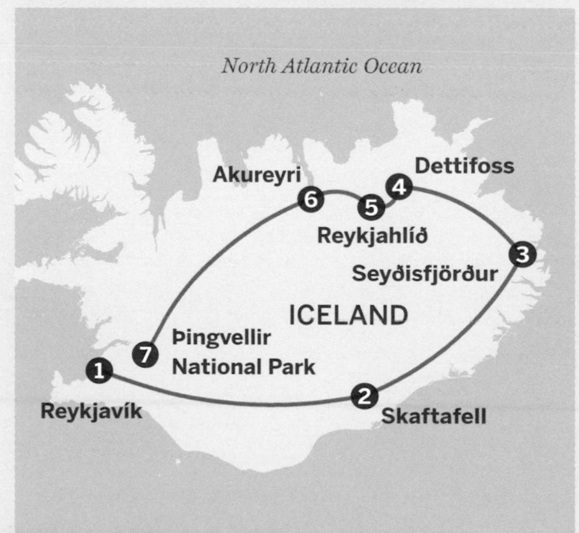

CLASSIC NORWAY DISCOVERER

Oslo – Hardangerfjord

Explore Norway's quintessential sights – cities, fjords, forests and ferries – by a combination of rail and sea.

FACT BOX

Carbon (kg per person): 71
Distance (km): 1280
Nights: 5
Budget: $$$
When: May-Sep

❶ Oslo

Norway's capital is a compact, cultured and fascinating little city. It has some stellar museums, not least the Nasjonalmuseet (National Museum), home to the nation's principal artistic treasures; and the captivating Vikingskipshuset (Viking Ship Museum), where you can see three real-life Viking ships that were used to bury important chieftains. The city also has a growing reputation for adventurous architecture, epitomised by the striking new opera house and the Renzo Piano-designed Astrup Fearnley Museet.

ABOVE: BERGEN

Catch an early train on the Bergensbanen, Norway's most scenic railway, to Myrdal (5 daily; 4hr 45min).

❷ Myrdal

The mountain town of Myrdal has one compelling reason to stop: the chance to ride the Flåmsbana Railway, one of the world's steepest railway lines, at a gradient of 1:18. This 20km-long engineering wonder hauls itself up 866m, passing deep ravines and thundering waterfalls, and chundering through 20 tunnels. It takes 45 minutes to descend from Myrdal to Flam. The railway runs up to ten times daily, so you can make the up-and-down journey in an afternoon before catching an onward train to Bergen.

Take the train to Bergen (4 daily; 2hr 30min).

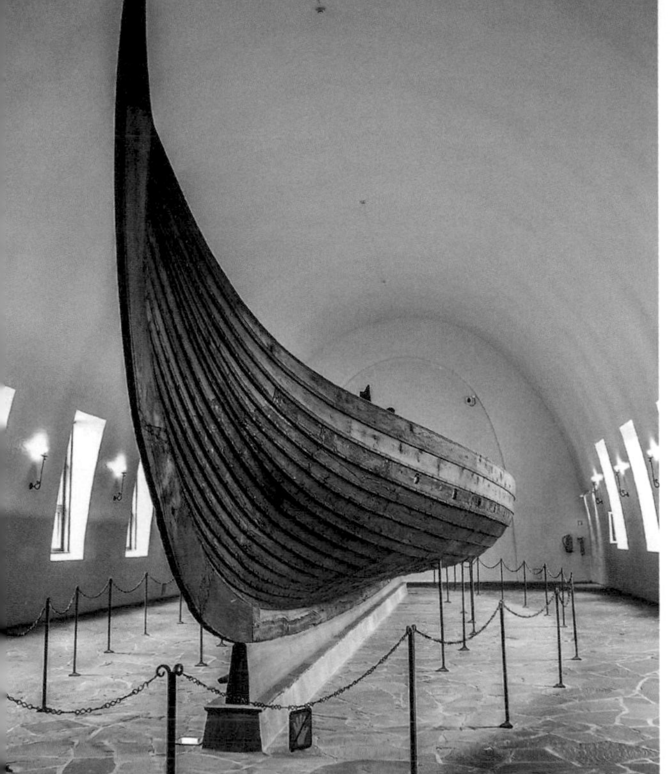

ABOVE: VIKINGSKIPSHUSET, OSLO

❸ Bergen

Bergen was an important seaport during the Hanseatic era – a heritage that can still be seen in the colourful wooden buildings of Bryggen, a Unesco World Heritage Site. It's a lovely city: clapperboard houses creep up the seven hillsides, ferries flit around the seven magnificent fjords and excellent art museums provide culture. It's worth riding the cable car to the top of Mt Fløyen for a pan-Bergen view.

Several ferries run to Hardangerfjord from Bergen, including Hardangerfjordekspressen passenger service (2hr).

❹ Hardangerfjord

Bergen is an ideal launchpad for a fjord cruise. Two hours' south of the city, Hardangerfjord boasts all the classic Norwegian features you could ask for: sheer cliffs; icy blue waters; misty mountain tops. The Bergen tourist office offers organised excursions which include the Vøringsfossen waterfall and the Hardangervidda Nature Centre in Eidfjord. You can be back in Bergen in a day, but with more time, you could also take a guided glacier walk in Folgefonna National Park.

Once back in Bergen, hop on the Bergensbanen back to Oslo for onward connections.

ALTERNATIVES

Extension: The Hurtigruten

So much more than a mere ferry, the iconic Hurtigruten provides one of the most spectacular boat journeys on earth. Day in, day out, one of 12 ferries heads north from Bergen, pulling into 35 ports on its six-day journey to Kirkenes. Along the way, it dips into coastal fjords, docks at isolated villages barely accessible by road, draws near to dramatic headlands and crosses the Arctic Circle, only to return a few days later. It's the best way to see Norway's coast in all its glory: the return journey takes 11 days and covers 5200km. If you're lucky (especially if you're travelling in winter) you may even be treated to a display of the Northern Lights – and there aren't many ferry trips which can offer that, now are there?

Ferries sail from Bergen to Kirkenes, stopping at 32 other ports en route; the full journey takes from seven to 11 days, with at least one sailing per day year-round.

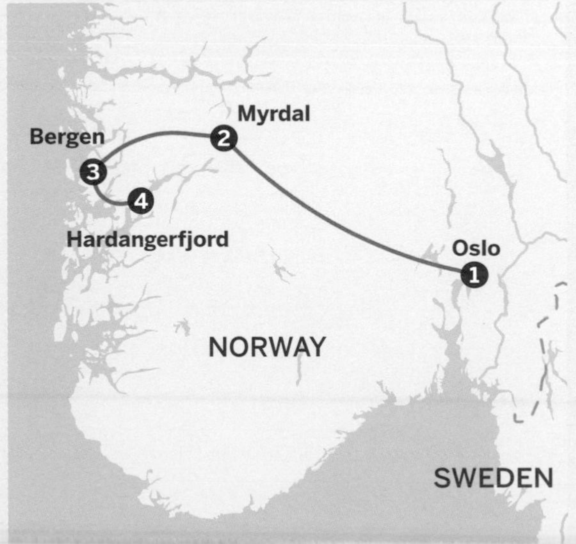

FOLLOW THE ALVAR AALTO TRAIL

Helsinki – Villa Mairea

Pay homage to the masterworks of Finland's greatest architect, Alvar Aalto, on this cross-country train trip.

FACT BOX

Carbon (kg per person): 20
Distance (km): 1017
Nights: 4-5
Budget: €€€
When: Jun-Aug

ABOVE: AALTO'S CAPITOLIUM BUILDING, JYVÄSKYLÄ UNIVERSITY

❶ Helsinki

Finland's capital is proof that good things often come in small packages. Though compact, the city is full of wonderful architecture and design, including significant Aalto buildings and spaces. When here, visit the great man's home and studio (northwest of the city centre), take in a concert at his marble-clad Finlandia Hall, browse his furniture designs at Artek and linger over dinner at the Savoy restaurant, which has retained its original Aalto-created interior. Before moving on, enjoy other city highlights including the mighty fortress of Suomenlinna, a short ferry ride away; the Design Museum, with some of the works of Finland's other top designers; and the Museum of Finnish Architecture.

Take the train from Helsinki's central train station to Jyväskylä (every 2hr; 3hr 30min).

❷ Jyväskylä

Aalto spent much of his childhood and early career in this town on the western edge of Finland's glistening Lakeland region. Visit the Alvar Aalto Museum, which focuses on his major buildings and furniture designs, as well as the Museum of Central Finland and the University of Jyväskylä campus (don't miss the swimming hall).

ABOVE: FINLANDIA HALL, HELSINKI

Get back on the train and head to Turku, with a change in Tampere (every 2-3hr; 4hr).

3 Turku

Finland's second city, Turku is renowned for its experimental art, design and music scenes. It's also where Aalto's work base was located between 1927 and 1933, and where he traded Nordic classicism for the functionalism that was to become his trademark. His best-known work here is the office building designed for the Turun Sanomat Newspaper at Kauppiaskatu 5.
Continue on the train to Pori, changing at Tampere (4 daily; 3hr 30min-4hr 45min).

4 Villa Mairea

Built in 1939 for wealthy art patrons Maire and Harry Gullichsen, this villa is perhaps Aalto's most influential and beloved building. A mash-up of Finnish vernacular, Japanese and Modernist influences, the result is sublime, both inside and out. Contact the property to organise a guided tour including the interior, which was designed by Aalto's wife and architectural partner, Aino. The villa is 16km from Pori, around 40min on local bus 66 or 64.
Head back to Helsinki on the train (every 2hr via Tampere; 3hr 30min-4hr) for onward international connections.

ALTERNATIVES

Extension: Säynätsalo Town Hall

One of Aalto's most significant designs, this complex on a forested island near Jyväskylä was constructed between 1949 and 1952. Its grassy inner courtyard bathes the interior with light, and its various spaces (library, council chamber etc) feature purpose-designed furniture. Overnight accommodation in the building is available too.
To get there, take the regular bus from Jyväskylä (20min).

Extension: Paimio Sanatorium

Aalto garnered global attention with his 1933 design for this Modernist former tuberculosis sanatorium, 30km east of Turku on the outskirts of Paimio. Its furnishings were custom-designed and included the famous Paimio chair, made with a single piece of undulating bent plywood.
Turku-based travel agent Magni Mundi offer guided tours, or you can take a bus from Turku (45min–1hr).

BELOW: HELSINKI'S WATERFRONT

Trip Builder

AFRICA

Want a one-on-one with a mountain gorilla in Uganda or Rwanda? Lemurs in Madagascar, or the Big Five in big-name national parks like Tanzania's Serengeti or Namibia's Etosha? Check, check and check – but don't miss road-tripping the world's highest country in Lesotho, discovering medinas and mosques in Morocco or vineyard-hopping South Africa's Cape Winelands. And then there's a daring dip in Victoria Falls' Devil's Pool or just lazing beachside in Mauritius, Mozambique and the Cabo Verde islands.

CONTENTS

ABOVE: WILDEBEEST IN SERENGETI NATIONAL PARK, TANZANIA

TAKE IT SLOW IN NORTHERN MOZAMBIQUE

Pemba – Gurúè

Slow travel is par for the course on this zigzag around northern Mozambique, taking in the fabled Swahili coast and its lost-in-time islands.

FACT BOX

Carbon (kg per person): 123
Distance (km): 1659
Nights: 9-10
Budget: $$
When: May-Nov

❶ Pemba

Hub of Mozambique's north coast, Pemba is a city of three parts: a Portuguese-flavoured port; a lively African township; and a generously wide beach (Wimbi) backed by an assemblage of budget guesthouses. It's a great place to slip into the relaxed mood of coastal Mozambique.

The easiest way to get to Ibo Island from Pemba is to book a private transfer or direct boat via your Ibo accommodation.

❷ Ibo Island

Arguably the finest undiscovered beauty spot in sub-Saharan Africa, Ibo is the largest island in the Quirimbas archipelago. Heavy with the ghosts of centuries past, it holds a mildewed ensemble of Portuguese-era architecture in a lush palm-fringed Indian Ocean setting. Dhows navigate the surrounding seas and islanders make a living from fishing, crafting silver and growing their own coffee.

Organise a car transfer to Mozambique Island from your Ibo hotel, then a dhow to Tandahangue village from where it's a full day's ride to the Ilha.

❸ Mozambique Island

Attached to the mainland by a narrow one-lane bridge, the Ilha, as Mozambique Island is known, is a magical place with narrow history-filled streets and food and culture infused by multiple religions, races and languages. Explore the massive Portuguese fort, the oldest church in sub-Saharan Africa and placid bays.

Regular *chapas* head west from Mozambique Island to Nampula (3hr).

ABOVE: NAMPULA CATHEDRAL

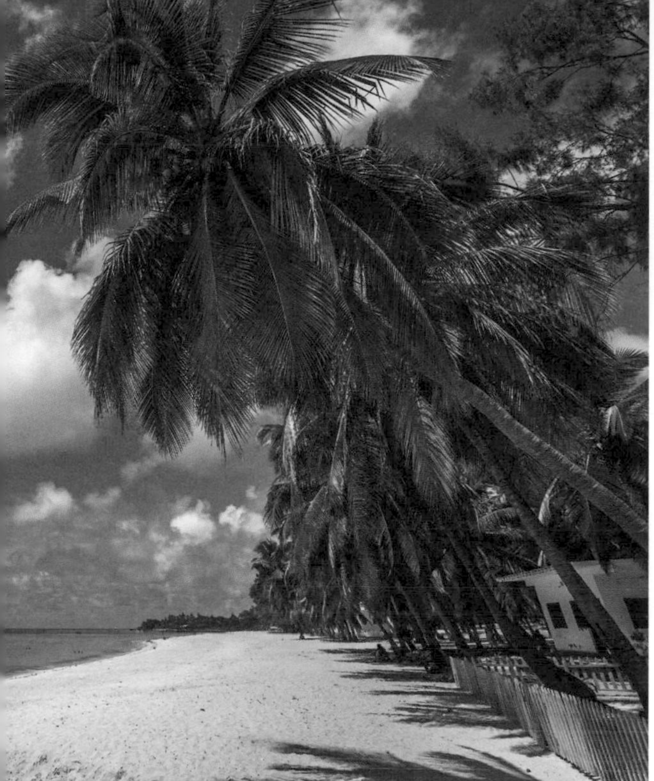

ABOVE: WIMBI BEACH, PEMBA

❹ Nampula

After the wonderful chaos of Mozambique Island, modern Nampula makes a welcome breather. Visit its dusty museum and oversized cathedral, stay in an air-conditioned hotel and grab a Portuguese *galão* (white coffee) before taking the train west.

Take the train to Cuamba (2 weekly Tue & Sat; 13hr).

❺ Cuamba

The only reason to make the lengthy trip to Cuamba is for the train ride. Absorb the dusty essence of southeast Africa and its bustling trackside commerce before getting the first bus south through arid, hilly countryside to Gurúè.

There is usually one daily bus to Gurúè (5hr). Ask around at the Maçaniqueira market.

❻ Gurúè

Floating in a sea of tea plantations and guarded by misty Mt Namúli, Gurúè is worth the hot, ponderous slog to get there. Lodging is simple, the town itself is unremarkable, but the hiking through hillsides packed with tea and local village life is sublime.

Taxi back to Nampula (6hr) and fly to Maputo for onward connections.

ALTERNATIVES

Extension: Mt Namúli

Mozambique's second-highest peak inhabits a little-penetrated region that was all but cut off until the late 1990s by civil war. Summiting it is a long, tough day trip; you'll need a guide, transport to the start-point 12km from Gurúè, and gifts for the local Makua tribe. It's an invigorating ascent through tea plantations and pockets of biodiverse indigenous forest.

Arrange guides and transport at Gurúè's Pensão Gurúè.

Diversion: Maputo

Welcome to one of Africa's most pleasant capitals, a strollable city with Mediterranean airs, Portuguese-quality cafes and pockets of eclectic architecture. Since you're highly likely to be entering and exiting Mozambique via Maputo, a short layover in the capital is recommended. Essential sights include a fort, a national art museum and a grand railway station (which is a piece of art in its own right).

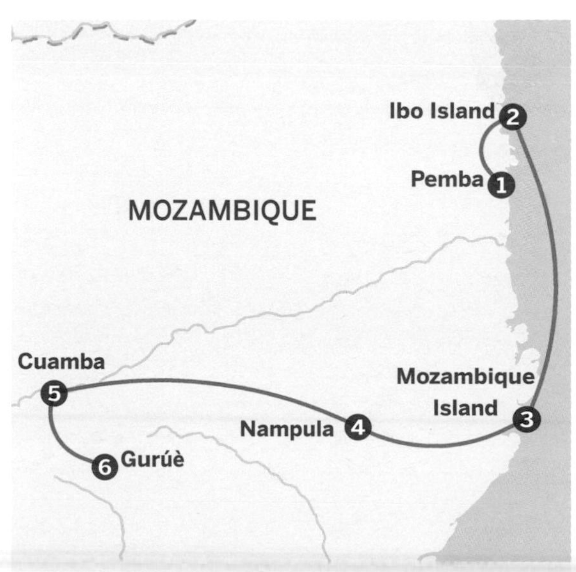

BEYOND CAPE TOWN: VINEYARDS, DESERT AND OCEAN

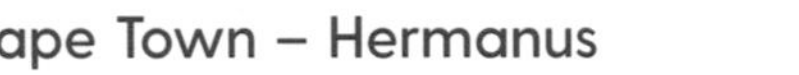

Cape Town – Hermanus

Cosmopolitan Cape Town is the gateway to a dramatic road trip through South Africa's lush vineyards, wild desert and Atlantic Ocean shores.

FACT BOX

Carbon (kg per person): 91
Distance (km): 731
Nights: 7
Budget: $$
When: Sep-May

ABOVE: SOUTHERN RIGHT WHALE, HERMANUS

❶ Cape Town

Start by taking the vibrant pulse of South Africa's Mother City at an al fresco restaurant on historic Victoria Harbour, from where boats leave to visit Mandela's Robben Island prison. Outdoor enthusiasts will want to trek up Table Mountain, while food and art lovers head to Woodstock's Neighbourgoods Market for artisan brewers, gin distillers and evocative street art. There's great accommodation in the kaleidoscope-coloured houses of Bo-Kaap, the traditional Cape Malay quarter.

Drive 95km east via Stellenbosch.

❷ Franschhoek

The countryside of the Cape Winelands is an unforgettable mix of vineyards set against dramatic mountain backdrops. The best base for exploring is the former French settlement of Franschhoek where most wine estates – Chamonix, La Motte, Haute-Cabriere – offer restaurants, picnics, kids activities, vineyard tours and tastings.

Continue 140km northeast.

❸ Aquila Private Game Reserve

The mountain landscapes disappear abruptly on entering the desolate emptiness of the semi-desert Great Karoo. In the middle of this emptiness lies a

ABOVE: CAMPS BAY OCEAN POOL, CAPE TOWN

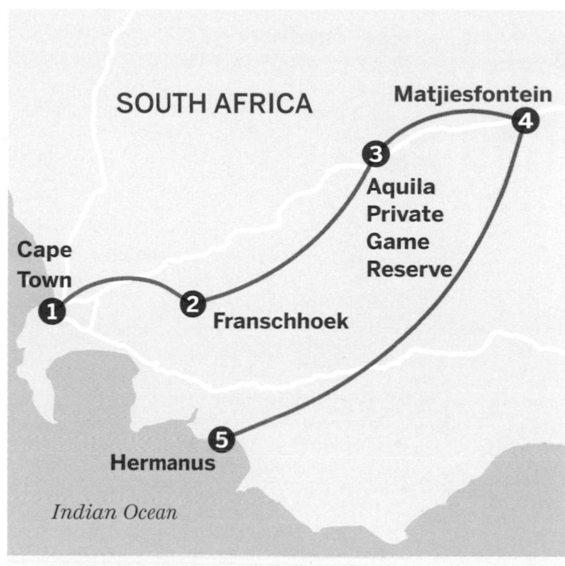

small family-friendly game reserve, offering jeep or quad excursions to get up close to zebras, giraffes, rhinos and elephants.

Pick up the N1 at Touws River and drive northeast for 66km.

4 Matjiesfontein

Settlements are rare in the arid Karoo until you reach the oasis-like Matjiesfontein, a perfectly preserved, time-warp Victorian ghost town that lives on through its splendid Lord Milner Hotel.

Head back along N1 to Touws River, then take the picturesque R318 and 326 to Hermanus (310km).

5 Hermanus

Hermanus sits on the Atlantic's Walker Bay, a natural park and marine paradise that is one of the world's prime spots for whale watching, as well as the more dangerous sport of diving with great white sharks. From its founding in the early 1800s as a commercial whaling port, Hermanus has successfully transformed into an ecotourism destination. It's also the starting point of an up-and-coming wine route into the lush Hemel-en-Aarde (Heaven and Earth) valley, producing distinctive Pinot Noir and Chardonnay.

Head 120km back to Cape Town.

ALTERNATIVES

Day trip: Paternoster

This tiny fishing village of whitewashed cottages along a sandy Atlantic beach is a favourite getaway for Cape-dwelling locals. More than a seaside resort, Paternoster has recently become a gourmet hotspot, with the 20-seater Wolfgat winning top prize in the 2019 World Restaurant Awards for its sustainable 'strandveld' cuisine, based on beach and rock-pool foraging.

The R27 road runs 160km from Cape Town up to Paternoster.

Extension: Johannesburg

A quick flight transports you to South Africa's largest city, a Rainbow Nation capital and a vibrant cultural hub. Chaotic and sometimes anarchic, but today overcoming economic and social problems, Jo'burg buzzes. After classic must-sees of a Soweto tour and the Apartheid Museum, discover local art and fashion, creative restaurants and a funky music scene.

There are around 20 nonstop daily flights from Cape Town to Johannesburg (2hr).

LESOTHO'S MOUNTAIN ROADS

Malealea Lodge – Sani Pass

Road-trip across Lesotho, crossing Southern Africa's highest passes, stopping to admire the Mountain Kingdom's peaks and staying in the country's historic trading-post lodges.

FACT BOX

Carbon (kg per person): 107
Distance (km): 865
Nights: 9-13
Budget: $$
When: Sep-Apr

❶ Malealea Lodge

Begin your journey into the mountains with one of Lesotho's best experiences, the community-run lodge in a former trading post at the remote village of Malealea. Cross the Gates of Paradise Pass (2003m), built over a century ago by the trading post's intrepid founder, and follow the sign's advice to 'pause and look upon a gateway of paradise' as the Thaba Putsoa Mountains rise before you. Enjoy hiking, trekking on the local Basotho ponies and unforgettable sunset performances by the village choir.

Drive 90km north.

❷ Roma Trading Post Lodge

This lowlands lodge is a tranquil place to rest in Roma, home of the country's only university, before another mountain ascent. Established in 1903, the lodge offers thatched rondavels (traditional round huts) and tours to some *minwane* (dinosaur footprints).

Continue 210km east to Katse.

❸ Katse Dam

God Help Me Pass (2332m) and several others climb to the Katse Dam, an engineering feat that generates hydroelectricity and supplies Johannesburg with water. With a visitor centre, guided tours, a midrange lodge and a verdant botanical garden, Katse makes a relaxing stopover.

Drive 175km north to Ts'ehlanyane.

❹ Ts'ehlanyane National Park

Return to the lowlands, stopping atop Mafika-Lisiu Pass (3090m) for views that go on forever, and follow Lesotho's northern border to the hiking and horse-riding trails in Ts'ehlanyane National

ABOVE: RONDAVEL HUT, NORTHERN LESOTHO

ABOVE: KATSE DAM

Park, an accessible chunk of the Maluti Mountains. Ts'ehlanyane is also home to Maliba Lodge, Lesotho's only five-star accommodation.

Head 120km east to Afriski.

5 Afriski Mountain Resort

One of Southern Africa's highest road passes, Mahlasela (3222m) leads to one of the continent's few ski resorts. If it's winter (June to August), the slopes await. Otherwise, there's hiking, ziplining and eating at Sky Restaurant, Africa's highest restaurant (and pizza oven) at 3010m.

Drive 140km southeast.

6 Sani Pass

You've made it across the world's highest country (Lesotho also has the highest low point, at around 1400m, in the so-called lowlands) to the top of the famous Sani Pass (2876m), which helter-skelters through the Drakensberg range to the South African border. Check into Sani Mountain Lodge and toast the mountains with an altitudinous ale in Africa's highest pub; and hike up Thabana-Ntlenyana (3482m), the continent's highest peak south of Kilimanjaro.

Head 260km east to Durban and its international airport.

ALTERNATIVES

Extension: Semonkong

Semonkong, 'Place of Smoke', takes its dramatic name from the nearby 'smoking' Maletsunyane Falls (204m). The excellent Semonkong Lodge offers the world's longest commercially operated, single-drop abseil down the misty falls, as well as pub crawls by donkey. Stop on the journey into the highlands at the clifftop Ramabanta Trading Post Lodge, dating from 1939.

It's 85km from Roma to Semonkong via Ramabanta.

Diversion: Morija

Spend a night in sleepy Morija, the site of Lesotho's first European mission. It's also home to the country's unofficial national museum, the Morija Museum & Archives, which gives the lowdown on the culture of the Basotho people. The town's treasures continue with one of Lesotho's loveliest questhouses, the stone-and-thatch Morija Guest Houses. Hiking and pony trekking are on offer.

Morija is 45km from Malealea and 50km from Roma.

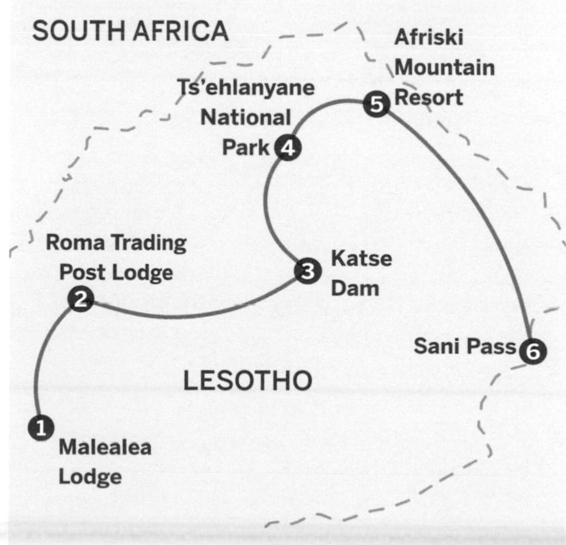

MALAWI FROM MOUNTAINS TO LAKE

Viphya Plateau – Liwonde National Park
This grand tour of the friendly 'warm heart of Africa' takes in mountains, lakeside towns, wildlife reserves, mission stations and hiking trails.

FACT BOX

Carbon (kg per person): 145
Distance (km): 1170
Nights: 8-15
Budget: $$
When: May-Aug

❶ Viphya Plateau

Stop at welcoming Luwawa Forest Lodge on the drive north from the Malawian capital, Lilongwe, for an experience of rural life at 1585m. Malawi may be famous for its lake, but Viphya is a lesser-known highlight, one of several plateaus offering cool-climate forest hikes.

Drive 120km to Mzuzu.

❷ Mzuzu

Malawi's third city, complete with a single traffic light, offers a glimpse of unhurried, everyday life on its dusty streets. There's a museum with an exhibit on your next stop and the nearby Mzuzu Coffee Den for a cup of the strong, sweet local coffee. Stay, relax and enjoy homemade pasta at Italian-owned Macondo Camp.

Continue north for 145km.

❸ Livingstonia

Organise a 4WD in lakeside Chitimba to ascend the tortuous hairpin bends known as the Gorode to the mountaintop Livingstonia mission station, established by the Free Church of Scotland in 1894. The two eco-lodges perched atop the escarpment, Mushroom Farm and Lukwe EcoCamp, and the old mission buildings, including the church with its stained-glass window of namesake David Livingstone, make this sleepy spot hard to leave.

Drive 160km south along Lake Malawi to Nkhata Bay.

❹ Nkhata Bay

Lake Malawi is the world's most biodiverse freshwater lake – and feels more like the Caribbean Sea than a lake, with its endless blue vistas and Nkhata Bay's reggae-playing lodges. Aqua Africa can take you diving to meet schools of iridescent cichlids, followed by some hammock time at the likes of Mayoka Village.

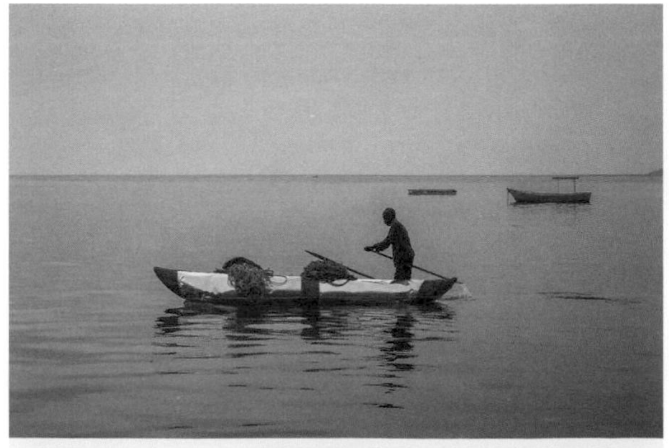

ABOVE: LAKE MALAWI AT CAPE MACLEAR

ABOVE: SHIRE RIVER ELEPHANTS, LIWONDE NATIONAL PARK

Continue 195km south.

5 Nkhotakota Wildlife Reserve

African Parks NGO has turned Malawi into an up-and-coming safari destination, and Nkhotakota benefitted from a historic translocation of 500 elephants from two southern parks. It's rough and real African bush, with some beautiful lodges alongside the croc-inhabited Bua River.

Drive 280km south to Cape Maclear.

6 Cape Maclear

Cape Mac is the original lakeside hangout, with fishing nets drying on powdery sand overlooked by beach cabanas and baobabs. Dive, kayak to Domwe or Mumbo Island for the night, visit missionary graves and clink a cold beer at a beachfront lodge.

Continue 150km south.

7 Liwonde National Park

Liwonde is one of Africa's best spots for river-based wildlife watching. The lodges offer boat and canoe trips to see the Shire River's healthy population of hippos, crocodiles and elephants.

Finish the trip 120km south in second-city Blantyre, which has onward connections.

ALTERNATIVES

Extension: Mua

Established in 1902, this red-brick Roman Catholic mission seems transposed from Tuscany. Opposite the tin-roofed church (services in Chichewa) is the mission's remarkable Kungoni Centre of Culture & Art, which preserves Chewa, Yao and Ngoni heritage. The Chamare Museum here has a striking 'tree of spirits' hung with 280 traditional Gule Wamkulu masks.

Mua is 190km from Nkhotakota, 90km from Cape Maclear.

Diversion: Mt Mulanje

On a misty day, this massif of some 20 gnarly granite peaks, rising to over 2500m, lives up to its 'Island in the Sky' local name. Hiking trails and forestry huts allow access to green valleys with rivers, waterfalls, endemic Mulanje cedars, and wildlife from klipspringers to black eagles. Mulanje is surrounded by emerald-green tea plantations, including the fascinating Satemwa estate.

The area is 155km from Liwonde, 70km from Blantyre.

EASE THROUGH EGYPT FROM THE DESERT TO THE NILE

Cairo – Abu Simbel

Egypt isn't only pyramids, temples and tombs. Veer off the normal Nile-hugging tourist route to explore the Western Desert oases beyond.

FACT BOX

Carbon (kg per person): 65
Distance (km): 2079
Nights: 14
Budget: $-$$
When: Nov-May

❶ Cairo

The country's capital is one of the best places to get your fill of Ancient Egypt at the Pyramids of Giza and Egyptian Museum. Then move forward in time to explore the city's Coptic Christian heritage, and get lost amid the Mameluke era's muddle of mosques and mausoleums.

Take an Upper Egypt bus from Cairo Gateway (Turguman) Bus Station to Bawiti, in Bahariya Oasis (2 daily; 5hr). Alternatively, microbuses leave when full from El-Monieb Bus Station in Giza (4hr).

❷ Bahariya Oasis

Bahariya Oasis is the gateway to the White Desert's lunarscape of surreal, wind-sculpted chalk-rock cliffs and spires, and the glistening, rock-pitted plateau of the Black Desert.

Take an Upper Egypt bus to Mut in Dahkla Oasis (2 daily; 4hr).

❸ Dahkla Oasis

Surrounded by dunes and sprinkled with hidden hot springs and palm-shaded villages, Dahkla is quintessential slow-paced oasis life. Don't miss the fortified mudbrick town of Al-Qasr and the painted tombs of Qarat Al-Muzawwaqa.

Take an Upper Egypt bus to Al-Kharga (2 daily; 3hr).

❹ Al-Kharga Oasis

The main town of Al-Kharga Oasis isn't a looker, but beyond stand the mural-covered tombs of Al-Bagawat's early-Christian necropolis, and a dunescape overlooked by Roman forts and Ancient Egyptian temples.

Take an Upper Egypt bus to Asyut (3 daily; 3hr 30min) then a train from Asyut Station to Luxor (8 daily; 4hr 30min).

ABOVE: NILE FELUCCAS, ASWAN

ABOVE: VALLEY OF THE KINGS, LUXOR

❺ Luxor

Temples and tombs galore await in famed Luxor. Marvel at burial-chamber art in the Valley of the Kings and gaze at giant-sized wall reliefs depicting pharaonic triumph at Medinat Habu. Afterwards, crane your neck at colossal columns in Karnak's Hypostyle Hall.

Take the train to Aswan (7 daily; 3hr 15min).

❻ Aswan

Aswan is Egypt's most laidback town, a place to kick back and toast Nile sunsets with *shisha* (water-pipe) and *karkaday* (hibiscus drink) after strolling the Nubian villages of Elephantine Island and relaxing on a felucca boat trip.

Take the bus to Abu Simbel (8am daily; 4hr).

❼ Abu Simbel

Most travellers only come on a day trip, but overnight in sleepy Abu Simbel and you get Ramses II's triumphant temples to yourself. Salute the pharaoh's colossal likeness standing sentinel, then wander column-studded halls wrapped in boastful reliefs relaying his victories.

Take the bus back to Aswan (daily; 4hr), then train to Cairo (9 daily; 13hr). Alternatively, fly to Cairo from Aswan (3 daily; 1hr 20min).

ALTERNATIVES

Extension: Dendara

Tag on an extra day in Luxor and visit Dendara, 4km southwest of Qena. Walk through the Temple of Hathor's interior of blue-tinged reliefs, glared down upon by Hathor-head column capitals, and you'll understand why Egypt's ancient gods were so revered.

Take a train from Luxor to Qena (10 daily; 1hr 15min) then a taxi to the site. Or arrange a taxi all the way from Luxor.

Diversion: Felucca to Kom Ombo from Aswan

See Egypt at its most lush and tranquil and get a close-up of Nile life on a multi-day felucca trip. Sail downriver from Aswan, passing mudbrick Nubian villages and palm-fringed banks backed by desert dunes. It's two days/one night to reach the Temple of Kom Ombo, dedicated to gods Horus and Sobek.

Felucca trips are usually arranged directly with felucca captains in Aswan.

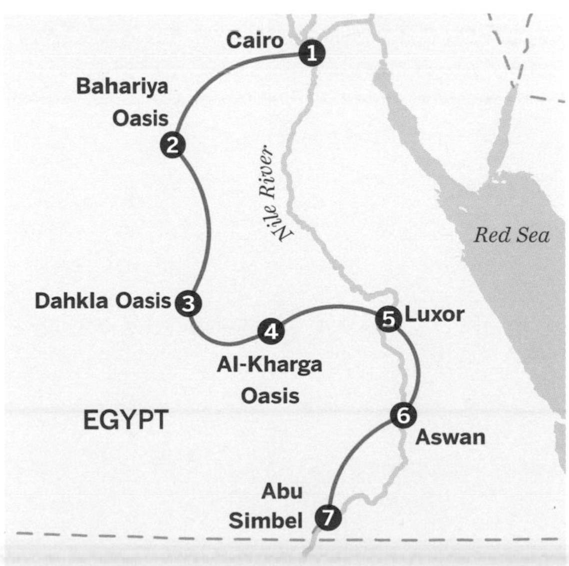

TREASURES OF ANCIENT NUBIA IN EGYPT AND SUDAN

Aswan – Meroë

Follow the Nile into Sudan through the territory of Ancient Nubia, where toppled temples and pyramids half swallowed by dunes come without the crowds.

FACT BOX

Carbon (kg per person): 47
Distance (km): 1690
Nights: 12-14
Budget: $-$$
When: Nov-Mar

❶ Aswan

Aswan sits on the border between Ancient Egypt and Ancient Nubia. Once you've got here from Cairo – either by train (9 daily; 13hr) or plane (3 daily; 1hr 20min) – visit the Nubian Museum for some historical insights, then sail by felucca to Seheyl Island where pharaonic officials chiselled inscriptions into the cliff before taking Nubia journeys.

Take a bus to Wadi Halfa in Sudan (4-6 daily Sat-Thu; 7hr including the border crossing).

ABOVE: MEROË PYRAMIDS, SUDAN

❷ Wadi Halfa

The distance isn't lengthy but the border crossing can take hours, so plan to be stuck in scruffy Wadi Halfa for the night.

Take a microbus heading to Dongola and get off at Abri (roughly hourly; 3hr).

❸ Abri

From Nile-side Abri, surrounded by fava bean fields, visit Sai Island's rubble-strewn fort and church remnants, and the half-tumbled temple ruins of atmospheric Soleb.

Microbuses leave Abri when full, heading to Kerma (2hr 45min). Alternatively, all Dongola-bound microbuses can drop you at Kerma's highway turnoff, from where it's a 4km walk into town.

❹ Kerma

This small village was the site of the Ancient Nubia's first great civilisation. Climb up the Western Deffufa, a slumping, mudbrick temple, to view the excavated remains of Ancient Kerma below.

Microbuses to Dongola leave when full (1hr). Transfer in Dongola for microbuses to Karima (regular; 2hr 45min).

ABOVE: SOLEB TEMPLE RUINS, SUDAN

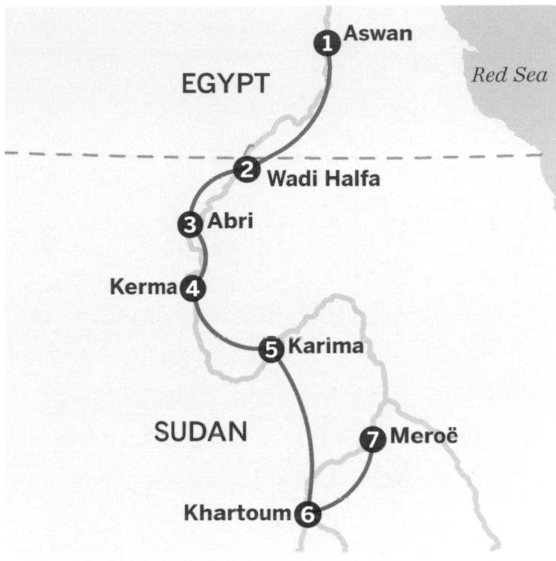

5 Karima

Scrabble up Jebel Barkal for views of Karima's pyramid field and sprawling desert beyond. Delve into painted tombs at El Kurru then beeline to Nuri, where pyramids poke out between dunes.

Take a bus to Khartoum (every 1-2hr; 6hr).

6 Khartoum

The Sudanese capital is where the waters of the White and Blue Niles meet. It's also where you can view early-Christian Faras frescoes at the National Museum (salvaged from the rising waters of Aswan's High Dam) and, if here on a Friday, take in the vivid dervish *dhikr* (Sufi ritual) at Hamed el-Nil Tomb.

Private transfers are the easiest option to Meroë (4hr) and allow stops at Naqa and Musawarat es-Suffra along the way.

7 Meroë

Last capital of Nubia's Kingdom of Kush and Sudan's greatest treasure, Meroë presents the visitor with a field of broken-tipped pyramids fighting against surrounding orange-hued sand dunes. Sunset is a magical time to be here.

Once back in Khartoum, the airport has onward international connections.

ALTERNATIVES

Extension: Abu Simbel

Ramses II's most triumphal temple was built on Ancient Egypt's border. Hewn out of a cliff, the Great Temple of Abu Simbel kept a watchful eye on Egypt's Nubian neighbours to the south. Pass under the pharaoh's colossal stone effigies to enter halls where boastful reliefs recount tales of Ramses II's might.

Hotels and travel agencies in Aswan can organise day trips to Abu Simbel.

Diversion: Naqa and Musawarat es-Suffra

Between Khartoum and Meroë, you can veer off the main highway to two Kingdom of Kush temple sites in the scrub-scattered desert plateau. Naqa's Lion Temple is wrapped in ornate reliefs of Kushite rulers and gods, while Musawarat es-Suffra is a vast complex with detailed elephant carvings.

Khartoum hotels and travel agencies can organise Meroë trips which include stops at both sites.

JOURNEY TO THE SERENE LAKES AND RIVERS OF GABON

Libreville – Port Gentil

From Libreville, pass myriad villages to Lambaréné, home of Dr Schweitzer's famous hospital. The town is set in a watery paradise teeming with hippos and birds.

FACT BOX

Carbon (kg per person): 52
Distance (km): 567
Nights: 3-4
Budget: \$-\$\$
When: Dec & Jan, Jul-Sep

❶ Libreville

Gabon's capital is a vibrant, cosmopolitan city with some brilliant bars and clubs that really get going after midnight. For an enlightening view of Gabonese culture, don't miss a guided tour at the National Museum of Arts and Traditions on the seafront.

To reach Lambaréné, take a shared taxi to the SOGATRA bus station, 8km east of Libreville, to pick up the bus (2 daily at 1 and 3pm; 4hr) that travels through rural villages and over the equator.

❷ Lambaréné

With its glossy lakes, fast-flowing rivers and thick green foliage, Lambaréné has a peaceful, enticing air, as if infused with the grace of Dr Schweitzer and his humanitarian legacy. Most of the town is located on the south bank of the Ogooué River, with a limited number of places to stay.

The Albert Schweitzer Hospital is a short walk from the centre; cross the river to the north bank.

❸ Albert Schweitzer Hospital

This famed hospital was set up in 1913 by Nobel Prize-winning Dr Schweitzer and his wife Hélène to treat leprosy patients. No one was ever turned

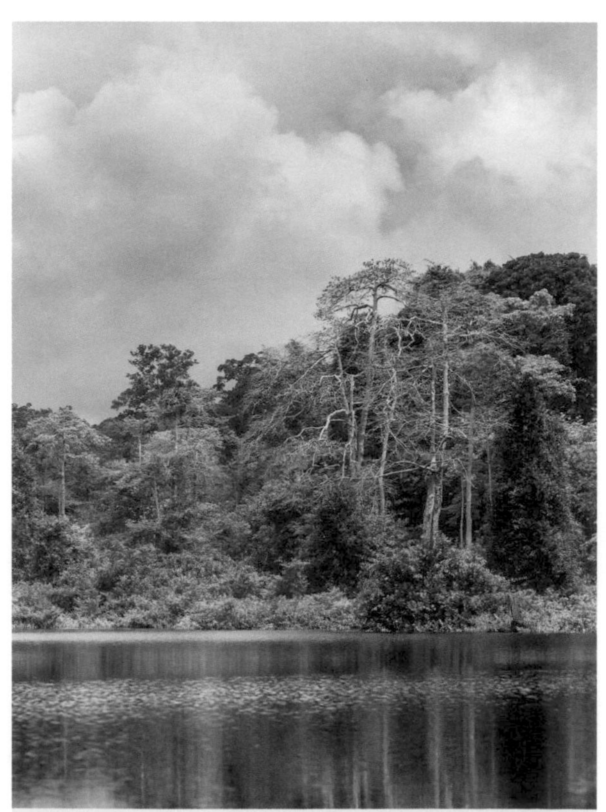

ABOVE: OGOOUÉ RIVER NEAR LAMBARÉNÉ

ABOVE: ALBERT SCHWEITZER HOSPITAL, LAMBARÉNÉ

away, whatever their religion or ability to pay. The hospital is still in operation and has expanded to treat tropical diseases such as malaria, with obstetrics and dentistry on offer too. Explore the museum in the old hospital building on a guided tour to see photos, paintings and the impeccable house, and visit the little zoo which has animals descended from the original roster of inhabitants. Dr Schweitzer, his wife and some staff members are buried in the grounds. You can stay here, too, in the atmospheric old staff quarters, now revamped.

Be sure to book your onward boat journey down the Ogooué River at the port a day or two before departure to Port Gentil. Arrive at the jetty at least an hour in advance for the trip (7am daily; 7hr 30min). Lifejackets are provided, and basic sandwiches and plenty of beer are available.

4 Port Gentil

Gabon's second city is its economic and industrial hub with gargantuan oil rigs out at sea. There's a laidback air with some pleasant hotels, a few cool places to eat and drink and a stretch of beach ideal for seaside strolling.

Take a short flight back to Libreville for onward connections (3-5 daily; 35min).

ALTERNATIVES

Extension: Lambaréné lakes

Hire a manned *pirogue* (dug-out canoe) to explore the serene river and lakes surrounded by luxuriant rainforest. You'll come across a few seine fisherfolk and perhaps hippos, but otherwise it's just water, trees and sky. On a half-day trip, you can reach Lake Zilé with its 30 islands full of birds – look out for one island populated exclusively by white cormorants. If you choose a full day's *pirogue* trip, stock up with provisions for a picnic lunch in the market or small shops at Lambaréné, and then head for remote Lake Azingo, northeast of town, where there is a larger population of hippos, crocodiles and birds. On your return to port, you'll pass the Schweitzer Hospital and glide under a bridge where thousands of bats hang out.

Hire manned *pirogues* at Lambaréné port or from Hotel Ogooué Palace.

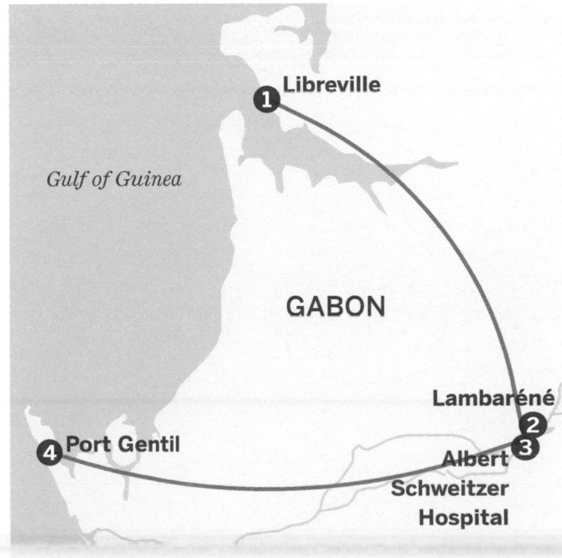

EXPLORE EQUATORIAL GUINEA'S RAIN-DRENCHED URECA

Malabo – Ureca

If remote, untamed rainforest fringing black-sand beaches where turtles lay eggs appeals to you, then Ureca on Equatorial Guinea's Bioko Island is a trip worth making.

FACT BOX

Carbon (kg per person): 27
Distance (km): 219
Nights: 3-4
Budget: $$
When: Nov-Mar

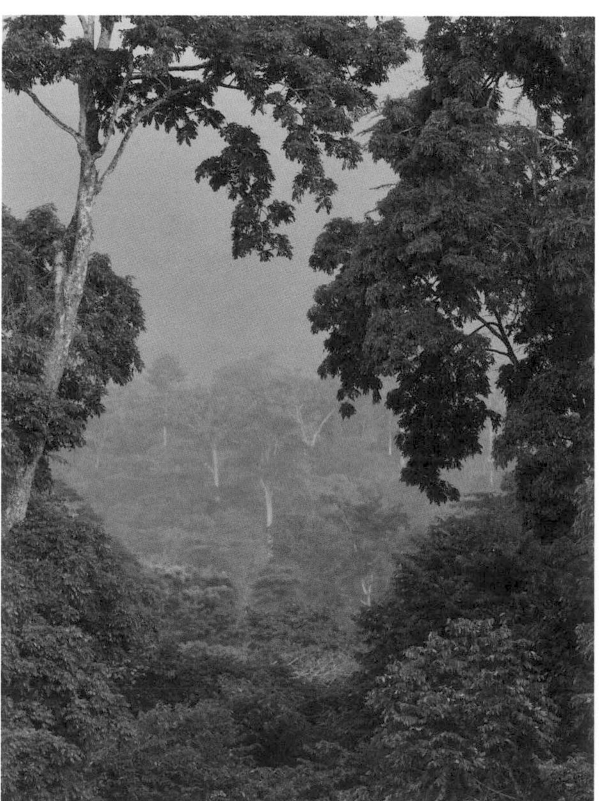

ABOVE: TROPICAL RAINFOREST NEAR LUBA

❶ Malabo

The capital of Equatorial Guinea, Malabo sits on the northern tip of Bioko Island off the coast of Cameroon. Most of the city has multi-lane highways, high-rise office blocks and upmarket hotels for oil executives. But tucked away around the port are beautiful vestiges of Spanish colonial architecture like the Gaudí-inspired Catedral de Santa Isabel.

Organise a 4WD vehicle and driver in Malabo and head 85km south along the coast to Moka.

❷ Moka

At 1380m, Moka offers welcome cooler temperatures. It's also a great place to stop off and take in the spectacular views from the mirador (viewpoint). A pretty pink church sits above the small village. Look out for stately ceiba, the national tree, on the edge of the rainforest.

Continue the drive for 30km west to Luba.

❸ Luba

The second-largest town on Bioko Island, Luba has an important port and is a good place for a fishy lunch. Mostly though, it's the jumping-off point for the trip's highlight, Ureca.

The steep road from Luba to the tiny village of

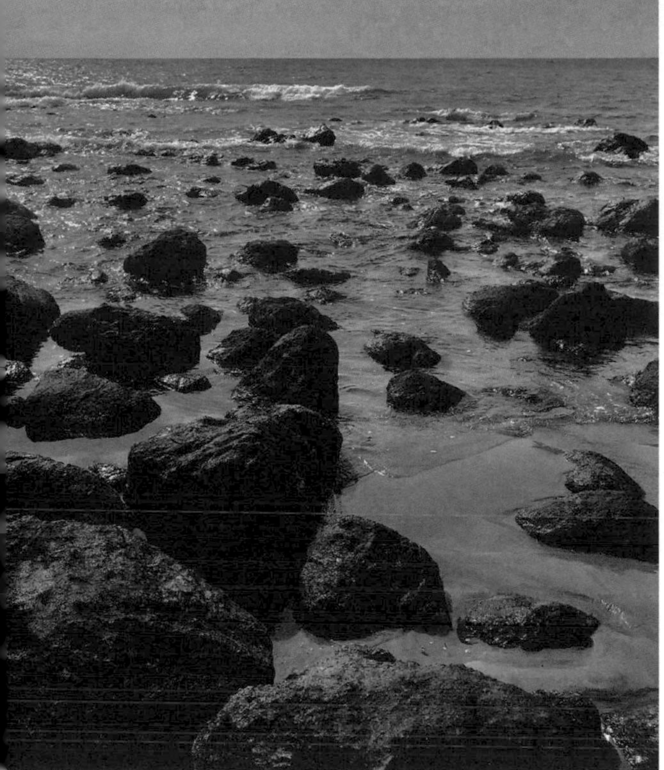

ABOVE: LUBA'S ARENA BLANCA COASTLINE

Ureca on the southern coast of Bioko Island is 26km long and will take up to an hour to drive.

4 Ureca

Ureca is one of the wettest places on earth, receiving an annual 10.45m (yes, metres) of rain: expect to get wet, even in the dry season. The Bioko Biodiversity Protection Program (BBPP), a partnership between Drexel University in Philadelphia and Malabo's University, operates a seasonal research camp, Moaba, to conserve critically endangered primates and nesting marine turtles; you can learn about the conservation work, protect biodiversity and help promote a sustainable livelihood for local people. You can also join night-hikes to remote beaches to monitor the turtles that come ashore to lay their eggs between November and January, and hike into the forest around a caldera to spot birds and various monkeys, including the highly endangered, endemic Pennant's Red Colobus. Accommodation is in comfortable cabins on the beach; meals, water and guides are provided. Be sure to check with the BBPP about what equipment to bring.

For the return journey to Malabo (78km), it's a tough drive from the coast through the rainforest, but once you get to Luba, it's a straightforward trip, passing the university and an impressive mosque.

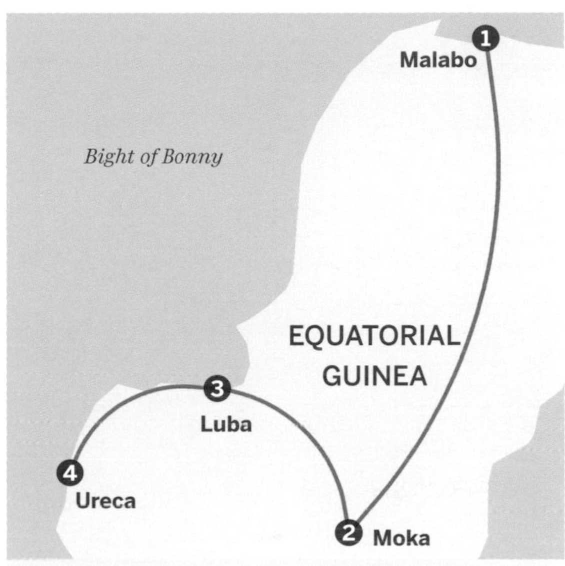

ALTERNATIVES

Extension: Pico Basilé

Equatorial Guinea's highest mountain is Pico Basilé on Bioko. The excellent road up starts 6km after Rebola village and climbs to the top at 3011m – but it's worth considering hiking part of the way. At 2800m is an impressive church with a statue of the Virgin Mary, while at the summit spectacular views reach as far as Nigeria and Mt Cameroon.

Arrange a car and driver in Malabo for the trip.

Diversion: Moka Wildlife Center hiking trails

With one day to explore Bioko Island and its animals, drive from Malabo to Moka and the BBPP's Wildlife Center for exciting hikes suitable for all fitness levels. The 1km Nature Trail is perfect for families, or try a half-day into the forests to magnificent river valleys with waterfalls. A more challenging route climbs through forests to a crater and volcanic Lago Biao.

Arrange a car and driver in Malabo to get to Moka.

MADAGASCAR'S NATIONAL PARKS AND LEMURS

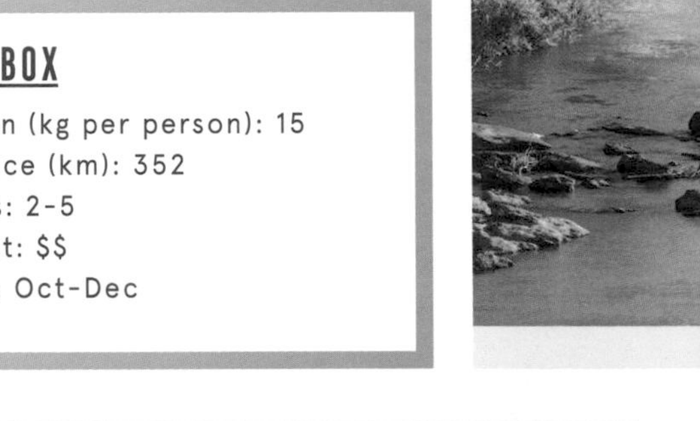

Antananarivo – Parc National Mantadia
Dense rainforest, rushing rivers and sacred waterfalls teeming with birds, frogs and lemurs are all within reach by *taxi-brousse* (bush taxi) from Antananarivo.

FACT BOX

Carbon (kg per person): 15
Distance (km): 352
Nights: 2-5
Budget: $$
When: Oct-Dec

❶ Antananarivo

Madagascar's lively capital has a traffic-congested centre surrounded by leafy hills full of traditional and colonial buildings housing excellent restaurants and shops. A visit to the Rova is a must: the royal palace perched on the highest hill has commanding views across the city.

🚕 *Taxis-brousses* heading east from Antananarivo leave from the Ampasapito bus station. The first leg of the journey is to Moramanga (2hr 45min), where you pick up another *taxi-brousse* to Andasibe (45min). Taxis leave when full, and there are departures all day.

❷ Andasibe

Andasibe is a large village surrounded by two national parks and other reserves, making it the perfect base for exploring the region.

Parc National Analamazaotra is an easy 1.7km walk north of Andasibe. Buy tickets for the Analamazaotra and Mantadia parks and hire the obligatory guide at the office in Andasibe.

❸ Parc National Analamazaotra

The park has 8 sq km of fairly flat rainforest, making for easy accessibility. Four easy-going walking trails and 60

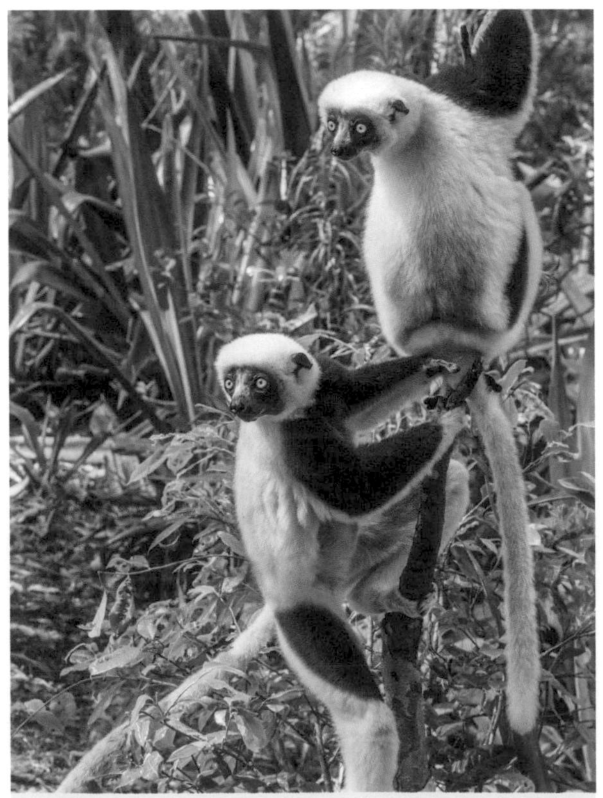

ABOVE: DIADEMED SIFAKA LEMURS, PARC NATIONAL MANTADIA

ABOVE: ANDASIBE OUTSKIRTS

resident indri lemur groups of up to four individuals mean you are highly likely to see them – and will definitely hear them. You'll also wake to their eerie calls as soon as the sun rises. Other lemurs here include black-and-white ruffed, red-fronted, woolly, grey bamboo and diademed sifakas. Look out for chameleons, too, particularly the huge, colourful Parson's chameleon.

Rent a bicycle from your hotel or take a taxi from Andasibe to the Parc National Mantadia, 17km north of the village.

4 Parc National Mantadia

More remote and less visited than Analamazaotra, this park has spectacular landscapes with magical waterfalls, rivers and rougher terrain. There is one easy circuit of 1.5km where you're likely to see indris and orchids, and you can swim in a natural pool under a waterfall. Other circuits are longer and more strenuous but highly rewarding, with possible sightings of indri, brown, woolly and bamboo lemurs and diademed sifakas. Parson's chameleons lurk, rare frogs abound and birds are prolific.

Begin your journey back to Antananarivo early with a *taxi-brousse* from Andasibe to Moramanga (departures all day; 45min). Change there for Antananarivo (departures all day; 2hr 45min).

ALTERNATIVES

Extension: Parc des Orchidées

An easy stroll along the road between Andasibe village and the National Park Office is this small park full of exquisite orchids, in flower from October to December. Over 100 species of orchids grow here, many of them epiphytes and all endemic. Hiring a guide is useful as there is no signage. This is a perfect spot to find chameleons, too.

The Parc des Orchidées is free to enter.

Diversion: Parc Mitsinjo Night Walk

Don't miss a night walk in the Parc Mitsinjo, opposite the National Park office, to see rare leaf-tailed lizards, sleeping chameleons and nocturnal lemurs, including the endearingly named hairy-eared dwarf lemur and the elusive aye-aye with its long finger for digging out grubs beneath bark.

Take a torch: the walk back to your hotel has no street lighting and is pitch black on a moonless night.

NORTHERN MOROCCO DISCOVERER

Casablanca – Moulay Idriss Zerhoun

Weave between mountains, valleys and the sea on this historic pilgrimage around the top of Morocco, taking in medieval bazaars, mosques and epic views.

FACT BOX

Carbon (kg per person): 31
Distance (km): 740
Nights: 10
Budget: $
When: Apr-Jun, Sep-Oct

❶ Casablanca

Morocco's most modern city, and the country's major international flight hub, has Moorish Art Deco architecture and a small, friendly medina. It's a good place to acclimatise before heading north. Join locals taking a sunset stroll along the seafront promenade and book a tour of the imposing Hassan II Mosque – the largest functioning mosque in Africa and the only one in Morocco that non-Muslims can enter.

Take the fast train from Casa Voyageurs station to Tanger Ville (several weekly; 2hr 10min).

ABOVE: CHEFCHAOUEN *DERB*

❷ Tangier

Less than 35km from Europe, Tangier was a mid-20th-century hangout for foreign writers, artists and political dissidents. Its whitewashed medina, recently spruced up, climbs to a crenellated kasbah overlooking the Straits of Gibraltar. Follow lively chatter to old beatnik cafes once frequented by rock stars and authors.

Book ahead for the CTM bus to Chefchaouen (daily; 3hr).

❸ Chefchaouen

Winding painted *derbs* (alleyways) lined with colourful flowerpots are a signature feature of Morocco's achingly pretty blue town. The Rif Valley backdrop and riverside setting add to the allure. Instagrammers love it, but Chefchaouen still has a sleepy feel – come here to wander and relax.

Take a CTM bus to Fez (daily; 4hr).

❹ Fez

Prepare yourself for the sensory overload of Fez's medina. It's the world's largest car-free urban area, where donkeys still lug goods down medieval streets and artisans work in cubby-hole studios beneath a sea of minarets. It's intoxicating, magical and a great place to shop. Hire a guide to explore.

ABOVE: BAB EL MANSOUR, MEKNES

Take the ONCF train to Meknes (roughly hourly; 40min).

Meknes

Fez's little brother, this former capital of Morocco is more chilled and less touristy. Stay a day to tour 17th-century relics – the giant granary of Heri Es Souani and glorious Bab El Mansour gate – from the reign of Moulay Ismail, Morocco's most fearsome, war-mongering sultan.

Take a grand taxi from Meknes' Institut Français – they leave when they're full (40min).

6 Moulay Idriss Zerhoun and Volubilis

Cascading over two green humps in the mountains north of Meknes, Moulay Idriss Zerhoun is an important pilgrimage site, and was closed to non-Muslims until 1912. You won't find endless carpet shops here; instead, you'll mix with throngs of devout pilgrims en route to the Mausoleum of Moulay Idriss (off-limits to non-Muslims). The town also promises excellent grilled *kofta* (spiced mincemeat skewers) and easy access to the important Roman ruins of Volubilis down the road.

Backtrack to Fez for onward trains or domestic and international flights.

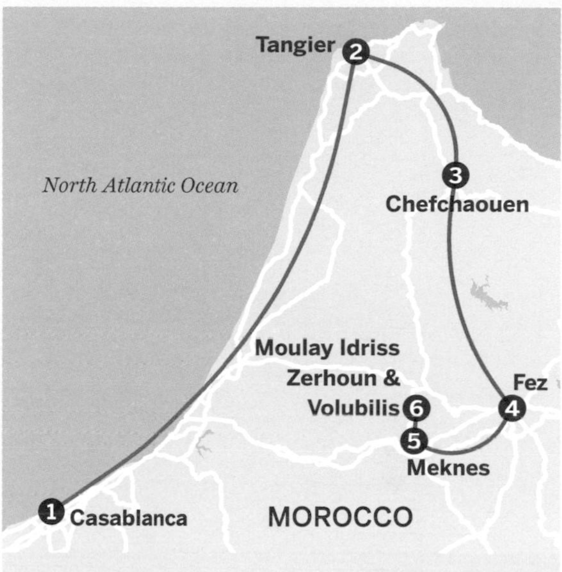

ALTERNATIVES

Extension: Tetouan

With an extra day or two, stop in little-visited Tetouan between Tangier and Chefchaouen. Tucked between the sea and the Rif Mountains, this former capital of northern Morocco's Spanish protectorate offers unique Hispano-Moorish architecture, a Unesco-listed whitewashed medina, and the big-draw Royal Artisan School of Moroccan arts and Tetouan Museum of Modern Art.

Catch a CTM bus from Tangier (frequent; 45min–1hr 30min).

Diversion: Asilah

Asilah has an artsy vibe, sea breezes, a quiet medina – and few visitors. Its white walls are splashed with colourful murals, painted afresh each year during the Asilah Festival (June/July), and the medina is crammed with small galleries. Just outside town lies a string of sandy beaches – busy with local visitors during summer, but nearly deserted at other times of year.

Trains run to Asilah from Tangier (several daily; 45min).

A CIRCUIT OF LAKE VICTORIA FROM UGANDA TO TANZANIA

Kampala – Kakamega Forest Reserve
Take a self-drive road trip around the greatest of Africa's Great Lakes, passing through four countries and some of the continent's most treasured national parks.

FACT BOX

Carbon (kg per person): 286
Distance (km): 2300
Nights: 20
Budget: $$$
When: Jan-Feb, Jun-Sep

❶ Kampala, Uganda

Kampala is one of the most likeable and walkable capitals in Africa, set over gently contoured hills to the north of Lake Victoria, not far from the source of the White Nile. To learn about modern Ugandan history, head to Mengo Palace, a former residence of Buganda kings, where you'll find subterranean prisons synonymous with dictator Idi Amin.

Allow around two days to drive the 545km south to Volcanoes National Park – Lake Mburo and Lake Bunyonyi are excellent stops en route.

❷ Volcanoes National Park, Rwanda

Set on the mountainous frontier with the DRC, the lush forests of Volcanoes National Park are the domain of golden monkeys, hyenas and bushbucks. The headline act is, of course, the mountain gorilla: join a guided trek (and pay for an expensive permit) to come face to face with a silverback.

Allow two days to drive 540km southeast, crossing the border at Rusumo Falls and reaching the Tanzanian shore – consider a stop in Kigali. Liaise with Rubondo Island National Park authorities to organise transfers from the mainland – self-drivers normally catch the ferry from Kigongo.

❸ Rubondo Island, Tanzania

One of Tanzania's most sublime (and also secret) spots, Rubondo is an island reserve on Lake Victoria that's home to elephants, giraffes and sitatunga, an amphibious antelope. Humans are comparatively

ABOVE: VOLCANOES NATIONAL PARK GORILLAS, RWANDA

ABOVE: WILDEBEEST AT SERENGETI NATIONAL PARK, TANZANIA

scarce – there's simple accommodation near the park HQ and just one luxury camp. Stays should be arranged well in advance.

Allow another two days to drive 515km east to the western edges of the Serengeti National Park.

4 Serengeti National Park, Tanzania

The national park to end all national parks, Serengeti sprawls across the vast grasslands of the Rift Valley. Its greatest spectacle is the wildebeest migration – with herds following the cycle of the rains, dodging a predatory cast of lions, leopards and hyenas as they go.

It's a full day's journey to drive the 415km north to Kakamega Forest National Park in Kenya.

5 Kakamega Forest National Reserve, Kenya

Kakamega counts as one of the last pockets of a mighty rainforest that once extended across East Africa. Primates resident include colobus, de Brazza's and Sykes monkeys – scour the canopy as you hike the half-day Isiukhu Trail.

It's 285km back to Kampala, or roughly the same to Nairobi to the east. Both have airports with international connections.

ALTERNATIVES

Extension: Ssese Islands, Uganda

A serene archipelago in Lake Victoria's northern reaches, the Ssese Islands' key appeal is as a place to detach from the modern world, treading white-sand beaches and lazing in hammocks by day, spending nights at lakeside bonfires under starlit skies.

Buggala Island is the main hub, reached by a ferry (daily; 3hr 30min) from the port of Entebbe on the mainland, not far from Kampala.

Diversion: Akagera National Park, Rwanda

Akagera National Park is a rare conservation success story. This region of wetlands and savannah was ravaged during the turmoil of 1990s Rwanda; three decades on, wildlife has bounced back, with lions and black rhinos reintroduced, along with resident populations of leopard, buffalo and elephant.

From the capital, Kigali, the park is 105km east.

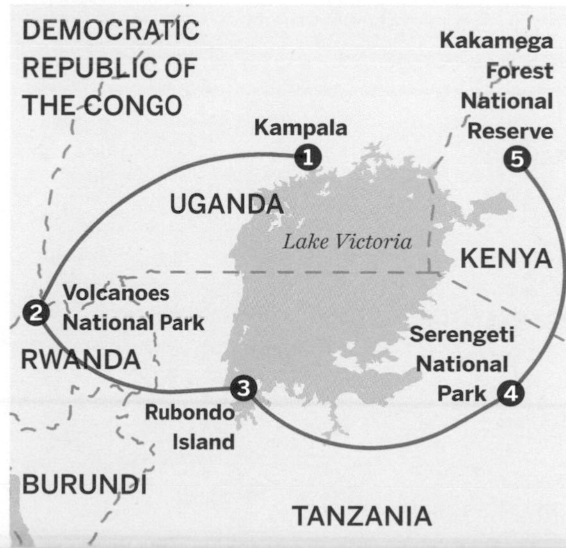

ANCIENT TUNISIA EXPLORER

Tunis – Mahdia

Far away from the busy beach resorts, see the historical side to this North African country, travelling between ancient cities by train, bus and taxi.

FACT BOX

Carbon (kg per person): 26
Distance (km): 420
Nights: 12-14
Budget: $
When: year-round

❶ Tunis

Tunis has onion-layers of history from the twisting alleyways of its medina to the artists' haunt of Sidi Bou Said. Its oldest incarnation, however, is Carthage, where you can stroll the atmospheric ruins of Hannibal's capital, razed to the ground by invading Romans in 146 BCE.

ABOVE: BIZERTE HARBOUR

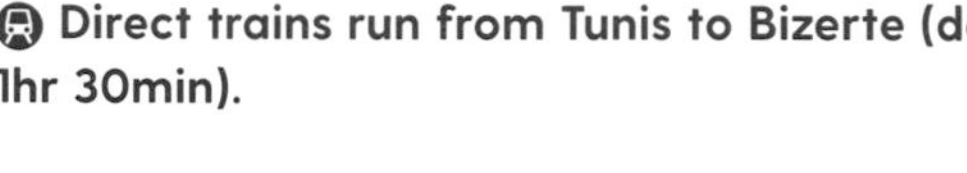

Direct trains run from Tunis to Bizerte (daily; 1hr 30min).

❷ Bizerte

A quieter counterpoint to its southern neighbour Tunis, Bizerte is a place where old mosques and storied ramparts huddle around a stone harbour. Idle away a day or two at its waterfront cafes, watching fishing boats puttering home with the morning's catch.

Return to Tunis and catch a bus to Teboursouk (11 daily; 2hr), from where it's a 10min taxi ride to Dougga.

❸ Dougga

The best-preserved Roman town in North Africa, Dougga's tumbledown temples and grand bathhouses are strewn over an Arcadian landscape of rolling hills and olive groves. The highlight is standing amidst soaring columns in the evocatively named Square of the Winds.

Retrace your steps to Tunis and catch a bus on to Kairouan (hourly; 2hr).

❹ Kairouan

Kairouan is cited as Islam's fourth holiest city, with its Great Mosque, founded in the 7th century, its

ABOVE: EL JEM COLOSSEUM

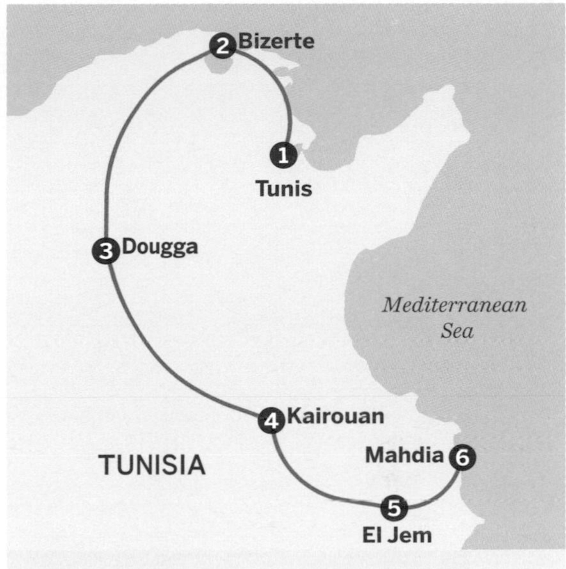

centrepiece. Non-Muslims can peek into the prayer hall from the courtyard, or join pilgrims exploring the thronging bazaars, backstreets and *babs* (gateways) of the medina.

From Kairouan, catch a taxi to Sousse (1hr), then a train on to El Jem (7 daily; 1hr).

5 El Jem

The colosseum of El Jem is almost as impressive as its sibling in Rome. In later life it doubled as a fortress and a marketplace, but despite the years and uses it still looks as if gladiators have only just departed. Channel your inner Russell Crowe exploring its lion pits.

From El Jem, take a taxi to Mahdia (1hr).

6 Mahdia

Perhaps the least touristy town on Tunisia's central coast, Mahdia is a fine place to unwind after a journey among ancient empires, with whitewashed houses perched on a breezy peninsula and sandy beaches overlooked by stout Ottoman battlements. It's also the centre of a rich weaving culture – drop in on artisans at work.

From Mahdia, direct trains run to the airport at Monastir (hourly; 1hr), which has European connections.

ALTERNATIVES

Extension: Tozeur and the Jerid

Tunisia's diverse landscapes include Med beaches and Saharan dunes, but its wildest corner is the Jerid, a territory of oasis towns, lunar-like salt flats and film locations from the original *Star Wars* movies. Tozeur is a good base from which to arrange excursions to shady canyons and hidden waterfalls, as well as the places that doubled as the planet of Tatooine.

You can fly to Tozeur from Tunis (4 weekly; 1hr).

Diversion: Djerba

Djerba feels like a tropical island that has mistakenly wandered off the Tunisian coast, with tempting white-sand beaches and turquoise tides. Step inland and you'll find a deeply distinctive culture, with one of North Africa's last remaining Jewish communities. Visit the El Ghriba synagogue – the oldest on the continent, adorned with wonderfully luminous blue tiles.

From Mahdia, take a taxi to Sfax (1hr 30min), then a bus to Djerba (daily; 5hr).

KWAZULU-NATAL ROAD TRIP

Durban – Royal Natal National Park
Wildlife, beaches, battlefields, mountain hikes, Zulu culture – pack in the best of South Africa on a road trip through its easternmost province.

FACT BOX
Carbon (kg per person): 123
Distance (km): 989
Nights: 7-10
Budget: $$
When: Apr-May, Sep-Oct

❶ Durban

The northeastern patch of KwaZulu-Natal, sometimes dubbed the Elephant Coast, offers superb wildlife-watching, historic battlefields and the Drakensberg Mountains. Begin in Durban, South Africa's third-biggest city. It's really just a place to pick up your car, though the international airport sits alongside a series of fabulous beaches, with great surf and opportunities for tasting Durban's signature bunny chow – curry, typically mutton or lamb, served in a hollowed-out white-bread loaf.

Rent a car at the airport and drive 210km north.

❷ St Lucia and iSimangaliso Wetland Park

Friendly St Lucia is gateway to the 'land of wonder', as the Zulu name iSimangaliso translates. A vast mosaic of lakes, estuaries, forest, savannah and coast, this wetland park's animal highlights include the Big Five – lion, leopard, elephant, rhino and buffalo – plus hundreds of bird species, zebra and plentiful hippos. And on top of all that, between September and March, thousands of humpback whales congregate offshore.

Continue your drive 73km west to Hluhluwe-iMfolozi Park.

❸ Hluhluwe-iMfolozi Park

One of Africa's oldest protected areas is known for the flagship conservation project that rescued the southern white rhino from the brink of extinction. Self-drive safaris yield plentiful game sightings of elephant, buffalo, zebra and black and white rhino. But best are guided walking trails through the iMfolozi wilderness area – truly immersive, back-to-basics experiences shepherded by expert guides and scouts.

Drive 170km southwest to reach Isandlwana/Rorke's Drift.

ABOVE: BUNNY CHOW, A DURBAN SPECIALITY

ABOVE: AFRICAN ELEPHANT, HLUHLUWE-IMFOLOZI PARK

❹ Isandlwana/Rorke's Drift

Dozens of battle sites, fortifications, museums and memorials scattered across central KwaZulu-Natal provide reminders of various conflicts that ravaged the region during the 19th and early 20th centuries. Among the most commonly visited and atmospheric sites are Isandlwana and Rorke's Drift, where Zulu warriors and British soldiers clashed in November 1879.

Continue the drive 224km west to Royal Natal National Park.

❺ Royal Natal National Park

The sheer escarpments of the Drakensberg Mountains certainly have something of the dragon about them (their name comes from the Afrikaans for 'dragon mountains'), and the stretch encompassed by Royal Natal National Park is among the most dramatic. With jaw-dropping views of the rock wall known as the Amphitheatre and access to superb hiking along the Tugela Gorge, Thendele Camp must be one of the most scenic stays on the planet.

Drive 312km back to Durban's international airport, or 350km north to Johannesburg for onward connections.

ALTERNATIVES

Extension: Sodwana Bay

At the northern end of iSimangaliso, shipwreck-dotted Sodwana Bay is one of the world's great dive spots. Africa's southernmost coral reef, it's home to a dizzying 1200 fish species and sees seasonal visits from whale sharks, humpbacks and leatherback and loggerhead turtles. There's also great surfing and kiteboarding, game fishing and fine sandy beaches.

Sodwana is 161km north of St Lucia.

Diversion: Giant's Castle Game Reserve

Southern Africa's highest range, the Drakensberg Mountains form a jagged arc around Lesotho's eastern border, with plentiful hiking across several protected areas. Giant's Castle mixes terrific trekking with wildlife watching – notably lammergeier (bearded vulture) and eland antelope – plus intriguing ancient San rock-art sites.

Giant's Castle Camp is a 154km drive southeast from Royal Natal NP.

ISLAND-HOPPING IN CABO VERDE

São Vicente – Fogo

Take in lush valleys, soaring peaks and untouched beaches on a ferry trip around the islands of Cabo Verde, sitting pretty off the coast of West Africa.

FACT BOX

Carbon (kg per person): 33
Distance (km): 440
Nights: 14
Budget: $$
When: Oct-Jun

❶ São Vicente

In the northern reaches of the Cabo Verde archipelago lies the island of São Vicente. It has attractive beaches at Calhau to the south and Baía das Gatas in the east, and there's also a fine lookout over the island atop the 750m peak of Monte Verde. São Vicente's biggest draw, however, is the pretty harbourfront town of Mindelo. Set with cobblestone streets and colourful colonial-era architecture, this place is well known for its live music scene (the famed late singer Cesária Évora was born here), and you can catch the melodic strains of traditional *morna* all across town.

Take the CV Interilhas ferry from Mindelo to Porto Novo on Santo Antão (daily; 2hr).

❷ Santo Antão

Home to Cabo Verde's most dramatic scenery, Santo Antão has soaring peaks, greenery-filled canyons and mist-covered valleys. You'll find stunning walks here: a classic route follows the coastal track from Ponta do Sol to the hidden cliffside town of Fontainhas, taking you up to craggy heights with sweeping views over the Atlantic. Good bases for island exploring include the fertile valley settlement of Paúl and the seaside town of Ponta do Sol.

Catch the ferry back to Mindelo, and then fly to Praia on Santiago (3 weekly; 55min).

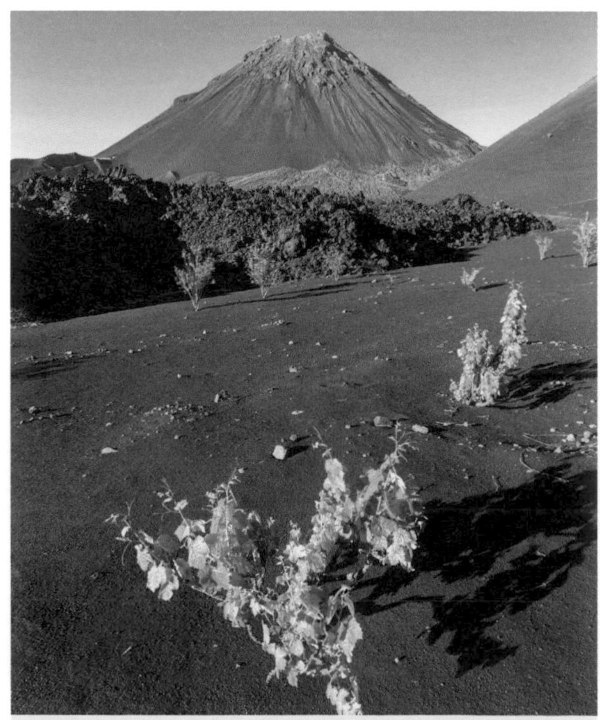

ABOVE: VINES IN THE SHADOW OF PICO DO FOGO

ABOVE: FONTAINHAS, SANTO ANTÃO ISLAND

❸ Santiago

The largest island of Cabo Verde has a little of everything: sandy beaches, desert plains, verdant valleys and a mountainous interior. The buzzing capital city Praia is a fine launchpad for a day trip to Cidade Velha, a Unesco World Heritage Site that contains the ruins of a 16th-century fort and 17th-century cathedral. On the north side of the island, tranquil Tarrafal has a pretty beach and relaxing B&Bs, with good walks into the surrounding countryside.

CV Interilhas ferries sail to São Filipe on Fogo (several times weekly; 6hr).

❹ Fogo

Aptly named Fogo (fire) is home to the archipelago's only active volcano, an eponymous, perfectly conical peak (2829m) visible from across the island. The charming town of São Filipe has flower-draped plazas and a black-sand beach. Beneath the smouldering volcano, the village of Chã das Caldeiras has a small but vibrant community (catch nightly singalongs at the village store), plus half-buried houses (destroyed in a 2015 eruption) and nearby vineyards.

Fly back to Praia where the airport has onward international connections.

ALTERNATIVES

Extension: Brava

In the archipelago's southwest corner, tiny Brava's dramatic coast and mountain-filled hinterland make it a delight for walking or just soaking up the beauty of the shoreline. The quaint town of Vila Nova Sintra in the middle of the island is a good starting point for walks, though for sheer beauty, the hillside-backed coastal village of Fajã de Água is unmatched.

CV Interilhas ferries sail between Fogo and Brava (3 weekly; 50min).

Diversion: Maio

Glittering like a white crystal in a sea of turquoise, Maio is an island of squeaky-clean beaches, an acacia-dotted interior and a string of 13 villages. The relaxed pace makes it ideal for unwinding after days of travel; you can find a few ocean-facing B&Bs and seafood restaurants perched on the beach.

CV Interilhas ferries link Praia with Maio (3 weekly; 2hr 30min).

THE BEST OF SENEGAL

Dakar – Cap Skirring

One of West Africa's most dynamic destinations has a captivating blend of cultures amid striking tropical landscapes, easily visited by ferry and bus.

FACT BOX

Carbon (kg per person): 58
Distance (km): 1050
Nights: 12
Budget: $$
When: Nov-May

❶ Dakar

Senegal's capital is a dizzying whirl of vibrant markets, dynamic nightlife and fascinating neighbourhoods spread across Africa's westernmost peninsula. There's much to discover, including perusing cultural treasures at the IFAN Museum of African Arts along with surfing and beachcombing on the nearby island getaway of Île de N'Gor. Get the lay of the land from the Phare des Mamelles, a 19th-century lighthouse which also doubles as a restaurant and live music spot, with excellent views out across the Atlantic.

Take a ferry from Dakar's passenger port to Île de Gorée (every 1-2hr; 20min).

ABOVE: ÎLE DE GORÉE

❷ Île de Gorée

The beauty of this peaceful, vehicle-free island is obvious: narrow lanes wind past colonial-era brick buildings and wrought-iron balconies festooned with bougainvillea. But Gorée has a ghastly history, with several key sites serving as a memorial to the island's part in the horrific transatlantic slave trade; learn about the miseries of the past at the 18th-century Maison des Esclaves (Slave House). Today, Gorée is home to a vibrant artists' community; you can visit studios dotted around the island.

Take the ferry back to Dakar and catch a Dakar Dem Dik bus from Gare Liberté 5 to Saint-Louis (frequent; 5hr).

❸ Saint-Louis

A Unesco World Heritage Site, Saint-Louis is an historic river-island town of crumbling colonial buildings, horse-drawn carts and peaceful ambience. Intriguing

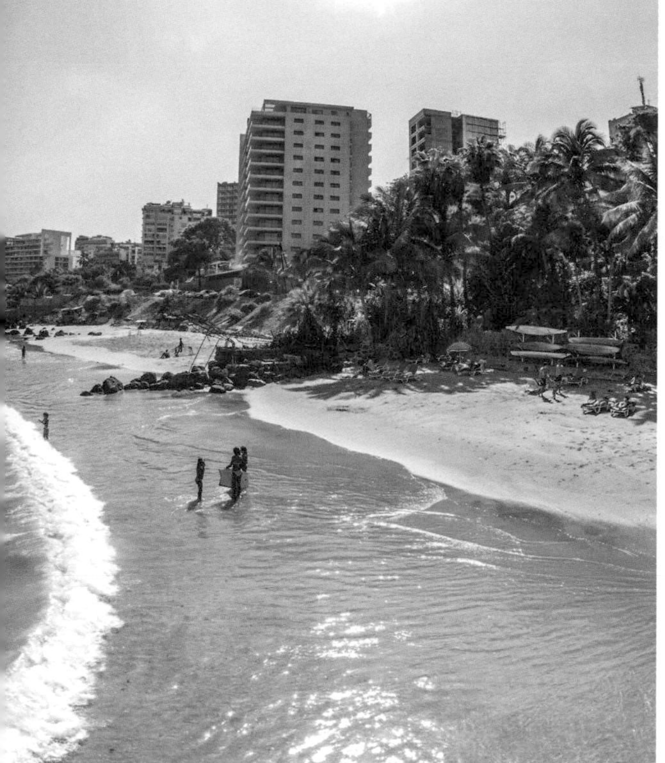

ABOVE: DAKAR

architecture abounds, such as the 19th-century cast-iron Pont Faidherbe linking the island to the mainland. There's also great birdwatching nearby at the Langue de Barbarie National Park.

Return to Dakar by bus and catch an overnight ferry to Ziguinchor (4 weekly; 15hr).

4 Ziguinchor

Ziguinchor is a former colonial centre of tree-lined streets and busy markets, with tiny mangrove-fringed river communities just beyond the city limits. The town is known for its festivals and you can catch live performances, art exhibitions and other events year-round at the Alliance Franco-Sénégalaise.

Take a minibus west to Cap Skirring from the Gare Routière, 1km east of the centre (frequent; 2hr).

5 Cap Skirring

Home to some of West Africa's finest beaches, Cap Skirring makes a great final stop on your Senegal journey. You can unwind on the sands, explore traditional Diola villages to the east and indulge in good restaurants and nightlife.

Cap Skirring's small airport has direct flights to Dakar for onward international connections.

ALTERNATIVES

Extension: Lake Retba

Some 30km northeast of Dakar, Lake Retba (aka Lac Rose) has a distinctly Martian aspect. Its flamingo-pink hue results from the cyanobacteria that feed on its high salt content; the high dunes of salt ringing the shore add to the otherworldly setting. Lakeside guesthouses can arrange boat trips and horse rides.

Dakar Dem Dik bus 11 goes from Terminus Lat Dior to Keur Massar; from there take DDD bus 311 to the lake.

Diversion: The Gambia

Africa's smallest country, the Gambia is entirely surrounded by Senegal and has a photogenic coastline of golden beaches backed by fishing villages and verdant coastal reserves. After taking in the beaches of Kololi, head inland to the River Gambia National Park, home to over 600 bird species, plus manatees, hippos, crocodiles and primates.

Direct flights connect Dakar and Banjul, the Gambian capital (2 daily; 40 min).

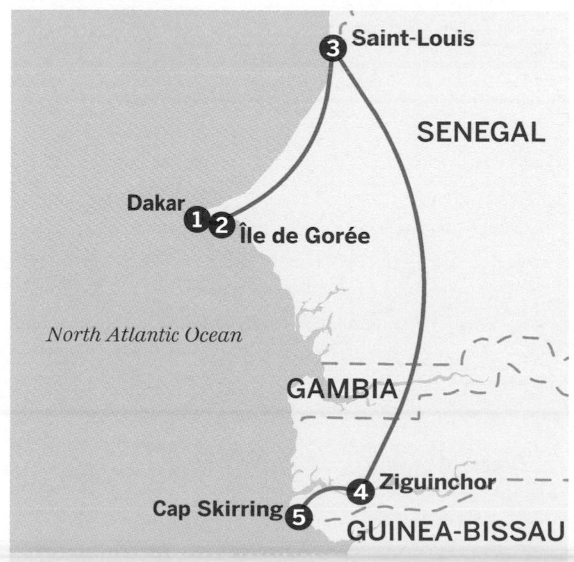

NAMIBIA CLASSIC ROAD TRIP

Windhoek – Etosha National Park

Take a wild, wildlife-filled road trip, making a classic loop around the highlights of this southwest African country while steering away from the crowds.

FACT BOX

Carbon (kg per person): 232
Distance (km): 1860
Nights: 14
Budget: $$
When: Apr-Oct

❶ Windhoek

Most travellers arrive in Namibia's German-flavoured capital on flights via Johannesburg. Rent a car at the airport (though roads are largely gravel, a 4WD isn't necessary), rest up then move on – your time is better spent elsewhere.

Drive 300km southwest into the Namib-Naukluft National Park.

❷ Namib-Naukluft National Park

Rugged mountains meet the ancient ripples of the Namib Desert and the wild Atlantic Ocean in Namibia's west. The headline act in the Namib-Naukluft National Park is Sossusvlei, where apricot-coloured dunes tower over cracked white clay pans, but the park offers much more besides: amazing hiking, biking and horse-riding across the Naukluft escarpment; farm stays and luxury lodges; star-gazing that's out of this world. There are numerous accommodation options in the area around the settlement of Sesriem.

Drive 350km northwest to Swakopmund.

❸ Swakopmund

Bavaria-by-sea, in southern Africa – Swakopmund is a delightful oddity, dotted with timbered Germanic architecture. It's also the country's unofficial adventure capital. Options here include mountain- and quad-biking, 4WD-ing to the Skeleton Coast and sandboarding down the dunes. Boat and kayak trips from nearby Walvis Bay might encounter dolphins, Cape fur seals, flamingos, even whales. Be sure to pick up a few German pastries from one of Swakopmund's sweet-smelling *bäckerei* (bakeries) too.

Continue 350km north into Damaraland.

ABOVE: SOSSUSVLEI, NAMIB-NAUKLUFT NATIONAL PARK

ABOVE: ETOSHA NATIONAL PARK

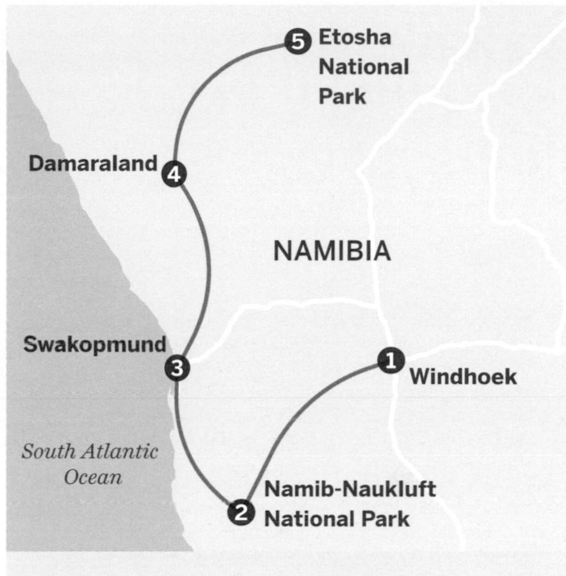

4 Damaraland

With stiff competition, this swathe of semi-desert, rolling hills and rocky outcrops might be Namibia's most outstanding wilderness. The whole region, though barren in appearance, is teeming with life, while Twyfelfontein's treasure-trove of ancient rock art is Unesco-listed. Remote lodges here (some owned in partnership with local Damara communities) offer safaris to spot desert-adapted elephants, gemsbok, greater kudu and eland, plus insight into unusual plants and the chance to track black rhino on foot.

Drive 390km northeast to Etosha National Park.

5 Etosha National Park

This national park, centred on a vast, blinding-white salt pan, is Namibia's Kruger. An expansive reserve, it's ideal for self-drive safaris and is stuffed to the gills with wildlife. Among the species drawn to its waterholes, thorn-bush savannah and mopane woodland are black rhinos, large elephant herds, plentiful lions, leopards and cheetahs, innumerable antelopes, plus giraffes, zebras and 340 species of birds. Camps within the park have floodlit waterholes so you can continue spotting after dark.

Round off the trip with a 470km drive south back to Windhoek for onward connections.

ALTERNATIVES

Extension: Fish River Canyon and Luderitz

For a fuller Namibia circuit, drive south from Windhoek, via Keetmanshoop's quiver tree forests, to Fish River Canyon, the world's second-largest. Then veer west to time-warp Luderitz to see penguins on the coast and a sand-swallowed ghost town just inland.

Keetmanshoop is 500km south of Windhoek; Ais-Ais (for Fish River Canyon) is a further 200km south, Luderitz 340km west.

Diversion: Waterberg Plateau

Rather than rushing from Etosha back to Windhoek, pause mid-way on the Waterberg Plateau, a striking upland of rich-red sandstone cliffs, deep ravines, open grasslands and mixed forest. It's excellent for hiking, is dotted with San art and harbours numerous species, from rare Cape vulture to leopard and cheetah.

Waterberg Plateau is around 280km south of Etosha NP on the way to Windhoek.

THE BACKROADS OF MOROCCO'S HIGH ATLAS

Azilal – Demnate

Take a deep drive into stunning landscapes and traditional cultures on a remote road trip in Morocco's interior to find high adventure in the High Atlas.

FACT BOX

Carbon (kg total): 48
Distance (km): 387
Nights: 6–8
Budget: $$
When: Jun–Oct

❶ Azilal

Time your trip to start on a Thursday so you can visit Azilal to coincide with one of central Morocco's less-touristed traditional weekly markets.

Drive 98km on progressively rougher roads along the R304 to panoramic views of Bin el Ouidane Reservoir, turning east onto the R306 and south to the R302 into Reserve Naturelle de Tamga.

ABOVE: DEMNATE CERAMICS

❷ Cathédrale des Rochers

Park the car and hit the trails – hiking routes wind throughout the Reserve Naturelle de Tamga, including around and up the Cathédrale des Rochers massif that defines the horizon. Make sure to stop in at the impressive local botanical gardens.

Continue 45km south to Zaouiat Ahansal.

❸ Zaouiat Ahansal

Traditional High Atlas architecture is better preserved here than anywhere else in the region, thanks to the restoration efforts of the Atlas Cultural Foundation. Zaouiat Ahansal's four distinct villages boast towering 17th-century *igherm* (traditional communal granaries) and 18th-century religious structures connected to the local saint – all of it surrounded by awesome mountain landscapes.

Follow the R302 36km to a marked turn-off, continuing on the dirt road for 26km from there to Ait Bougmez.

❹ Ait Bougmez

The 'Happy Valley' will put a smile on the face of travellers keen to explore rural villages on easy walks; alternatively, head into the M'Goun Mountains for more adventurous treks.

ABOVE: AIT BOUGMEZ

Continue on the valley road once it rejoins the R302, a total of 18km west to Ait Bououli.

5 Ait Bououli

Along the High Atlas' most rugged roads, the cliffside villages and ochre-shaded escarpments of Ait Bououli testify to both the beauty of the terrain and difficulty of living in these precipitous valleys.

Continue 57km northwest on the R302 through the hilltop village of Ait Blel to Imi n'Ifri.

6 Imi n'Ifri

Walk the short path down through Imi n'Ifri's natural travertine bridge into a narrow gorge, stopping for a quick dip in natural mineral springs popular with local families.

Drive 7km north to Demnate.

7 Demnate

Pass through Demnate on a Sunday and you'll have the opportunity to buy the town's famous olives, almonds and honey at the weekly bazaar. For pottery lovers, any day is great to visit nearby Boughlou village to shop directly with local artisans.

Drive 100km west to Marrakesh for onward international connections.

ALTERNATIVES

Extension: Taghia village

Below the distinctive peak of Jebel Oujdad (2695m), climbers have long made bucolic Taghia a base for scaling the surrounding cliffs. The village is only just becoming popular among tourists interested in local culture and day-hikes – get there before everyone cottons on to its appeal.

From Aguddim village in Zaouiat Ahansal, follow the 2km dirt road past Tighanimin to the turnoff for the 5.5km hike to Taghia.

Diversion: Cascades d'Ouzoud

Among the most popular natural attractions in central Morocco, the 110m-high multi-tiered waterfalls at Ouzoud deserve every bit of hype – as do the local troops of Barbary macaques. Most visitors come on a day trip from Marrakesh, but stick around until the quieter late afternoon, or overnight here and enjoy the morning hours nearly free from crowds.

The cascades are 28km from Azilal.

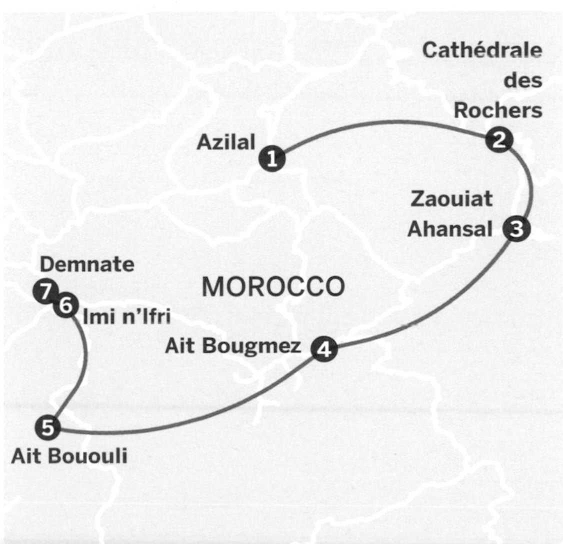

MAURITIUS BEYOND THE RESORTS

Mahébourg – Grand River South East

Take a beach-studded road trip right around the coast of Mauritius, eschewing the island's resorts for a more immersive travel experience.

FACT BOX

Carbon (kg per person): 28
Distance (km): 223
Nights: 7
Budget: $$
When: May-Dec

❶ Mahébourg

Most visitors step off the plane at the island's international airport and into a resort shuttle, but the beauty of picking up a rental car instead means you can get your first taste of an authentic Mauritian fishing village just a 15min drive to the east in Mahébourg. Recover from your flight at a waterfront guesthouse and explore the town's Monday market and natural history museum before hitting the road.

It's a 60km drive west from Mahébourg to La Gaulette.

❷ La Gaulette

It's not a long drive to the west coast seaside town of La Gaulette, but you'll want to take all day to do it, whichever route you take. Drive the coast road and stop for a swim at the string of idyllic beaches you pass along the way, or head north on the B89 from Surinam to check out the lookouts and hiking trails of Black River Gorges National Park, and the unusual 'seven-coloured earth' of Chamarel. When you arrive in relaxed La Gaulette, sign up for a kitesurfing lesson (or go for another swim) at nearby Le Morne's dazzling turquoise lagoon, or take a boat to nearby Île aux Bénitiers for a beach day.

Drive 71km north to Trou aux Biches.

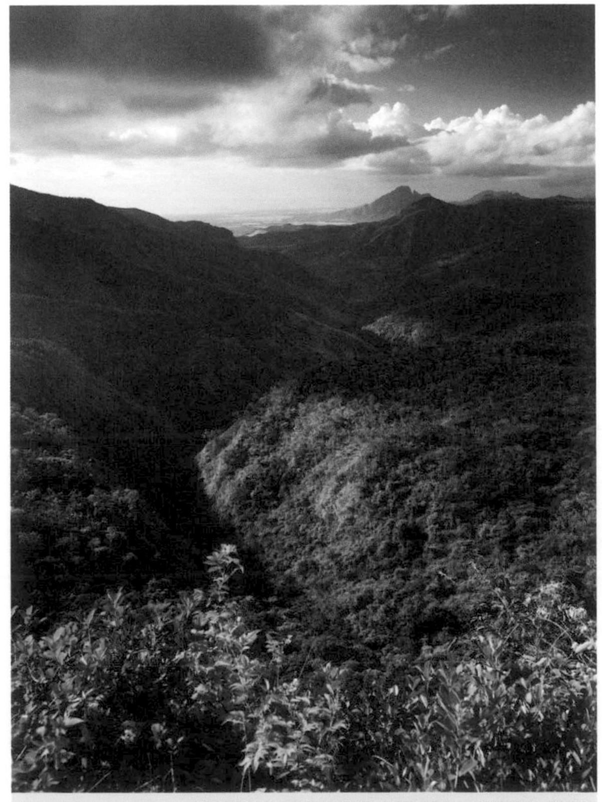

ABOVE: BLACK RIVER GORGES NATIONAL PARK

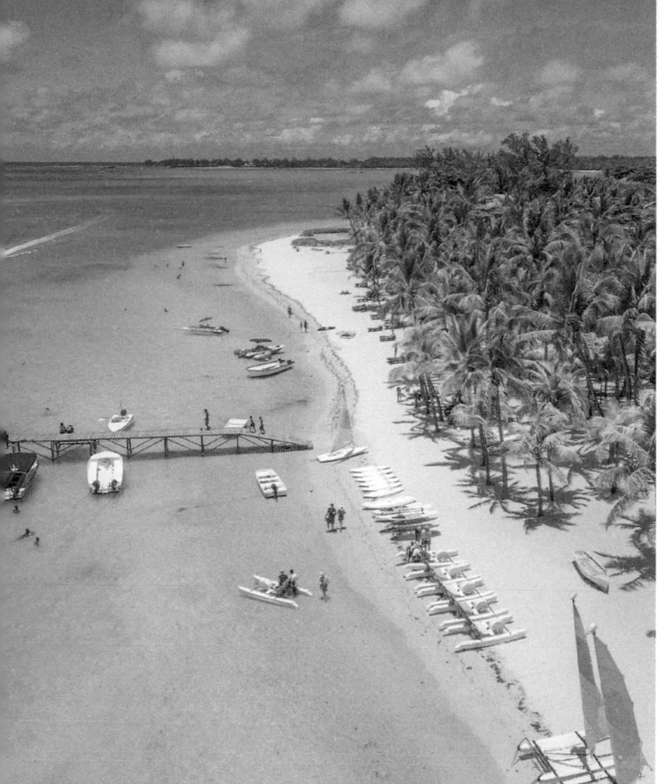

ABOVE: TROU AUX BICHES BEACH

3 Trou aux Biches

Home to one of the island's loveliest beaches, as well as some great street food, low-key Trou aux Biches is a top base for exploring the northwest coast, and for making a day trip by local bus to the capital, Port Louis. On your drive over from La Gaulette, avoid Port Louis' notorious traffic by taking the inland road to Trou aux Biches, taking a right off the A3 at Cascavelle.

Drive 60km southeast to Grand River South East.

4 Grand River South East

You'll want to keep your swimwear close at hand for the spectacular drive along the island's east coast. Pause in sleepy Poudre d'Or to watch local fisherfolk hauling in their catches before making the most difficult decision of your day – which beach to spend it on (try Belle Mare or Palmar). Once in Grand River South East, stay at Otentic, the island's only glamping experience, and sign up for the free return boat trip to nearby Île aux Cerfs, which gets you back to Otentic in time for a delicious Creole lunch.

Head 32km to the airport and onward connections.

ALTERNATIVES

Extension: Port Louis

Sidestep the stress of driving in Mauritius' capital by taking the local bus to admire its heritage architecture and explore its vibrant street food and art scenes. Don't miss the Aapravasi Ghat World Heritage Site on the seafront, which served as an immigration depot for indentured labourers from India.

Regular buses run between Trou aux Biches and Immigration Square in Port Louis (1hr).

Diversion: Île aux Aigrettes

The Mauritian Wildlife Foundation runs day trips to this small island and research station. A kind of Noah's Ark for rare and endangered flora and fauna, its residents include Aldabra giant tortoises brought in from the Seychelles to take over the ecological role of the extinct Mauritian giant tortoise.

Point Jerome Embarkation Point (where you can leave your car) is on the southern fringe of Mahébourg.

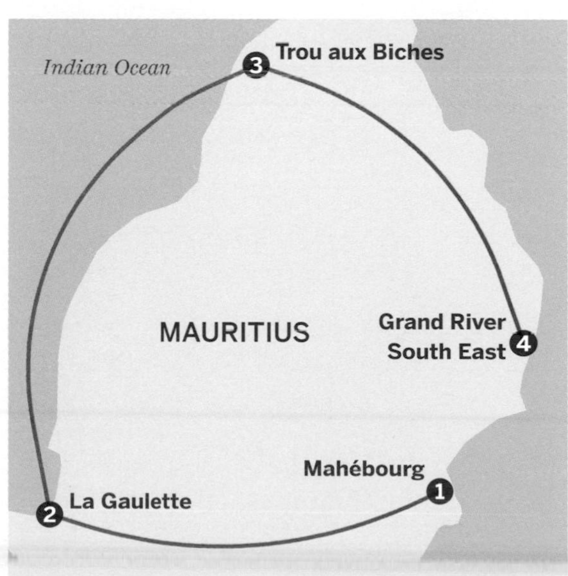

UGANDA, THE PEARL OF AFRICA

Kampala – Entebbe

Trek Uganda's lush jungles to find mountain gorillas in between wildlife safaris and adrenaline-packed adventure. For safaris, you'll need to hire a 4WD or book a tour.

FACT BOX

Carbon (kg per person): 202
Distance (km): 1186
Nights: 14
Budget: $$$
When: Jun-Feb

❶ Kampala

The first port of call for most travellers to Uganda is Kampala. In between arranging park permits, be sure to tap into the pulsating energy of this modern, cosmopolitan East African city and embrace its vibrant art scene, contemporary dining and raucous nightlife.

ABOVE: KAMPALA

From Kampala take an express shuttle bus east to Jinja (daily; 3hr).

❷ Jinja

Prepare for any prior expectations you had of Africa to be wiped with a sudden rush of white water as you plunge headfirst into Jinja's Grade-V rapids. Home to a thriving outdoor adventure scene, Jinja is where you'll find world-class rafting to go with bungee jumping and tubing. Afterwards, slow things down with a calming sunset cruise along the source of the River Nile.

From Jinja it's a 460km drive to Queen Elizabeth National Park; either arrange a tour or rent a 4WD.

❸ Queen Elizabeth National Park

No trip to East Africa is complete without going on safari – and Queen Elizabeth National Park offers one of the finest, with a memorable setting across varied ecosystems. Among opportunities to spot elephants, leopards, chimpanzees and hippos, the park's most famous residents are perhaps its tree-climbing lions – seen lazing high in the limbs of fig trees.

Arrange a tour or private car to Bwindi Impenetrable Forest National Park (300km).

ABOVE: MOUNTAIN GORILLA, BWINDI

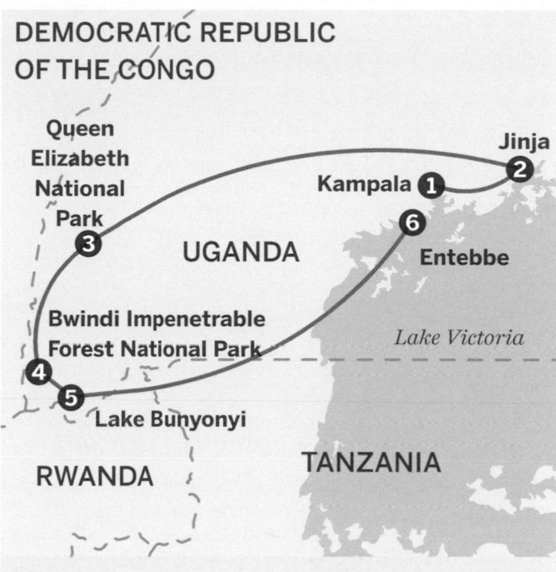

4 Bwindi Impenetrable Forest National Park

High on the bucket list of animal lovers the world over is the opportunity to see mountain gorillas in their natural habitat. In Bwindi you'll get a close encounter like no other, as you stand mere metres from these gentle, awe-inspiring and critically endangered creatures as they go about their forest business.

Hire a private car for the 78km drive from Bwindi (Buhoma) to Kabale town, gateway to Lake Bunyonyi.

5 Lake Bunyonyi

If Uganda is the Pearl of Africa, Lake Bunyonyi is its polished gleam, with an astonishing lakeside setting speckled with islands and green terraced hills. Relax by its sparkling waters while dining on freshwater crayfish and enjoying its famed sunsets.

It's a 400km drive to Entebbe.

6 Entebbe

Finish lakeside by Africa's largest and most famous body of water, vast Lake Victoria. Green and relaxed, Entebbe is a pleasant town to unwind before a flight and is popular for its chimpanzee wildlife refuge.

Entebbe's international airport offers onward international flights.

ALTERNATIVES

Extension: Mgahinga Gorilla National Park

Offering a less touristy alternative to Bwindi, Mgahinga gorilla permits are easier to obtain and provide a more dramatic setting among the volcanic Virunga mountain range. Gorillas are the stars of the show, but the endemic golden monkeys here also impress, as do the jungle treks to scale its volcanos.

Kisoro is the park's gateway town, a 48km drive from Lake Bunyonyi, or a full day's drive from Kampala (484km).

Diversion: Kibale National Park

While most folk come to Uganda for mountain gorillas, encountering chimpanzees in Kibale is also a special experience. It can easily be tagged on to a Queen Elizabeth NP itinerary and is within close proximity to Fort Portal's stunning Crater Lakes.

The park is 105km north of Queen Elizabeth NP and can be visited en route between the latter and Jinja.

VICTORIA FALLS FROM ZIMBABWE AND ZAMBIA

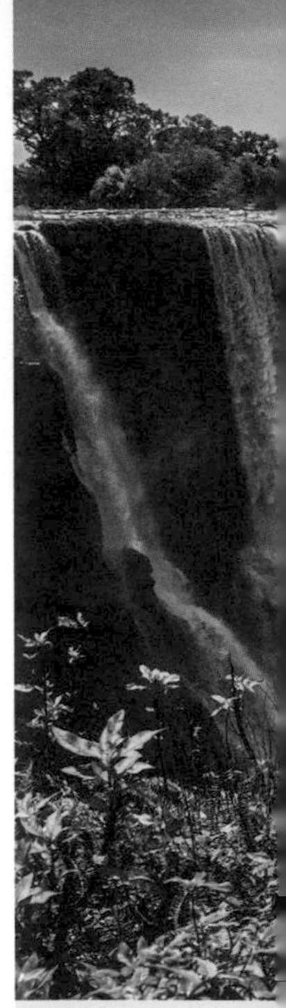

Victoria Falls town – Livingstone

Come for one of the world's natural wonders and stay for outdoor adventure and wildlife on this two-country trip – dual visas allow easy access back and forth.

FACT BOX

Carbon (kg per person): 12
Distance (km): 73
Nights: 4
Budget: $$
When: year-round

❶ Victoria Falls town (Zimbabwe)

Starting on the Zimbabwe side, spend a night or two in this tourist town – a relaxed base within walking distance of all the action. Take in the culture of the indigenous Shona and Ndebele people, sample local fare with a contemporary twist, and get a taste of colonial history at Victoria Falls Hotel.

Walk 2km to Victoria Falls.

ABOVE: ZAMBEZI RAFTERS, VIC FALLS BRIDGE

❷ Victoria Falls

Forming one of the world's most spectacular international borders are these World-Heritage listed falls that join Zimbabwe and Zambia with a curtain of raging water. Known as 'the smoke that thunders', at full force the power and fury of these falls is exhilarating – often leaving visitors soaked as they stand in awe before them.

Back in Vic Falls, arrange a safari to Zambezi National Park.

❸ Zambezi National Park

Stretching along the Zambezi River on the Zimbabwe side, a safari through this national park gives visitors the chance to encounter elephants, lions, giraffes, sables and leopards.

From Vic Falls town, walk 3km to the next stop, the bridge, but most operators include transport in their packages.

❹ Vic Falls Bridge

Arching spectacularly across the Zambezi River is

ABOVE: VICTORIA FALLS

the iconic Victoria Falls Bridge. It's here where the outdoor adventure scene comes into its own – from leaping off the bridge on a bungee jump to taming those white-water rapids beneath, there's a heap of adrenaline-packed activity to keep you busy.

Back in Vic Falls town, arrange a tour (transport included) to Livingstone Island on the Zambian side.

5 Livingstone Island and the Devil's Pool

Livingstone Island is where the namesake doctor and explorer first viewed the falls. To imagine the thrill he must have felt, visitors today can take a dip here in the Devil's Pool. This precarious rockpool perched at the top of the falls allows you to swim up and peek directly over the pouring torrent. Strictly for daredevils.

Take a taxi for the 13km ride to Livingstone.

6 Livingstone, Zambia

For those wanting to stay in Zambia, Livingstone is the closest town to the falls. It's a delightful base from which to explore the region, offering a wonderfully sleepy and relaxed atmosphere, rich in local culture, history and cuisine.

Livingstone's international airport is 5km northwest of town.

ALTERNATIVES

Extension: Hwange National Park (Zimbabwe)

While not as famous as the likes of Kruger or Serengeti, Zimbabwe's Hwange National Park, 180km south of Vic Falls, has many of the same animals – minus the crowds. Come to see some of the world's largest herds of elephants, plus lions, cheetahs, leopards, African wild dogs and giraffes, plus some 400 species of birds.

Take the bus down from Vic Falls (daily; 2hr), en route to Bulawayo.

Diversion: Sinazongwe and Lake Kariba (Zambia)

Take a break from the rigours of life on the road for some R&R at this laidback and very local-style resort town on the shore of manmade Lake Kariba. Here it's all about sunsets, beaches, canoeing and excellent fishing.

You'll need to arrange a private car/driver to get from Livingstone to Sinazongwe on the lake's northern shore (4hr 30min).

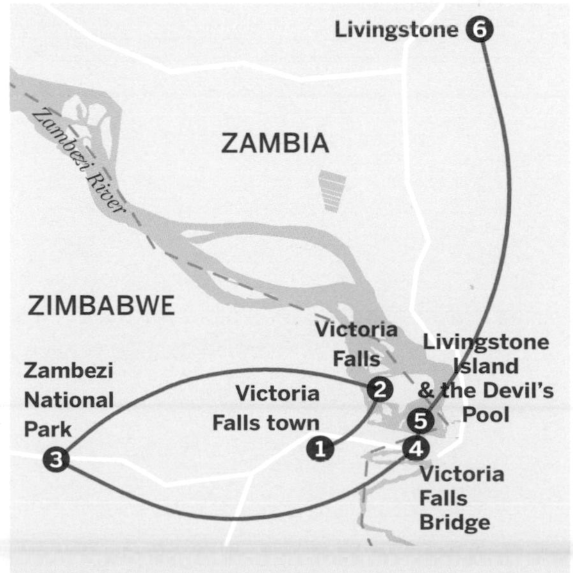

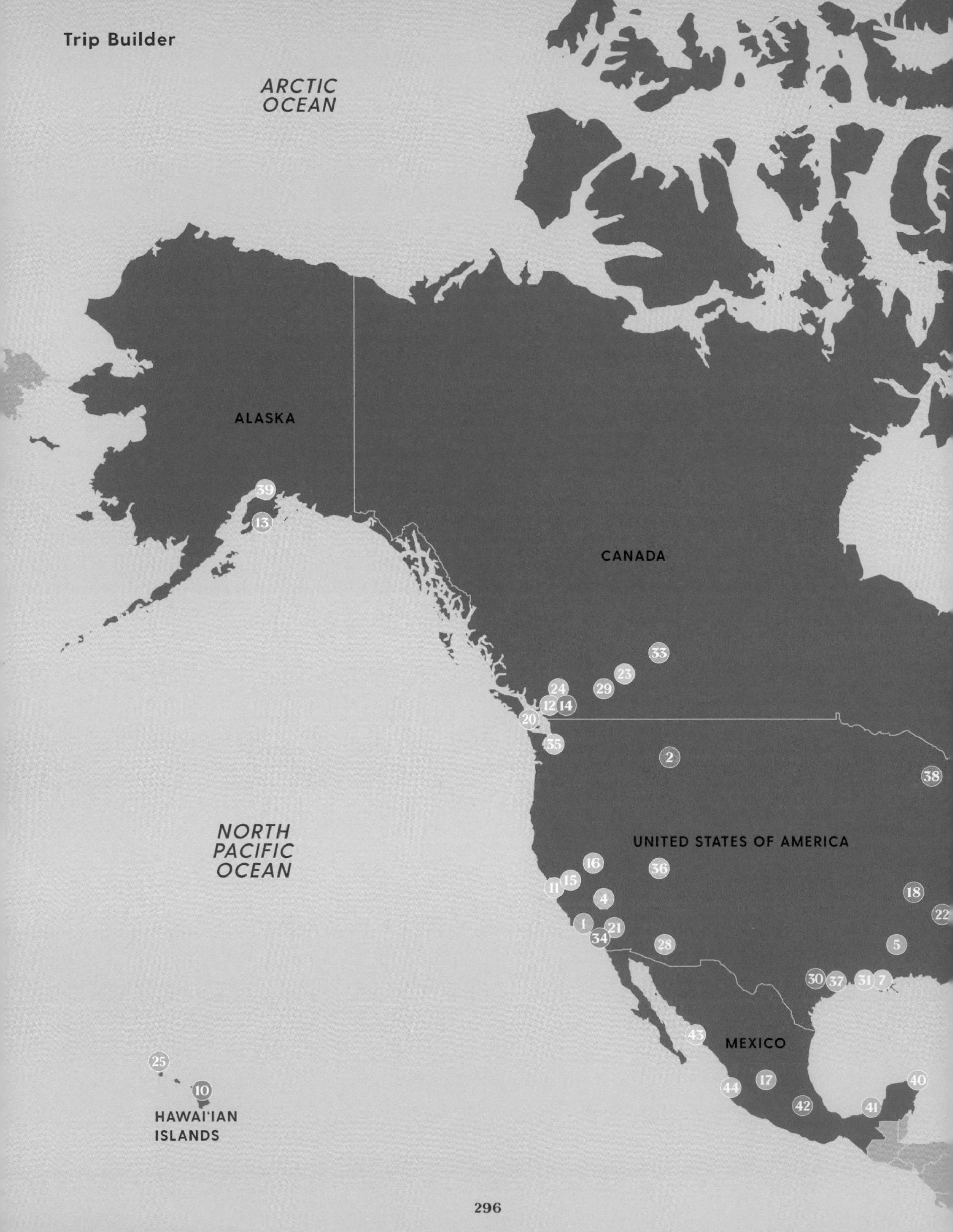
ARCTIC OCEAN
ALASKA
CANADA
NORTH PACIFIC OCEAN
UNITED STATES OF AMERICA
MEXICO
HAWAI'IAN ISLANDS
39
13
33
23
24
29
12
14
20
35
2
38
16
36
11
15
4
18
22
1
21
34
28
5
30
37
31
7
43
17
44
42
40
41
25
10

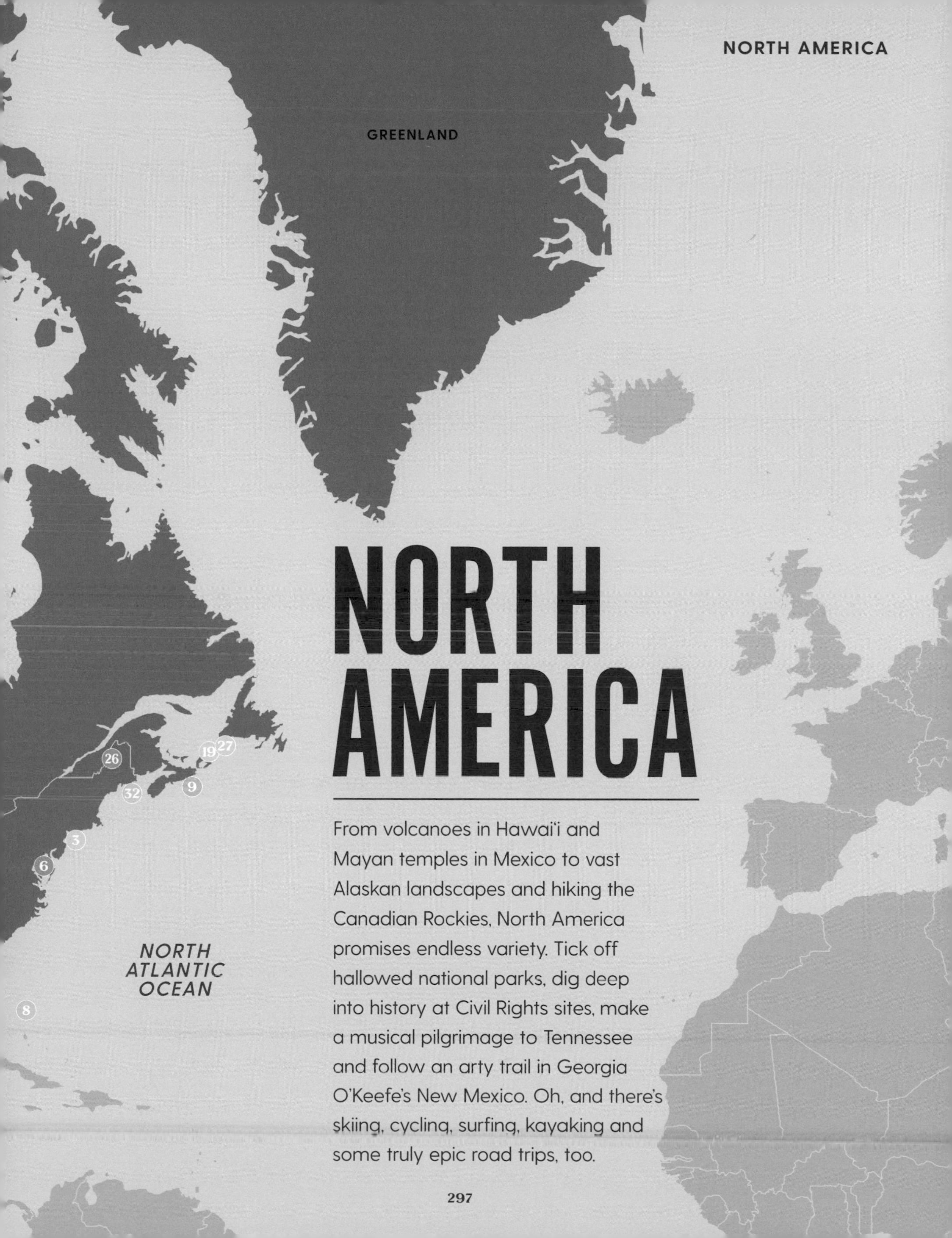

GREENLAND

NORTH AMERICA

From volcanoes in Hawai'i and Mayan temples in Mexico to vast Alaskan landscapes and hiking the Canadian Rockies, North America promises endless variety. Tick off hallowed national parks, dig deep into history at Civil Rights sites, make a musical pilgrimage to Tennessee and follow an arty trail in Georgia O'Keefe's New Mexico. Oh, and there's skiing, cycling, surfing, kayaking and some truly epic road trips, too.

NORTH ATLANTIC OCEAN

26 19 27 9 32 3 6 8

CONTENTS

ABOVE: DOWNTOWN MIAMI, FLORIDA

AROUND LOS ANGELES BY RAIL

Pasadena – Santa Monica

Though Los Angeles practically defines US-style car culture, many highlights – arts and architecture, world foods and beaches – can be explored car-free via light-rail.

FACT BOX

Carbon (kg per person): 1.8
Distance (km): 43
Nights: 3-5
Budget: $$
When: year-round

❶ Pasadena

Exit Memorial Park Station to explore lovely, lively Pasadena's trove of California architecture from Craftsman (Gamble House) to Spanish Colonial (Civic Center), world-class art and antiquities (Norton Simon Museum) and Asian flavours (all around town). Old Pasadena offers a mix of fun and funky shops.

Ⓜ Take Metro L Line to Union Station (15min).

❷ Downtown Los Angeles

From Union Station's soaring, Spanish-style main hall, emerge onto Olvera St, LA's first settlement and historic heart of its Mexican-American community. Walk to Frank Gehry's gravity-defying Walt Disney Concert Hall, visit the stunning Broad Museum and Museum of Contemporary Art, then grab a bite from the gourmet-chic stalls of the circa-1917 Grand Central Market. Spend the night in 'DTLA' – there's plenty more to see.

Ⓜ From 7th St/Metro Center Station, take Metro E Line to Expo Park/USC station (11min).

❸ Exposition Park

Exposition Park is LA's family-friendly go-to. Live out astronaut fantasies via the California Science Center's Space Shuttle, discover dinosaurs and diamonds at the Natural History Museum, and explore African American art and history at the California African American Museum. Upcoming attractions? The Lucas Museum of Narrative Art (yes, *Star Wars* creator George Lucas).

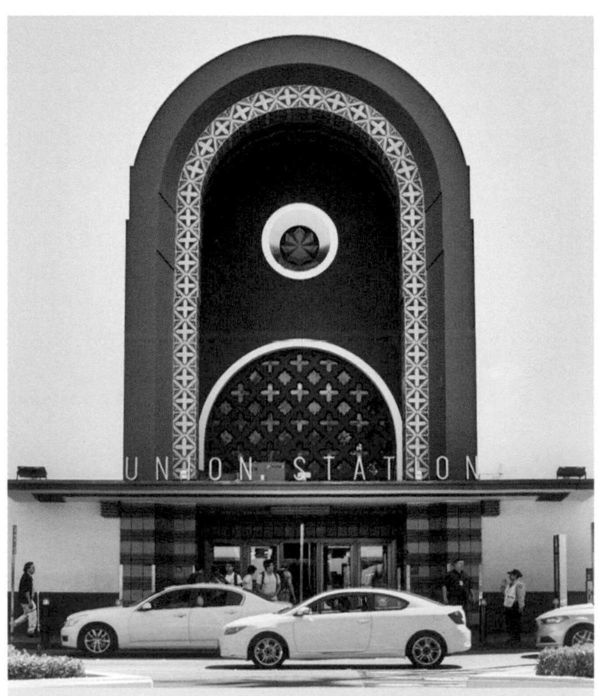

ABOVE: LA'S ART DECO UNION STATION

ABOVE: SANTA MONICA BEACH

Jump on Metro E Line to Culver City (18min).

4 Culver City

Self-proclaimed 'Heart of Screenland'– Sony (formerly MGM) Studios are here – and one-time sleepy bedroom community, Culver is now ready for its close up. In the historic downtown, check out the Culver Hotel (where the Munchkins stayed during the filming of *The Wizard of Oz*), Art Deco Kirk Douglas Theatre and new Culver Steps plaza. Browse designer chic in the Helms Bakery District and nearby art galleries, and mind-blowing contemporary architecture in the Hayden Tract neighbourhood.

Take Metro E Line to Downtown Santa Monica Station (18min).

5 Santa Monica

LA's seaside belle, Santa Monica is the end of train line and the iconic Route 66. Its beachside bicycle path and Ferris-wheel-topped pleasure pier grace countless postcards, and sunsets over the Pacific are legendary. Buskers and pleasure-seekers flock to Third Street Promenade, while fashionistas groove to the local vibe of Main St and chichi Montana Ave.

Santa Monica's a popular (though pricey) place to stay and close to LAX airport.

ALTERNATIVES

Extension: Huntington Library and Gardens

The Huntington offers a unique mix of art, literary history and inspirational gardens. Bookish treasures include a 15th-century *The Canterbury Tales*, while the art spans Gainsborough to Warhol. Allow plenty of time for the gardens – the Chinese, Japanese and desert sections all impress – and for high tea in the Rose Garden's tea room.

It's a 10min ride from downtown Pasadena; book a ride-share.

Diversion: Venice Beach

Take a walk on LA's wild side, namely Venice's Ocean Front Walk, a compelling (if often trashy and occasionally gritty) whirlwind of sun seekers, street performers, cyclists, in-line skaters, iron pumpers, fortune tellers, graffiti artists, drum circlers and beachfront-bar party peeps. Just inland, find Abbot Kinney Blvd's cool shops and restaurants, or wander the Venice Canals.

Venice is a beachfront stroll or cycle south from Santa Monica.

NATIONAL PARKS OF NORTHWEST WYOMING

Mammoth Hot Springs – Jenny Lake
Road-trip round Yellowstone, the US' (and the world's) first national park; after admiring its geysers and wildlife, head high into the mountains of adjacent Grand Teton.

FACT BOX

Carbon (kg per person): 26
Distance (km): 206
Nights: 2
Budget: $$$
When: early May–early Oct

❶ Mammoth Hot Springs

Mammoth Hot Springs' massive, otherworldly geothermal and geological wonders offer you a literally warm welcome to Yellowstone National Park – limestone pools, yellow springs and travertine terraces. Detour east to the Lamar Valley (40km) to appreciate the park's vastness (bigger than some states) and, with luck, spot bison and elk – maintain a safe distance of at least 100m and be prepared for travel delays due to animal-gawker traffic jams.

Drive south for 34km.

❷ Norris Geyser Basin

Wander miles of wooden walkways that cover Norris Geyser's two sections. Porcelain Basin, Yellowstone's warmest location, boasts colourful geysers flowing green, blue, white and gold; in larger Back Basin, the best-known sight is Steamboat Geyser, the world's tallest – if you're lucky to catch it when it's active it can spout over 100m.

Back in the car, head 30km south.

❸ Grand Prismatic

Yellowstone's deepest hot spring (37m) and often called its most beautiful natural feature, Grand Prismatic has a surface made up of concentric rings of bright orange, yellow and green waters around an eye-popping blue centre. Walk 1km of boardwalk for close-ups or drive to Fairy Falls and hike 1.5km for aerial views from an overlook.

Continue 11km east.

ABOVE: GRAND PRISMATIC, YELLOWSTONE

ABOVE: OLD FAITHFUL, YELLOWSTONE

❹ Old Faithful

America's most famous geyser owes its name and fame to the regularity of its eruptions – about every 90 minutes (staff and an app announce the next time). Escape the crowds on miles of paved paths which connect with dozens of smaller geysers, many with ethereal, architectural forms. Book ahead for a night at 1904 Old Faithful Inn, the world's largest log-built building.

Drive 105km south, crossing the Continental Divide and into Grand Teton National Park.

❺ Signal Mountain

Near the centre of Grand Teton, Signal Mountain's 275m summit offers a breathtaking panorama of five surrounding mountain ranges: Yellowstone Plateau; Teton; Snake River; Gros Ventre; and Absaroka. Bring binoculars to spot wildlife on the valley below.

Head 26km southwest.

❻ Jenny Lake

This glacier-carved lake is Grand Teton's most visited attraction. Hiking trails through coniferous shoreline forests lead to towering waterfall cascades. For a shorter trip to the falls, scheduled boats cross the lake. Reward yourself with a stay at Jenny Lake Lodge.

Drive 40km south to Jackson for onward flights.

ALTERNATIVES

Extension: Jackson, Wyoming

Jackson packs an Old West town vibe into one of America's ritziest communities. It's a popular base for climbers, riders and cyclists (and, in winter, skiers), and in town you'll find tasty food, lively bars and upscale lodging. The National Museum of Wildlife Art has over 5000 wild-animal-related works.

Jackson is just 8km south of Grand Teton National Park.

Diversion: Craters of the Moon National Monument and Preserve, Idaho

Vast lava fields form a mystical, if not exactly lunar, landscape in the Craters of the Moon. A short but steep hike up Inferno Cone provides views of the Snake River Plain and, in clear weather, the Grand Tetons. Several park hiking trails offer geological tours through shallow caves and lava tubes.

The park is about 460km southwest of Yellowstone.

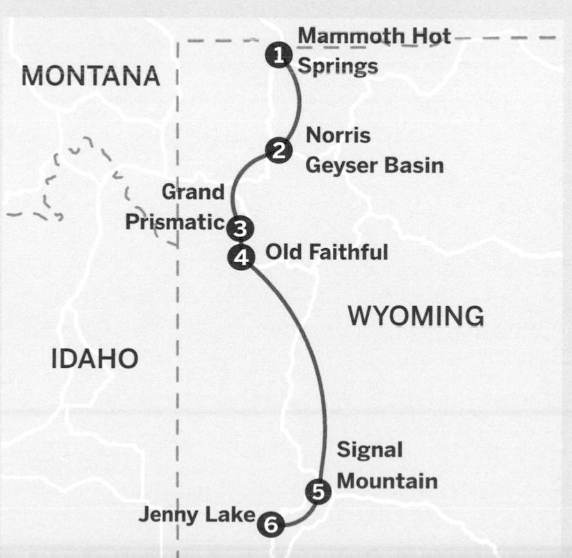

THE USA FROM COLONY TO COUNTRY

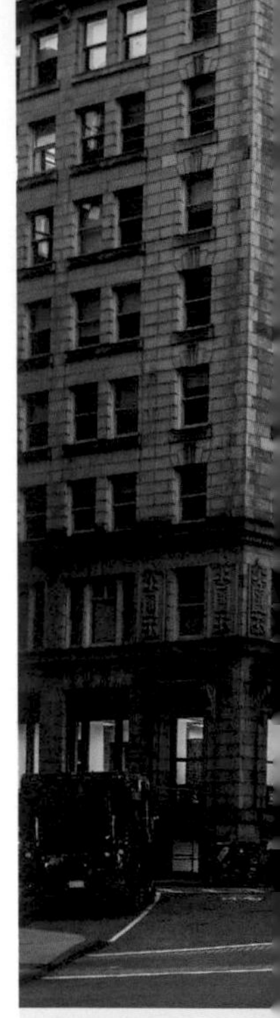

Washington, DC – Boston

Take a train trip into the story of the USA's formation via a journey north from the nation's capital through the coastal mid-Atlantic region to Boston.

FACT BOX

Carbon (kg per person): 14
Distance (km): 724
Nights: 8
Budget: $$$
When: year-round

❶ Washington, DC

The US capital celebrates the founding of the country with majestic monuments and free museums galore along the two-mile-long National Mall. The Washington Monument and the Jefferson Memorial honour the United States' first and third presidents respectively. The National Museum of American History and the US Capitol spotlight the country's beginnings too. The White House, Lincoln Memorial and Smithsonian collections continue the story.

Take the train from Union Station to Philadelphia (several daily; 2hr).

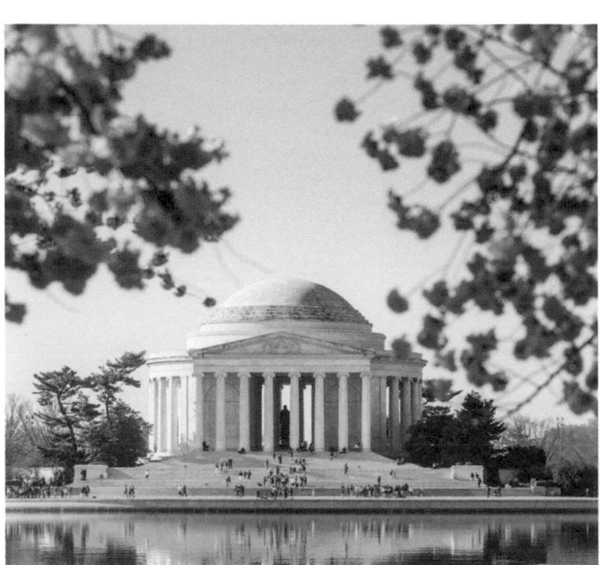

ABOVE: LINCOLN MEMORIAL, WASHINGTON, DC

❷ Philadelphia

Dotted with gracious squares and lined by cobbled alleys, Philly is a beautiful place. As the 'birthplace of American government' – where the Founding Fathers signed the Declaration of Independence in 1776 – history abounds: Independence Hall, Independence National Historical Park, the Museum of the American Revolution and the Liberty Bell are highlights. For post-exploration refuelling, the city's dining scene has progressed beyond the famed cheese-steak sandwich with great restaurants tempting from many a street corner.

From neoclassical 30th St Station take the train to NYC (several daily; 1hr-1hr 30min).

❸ New York City

The Big Apple is a world-class destination for food, music, theatre, fashion, publishing, advertising and finance. It's also home to many an historic site connected to the earliest days of what would

ABOVE: OLD STATE HOUSE ON BOSTON'S FREEDOM TRAIL

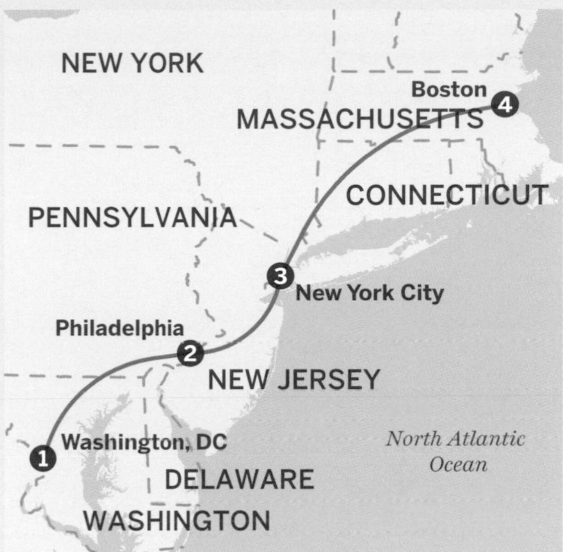

become the US. Skirting the edge of Manhattan and now home to artworks, gardens and museums, 12-acre Battery Park is the spot where the Dutch first settled in 1623. It was also here that the first battery of cannon was erected to defend the fledging settlement of New Amsterdam. Though unveiled much later, in 1886, the iconic Statue of Liberty symbolises on a grand scale the ideas of opportunity and freedom established by the Founding Fathers.

Penn Station in Midtown is the departure point for trains to Boston (several daily; 3hr 45min to 5hr 30min).

4 Boston

Arguably the USA's most historic city, Boston was the site of many events intimately connected to the country's founding: the Boston Tea Party; Paul Revere's Ride; the first battle of the Revolutionary War. Trace these world-changing moments on the 2.5-mile red-brick Freedom Trail, starting at Boston Common and wending its way to Charlestown across the Charles River. Then settle into Boston's oyster houses and cafes for sustenance before letting loose at the edgy music clubs near Harvard University, founded in 1636.

Boston has onward national and international connections.

ALTERNATIVES

Extension: Mount Vernon

The home of George and Martha Washington, the nation's original First Family, Mount Vernon is one of the most visited historic shrines in the nation. Tours of the estate offer glimpses of 18th-century farm life and do not gloss over Washington's ownership of enslaved people, taking in the living quarters and burial grounds of the enslaved.

A car is the easiest way to get to Mount Vernon, 28km south of DC.

Diversion: Valley Forge National Historical Park

After the Battle of Brandywine defeat, George Washington withdrew his army to Valley Forge; today, the site symbolises his endurance and leadership. Historical Park marks the former encampment, its scenic beauty contrasting with the knowledge that 2000 soldiers perished here from freezing temperatures, hunger and disease.

Valley Forge is 35km northwest of Philly, accessible by car or bus.

LOOP THE GRAND CANYON ON A CLASSIC US ROAD TRIP

Las Vegas – Hoover Dam and Lake Mead
From Vegas, this red-rock driving loop circles the Grand Canyon, shooting northeast across Nevada and Utah, dropping south into Arizona and cruising back to Sin City.

FACT BOX

Carbon (kg per person): 272
Distance (km): 1360
Nights: 10
Budget: $$
When: Apr-Oct

❶ Las Vegas

This classic road trip kicks off beneath the neon lights of the Strip, front row for Vegas' dazzling show – dancing fountains, a spewing volcano and the Eiffel Tower – all attached to tempting casinos. Live acts, the Mob Museum and plenty of top-notch dining round out the fun.

Drive 74km northeast to Valley of Fire State Park.

ABOVE: LAKE MEAD

❷ Valley of Fire State Park

The park's psychedelic outcroppings will electrify your social media feed, especially if shot at sunrise and sunset when the desert light hits the sandstone just right.

Continue 200km east, crossing into Utah.

❸ Zion National Park

Zion's soaring red-rock cliffs hide graceful waterfalls, narrow slot canyons and hanging gardens. It's a surprisingly lush wonderland lying in the shadow of the head-for-heights-needed Angels Landing Trail – walk it if you dare.

Drive 200km southeast.

❹ Grand Canyon National Park – North Rim

The primary distractions at the remote and lofty North Rim are the canyon-edge national park lodge, a campground, a motel, a general store and miles of trails carving through sunny meadows thick with wild flowers, willowy aspens and ponderosa pines.

It's a 340km drive from the North to the South Rim.

❺ Grand Canyon National Park – South Rim

The sheer immensity of the canyon, carved by the

ABOVE: GRAND CANYON

Colorado River over six million years, grabs you first when viewed from the South Rim. Next you notice dramatic layers of rock and then the artistic details – rugged plateaus, crumbly spires, maroon ridges – that catch your eye as shadows flicker across the stones. Walks along the rim are breathtaking, but to get a true sense of its scale, hike down into the canyon – the South Kaibab Trail is rightly popular.

Head 126km south.

❻ Flagstaff

A gateway town for the South Rim, Flagstaff features a pleasant, walkable downtown crammed with eclectic vernacular architecture, vintage neon, craft breweries and a restaurant scene that is small but high quality. The surrounding countryside boasts myriad high-altitude pursuits.

Drive 360km west to the Hoover Dam.

❼ Hoover Dam and Lake Mead

Straddling the Arizona-Nevada state line, the graceful curve and Art Deco style of the 221-m-high Hoover Dam contrasts sharply with the stark desert. A scenic drive winds beside dam-created Lake Mead.

Round off the trip with the 60km drive back to Vegas.

ALTERNATIVES

Extension: Bryce Canyon National Park

Pastel-coloured spires and hoodoos pop like creations in a Dr Seuss book. Though the smallest of Utah's national parks, Bryce is perhaps the most immediately visually stunning. This is particularly true at sunrise and sunset when an orange wash sets the rock formations ablaze. Steep trails descend into an amphitheatre of otherworldly spires.

Bryce Canyon is 116km east of Zion National Park.

Diversion: Horseshoe Bend

The bird's-eye view of Horseshoe Bend is epic – or perhaps alarming. There's no barrier between you and the edge, with the Colorado River, 300m below, carving a perfect U through the Navajo sandstone cliffs: simultaneously beautiful and terrifying. The trail from the parking to the overlook is around 1km.

From the Grand Canyon North Rim, it's a 200km drive northeast to Horseshoe Bend.

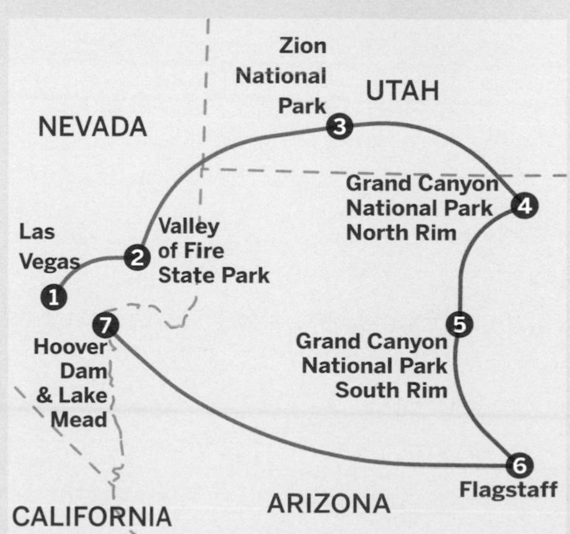

SOUTHERN CIVIL RIGHTS SITES

Montgomery – Memphis

Learn the history of the American Civil Rights Movement on a drive through pivotal cities across Alabama, Mississippi and Tennessee.

FACT BOX

Carbon (kg per person): 92
Distance: 740
Nights: 6
Budget: $
When: year-round

❶ Montgomery

The Civil Rights Movement began in earnest in 1955 in the capital of Alabama when Rosa Parks refused to give up her seat for a white passenger on a city bus – the moment is recreated at the Rosa Parks Museum in downtown Montgomery. Martin Luther King Jr led the subsequent year-long bus boycott here and was pastor of the city's Dexter Avenue Baptist Church, a short walk east of the museum. Just south, the Civil Rights Memorial Center poignantly honours the movement.

Drive 80km west following the route of the Selma to Montgomery National Historic Trail.

❷ Selma

On 7 March 1965, aka Bloody Sunday, Alabama state troopers and recently deputised local white men attacked 500 peaceful marchers on Selma's Edmund Pettus Bridge with clubs and tear gas. Film of the horrifying attack was shared across the nation and garnered large support for the Civil Rights Movement. The National Voting Rights Museum is near the southern side of the bridge.

Continue 320km west, crossing into Mississippi.

❸ Jackson

Mississippi's capital contains a striking mix: handsome

ABOVE: LORRAINE MOTEL, MEMPHIS

ABOVE: COMMEMORATION AT EDMUND PETTUS BRIDGE, SELMA

residential areas; hip and food-minded Fondren District, and swathes of urban blight. At the acclaimed Mississippi Civil Rights Museum downtown, eight exhibit halls pack a visceral punch explaining the national Civil Rights Movement through the lens of the fight for equality in Mississippi. It's hard not to be profoundly moved when confronted with the rifle that shot and killed activist Medgar Evers in 1963.

Drive 340km north.

4 Memphis

Martin Luther King Jr's crusade was abruptly halted here in Memphis in April 1968 when he visited in support of a Black sanitation workers' strike. He was shot on the balcony outside room 306 at the Lorraine Motel, now part of the impressive and essential National Civil Rights Museum. Elsewhere, downtown's Beale St transformed into one long music venue filled with nightclubs and good times in the 1890s and later became an epicentre for blues greats. The city's eclectic but uniformly delicious contemporary dining scene is a welcome addition to Memphis' attractions and includes weathered burger and BBQ joints as well as trendy chef-driven hotspots.

Memphis International Airport has multiple onward connections.

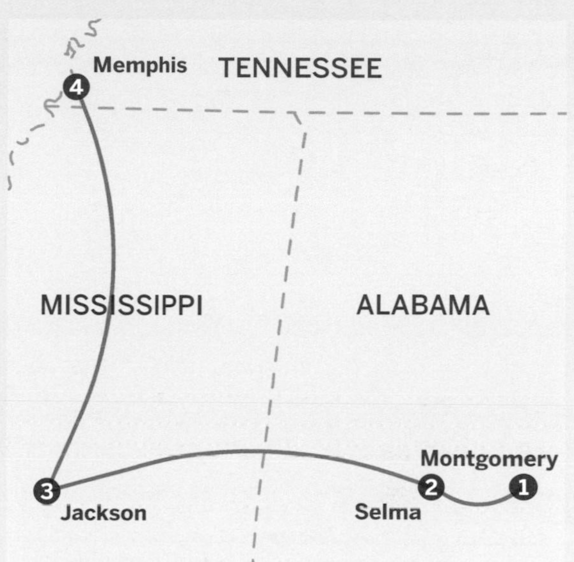

ALTERNATIVES

Extension: Birmingham

Civil rights campaigners embarked on a desegregation campaign in downtown Birmingham, Alabama that employed masses of local activists, often high school students. Police responded with water cannon and attack dogs, and the events are immortalised in sculpture at Kelly Ingram Park.

Birmingham is 146km north of Montgomery.

Diversion: Mississippi Delta

Stretching south along the river, the Delta is where songs of labour, suffering and love developed into the blues. Hear live blues at Red's, a late-night juke joint in Clarksdale, the Delta's scrappy focus. The Emmett Till Historic Intrepid Center in nearby Glendora spotlights Till's murder by two white men, and their acquittal by an all-white jury – a Civil Rights Movement flashpoint.

Clarksdale is 120km south of Memphis.

VIRGINIA CAPITAL TRAIL: CYCLING THROUGH US HISTORY

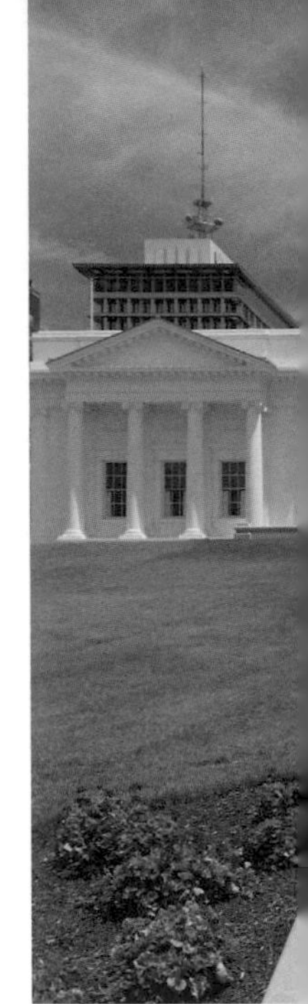

Richmond – Colonial Williamsburg
Pedal into the past along the Virginia Capital Trail, following a mostly flat route that takes in a 17th-century plantation, historic Jamestown and Colonial Williamsburg.

FACT BOX

Carbon (kg per person): 0
Distance (km): 121
Nights: 2
Budget: $
When: May-Oct

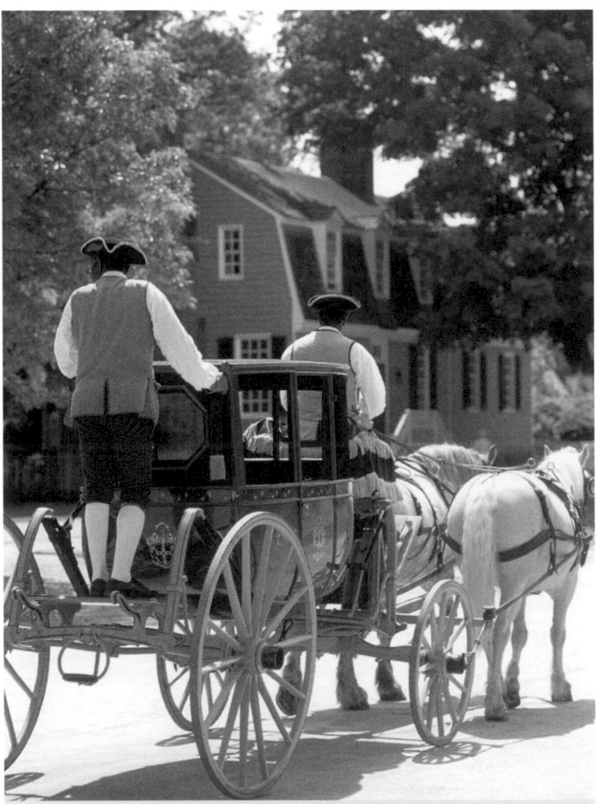
ABOVE: COLONIAL WILLIAMSBURG

❶ Richmond

Richmond, capital of Virginia, wins kudos for its vibrant restaurant and craft brewery scene, especially in the emerging Scott's Addition area. The Capital Trail officially starts at Great Shiplock Park in the city's downtown. From here, cycle to the American Civil War Museum and Belle Isle, former home to a Civil War quarry and POW camp. The State Capitol building, designed by Thomas Jefferson, and St John's Church, where Patrick Henry gave his Liberty or Death speech, are nearby.

From mile marker 51.2, the start of the trail, cycle southeast to mile marker 30 and the turnoff to Shirley Plantation.

❷ Charles City County

Built using riches gathered from ownership of enslaved people, the grand homes of Virginia's first families were a clear sign of the era's race and class divisions. A string of them line the Capital Trail on the north side of the James River: Shirley Plantation, dating from 1613, is the largest intact colonial estate in the country. Upper Shirley Vineyards is a picturesque winery nearby. Berkeley Plantation, a short distance to the east, was the location of the first official Thanksgiving in 1619, and the home of a signatory of the Declaration of Independence.

Pedal to mile marker 0 to reach Jamestown.

ABOVE: VIRGINIA STATE CAPITOL, RICHMOND

❸ Jamestown

In 1607 a group of 104 English men and boys settled on this swampy island bearing a charter from the Virginia Company of London to search for gold and other riches; by early the next year only 40 were still alive. The survivors ultimately established the first permanent English settlement in North America; initially assisted by the area's Native Powhatan people, they later massacred and almost obliterated them. Today's Jamestown Settlement reconstructs the fort, a Native American village and the ships that carried the settlers. Nearby Historic Jamestowne is the original Jamestown site.

Cycle 11km north from the end of the trail to get to Colonial Williamsburg.

❹ Colonial Williamsburg

The restored capital of England's largest colony in the New World is a living, breathing, working history museum. The 30-acre historic area contains 88 original 18th-century buildings where costumed interpreters in period dress go about their jobs, giving visitors a glimpse into colonial life. Old-timey taverns serve appropriately old-timey fare or modern options abound in adjacent Merchants Square.

Cycle or catch the train back to Richmond.

ALTERNATIVES

Day trip: Monticello

Designed by Thomas Jefferson, founding father and third US president, Monticello is an architectural masterpiece. Located in Charlottesville, it is the only home in the US designated a World Heritage site. Tours do not shy away from addressing the contributions of the enslaved people who lived and laboured here.

A car or tour is the easiest way to get to Monticello, 112km west of Richmond.

Day trip: Yorktown

On 19 October, 1781, British General Cornwallis surrendered to George Washington at Yorktown, effectively ending the War of Independence. Multimedia exhibits at the Museum of the American Revolution vividly describe the build up, the war itself and daily life on the home front. George Washington's original field tent is in the visitor centre.

Yorktown is a bikeable 19km southeast of Colonial Williamsburg.

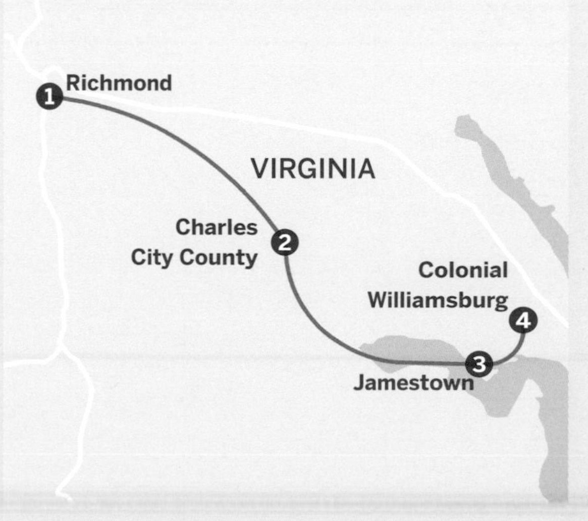

FABULOUS FOOD FINDS ON THE US GULF COAST

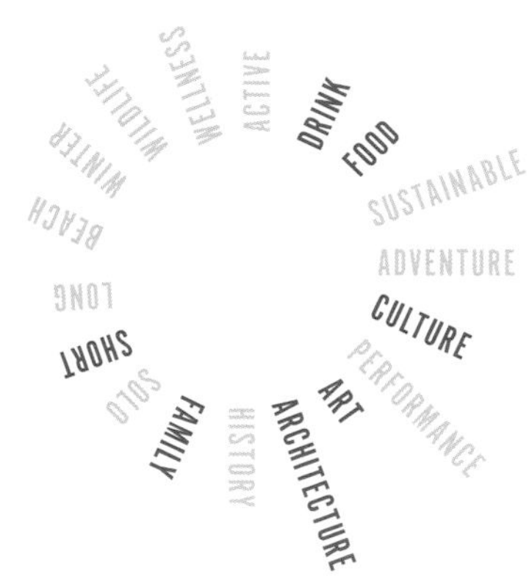

New Orleans – Houston

The melting humidity in Louisiana and Texas combines with the regions' melting pot of food and people on this celebratory, gastronomic road trip.

FACT BOX

Carbon (kg per person): 110
Distance (km): 550
Nights: 3
Budget: $$
When: Feb-Apr

❶ New Orleans

There is no city quite like New Orleans, where a rainbow of exquisite colonial architecture blends perfectly with a culturally liberal approach to life. The fantastic, eclectic food scene includes mainstays of classic Creole cuisine such as Dooky Chase and some of the world's best fried chicken at Willie Mae's. Less traditional, but no less delicious, are restaurants like Marjie's Grill where Southeast Asian influences are fused with Deep Southern food while retaining respect for both.

Drive 20km west to Kenner.

❷ Kenner

This suburb of New Orleans might have few specific sights, but the reason you're here is because it's home to many of the more recently arrived immigrants who form the backbone of the regional economy – and who have introduced a cosmopolitan mix of new flavours to the established food options. Excellent, budget friendly restaurants cover the gamut of international fare, many lining industrial Williams Blvd. Head to Little Chinatown for the chili-pepper chicken, Brazilian Café for *pastel de carne* (meat pies) or fill up on a *fritanga*, a Nicaraguan behemoth of fried cheese and roasted meats, at Nola Nica.

Drive 200km west into Cajun Country, past the primeval swamp of the Atchafalaya Basin, to Lafayette.

ABOVE: FRENCH QUARTER, NEW ORLEANS

ABOVE: NEW ORLEANS CRAWFISH

❸ Lafayette

Cajun Country, also known as Acadia, is a region steeped in American folklore. The area was settled by French-Canadians escaping British rule in the 18th century, and their descendants have developed a culture deeply rooted in the swamps and prairies of southern Louisiana – as well as distinctive food and music. Catch a foot stomping show at the Blue Moon Saloon, Lafayette's most storied music venue, then grab some eggs and *boudin* (local sausage) at the magnificent French Press.

Drive 330km west to Houston, crossing from Louisiana into Texas.

❹ Houston

Houston is one of the most international cities in the world, and amid the dense urban sprawl there's plenty of gastronomic evidence of its multiculturalism. Try the straight-out-of-Hyderabad *biryani* (an all-in-one rice dish) at Biryani Pot; the simple elegance of vegetarian Vietnamese food at Duy Sandwiches; *jollof* (smoky West African rice) at Baba Jollof; or the transcendent crawfish swimming in garlic butter sauce at Crawfish & Noodles.

Houston's airport has global onward connections.

ALTERNATIVES

Extension: Barataria Preserve

Just a little way south of New Orleans sits the Barataria Preserve, a part of Jean Lafitte National Historical Park. From the visitor centre, boardwalk trails crisscross a haunting wetland-scape of flooded cypress forests and blackwater swamps. There are plenty of alligators out here, although they're sometimes concealed by the park flora.

Barataria Preserve is 30km south of New Orleans.

Diversion: Chicot State Park

This region's culture is intimately tied to its natural setting, and this state park, which surrounds the shores of Lake Chicot, is a good spot for exploring the beauty of southern Louisiana's forest and swamp environments. Waterside cabins – they're actually built on stilts on the water – are an excellent accommodation choice.

Chicot State Park is about 80km north of Lafayette.

SOUTH FLORIDA SHUFFLE

Miami – Key West

Experience culture and kookiness in a US road trip around South Florida, from art and architecture in Miami to the fantastical pageant that is daily life in Key West.

FACT BOX

Carbon (kg per person): 62
Distance (km): 322
Nights: 5
Budget: $$
When: Mar–May

❶ Miami

The 'Magic City' offers an intoxicating mix of the Spanish-speaking Caribbean, European style and American attitude. Make sure to take time for the adult playground of Wynwood, a hub of galleries and street art, and stroll around Little Havana, where the original Cuban residents have now been added to by arrivals from across the world.

ABOVE: EVERGLADES NATIONAL PARK

Rent a bike and cycle (or take a taxi) from downtown across MacArthur Causeway to Miami Beach (12km).

❷ Miami Beach

Miami Beach is both a barrier island and a separate city from mainland Miami. In terms of iconic travel images – long stretches of sandy beach, eye-catching Art Deco architecture and the neon nights of Ocean Drive – the beach is more 'Miami' than Miami proper. Check out the gorgeous hotels lining Collins Ave from Lincoln Rd up to 24th St, and stroll the seafront along South Beach's Ocean Drive.

Rent a car and head 90km south of the city to the Everglades.

❸ Everglades National Park

Ecologically, South Florida sits on a seasonally flooded prairie interspersed with hammocks (groves) of subtropical forest. This 'river of grass' is best preserved in Everglades National Park, a gem of the National Park Service that is rife with wetlands, long, low horizons and a zoo's worth of wildlife, including

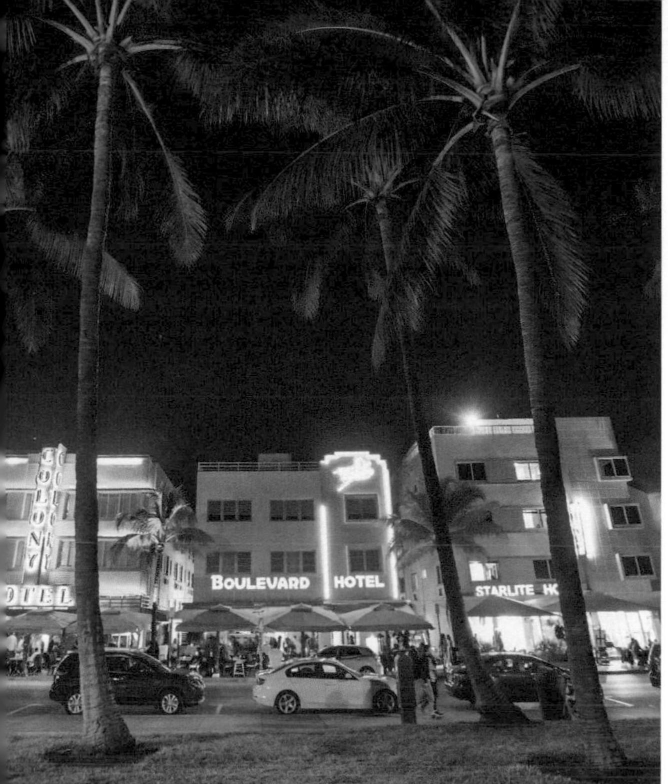

ABOVE: OCEAN DRIVE, MIAMI BEACH

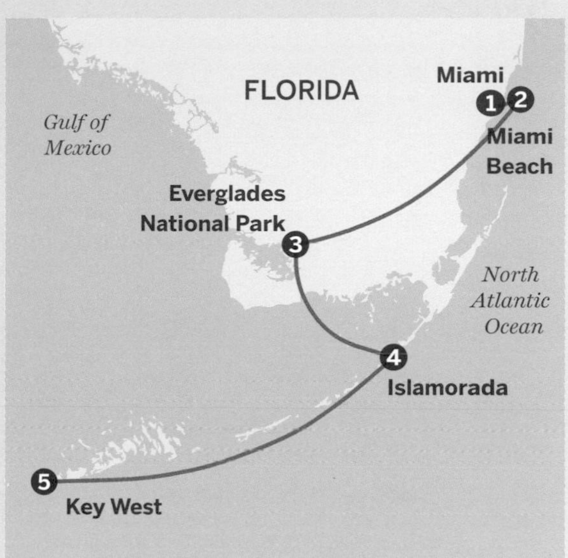

dozens of alligators at the Royal Palm Visitor Center.
Drive 90km south, joining the Overseas Hwy to reach the Florida Keys.

4 Islamorada

The Florida Keys are an archipelago of over 880 islands of which only around 30 are inhabited. The biggest settlements are connected to the mainland by the Overseas Hwy. In the middle of the Keys, quirky Islamorada encompasses the mangroves and calm mudflats of Anne's Beach and an arts-market-fishermen camp at Robbie's Marina, making it a fine place to break up the journey.
It's another 130km to end of the road Key West.

5 Key West

The southernmost city in the continental USA is a beautiful place, all heritage architecture and swaying palm trees, with the sunset painting everything a rich pink, red and orange as evening comes on. Duval St is the main drag, where you'll find tons of art galleries, restaurants and rum-soaked nightlife. And don't miss trying some Key lime pie in the place it was – allegedly – invented.
Key West airport has flights to several US cities, or drive back to Miami.

ALTERNATIVES

Extension: Shark Valley

Everglades National Park covers an enormous area that extends across the Florida peninsula to the Gulf of Mexico. If you head west of Miami on the Tamiami Trail (US-41) you'll leave the wet prairie for a flooded forest setting where, at Shark Valley, you can take a tram tour and spot more swamp wildlife.
Shark Valley is about 65km west of Miami.

Diversion: Key Largo

The northernmost of the major inhabited Keys, Key Largo is also home to John Pennekamp Coral Reef State Park. This is one of the finest spots in the continental USA for diving and snorkelling, and the park's infrastructure is largely geared towards getting people underwater. There are glass-bottom boat tours as well, if you don't want to strap on flippers.
The park is about 100km south of Miami and just off the Overseas Hwy.

A CANADIAN ATLANTIC ADVENTURE

Halifax – St John's

Enter a maritime world of salty winds and lashing storms, friendly people and epic views with a road trip through Canada's beautiful, wild Atlantic provinces.

FACT BOX

Carbon (kg per person): 228
Distance (km): 1458
Nights: 7
Budget: $$
When: Aug-Oct

❶ Halifax

The largest city in Canada's Atlantic provinces is a bustling town dotted with important architectural highlights such as the Citadel National Historic Site, a star-shaped fortress that played a key role in provincial history. In complete contrast is the Halifax Central Library, an ultra-modern hulk of terraces, Cubist-style angles and shining glass. The country's national diversity is celebrated and explored at the Canadian Museum of Immigration at Pier 21.

Drive 62km east to the town of Lake Charlotte.

❷ Lake Charlotte

Nova Scotia's heritage is especially evident at the Memory Lane Heritage Village, a twee recreation of a 1940s Eastern Shore community, complete with a schoolhouse, church and farmstead. The buildings here have either been relocated or painstakingly restored, and the atmosphere is warm and family-friendly.

Drive 250km along the Trans-Canada Hwy to Port Hastings, gateway to Cape Breton Island.

❸ Cape Breton Island

Rugged, raw and beautiful, Cape Breton Island is also infused with Scottish heritage, something that is obvious as you drive the Ceilidh Trail, home of the Glenora Inn & Distillery, the first single malt whisky distillery in North America. The Ceilidh connects to the Cabot Trail, a stunningly appealing road running right through the natural beauty of Cape Breton Highlands National Park.

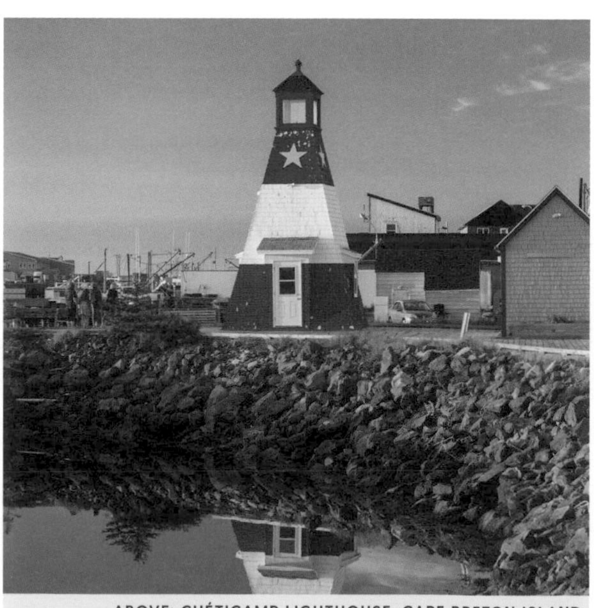

ABOVE: CHÉTICAMP LIGHTHOUSE, CAPE BRETON ISLAND

ABOVE: ST JOHN'S, NEWFOUNDLAND

Catch the ferry (3 weekly; 16hr) from North Sydney in Nova Scotia to Argentia in Newfoundland.

4 Irish Loop

Newfoundland is a wild, windswept place, filled with some of the friendliest folk in Canada. From Argentia, join the Irish Loop, an amazing drive that hugs the cliffs and green mountains of Newfoundland's Avalon Peninsula. In Witless Bay, consider taking a boat tour to get a close-up view of some of the largest seabird colonies in the world.

Following the Irish Loop, it's about 300km to St John's.

5 St John's

Newfoundland's provincial capital is a handsome collection of candy-coloured, hillside-covering homes. Make sure to take a trip to The Rooms, a finely executed museum that is a wonderful window into Newfoundland and Labrador. Then go for a windy hike around Signal Hill National Historic Site, a fortress that affords exceedingly dramatic views of St John's.

Flights connect St John's with other Canadian and international cities.

ALTERNATIVES

Extension: Mistaken Point Ecological Reserve

Located in the southeastern corner of the Irish Loop, Mistaken Point is raw even by the standards of Newfoundland. It also holds huge geological significance. Here you'll find the Mistaken Point Formation, embedded with a treasure-trove of Precambrian and Ediacaran fossils that are the oldest remains of multicellular life on Earth.

Mistaken Point is about 150km south of St John's.

Diversion: Cape Spear

The easternmost point of Canada (and North America – next stop Europe) is marked by Cape Spear Lighthouse National Historic Site, which perches above the rolling Atlantic Ocean. The views here are vast, the air is as fresh as you'd expect, and there are often icebergs to be spotted floating on the grey waves.

The lighthouse is 16km east of St John's via twisty Black Head Rd.

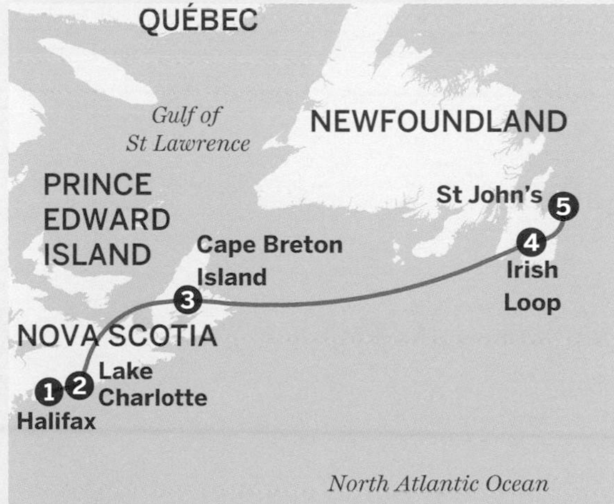

A TALE OF TWO HAWAI'IAN ISLANDS

Kailua-Kona, Big Island – Waimea, Kaua'i
Journey from the volcanic and historical highlights of the Big Island to the active thrills and natural spectacle of more compact Kaua'i.

FACT BOX
Carbon (kg per person): 102
Distance (km): 791
Nights: 8-12
Budget: $$$
When: May-Sep

❶ Kailua-Kona, Big Island

After arriving at Kona International Airport, explore the historical highlights of the Big Island's leeward side including the Hulihe'e Palace, a stately 1838 mansion used as a summer house by the Hawai'ian royal family. At the nearby Kaloko-Honokohau National Historical Park, ancient petroglyphs combine with sea turtle-spotting and a tranquil beach.

From Kailua-Kona, drive 200km east across the island to Volcano.

ABOVE: WAIMEA CANYON, KAUA'I

❷ Volcano, Big Island

After passing the spectacular Hamakua Coast and laidback Hilo, you arrive in forest-clad and arty Volcano, the base for exploring Hawai'i Volcanoes National Park. Embark on a day-hike along the Crater Rim Trail on Kilauea's summit then drive the Chain of Craters Rd, descending through lava fields to end at the spectacular Hōlei Sea Arch.

Continue 140km west around the Big Island's south coast and the black-sand beach at Punalu'u.

❸ Captain Cook, Big Island

En route back to the island's west side, stop for lunch at Hana Hou, the United States' southernmost restaurant. Continue to the hillside settlement of Captain Cook, surrounded by coffee plantations. This is a good base for excellent snorkelling straight off the rocks at Two Step Beach, and renting a kayak to paddle across Kealakekua Bay to the spot where British maritime explorer Captain James Cook was killed in 1779.

ABOVE: GREEN TURTLE, BIG ISLAND

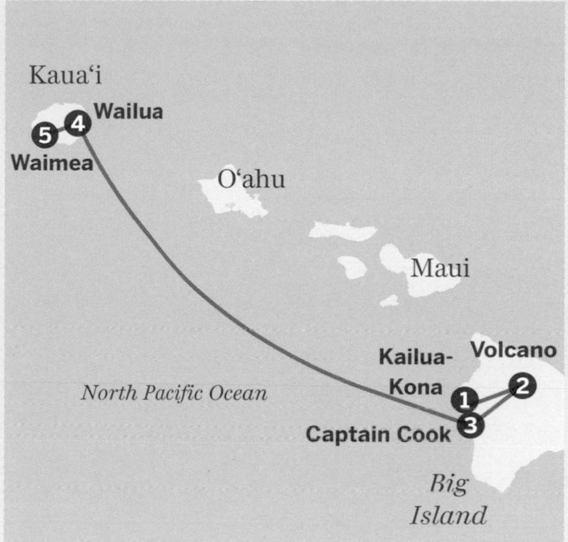

Drive 30km north back to Kailua-Kona; fly to Lihu'e on the island of Kaua'i (daily; 1hr); drive 10km north to Wailua.

4 Wailua, Kaua'i

Cottage and B&B options around Wailua make a good base for exploring the east coast of Kaua'i. Journey by kayak up the sacred Wailua River – taking in waterfalls and a short forest walk – and drive north to Kapa'a to rent a bike and negotiate the Ke Ala Hele Makalae Path to the sprawling surf beach at Paliku. Food trucks await back in Kapa'a for post-ride refreshment.

Drive west 48km to Waimea.

5 Waimea, Kaua'i

Using the historic Waimea Plantation Cottages as an accommodation base, drive the spectacular Waimea Canyon Drive, stopping at lookouts for waterfall vistas, and continuing for views of the Na Pali Coast from the end of the road at Pu'u o Kila. From Port Allen, embark on a catamaran sailing and snorkelling trip to see Na Pali's serrated pinnacles and verdant valleys from the ocean.

Return to Lihu'e and fly to Honolulu for onward connections.

ALTERNATIVES

Day trip: Hanalei

On Kaua'i's north coast, Hanalei's arcing bay has one of the island's best beaches. Take in the bohemian village vibe, incorporating food trucks and hip cafes, before getting active and tackling the two-day Kalalau Trail. A good one-day alternative is the shorter stretch of the trail from Ke'e Beach to Hanakapi'ai Beach. Either way, celebrate with drinks back in Hanalei at the raffish Tahiti Nui bar.

From Wailua it's around 40km north to Hanalei.

Day trip: Kipu Ranch Adventures

Head out in an All Terrain Vehicle on the red-dirt roads of the ranch where scenes from the *Jurassic Park* movies were filmed. Tours also take in the riverbank where Harrison Ford escaped on a rope swing in *Raiders of the Lost Ark*, and you can ascend a hill for views of the perfect beach from *The Descendants*.

Kipu Ranch Adventures departure point is around 18km southwest of Wailua.

CALIFORNIA COAST CRAFT BEER

San Francisco – Paso Robles
Follow one of the United States' most iconic coastal highways to discover the best of the brewing scene in California.

FACT BOX
Carbon (kg per person): 98
Distance (km): 792
Nights: 7-10
Budget: $$$
When: Jun-Sep

❶ San Francisco

Compact, liberal and wonderfully diverse, San Francisco is an easily navigated place to begin a Californian craft beer journey. Browse the farm-to-table stalls at the Ferry Plaza Farmers Market before exploring the beer scene around SoMa (South of Market). Brewery highlights include Cellarmaker, Black Hammer and 21st Amendment and, in nearby Dogpatch, Harmonic.
Drive 124km south to Santa Cruz along the scenic Pacific Coast Hwy (PCH, aka Hwy 1) via Half Moon Bay and Pescadero.

❷ Santa Cruz

Build up a thirst by biking along Santa Cruz's coastal West Cliff Drive or embarking on a guided sea kayaking tour leaving from the city's historic wharf. Then quench that thirst at Humble Sea Brewing with one of its excellent hazy IPAs, and at Sante Adairius Rustic Ales, specialists in Belgian beer styles.
Continue 68km south to Monterey along the PCH.

❸ Monterey

Explore the world-leading Monterey Bay Aquarium – the three-storey-high kelp forest features some surprisingly large sea creatures – before taking a whale-watching tour in the Monterey Bay National Marine Sanctuary. Blue and humpback whales cruise through the nutrient-rich bay from April to November. Visit Alvarado Street Brewery and Fieldwork Brewing

ABOVE: FIELDWORK BREWING, MONTEREY

ABOVE: SEA OTTERS ANCHORED IN KELP AT MONTEREY

after a big day out on the water.

Drive 220km along the PCH via Big Sur to San Luis Obispo.

4 San Luis Obispo

Travel down one of America's most spectacular roads via the dense forests and rugged, windswept coves and cliffs of Big Sur. Often dubbed 'SLO', San Luis Obispo combines a great local food scene with excellent craft beer and proximity to the vineyards of the Edna Valley. Dine along Higuera St during Thursday night's San Luis Obispo Farmers Market while planning your own craft beer trail including Libertine Brewing, There Does Not Exist and SLO Brew.

Head 50km north on Rte 101 to Paso Robles.

5 Paso Robles

Beer, wine and urban distilleries all feature in Paso Robles' Tin City precinct, a vibrant neighbourhood of industrial-style taprooms and tasting studios. Weekend live music and the city's best food trucks combine at BarrelHouse Brewing, while Silva Brewing, on the edge of the leafy downtown area, crafts excellent German-style brews and punchy American pale ales.

Head 330km north back to San Francisco on Rte 101 via Silicon Valley.

ALTERNATIVES

Day trip: Salinas

Salinas is the birthplace of Nobel Prize-winning author John Steinbeck, whose storied life is showcased at the National Steinbeck Center. Interactive exhibits chronicle the work of the author of *The Grapes of Wrath* and *East of Eden*; the museum also includes Rocinante, the campervan he travelled in during his own US road trip, detailed in *Travels with Charley*.

Salinas is 30km northeast of Monterey.

Day trip: Morro Bay

This raffish fishing town is dominated by Morro Rock, one of the Nine Sisters, a series of 21-million-year-old volcanic rocks punctuating the coastal landscape of the region. Rent a kayak, canoe or paddleboard to explore Morro Bay's giant estuary – wildlife includes sea otters and pelicans – before local beers at Three Stacks and a Rock Brewing.

Morro Bay is 21km west of San Luis Obispo.

EXPLORING CASCADIA BY RAIL

Vancouver – Crater Lake National Park
Journey from Canada to the US, visiting three dynamic cities by train before embarking on a road trip south to the spectacular peaks of the Cascade Range.

FACT BOX
Carbon (kg per person): 234
Distance (km): 1746
Nights: 7-10
Budget: $$$
When: Jul-Sep

❶ Vancouver

Begin this journey combining Canada and the United States in British Columbia's biggest city. Take in the otters and beluga whales at Vancouver Aquarium, and then walk or cycle around the Stanley Park seawall. The city's most diverse eating takes place along multicultural Commercial Drive, while historic 'Brewery Creek' along Main St is a hoppy hotspot for craft beer drinkers. Other Vancouver highlights include strolling the historic streets in Gastown, and grazing amid the flavour-packed food stalls of Granville Island Public Market.

Daily trains link Vancouver's Pacific Central Station to Seattle's King St Station (4hr 30min).

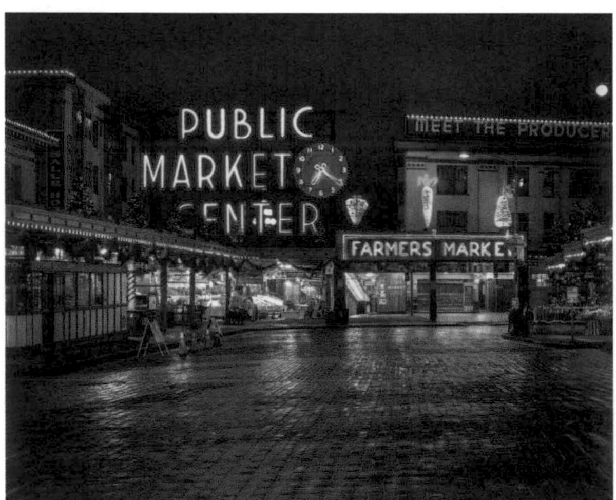

ABOVE: PIKE PLACE MARKET, SEATTLE

❷ Seattle

Experience the fun and theatre of the fish-throwing vendors at Pike Place Market before delving into the Museum of Pop Culture, where rock music history combines with sci-fi and fantasy TV and movies. The musical legacies of Seattle locals such as Jimi Hendrix and Kurt Cobain are all showcased. Next door, check out the retro *Jetsons* vibe of the Space Needle for great city views, and dive deep into the story of Boeing and aviation history at the Museum of Flight.

Continue on the daily Amtrak Cascades train to Portland's Union tation.

❸ Portland

Look forward to coffee, craft beer and the world's biggest independent bookstore in Oregon's capital of cool. Join a cycling tour to explore the Pearl District and along the Willamette River before getting pleasantly lost at Powell's City of Books. Try international flavours from good-value food trucks or

ABOVE: CRATER LAKE NATIONAL PARK, OREGON

be inspired by the cooking of some of America's most innovative chefs.

Rent a car in Portland and drive 180km south to Eugene.

❹ Eugene

Stroll the University of Oregon's leafy campus before browsing the Jordan Schnitzer Museum of Art's excellent Asian collection. Eugene's weekly Saturday market reinforces its alternative lifestyle and liberal credentials, while west of downtown the Whiteaker neighbourhood does its bit with artisan distilleries, craft breweries and urban wineries.

Continue south for 210km to Crater Lake National Park.

❺ Crater Lake National Park

Open from June to mid-October, the 53km Rim Drive around the serrated edge of Crater Lake is a brilliant way to experience the national park's stunning scenery. More than 30 viewpoints are dotted along the road, with the most spectacular vistas reserved for Cloudcap Overlook, more than 600m above the lake's intense cobalt waters.

Drive back to Portland and its well connected airport.

ALTERNATIVES

Day trip: Grouse Mountain

During spring and summer, Grouse Mountain swaps winter sports for warm-weather activities. Ride the Skyride gondola to check out the peak's popular Lumberjack Show, meet grizzly bears in their post-hibernation mountain refuge and explore the forested mountain on ropes and zipline adventures.

Take the SeaBus from downtown Vancouver to Lonsdale Quay, and transfer by bus to Grouse Mountain.

Day trip: Hood River and the Columbia River Gorge

Explore historic Columbia River Gorge, taking in brilliant views at Crown Point, before checking out the two-tiered Bridal Veil Falls and the 200m drop of the Multanomah Falls, the gorge's highest. Hood River is an ideal base to explore the gorge, relax in local wine-tasting rooms, go whitewater rafting or kayaking – and maybe drive the 100km south to spectacular Mt Hood.

It's around 100km from Portland to Hood River.

ACROSS ALASKA BY TRAIN AND BUS

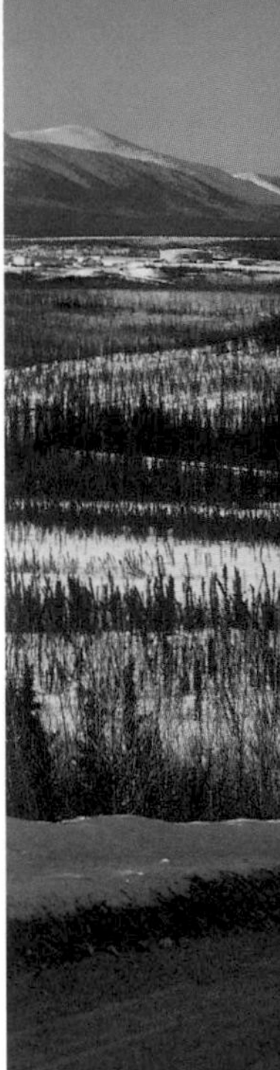

Seward – Prudhoe Bay

Traverse mainland Alaska south to north on its only railway, take a wilderness bus ride past the US' highest peak, then cross the Arctic Circle to visit an isolated oil outpost.

FACT BOX

Carbon (kg per person): 34
Distance (km): 982
Nights: 6-7
Budget: $$$
When: Jun-Aug

❶ Seward

A diminutive cruise port sandwiched between the Harding Ice Field and Resurrection Bay, Seward mixes a gritty downtown with a ritzier harbour. Refresh yourself on dive-bar beer and salmon sandwiches – and don't leave town without taking a shuttle to Exit Glacier, a shimmering tongue of landscape-altering ice patrolled by bears.

All aboard the Alaska Railroad Coastal Classic train to Anchorage (daily May-Sep; 4hr).

❷ Anchorage

In a state that's 99% wilderness, Alaska's biggest metro area presents a comforting pocket of civilisation with microbreweries, creative food and surprisingly good coffee. Get acquainted with Alaskan Native culture at the Anchorage Museum, before stretching your legs on the waterside Tony Knowles Trail with views of the ice-encrusted Alaska Range.

The Denali Star train rattles north to Denali National Park (daily May-Sep; 7hr 30min), offering dazzling views of North America's highest mountain en route.

❸ Denali National Park

Feral wilderness is viewable from the safety of a guided bus tour in Alaska's most accessible national park. Penetrate a stark, chilly tundra guarded by mountains on the 148km-long 'park road', where big

ABOVE: CARIBOU, DENALI NATIONAL PARK

ABOVE: DALTON HIGHWAY

fauna sightings are guaranteed and the vision of Denali (6190m) on clear days inspires hushed awe.

Back on the Denali Star continue to Fairbanks (daily May-Sep; 4hr).

4 Fairbanks

While not a pretty city, sprawling Fairbanks is the gateway to the Arctic and an essential supply centre for trips north. Spend a morning in the architecturally abstract University of Alaska Museum of the North, before loading up on snacks, bear spray and warm clothing for the coming bus ride.

Reserve ahead for the Dalton Highway Express, a minibus that runs the 500km from Fairbanks to Deadhorse/Prudhoe Bay (Jun-Aug; 16hr), crossing the Arctic Circle.

5 Prudhoe Bay

This greasy-oil-patch encampment is the terminus of an incredible journey across mountains, taiga and tundra to the top of the earth. Dip a toe in the Arctic Ocean, bed down with calendar-clocking oil workers in the utilitarian Prudhoe Bay hotel and enjoy the languid midnight sun.

Fly from Prudhoe Bay's tiny airport back to Anchorage.

ALTERNATIVES

Extension: Talkeetna

A strategic stop on the Alaska railroad that acts as both outfitting hub for assaults on Denali and home-base for people of an artistic bent, Talkeetna also has a healthy sense of humour – it once elected a cat as mayor. The town's small, easily walkable grid is punctuated with old wooden edifices hosting pubs, cafes and gift shops.

Day-trip from Anchorage using the train (2hr 45min).

Diversion: Whittier

In a state not short on weird, Whittier is extravagantly strange. A WWII military base expanded during the Cold War, it consists of a cruise port and two incongruous high-rises: the eerily abandoned Buckner Building and the jarring Begich Towers that houses 95% of the 'city's' 200-ish population. The saving graces: beautiful surroundings and an intriguing history.

To get here, take the daily bus from Seward (May-Sep; 2hr 15min).

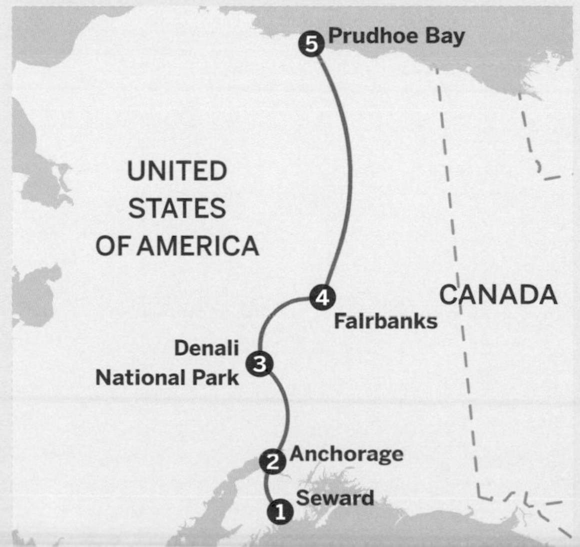

© Michael DeYoung | Design Pics/Getty Images; Noppawat Tom Charoensinphon | Getty Images

CANADIAN ISLAND-HOPPING IN BRITISH COLUMBIA

Vancouver – Victoria
Take to the water between British Columbia's two main urban hubs aboard the state ferry system, stopping off in the pastoral Southern Gulf Islands on the way.

FACT BOX
Carbon (kg per person): 9
Distance (km): 148
Nights: 5
Budget: $$
When: Apr-Oct

❶ Vancouver

Get a taste of rural beauty in the middle of BC's biggest city with a walk around the wilder parts of Stanley Park. Then head to Denman St to contemplate a United Nations of cuisines: more than 50 hole-in-the-wall restaurants in just five blocks. Round off this Vancouver segment with a hike (or cable-car ride) up Grouse Mountain for expansive views over the Strait of Georgia.

 **Take the Canada Line metro from City Centre to Bridgeport, then the 620 bus to Tsawwassen ferry terminal, followed by a ferry to Galiano (2 daily; 1 hr).**

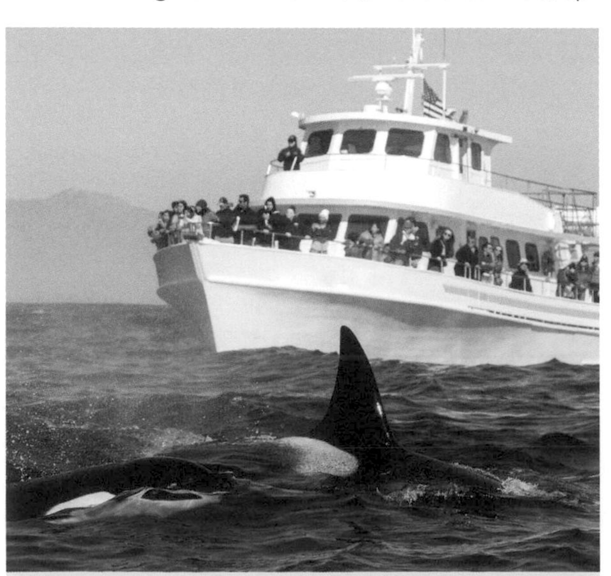

ABOVE: ORCA-SPOTTING, VICTORIA

❷ Galiano Island

A fine introduction to the dreamy tranquillity of the Gulf Islands, Galiano is a calm-inducing brew of quiet coves, wooded hills, driftwood-strewn beaches and esoteric art galleries. Despite no real town, there are several good places to stay and eat. Not much happens here. That's the point.

Take the ferry to Mayne (at least 3 daily; 30min).

❸ Mayne Island

Sleepy Mayne is ideal for gentle sea kayaking, curling up in a congenial B&B with a good novel and hitching rides using the island's unique car-stop service. Visit-worthy destinations include a quirky Saturday farmers market in the summer and a Japanese garden dedicated to the island's early pioneers.

All aboard the ferry to Salt Spring Island (2 daily; 1hr 15min).

ABOVE: STANLEY PARK TOTEM POLES, VANCOUVER

❹ Salt Spring Island

The largest Gulf Island has the archipelago's only real town, Ganges, and abounds with homemade businesses from wineries to cheesemakers to a rustic microbrewery. The highlight? Cycling up the steep 11km road to the top of Mt Maxwell for a Google-Earth view of the whole island chain floating in a calm sea.

Ferries shuttle between Long Harbour and Vancouver Island's Swartz Bay (daily; 1hr 15min). From here catch bus 70 or 72 into Victoria (1hr).

❺ Victoria

Once dubbed more British than Britain, Victoria is decidely cosmopolitan now with a diverse array of cafes and restaurants. But there are still ample opportunities to celebrate Anglophilia in tearooms, waterside fish-and-chips joints, carpeted pubs and (in summer) manicured cricket grounds. Stay close to the harbour to satisfy your Darjeeling cravings in the regal Empress Hotel before jumping on a boat for a windswept bout of whale-watching.

You can take a direct ferry back from Swartz Bay (Victoria) to Tsawwassen (Vancouver).

ALTERNATIVES

Extension: Pender Island

Two islands joined by a narrow bridge, the Penders, on the ferry route between Mayne and Salt Spring islands, display all of the classic Gulf Island calling cards: sleepy lanes, loose schedules and overgrown pastures inhabited by content-looking cows. To add some activity to this idyll, there's a vineyard, a small museum and a summer farmers market.

The island is reachable on ferries from Mayne or Salt Spring islands.

Diversion: West Coast Trail

Within striking distance of Victoria is one of the province's finest coastal hikes, a wild 75km foray along the eastern rim of the Pacific Ocean. The tough but rewarding trail through thick rainforest and past lonely beaches is usually spread over six days, and requires advance booking and a small payment. You'll also need to bring/rent camping gear.

The West Coast Trail Express provides pick-up and drop-off from Victoria (May-Sep).

CRUISE THE CALIFORNIA COAST BY TRAIN

San Francisco – San Diego

Gaze over the Pacific Ocean as you chug along California's magnificent slice of US coast, with both urban and beachside stopovers along your rail journey.

FACT BOX

Carbon (kg per person): 16
Distance (km): 800
Nights: 4-7
Budget: $$
When: year-round

❶ San Francisco

Begin your coastal odyssey in the City by the Bay. San Francisco is all about taking selfies in front of the Golden Gate Bridge, eyeballing the exhibits at the San Francisco Museum of Modern Art, hiking the hills for classic bay views and wandering Chinatown's atmospheric alleys. Every neighbourhood has excellent dining spots, whether you're craving Japanese-Peruvian, classic Italian, creative Californian or, of course, Chinese.

Head across the East Bay to Emeryville or Oakland and board the Amtrak train to San Luis Obispo (daily; 6hr 30min).

ABOVE: SEA LIONS NEAR LA JOLLA, SAN DIEGO

❷ San Luis Obispo

There's plenty of wine to sample in the tasting rooms around this central-coast university town. Try to arrive on a Thursday for the jam-packed evening farmers market – more a street party than a place to buy produce. An easy side-trip is astonishing Hearst Castle, built by publishing magnate William Randolph Hearst with no expense spared.

Take the train to Santa Barbara (several daily; 2hr 45min).

❸ Santa Barbara

With its historic mission church, sandy beaches, contemporary museums, gourmet restaurants and craft breweries, it's hard to be bored in this coastal town. Bonus: the train station is right by 'The Funk Zone' arts and bar district.

Back on the train, head to Los Angeles (several daily; 3hr).

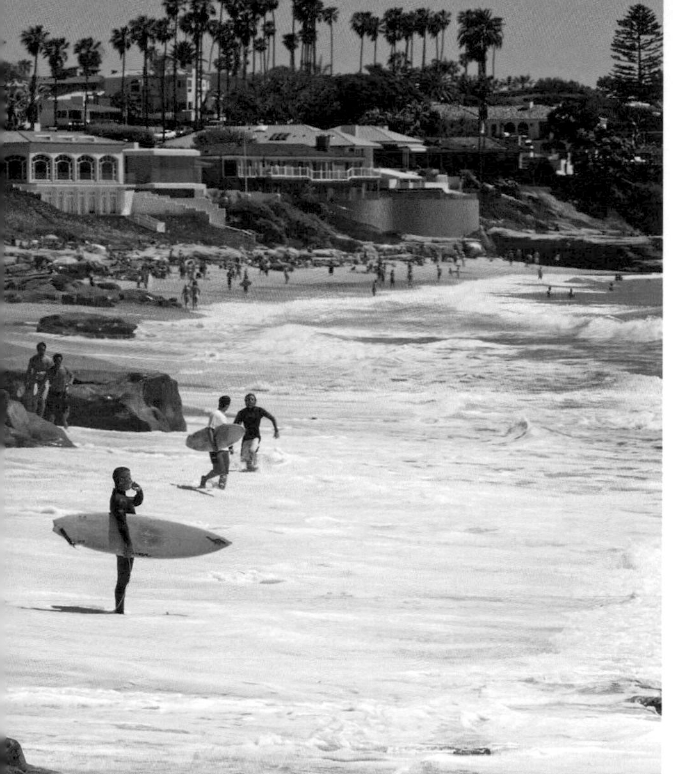
ABOVE: LA JOLLA'S WINDANSEA BEACH, SAN DIEGO

❹ Los Angeles

You could spend weeks exploring California's largest city. Spot stars (or at least their handprints) in Hollywood, watch the parade of people along quirky Venice Beach and look out over the metropolis from Griffith Park. Love art? Take in the contemporary galleries at The Broad, the hilltop Getty Center (amazing views too) and the massive Los Angeles County Museum of Art. And don't miss out on the huge range of international cuisine on offer, most notably Chinese, Japanese and Mexican.

Continue south on the train to San Diego (several daily; 3hr).

❺ San Diego

While beach culture is alive and well in San Diego, there's more to the city than sand. Its famous zoo in Balboa Park is a huge crowd-pleaser, and not just with children; the park also holds an eclectic selection of museums. Then check out the Latino art in Barrio Logan and visit posh La Jolla to dine overlooking the ocean or snap seaside photos in its sandy coves.

San Diego's airport has national and international connections.

ALTERNATIVES

Extension: San Diego's North County

Day-trip through the beach towns north of San Diego, learning some surfing history at Oceanside's California Surf Museum, listening to tunes past and present at Carlsbad's Museum of Making Music, or seeing what's on view at the contemporary Lux Art Institute in Encinitas. Have a fish taco on the sand before heading back downtown.

The coast-hugging Pacific Surfliner and local buses connect towns in the region.

Diversion: Portland, Oregon

Want a longer train journey? Head from SF to Oregon's coolest city. You never know quite what to expect in Portland, where a popular local slogan is 'Keep Portland Weird'. Hike rainforest trails in Forest Park, raise a glass at craft breweries and distilleries, tour the Japanese Garden and eat your way through the food trucks, clustered into 'pods' throughout the city.

Regular Amtrak trains run from SF to Portland (21hr 35min).

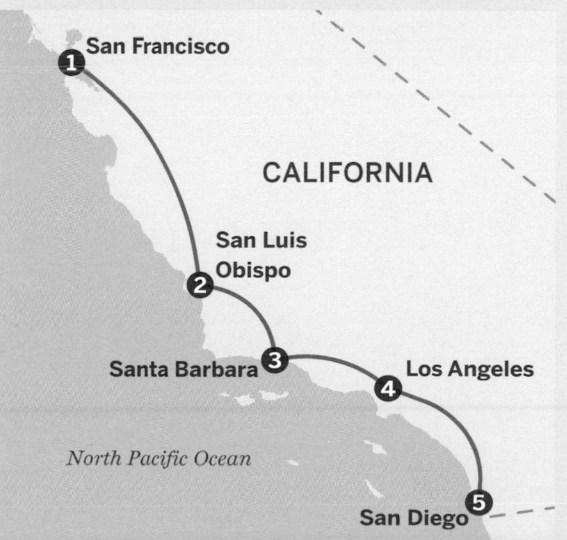

DRIVING CALIFORNIA'S HIDDEN HIGHWAY

Lake Tahoe – Las Vegas

California's Hwy 395 connects lakes with mountains with movie locations with deserts on a crowd-free, nonstop-highlights US road trip.

FACT BOX

Carbon (kg per person): 88
Distance (km): 655
Nights: 5–7
Budget: $$
When: May–Oct

❶ Lake Tahoe

Tahoe has year-round appeal. You can swim or kayak its clear water, hike or mountain bike its pine-covered hills (the Flume Trail is popular), ski its snow-covered peaks in winter, or just enjoy the views over a waterside drink.

Drive 180km south along Hwy 395.

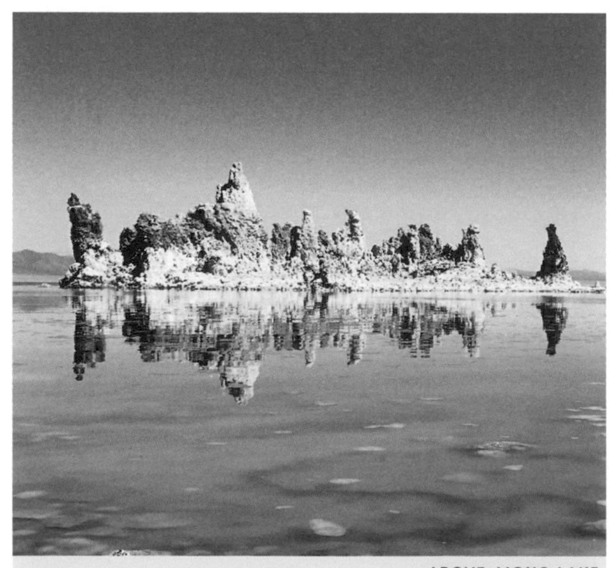

ABOVE: MONO LAKE

❷ Mono Lake

Perfect setting for a sci-fi movie, Mono Lake is an otherworldly waterscape where limestone tufa formations rise out of the shrimp-filled brine and millions of migratory birds make a temporary home to gorge on the alkali flies found in abundance here.

Continue 50km south along Hwy 395.

❸ Mammoth Lakes

Five swimmable lakes, multiple hiking trails, ski slopes aplenty and spectacular natural sights like Devils Postpile's basalt columns and Rainbow Falls are the daytime draws in Mammoth Lakes. And once the sun sets there are quality hotels and restaurants for end-of-the-day relaxation.

Continue 150km south along Hwy 395.

❹ Manzanar Internment Camp

A bleak place offering a bleak reminder of the USA's treatment of people of Japanese ancestry during WWII, Manzanar is where thousands of men, women and children were incarcerated from 1942 to 1945.

Drive 20km south on Hwy 395.

❺ Lone Pine

Think this town and those hills look familiar? You're

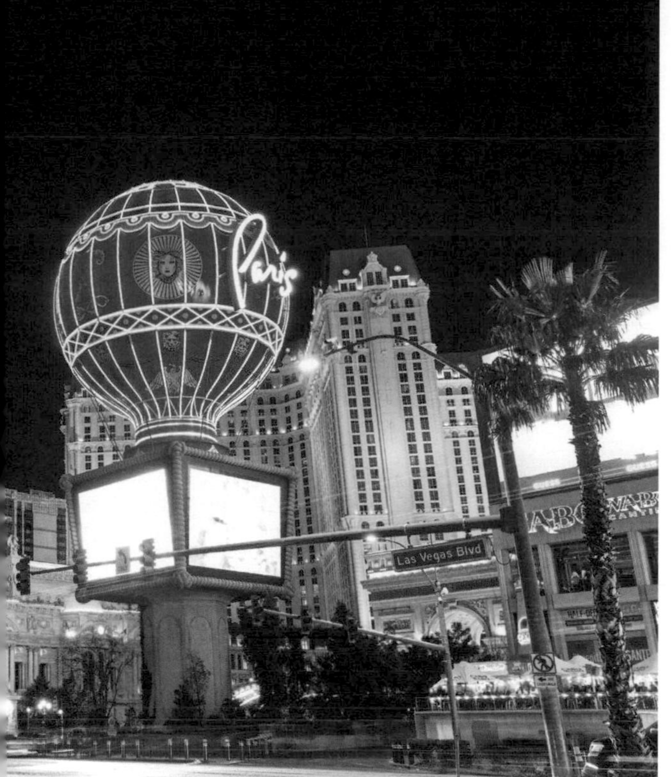

ABOVE: THE LAS VEGAS STRIP.

not wrong. Lone Pine and the Alabama Hills to its west were used in hundreds of movies, TV shows and commercials, as documented in the local museum. But the real star here is Mt Whitney, the highest peak in the contiguous US. Enter the permit lottery to climb it, or simply take in the cool air and views from the trailhead.

Head 85km east along Hwy 136/Hwy 190.

6 Death Valley National Park

The low-point of the trip, literally. Death Valley's Badwater Basin sits 86m below sea level, the lowest point in North America. The starkly beautiful national park holds another record – hottest place on earth (54.4°C), recorded in 2020.

Drive 170km southeast along Hwy 136/Hwy 190.

7 Las Vegas

And now for something completely different. The Las Vegas Strip is an adrenaline-packed jolt after the remoteness of the trip so far. But there's no denying the fun to be had amid the architectural pastiches, all-you-can-eat buffets, pulsating neon lights and chances to try your luck at the tables.

Vegas' McCarran Airport has dozens of national and international connections.

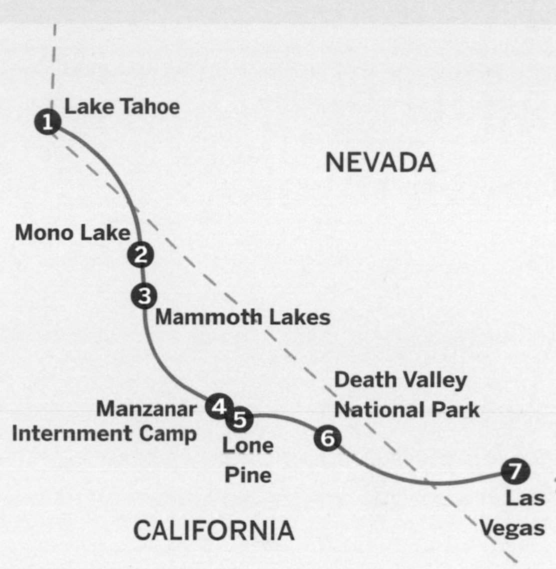

ALTERNATIVES

Extension: Bodie Ghost Town

In the hills east of the 395 lies Bodie, once a thriving gold-mining town, now a creepy-cool collection of abandoned homes and businesses. The road is bumpy and the elevation can make it chilly, but that just adds to the atmosphere as you peer through cracked windows into deserted houses and head underground to see mining life in all its harshness.

Bodie is 20km east of Hwy 395 on Hwy 270.

Diversion: Yosemite National Park

Turn west at Mono Lake and, snow permitting on Tioga Rd, you'll soon be in the impossibly majestic surroundings of Yosemite National Park. Take in the views from Glacier Peak; hike along the Merced River; try to capture Yosemite Falls with your camera – at every turn there's a natural wonder that makes the detour worthwhile.

Drive 120km west of Hwy 395 on Hwy 120.

MEXICO'S CRADLE OF INDEPENDENCE

Guanajuato – Dolores Hidalgo

Mexico's highlands have a history as rich as their silver mines, with three cities key to the country's independence.

FACT BOX

Carbon (kg per person): 20
Distance (km): 175
Nights: 4–5
Budget: $$
When: avoid Jun–Sep rainy season

1 Guanajuato

Though one of Mexico's most beautiful cities, Guanajuato is more than a pretty face. Yes, the brightly coloured buildings dotting the hills of the valley it sits in are undeniably photogenic, but those same hills contain silver that made this place rich – head underground to La Valenciana mine, once responsible for two-thirds of global silver. Guanajuato's elegant plazas and colonial architecture are peaceful today, but it was here that the first battle for Mexican independence took place in 1810 – a huge statue of local hero, El Pípila, looms over the valley. And cultural highlights run from the annual Cervantes Festival, celebrating Spain's great writer, and the museum-home of artist Diego Rivera, to nightly *callejoneadas*, fun tours through the city's alleys with singing, 17th-century-costume-wearing musicians.

Drive 80km east.

2 San Miguel de Allende

Now full of artists, visitors and expats who love the town's colonial buildings and excellent restaurants, San Miguel de Allende's past was much more tumultuous. Focus of the decades-long Chichimeca War between indigenous people and Spanish conquerors, it was later the birthplace of two independence martyrs, Juan Aldama and Ignacio Allende (his surname was

ABOVE: GUANAJUATO

ABOVE: SAN MIGUEL DE ALLENDE

added to the town's), who were executed in 1811. A post-independence decline was turned around with the arrival of US artists in the mid-20th century; many moved here, contributing to its cosmopolitan population, architectural preservation and cultural revival today.
Drive 40km north.

3 Dolores Hidalgo

Every Mexican knows 'El Grito de Dolores', the cry for independence declared from Dolores Hidalgo's main church on 16 September 1810. Local priest Miguel Hidalgo y Costilla (the town added his surname to its own in his honour) rang the church bells, summoned his congregation and urged them to revolt against the Spanish authorities. His exact words are unknown but the effect they had led to a long and ultimately successful fight for freedom (independence wasn't secured until 1821). These days the town is quiet – except each September when crowds celebrate in the plaza outside the famous church with its Churrigueresque façade; sometimes the Mexican president delivers a commemorative speech here. At other times, the museum next door provides an informative look at the town's history.
Drive 55km back to Guanajuato for bus connections to Mexico City and the airport at León.

ALTERNATIVES

Diversion: Parador Turístico Sangre de Cristo

This large, impressive complex in the hills west of Guanajuato houses a cafe, shops and three museums. One explores the region's mining heritage, another displays 36 mummies discovered in nearby churches. The third celebrates the Day of the Dead festival; exhibits include big-hatted La Catrina skeleton dolls.
Find the parador 23km west of Guanajuato.

Diversion: Cuna de Tierra vineyard

Guanajuato state's oldest vineyard is an architecturally stylish affair just north of Dolores Hidalgo. The owners reintroduced winemaking here in 2005, two centuries after Mexican vino was banned by the Spanish; today, they make some 80,000 bottles annually. Tour and taste the various reds and whites made on the premises and climb the tower for vineyard views.
The vineyard is 12km northeast of Dolores Hidalgo.

A TENNESSEE MUSICAL TRIBUTE

Nashville – Memphis

Titans of America's musical heritage, from country to blues, Nashville and Memphis are easily combined on a Tennessee two-city road trip.

FACT BOX

Carbon (kg per person): 43
Distance (km): 340
Nights: 4–5
Budget: $$
When: Apr, May, Sep & Oct

ABOVE: LIVE BLUES IN MEMPHIS

❶ Nashville

The genre most synonymous with Nashville gets a whole, bass-clef-shaped museum to itself in the Country Music Hall of Fame. Big names (Dolly, Johnny) are all covered, but it's the in-depth displays showing how country music became a phenomenon that impress most.

To the north stands the mother church of country music, the Ryman Auditorium. It was from here that the Grand Ole Opry radio shows were originally transmitted, with legends like Hank Williams and the Carter Sisters heard across the nation. It's still a favourite with performers and audiences.

South of downtown, plain looks disguise a distinguished musical legacy at RCA Studio B. Join a tour and expect a tingle down the spine standing in the spot where Dolly recorded *I Will Always Love You* and Elvis asked *Are You Lonesome Tonight?*

For live music, Nashville has three top options: brash Broadway and its neon-advertised honky-tonks (music bars); famed, intimate Bluebird Café where the likes of Taylor Swift honed their craft; and the Grand Ole Opry from where stage shows are still broadcast to a global audience.

Drive 340km southwest on Hwy 40.

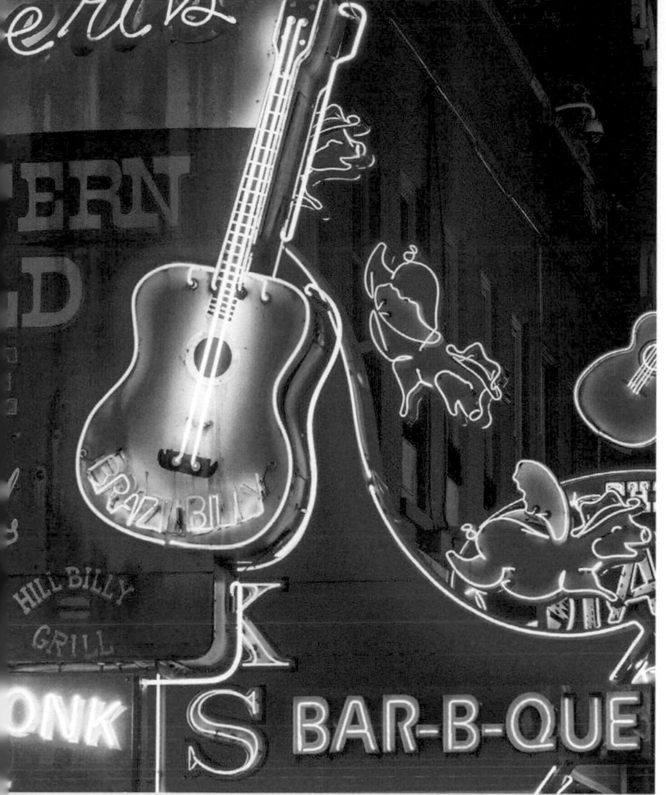

ABOVE: BROADWAY, NASHVILLE

❷ Memphis

Begin your exploration of Memphis' musical might at Sun Studio. It claims to be the birthplace of rock and roll, and with a roster of recording artists including Jerry Lee Lewis, Roy Orbison and Elvis (he cut his debut single, *That's All Right*, here in 1954), who are we to argue?

Stax Studio was to soul music what Sun Studio was to rock, and the Stax Museum offers insights into the development of soul and its most famous exponents, such as Otis Redding.

South of downtown lies Graceland, the place where Elvis lived, died and is buried. No matter how big (or little) a fan you are, this is a fascinating, fitting tribute to one of music's biggest stars. The house is a glimpse into his personal life (that Jungle Room!) while other displays chart his enormous cultural impact. Pay the extra to see his planes.

Thanks to past performers like BB King and Louis Armstrong, Memphis' Beale St is known as the Home of the Blues. Pedestrianised and well supplied with bars and restaurants, it guarantees a music-packed night out to remember.

Memphis has train, bus and plane connections to many US cities.

ALTERNATIVES

Diversion: Franklin

Franklin has plenty of small-town charm, its busy Main St lined with architectural beauties. But it's the bloodstained past that is the big draw: in November 1864, one of the US Civil War's bloodiest battles took place here. Three sites tell the story: Carnton, a plantation that became a field hospital; Carter House, scene of the fiercest fighting; and Lotz House, where cannonball damage is still visible.

Franklin is 35km south of Nashville on Hwy 65.

Diversion: National Civil Rights Museum, Memphis

Memphis' National Civil Rights Museum is an often-harrowing but necessary history of the struggles for equality by African Americans. Permanent collections examine Jim Crow laws and the Montgomery bus boycott, while temporary exhibitions cover topics like *The Green Book* guides that listed safe destinations for Black travellers.

The museum is a few blocks south of Beale St.

PRINCE EDWARD ISLAND EXPLORER

Charlottetown – Souris

A DIY driving tour around Canada's PEI goes beyond *Anne of Green Gables* for quirky experiences like climbing lighthouses, lobster fishing, shucking oysters and much more.

FACT BOX

Carbon (kg per person): 47
Distance (km): 378
Nights: 7
Budget: $$$
When: May-Sep

❶ Charlottetown

Charlottetown is touted as the birthplace of Canada. To see why you should start your visit at Province House, a handsome, Neoclassical building where in 1864 the 'Fathers of Confederation' met to plot out the creation of Canada. If you're into music, visit in summer and book ahead for the Charlottetown Festival, touted as 'the largest musical theatre festival in Atlantic Canada since 1965'.

Drive 35km west.

❷ Victoria-by-the-Sea

Victoria-by-the-Sea, located on the island's Red Sands Shore, is a pretty, arty village. The mint- and cream-coloured Victorian buildings house tea rooms, art galleries, a chocolate shop and great restaurants. The town is also an excellent spot from which to head out on a sea kayaking adventure.

Continue 66km northwest.

❸ Tyne Valley

Set on the Trout River, this quaint village and its surrounds is traditional Prince Edward Island at its best. Watch lobster boats unloading their baskets, oyster catchers with their tongs and farmers ploughing their fields. Don't miss Valley Pearl Oysters, where you can take a 'Get Shucked' tour.

Drive 15km northeast.

❹ Lennox Island First Nation

Head across a causeway to Lennox Island and you're in for a rich lesson in Mi'kmaq culture. Visit the cultural centre for information on Mi'kmaq history

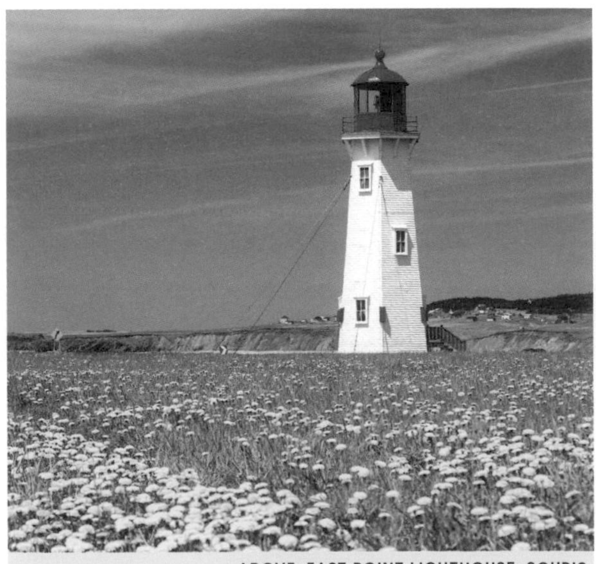

ABOVE: EAST POINT LIGHTHOUSE, SOURIS

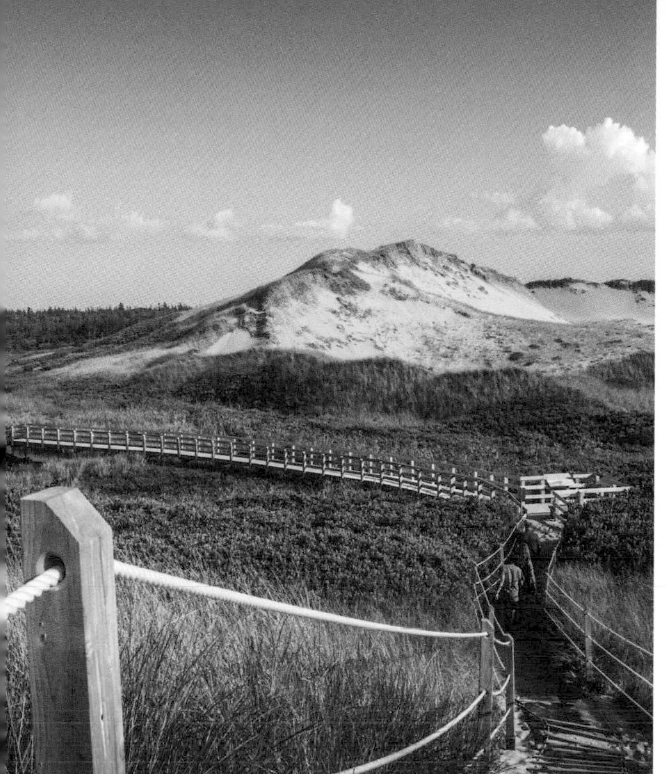

ABOVE: PRINCE EDWARD ISLAND NATIONAL PARK

and customs, then wander along the nature trail, The Path of Our Forefathers, through Pow Wow Grounds, blueberry fields and along the shore.

Head 100km east to Prince Edward Island National Park.

5 Prince Edward Island National Park

Lined by beaches, with forest behind the shoreline, PEI National Park has lots of outdoor activities just waiting for you, including swimming (for the brave) or hiring a bike to ride the 5km Robinsons Island Trail System.

Drive 84km to PEI's far east.

6 Souris

Souris is a small village (just 1000 inhabitants or so) but offers big things. Besides being a lovely place to walk around (the word 'cute' is highly applicable here), you can enjoy an attractive beach and historic lighthouse (take a tour to the top). Then wander around the artisans' stalls at Artisan on Main, and treat yourself to some of the island's best ice-cream. For yoga fans, there are studios galore.

Round off the trip with the 78km drive west back to Charlottetown.

ALTERNATIVES

Extension: Anne of Green Gables

PEI's literary claim to fame is much-loved *Anne of Green Gables.* Several sites devoted to the book can be found near Cavendish on the north side of the island. Within 20km of each other you'll find the home of author Lucy Maud Montgomery and a replica of Avonlea village, complete with a museum and souvenir shops.

The locations make an easy detour on the drive between Tyne Valley and PEI National Park.

Diversion: Point Prim Lighthouse

You can climb this historic brick structure, the oldest lighthouse on Prince Edward Island (built in 1845) and one of 50 in total on the island. The reward? A lovely view of Northumberland Strait and Hillsborough Bay. There's a virtual tour if you don't have a head for the 18m height.

The lighthouse is 50km south of Charlottetown.

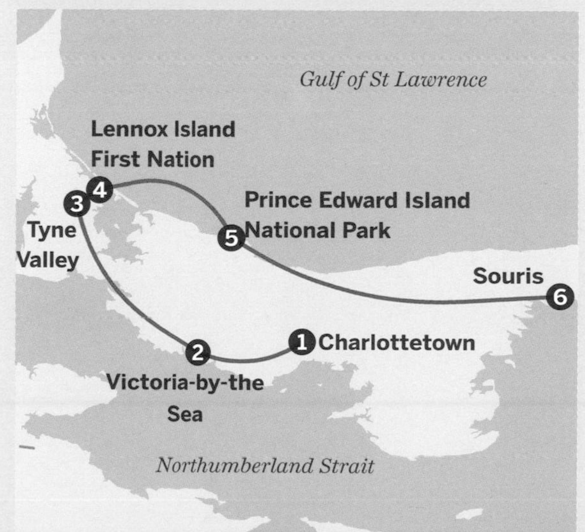

A CANADIAN ROAD TRIP ROUND VANCOUVER ISLAND

Victoria – Ucluelet

The largest of British Columbia's 6000 or so islands, Vancouver Island is also home to the province's capital and plenty of adventures for road-trip enthusiasts.

FACT BOX

Carbon (kg per person): 81
Distance (km): 649
Nights: 5-7
Budget: $$-$$$
When: Mar-Nov

❶ Victoria

Wander through the streets of BC's quaint capital, whose Victorian buildings and hanging flower baskets give it a British feel. Check out the Fairmont Empress Hotel for afternoon tea and the Gothic BC Parliament Buildings. Then it's time for a bike tour to some of Victoria's other sights: Canada's oldest Chinatown where the streets are lined with red and gold lanterns (don't miss the Gate of Harmonious Interest or Fan Tan Alley, one of the country's narrowest streets); Craigdarroch Castle, a perfect example of Victorian Baronial architecture with exquisite stained-glass windows; and the Royal BC Museum with engaging displays on local history and culture. Continue to Fisherman's Wharf and snap some pics of the colourful houseboats and playful seals, then cycle to Beacon Hill Park to admire the world's tallest freestanding totem pole (40m).

Drive 111km northwest.

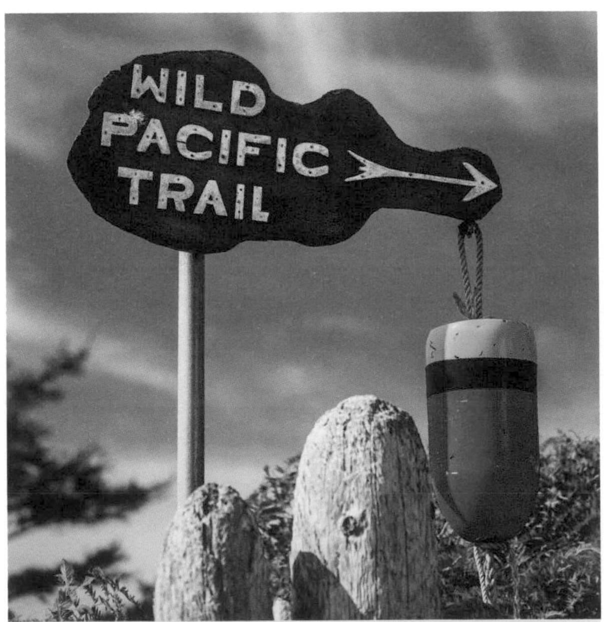

ABOVE: WILD PACIFIC TRAILHEAD, UCLUELET

❷ Nanaimo

Vancouver Island's second city, Nanaimo is the place to learn about local heritage, from First Nations to colonial, at the town museum; then sip on an ale at one of its craft breweries and nibble through a Nanaimo bar, the local chocolate speciality. Afterwards, take a 10-minute ferry hop to Saysutshun/Newcastle Island Marine Provincial Park where you can explore tidal pools and Snuneymuxw village sites, or hike or bike any of the island's 22km of trails.

Continue 207km west.

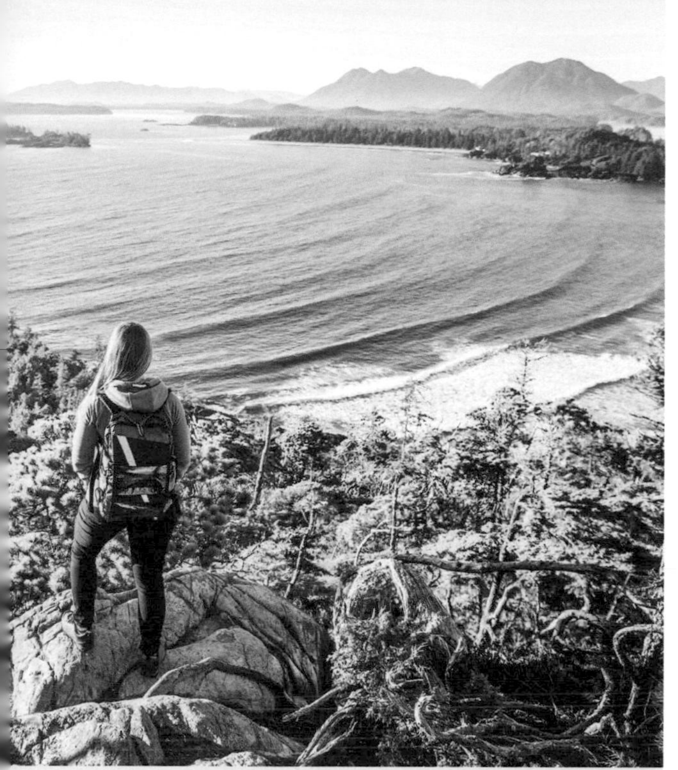
ABOVE: PACIFIC RIM NATIONAL PARK

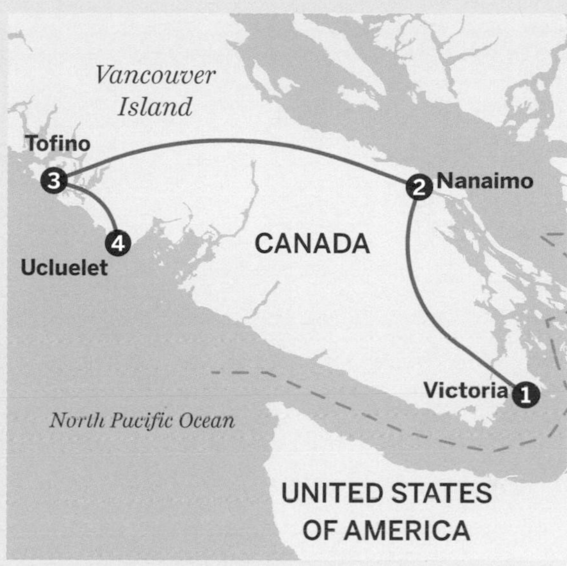

❸ Tofino

West-coast Tofino is the perfect base for exploring this stunning part of Vancouver Island, and the best way to familiarise yourself with the area is by hiring a Long Beach Nature Tour Guide. Then grab a surfboard (this is Canada's surfing capital), stand-up paddleboard or kayak, or get out on the water in Clayoquot Sound on a cruise. Head out on a bear or whale watching tour – chances of spotting the latter are high. Tofino itself is a growing gourmet hub; one excellent hotspot is Industrial Way, where you'll find charcuteries, brewing companies and bread-making outfits.

Drive 40km southeast.

❹ Ucluelet

Stroll along Ucluelet's Wild Pacific Trail, a scenic Pacific-side walk, before munching your way through delicious dishes at Pluvio, one of Canada's best restaurants. Clue up on your local marine species at the extraordinary Ucluelet Aquarium – all of its marine life is released back into the surrounding ocean at the end of the season.

Head 291km back to Victoria for onward connections.

ALTERNATIVES

Extension: Nanaimo to Tofino

Take your time driving between Tofino and Nanaimo. This across-the-island section of the trip is truly spectacular, passing over a dramatic mountain range and winding through ancient rainforests and past ink-coloured lakes. Detour to view massive trees in MacMillan Provincial Park's Cathedral Grove, hike the waterfall trails at Little Qualicum Falls Provincial Park, and pause at the viewpoints along the hairpin Hwy 4.

Diversion: Pacific Rim National Park

Be sure to go hiking in Pacific Rim National Park. Choose from trails that head along remote beaches or nature trails that weave through old-growth cedar-hemlock forests. One of the prettiest is the Willowbrae Trail, an historic track that used to link Ucluelet and Tofino before the establishment of an official road in 1942.

The park lies on the coast between Tofino and Ucluelet.

AMERICA'S MODERNIST MARVELS: LA TO PALM SPRINGS

Los Angeles – Palm Springs
Discover Modernist architectural masterpieces on a road trip from the City of Angels along the Californian coast to the coolest desert resort in the country.

FACT BOX

Carbon (kg per person): 58
Distance (km): 470
Nights: 10
Budget: $$$
When: year-round

❶ Los Angeles

Irresistible LA is a hedonistic home for the rich and famous, the wannabe rich and famous, the down and out, and everyone in between. It's home to the iconic Hollywood sign, Santa Monica's famous pier and eye-opening Venice Beach. It's also one of the best cities in the world to see classic examples of mid-century architecture – highlights include the Eames Case Study House, the Getty Center, Frank Lloyd Wright-designed Hollyhock House, the Stahl House by Pierre Koenig and the Schindler House by Rudolph Schindler.

From Santa Monica, drive the Pacific Coast Hwy (PCH) 58km south to Palos Verdes.

❷ Palos Verdes

Pop the top and head south for a spectacular drive along the PCH, hugging the coast and gaping at its string of enticing beaches. Past Redondo Beach, detour onto Palos Verdes Blvd and then right on Palos Verdes Dr for the scenic route along the stunning Palos Verdes Peninsula. Stop off at Wayfarers Chapel, a superb structure of glass and redwood designed by Lloyd Wright, son of Frank Lloyd Wright.

Continue 188km southeast on the PCH via Huntington Beach, Newport Beach and Laguna Beach to San Diego.

ABOVE: CLASSIC MID-CENTURY DESIGN, PALM SPRINGS

ABOVE: GIESEL LIBRARY, SAN DIEGO

❸ San Diego

Surf, sand and skateboards are what San Diego is all about. This relaxed city is the ultimate dropout for sun-seekers but is also home to world-famous sights such as the San Diego Zoo and the Hotel del Coronado. It's also an underrated hotspot for Modernist architecture – the UC San Diego's Geisel Library, Salk Institute for Biological Studies and the Timken Museum of Art in Balboa Park to name a few.

Turn inland and drive 224km northeast to Palm Springs.

❹ Palm Springs

Nowhere crams in more mid-century marvels than the Californian desert capital of cool. Back in the day, Palm Springs was a haven for the Rat Pack and Hollywood stars who partied here in homes designed by revered architects such as Albert Frey, Frank Lloyd Wright and Richard Neutra. Check in to a 1950s throwback motel with Sinatra on the record player, cocktails by the pool and soaring palm trees swaying overhead. Visit in February for the ultimate architecture celebration during Modernism Week.

Palm Springs International Airport has flights to national and international destinations.

ALTERNATIVES

Extension: Las Vegas

If you're craving more architecture and dramatic desertscapes, continue on to Sin City from Palm Springs. Las Vegas is known for its Postmodernist architecture – a reaction to Modernism in the late '60s, adding some colour and playfulness into the design – and also has a few mid-century delights, such as the Neon Museum and the Guardian Angel Cathedral.

Vegas is 375km northeast of Palm Springs.

Diversion: Joshua Tree National Park

The surreal landscape of Joshua Tree National Park is dotted with its famous tree-sized yuccas and otherworldly boulders. It's an easy day trip from Palm Springs so lace up your hiking boots and head out on any number of trails; or just find a perch and take in the trippy landscape and mystical atmosphere.

The park is 54km north of Palm Springs, on the way to Las Vegas.

BEST OF SOUTHERN APPALACHIA

Cloudland Canyon – Shenandoah

From Georgia's Cloudland Canyon head north through southern Appalachia to soak up national parks and forests, outdoor hubs and some of the eastern US' best camping.

FACT BOX

Carbon (kg per person): 172
Distance: 1381
Nights: 7-10
Budget: $$
When: Mar-Nov

❶ Cloudland Canyon State Park

Hike down to the gorge of this whimsically named state park to see babbling creeks, waterfalls and bafflingly large rock formations – keep an eye out for fantastical 'mushrooms' and sleepy 'tortoises'.

Drive 42km north, crossing from Georgia into Tennessee.

❷ Chattanooga

Known as the gateway to southern Appalachia, Chattanooga has plenty to detain you: indulge your kitsch craving at Rock City, learn about aquatic life at the Tennessee Aquarium or paddle the Tennessee River.

Continue 241km northeast.

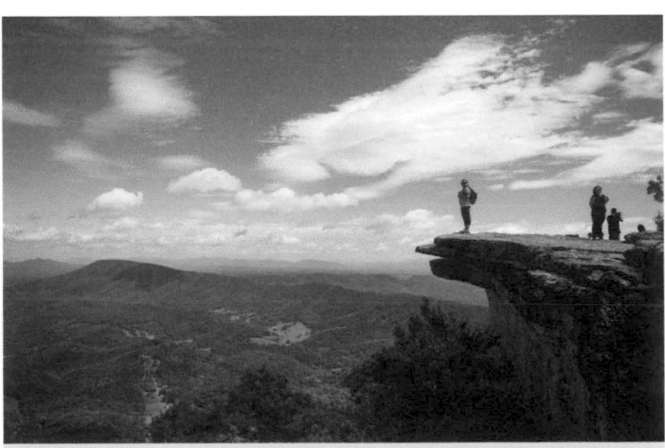

ABOVE: MCAFEE KNOB VIEWPOINT, ROANOKE

❸ Great Smoky Mountains National Park

The Smokies receive the country's highest number of national park visitors, but don't let that deter you – the park contains a ton of uncrowded options thanks to 1287km of trails that serve up spectacular views of the oldest mountain range in the US.

Drive 73km north.

❹ Knoxville

Knoxville has a buzzing downtown (home to three striking historic theatres) and its attractive Old City. Don't miss Knoxville's Urban Wilderness, a 1000-acre protected area that includes forest, parks, mountain-bike trails and more.

Continue 187km east into North Carolina.

❺ Asheville

Asheville has long been revered as a prime weekend break destination thanks to its restaurants and breweries – and the Biltmore, the spectacular Vanderbilt estate just outside town. Stay in the centre of the action or snag a cabin on the outskirts for a rural escape.

ABOVE: AUTUMN COLOURS, BLUE RIDGE PARKWAY

Head 10km southeast.

❻ Blue Ridge Parkway

Managed by the National Park System, the Blue Ridge Parkway winds 754km through the heart of Appalachia. Drive the route in autumn for eye-popping leaf colours, and walk at least a couple of trails along the way.

Follow the Parkway northeast for a beautiful 440km.

❼ Roanoke

Take a break from the road in Roanoke, the 'Capital of the Blue Ridge'. Hike a section of the famous Appalachian Trail and visit McAfee Knob, one of the best – and scariest – viewpoints in the region.

Drive 226km to Shenandoah National Park.

❽ Shenandoah National Park

The road ends in Shenandoah, a sprawling mountainscape encompassing 200,000 acres and 804km of trail. Set up at one of the park's many campsites, cruise the Skyline Drive and hike to numerous lookout points.

From the park it's 142km to Washington DC, the nearest major transport hub.

ALTERNATIVES

Extension: Blowing Rock and Boone

On your way through North Carolina via the Blue Ridge Parkway, stop off at the Moses H. Cone Memorial Park and hike one of its many trails, all easily accessible from the parkway; don't miss Blowing Rock for epic mountain views. Spend the night in the inviting university town just north, Boone, an eco minded community with a serious outdoorsy streak.

Boone is 136km northeast of Asheville.

Diversion: Lake Lure

Lovers of the movie *Dirty Dancing* should make the pilgrimage to Lake Lure, North Carolina, a small resort town located in Hickory Nut Gorge. Book a night at the historic Lake Lure Inn, where several scenes were filmed, or hike to nearby Chimney Rock. If you visit in September, you'll catch the Dirty Dancing Festival, which features dance lessons, watermelon races and lake-lift competitions.

Lake Lure is 45km east of Asheville.

GET ACTIVE IN THE CANADIAN ROCKIES

Calgary – Jasper National Park

Explore the mountains, lakes and glaciers of the Canadian Rockies on this active road trip, with breaks for hiking, paddling, climbing and other outdoor adventures.

FACT BOX

Carbon (kg per person): 103
Distance (km): 830
Nights: 6-8
Budget: $$
When: May-Oct

❶ Calgary

Before travelling through Banff and Jasper National Parks – the headline acts in Canada's Rocky Mountains – start in Calgary, the region's gateway city. Get an introduction to local history and culture at Glenbow Museum, stroll the riverfront and have coffee in the revamped East Village. Journey through Canada's musical heritage at the National Music Centre, then hit the clubs along the city's Music Mile.

Drive 125km west from Calgary to Banff.

❷ Banff National Park

More than 1600km of hiking trails wend through the mountains and along the glacier-fed lakes of Canada's first, and most visited, national park, Banff. Ride a gondola or brave a via ferrata to the peaks, soak in mineral-rich hot springs, or cruise the lakes by canoe or scenic boat tour. Visit the Cave and Basin National Historic Site to learn how the park got its start. Book accommodation and campsites well in advance for this popular park.

Continue 60km northwest.

❸ Lake Louise and the Icefields Parkway

A highlight of any Canadian Rockies trip is a stroll

ABOVE: BULL ELK, JASPER NATIONAL PARK

ABOVE: KAYAKING THE BOW RIVER, BANFF NATIONAL PARK

along much-photographed, turquoise-coloured Lake Louise. Hike into the glacier-topped mountains above the lake for stellar views. Canoe nearby Moraine Lake, then detour to Yoho National Park, where popular sites include the unmissable, jade-hued Emerald Lake and Takakkaw Falls, one of Canada's highest waterfalls (373m). Allow several hours to enjoy the drive north to Jasper on the Icefields Parkway – it ranks among Canada's most spectacular routes and rewards a leisurely pace. Capture the views above Peyto Lake, hike Wilcox Pass Trail for vistas across the glaciers, or take an excursion onto the Columbia Icefield, the Rockies' largest section of glacial ice.

Drive from Lake Louise to Jasper (230km), via Hwy 93/Icefields Parkway.

4 Jasper National Park

Among the attractions of Jasper National Park, the Canadian Rockies' largest, are boating on glacier-fed Maligne Lake, walking through the limestone gorges at Maligne Canyon and relaxing in Miette Hot Springs. You can hike for days, too. At night, take a virtual tour of the Jasper Dark Sky Preserve at the town's planetarium.

To return from Jasper to Calgary and its onward connections, retrace your route on the Icefields Parkway (415km).

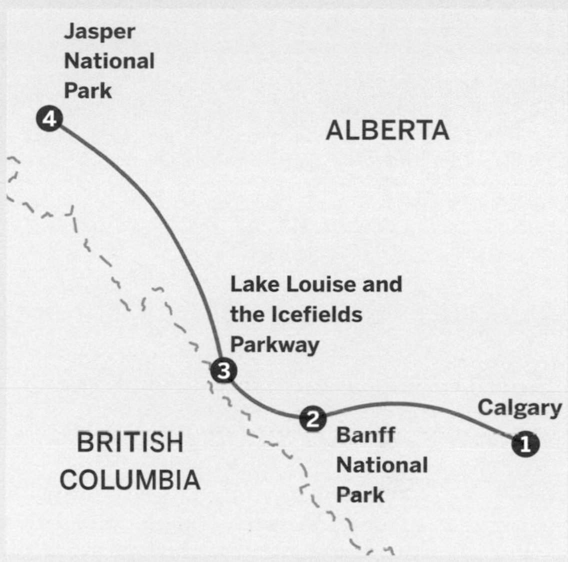

ALTERNATIVES

Extension: Edmonton

Detour east from Jasper for an urban culture weekend in Edmonton, with diverse neighborhoods and contemporary museums like the Art Gallery of Alberta or Royal Alberta Museum to explore. Outside town, watch bison roaming through Elk Island National Park.

Edmonton is 365km east of Jasper. Continue south for 300km to Calgary.

Diversion: Eastern British Columbia's parks

BC has some stunning parks within road-tripping distance of Jasper, Lake Louise and Banff. From Jasper, follow Hwy 16 to Hwy 5 for Wells Gray Provincial Park. Continue south, then east onto Hwy 1 for Mount Revelstoke, Glacier and Yoho national parks. A detour south on Hwy 93 leads to Kootenay National Park, before you return to Calgary.

Jasper to Calgary, via Wells Gray and the national parks, is a 1200km drive.

COAST-TO-COAST CANADA BY TRAIN

Vancouver – Toronto

From the west-coast rainforests to the Rocky Mountains, across the prairies to multicultural urban centres, this classic train trip shows off Canada's highlights.

FACT BOX

Carbon (kg per person): 89

Distance (km): 4466

Nights: 4-14 (depending how many stops along the way)

Budget: $$

When: year-round

❶ Vancouver

Vancouver is an outdoor enthusiast's paradise, a place where urban culture and nature mix. Cycle around Stanley Park, paddleboard from city-centre beaches or head into the North Shore mountains for hiking (or skiing) just 30min from downtown. Check out craft breweries, murals and indigenous art galleries, and enjoy top-notch seafood and Asian cuisine. Then board the *Canadian*, VIA Rail's flagship train, to start your iconic rail journey toward Toronto. It's a four-day, four-night trip straight through, but take your time and get off to explore some of the places on the way.

Take the VIA Rail to Jasper (2-3 weekly; 19hr).

❷ Jasper

The *Canadian* pulls into the centre of this Rocky Mountain town, surrounded by vast, spectacular Jasper National Park. Staff in the park visitor centre opposite the station can help you get oriented. Go for a hike, perhaps the Valley of the Five Lakes trail to five dramatically coloured bodies of water. In town, Jasper-Yellowhead Museum introduces you to local history, while Jasper Food Tours shows you what's good to eat. To day-trip through the mountains and glaciers along the Icefields Parkway, book a local tour or rent a car.

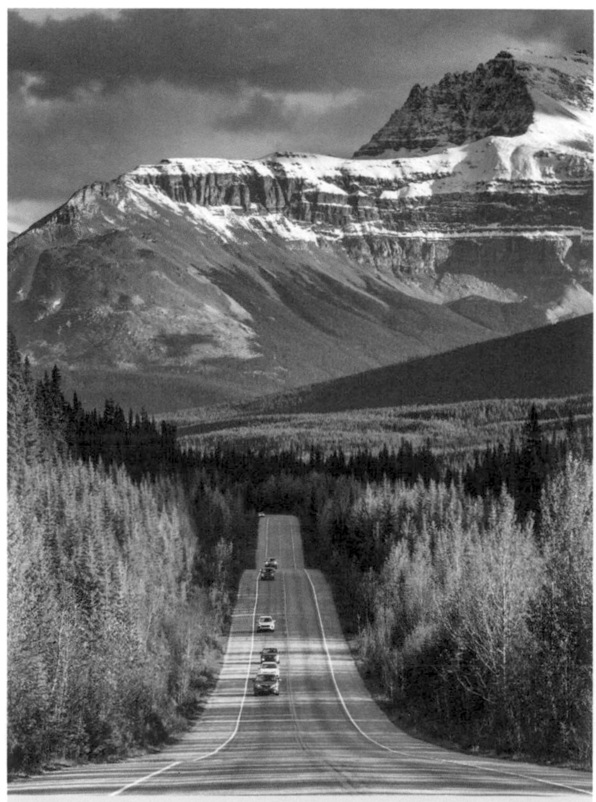

ABOVE: DRIVING THE ICEFIELDS PARKWAY

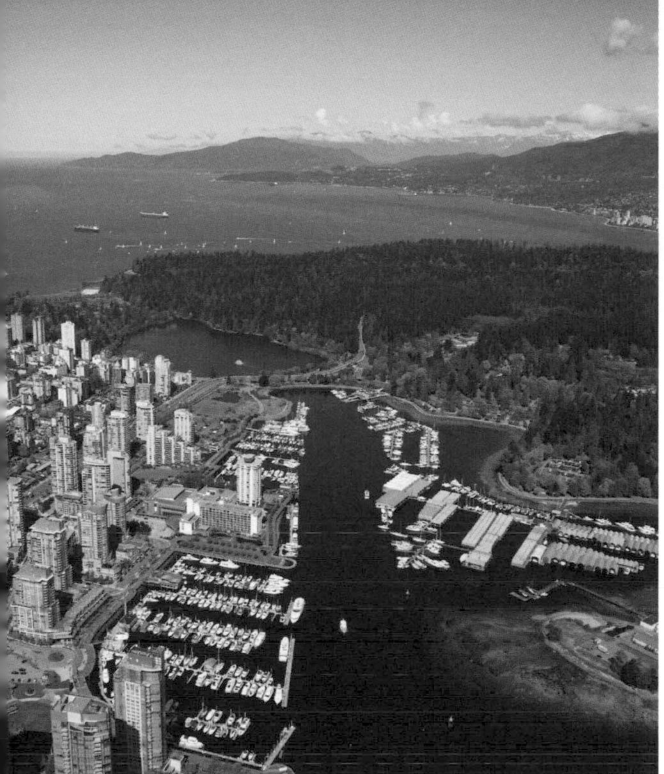
ABOVE: DOWNTOWN VANCOUVER AND STANLEY PARK

Continue on the VIA Rail train to Winnipeg (2-3 weekly; 33hr).

3 Winnipeg

It's a short walk from Winnipeg's train station to many of its attractions, including the Canadian Museum for Human Rights and Qaumajuq at the Winnipeg Art Gallery, which has the world's largest collection of contemporary Inuit art. The Forks, where the Red and Assiniboine rivers meet, is both an historic site and the home of a bustling international food hall.

Back on board your eastbound train to final stop, Toronto (2-3 weekly; 38hr).

4 Toronto

A short walk from Toronto's Union Station, the CN Tower is the best place to start in Canada's largest city, taking in the views (as far as Niagara Falls) from its 447m viewing platform. Then tour first-rate museums like the Art Gallery of Ontario, and walk the neighbourhoods in this multicultural metropolis. Catch the ferry on Lake Ontario for a nature escape to the Toronto Islands; if you want an up-close view of Niagara Falls, just 130km away, there are regular bus and train services from the city.

Toronto's airport has national and international connections.

ALTERNATIVES

Extension: Toronto, Montréal and Halifax

Extend your rail trip to the Atlantic coast. From Toronto, multiple daily VIA Rail trains travel to Montréal, with its art, street life and excellent restaurants. Continuing east, the *Ocean* train runs overnight through Quebec and New Brunswick, pulling into Halifax, Nova Scotia, 22 hours later. Halifax highlights include its historic waterfront, the Canadian Museum of Immigration at Pier 21, and anywhere you can eat lobster.

Diversion: Prince Rupert to Jasper

For an adventure in Canada's north, fly from Vancouver to Prince Rupert, on the northern British Columbia coast. Learn about the region's indigenous cultures at the Museum of Northern BC, then book a tour to Khutzeymateen Grizzly Bear Sanctuary to spot grizzlies in the wild. From Prince Rupert, a VIA Rail train runs 1160km southeast to Jasper, where you can transfer to the *Canadian* and continue east.

MAUI BEYOND THE BEACHES

Kahului – Wailuku

Beyond Maui's beautiful Hawai'ian beaches, dive into the island's diverse culture, road-tripping less touristed byways to uncover local heritage, arts and cuisine.

FACT BOX

Carbon (kg per person): 35
Distance (km): 278
Nights: 5-8
Budget: $$
When: year-round

❶ Kahului

Renting a car when you land at Kahului Airport gives you the most flexibility to explore. With big-box stores and rush hour traffic, Kahului looks like many American suburbs, but there's lively local life here too. See what's happening at Maui Arts & Cultural Center, find the latest cluster of food trucks and seek out crafts at Maui Swap Meet.

Drive 40km west to Lahaina.

❷ Lahaina

Make your way into the island's past amid the historic buildings in this west Maui town. Lahaina Heritage Museum gives you an overview, while Wo Hing Temple Museum highlights Maui's Chinese history. Check out the towering Buddha at Lahaina Jodo Mission and the sandy graves in adjacent Pu'upiha Cemetery.

Continue 74km east to Paia, via Kula and Makawao.

❸ Upcountry: Kula, Makawao and Paia

Head Upcountry to agricultural Maui. Sample cheeses and ice creams at Kula's Surfing Goat Dairy, visit Hawaii Sea Spirits for vodka distilled from organic sugarcane, then peruse the produce at Maui Nui Farm. In Makawao, try vodka brewed from pineapples at Hali'imaile Distilling Company. Then make your way

ABOVE: 'OHE'O GULCH ON THE ROAD TO HANA

ABOVE: PALM-LINED MAUI COASTLINE

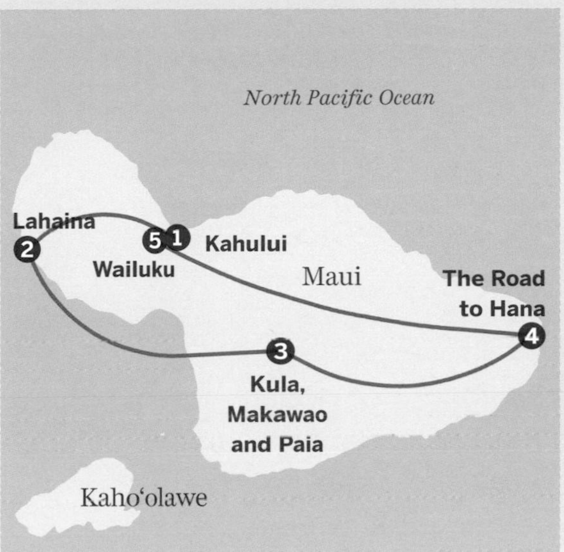

to the north shore to watch the windsurfers and kiteboarders at Ho'okipa Beach, before having dinner in one of the casually hip restaurants in Paia.

Hana is 72km southeast from Paia.

4 The Road to Hana

Maui's most famous drive is the Road to Hana, twisting and winding past waterfalls, across one-lane bridges and through tropical forests. Take your time as you navigate one switchback after another. Once you reach this rural town, poke around historic Hasegawa General Store, see if tiny Hana Cultural Center is open and grab lunch at a food truck or BBQ spot.

Retrace your route from Hana, and continue along the north shore to Wailuku, 84km from Hana.

5 Wailuku

If tourists come to this workaday town, it's often to hike to Iao Needle, a rocky pinnacle in the lush forests of Iao Valley State Park. But there's good eating here, too. Munch on *manju* (sweet pastries) at old-timey Home Maid Bakery, pick up a lunchtime bento at TJ's Warehouse and check the music schedule at Maui Coffee Attic. You'll feel at home in no time.

From Wailuku, it's only 8km back to the airport in Kahului.

ALTERNATIVES

Extension: Haleakalā National Park

Haleakalā volcano juts into the clouds, 3055m above sea level. Watching the sunrise from the summit is a special but extremely popular experience; reservations are required. You can hike along the rim or into the crater, too. Bring food, water and extra clothes to ward off the summit's chill.

It's 37km from Paia to the Haleakalā summit entrance. Allow an additional 30min to drive from the entrance gate to the summit.

Diversion: Kīpahulu

Most visitors who travel the Road to Hana simply turn around when they reach town. But continue along the Hana Hwy as it rounds the coast on Maui's east end to enter the Kīpahulu District of Haleakalā National Park. Enjoy the ocean views, hike through the humid forests to two waterfalls and camp – it's free – in the rustic drive-up campground overlooking the cliffs.

Kīpahulu is 19km from Hana.

ROAD-TRIPPING QUEBEC'S GASPÉ PENINSULA

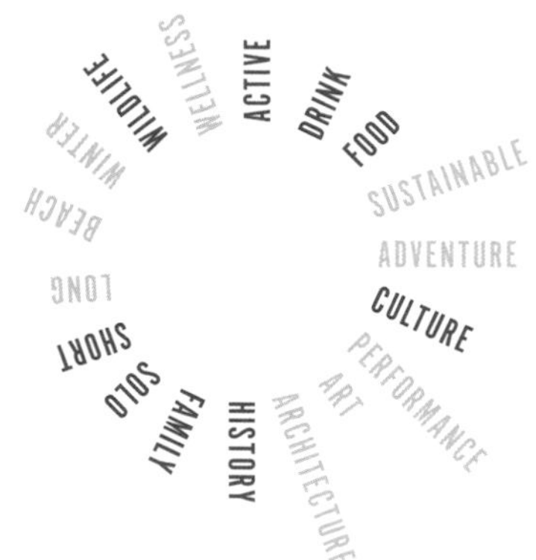

Quebec City – Sainte-Anne-des-Monts

Road-trip Canada's Gaspé peninsula to adventure through maritime national parks, whale-watch and experience the region's Acadian culture. Yes, there will be lobster.

FACT BOX

Carbon (kg per person): 189
Distance (km): 1524
Nights: 7
Budget: $$
When: Jun-Sep

❶ Quebec City

Begin your trip with a coffee and *pain au chocolat* in Quebec's Old Town; the city was founded in 1608. Explore Le Quartier Petit-Champlain above the waterfront, take a walk through the St-Jean-Baptiste and St-Roch districts, and pick up more pastries before you hit the road. You'll follow Hwy 132 around the Gaspé Peninsula that juts into the Gulf of St Lawrence. At Mont-Joli, when Hwy 132 splits, turn inland (southeast) to circle the peninsula counterclockwise and keep the ocean on your right, the better to enjoy the dramatic water vistas.

From Quebec City it's 555km northeast to Carleton.

ABOVE: NOTRE-DAME-DES-VICTOIRES, PLACE ROYALE, QUEBEC CITY

❷ Carleton

In this waterfront town, take a drive up to Mont St-Joseph for views across to New Brunswick, have dinner in one of the excellent restaurants and rest up for the night.

Continue 200km east to Percé, stopping en route for a visit to the Musée Acadien du Québec in Bonaventure to learn about the region's Acadian heritage.

❸ Percé

The seaside village of Percé is known for a rock, Rocher Percé. Boat tours take you around this massive slab with a photogenic archway beneath it, as well as to Percé's other main attraction, Île Bonaventure, North America's largest migratory bird refuge. The town is also an excellent option for chowing down on top-

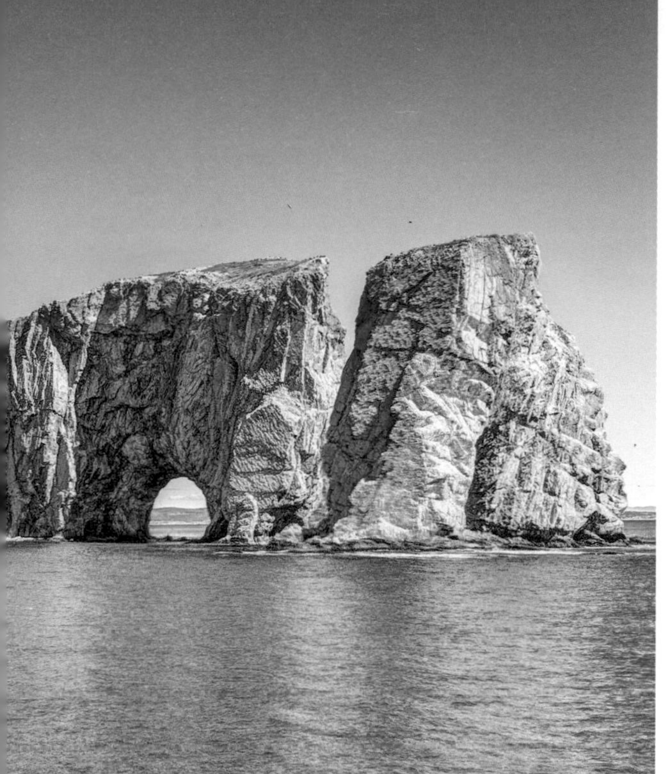

ABOVE: ROCHER PERCÉ, GASPÉ PENINSULA

quality seafood, particularly local lobster.
Drive 62km north to Gaspé.

4 Gaspé and Forillon National Park

Gaspé is the commercial hub for this northeast tip of the eponymous peninsula, with sleeping and eating options and a craft brewery. Forillon National Park, 30km east of town, is impossibly scenic, with cliffs plunging to the sea. Hike to Land's End, book a whale-watching cruise or fish for mackerel from Grande-Grave wharf.
Stay on the coastal highway, between the mountains and the ocean, for the 207km drive west to Sainte-Anne-des-Monts.

5 Sainte-Anne-des-Monts

Heading back towards Quebec City, it's worth breaking the journey in this waterfront town for more seafood. Detour inland to Parc National de la Gaspésie for a hike in the mountains, too. As you return to the city, pull off Hwy 132 at Sainte Flavie to see artist Marcel Gagnon's unusual sculpture, *Le Grand Rassemblement*, more than 100 stone figures marching out of the Gulf of St Lawrence.
From Sainte-Anne-des-Monts, it's 500km back to Quebec City.

ALTERNATIVES

Extension: Charlevoix

For a delicious detour, follow Hwy 138 northeast from Quebec City to the Charlevoix region. Base yourself in the cute town of Baie-Saint-Paul, where galleries and bistros line the main street, and follow La Route des Saveurs, a gourmet trail of more than 30 local bakers, cheesemakers, cider crafters and other food producers.
It's 94km from Quebec City northeast to Baie-Saint-Paul.

Diversion: Saguenay-Lac St Jean

Road-trip north of Quebec City to the Saguenay-Lac St Jean region to camp along Lac St Jean, hike to lookouts over Saguenay Fjord, and go whale watching from the town of Tadoussac. Then visit Musée Amérindien de Mashteuiatsh near Roberval, and buy monk-made goodies at Chocolaterie des Pères Trappistes de Mistassini.
Tadoussac is 215km northeast of Quebec City.

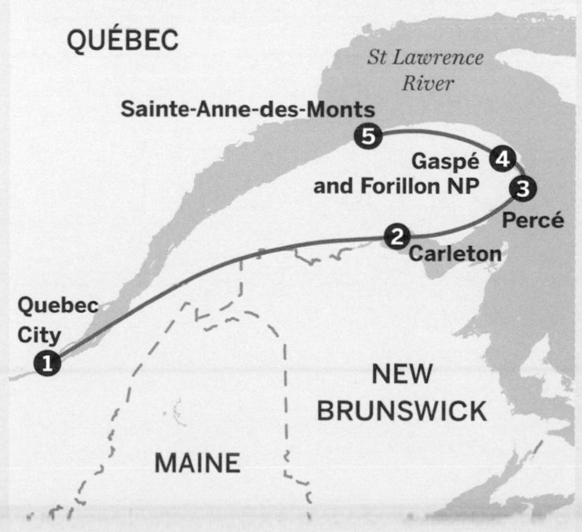

QUEBEC'S ÎLES DE LA MADELEINE

Souris – Île de Grande-Entrée

Canada's remote Îles de la Madeleine (Magdalen Islands) are known for dramatic red-sandstone cliffs, sandy beaches, Francophone culture and fresh seafood.

FACT BOX

Carbon (kg per person): 52
Distance (km): 257
Nights: 4-6
Budget: $$
When: Jun-Sep

❶ Souris

Although the Îles de la Madeleine belong to the province of Quebec, they're actually far closer to Prince Edward Island in Canada's Atlantic region, so start your trip on PEI's east coast in the fishing town of Souris, from where ferries to the Îles depart. Before setting off, climb East Point Lighthouse to take in the views and visit Basin Head Provincial Park, where the golden sands make a distinctive squeak as you walk.
Take the ferry to Île du Cap aux Meules (daily Jul-early Sep, 3-6 weekly in other seasons; 5hr).

❷ Île du Cap aux Meules

The ferry docks at Cap aux Meules, at the approximate centre of the island chain. Check out the distinctive red cliffs on the island's west side, and kayak from Parc de Gros-Cap. Take a lobster cooking class at Chef Johanne Vigneau's Gourmande de Nature or dine at her upscale restaurant, La Table des Roy. Relax over craft brews at À l'Abri de la Tempête.
From Cap aux Meules, drive 25km south to Havre-Aubert via Rte 199, which connects the six main islands.

❸ Île du Havre-Aubert

Musée de la Mer on the island's east coast takes you through the chain's maritime history. Next head to remote Sandy Hook beach where you might have the

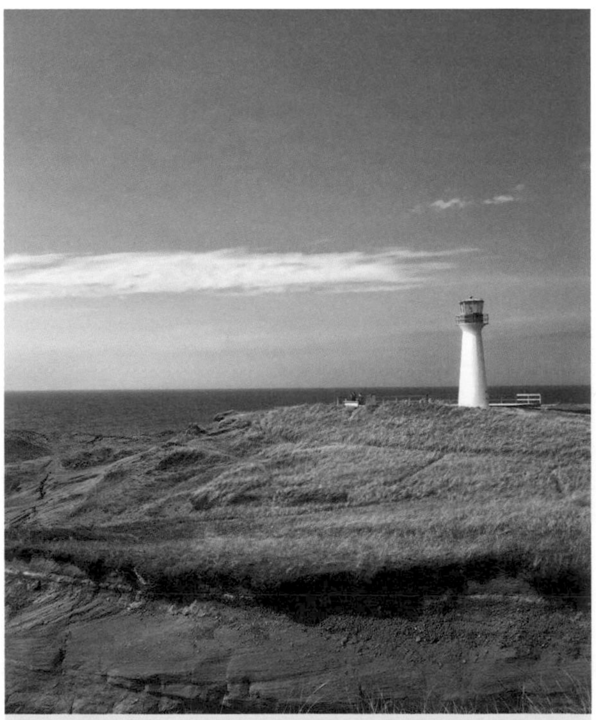

ABOVE: CAP AUX MEULES LIGHTHOUSE

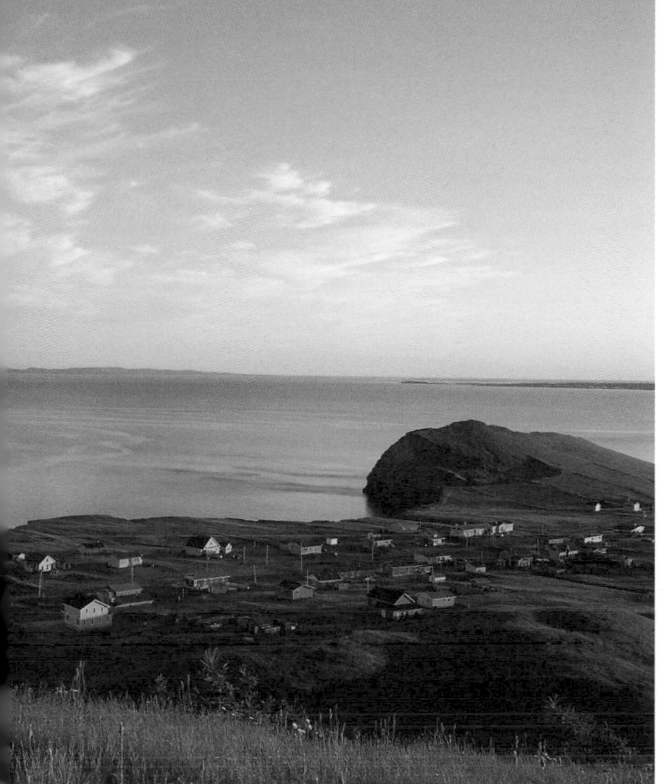
ABOVE: ÎLE DU HAVRE AUX MAISONS

place to yourself. Refuel with the rib-sticking fare at Café de la Grave, housed in a former grocery store.
Drive 33km north to Île Havre aux Maisons.

4 Île Havre aux Maisons
Loop back through Île du Cap aux Meules and on to Île Havre aux Maisons where you can sample more local foods, including smoked herring from Le Fumoir d'Antan and cheeses at Fromagerie du Pied-de-Vent.
Continue 40km north to Grosse Île.

5 Grosse Île
Follow Rte 199 along a narrow spit of land to the Madeleines' far north end to find the 10km of sand at beautiful Plage de la Grande Échouerie.
Drive 15km south to Grande-Entrée.

6 Île de Grande-Entrée
From Grande-Entrée wharf, the islands' main fishing port, you can watch the boats head out to sea (at least if you're up before dawn). Visit the Seal Interpretation Centre for exhibits about the chain's marine life.
Retrace your route from Grande-Entrée, continuing 57km to Cap aux Meules and the ferry port.

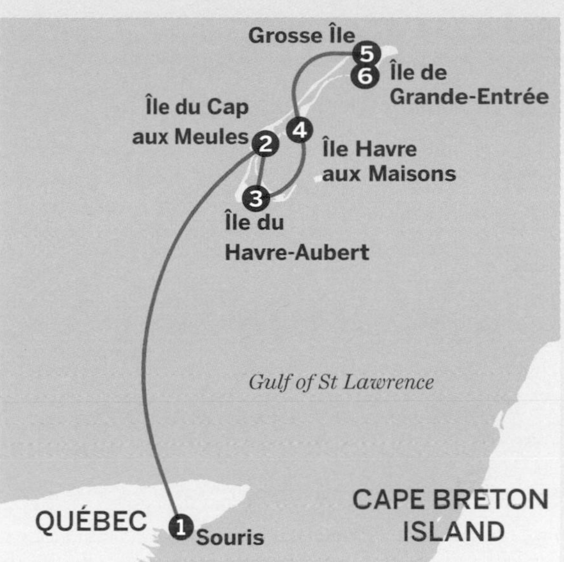

ALTERNATIVES

Extension: Charlottetown
Wander along the waterfront in PEI's pretty provincial capital, follow the boardwalk through Victoria Park and explore the heritage buildings in the historic core. Take a peek into the ornate St Dunstan's Basilica, or take a tour of COWS Creamery, PEI's well-known producer of ice cream and cheese, before digging into some local seafood.
From Souris, it's an 81km drive west to Charlottetown.

Diversion: Cavendish and vicinity
Cavendish was the hometown of *Anne of Green Gables* author Lucy Maud Montgomery; pay homage at Montgomery's Cavendish Homestead and the Anne of Green Gables Museum in nearby Park Corner. Also nearby is Prince Edward Island National Park, where you can stroll the dune-backed beaches along the coast.
Cavendish is 112km from Souris or 40km from Charlottetown.

DRIVING URBAN ARIZONA

Phoenix – Mesa

This US urban escape highlights Arizona's contemporary art, innovative architecture and Native American, Mexican and Spanish roots, from heritage buildings to cuisine.

FACT BOX

Carbon (kg per person): 84
Distance (km): 673
Nights: 4-6
Budget: $$
When: Oct-May

❶ Phoenix

Start your trip in Arizona's capital city, the region's main transportation gateway. Learn about the Southwest's Native peoples at the Heard Museum, stroll among the wildflowers at Desert Botanical Garden, examine the Western Gallery at Phoenix Art Museum and listen to recordings from more than 200 countries at the Musical Instrument Museum. Downtown, explore the galleries and street art in the the Roosevelt Row arts district. Hungry? Find the Phoenix Street Food Coalition's food trucks, or take yourself on a taco tour.

Drive 21km from downtown Phoenix to Scottsdale.

ABOVE: HEARD MUSEUM, PHOENIX

❷ Scottsdale

This upscale suburb has a thriving art scene, from its excellent Museum of Contemporary Art and an extensive public sculpture programme to the quirky Cattle Track arts collective where you can tour artists' studios. Scottsdale is also a centre for architectural innovation, a highlight of which is Frank Lloyd Wright's Taliesin West, the architect's winter home and school.

It's 185km southeast from Scottsdale to Tucson.

❸ Tucson

In this university town, you can hike among cacti in Saguaro National Park, explore the 18th-century Mission San Xavier del Bac then look for street art, like the unusual *Bike Church* sculpture, in its diverse

ABOVE: TUCSON SKYLINE AND SENTINEL PEAK

neighborhoods. Unwind in the bars on funky 4th Ave or graze through the Mexican restaurants along South 12th Ave. And don't leave without sampling the almost too-pretty-to-eat artisanal bonbons at Monsoon Chocolate.

Continue 133km east to Willcox.

4 Arizona Wine Country

Two of Arizona's small but growing winemaking regions are within easy drives of Tucson. Sonoita-Elgin, with about a dozen wineries, is 80km southeast, while other vineyards cluster around Willcox, further east. Bring a designated driver or book a wine tour. On your way to Willcox, pull off in Dragoon to see the exhibits of Native American art, history and contemporary culture at the Amerind Museum.

Drive 315km northwest from to Mesa.

5 Mesa

Returning to metropolitan Phoenix, stop in Mesa, an agritourism hub. Visit small farms, dairies, orange groves, even an olive oil producer. See what's on view at Mesa Contemporary Arts Museum, then for post art- and food-tour exercise, hike into the desert, go tubing on the Salt River or paddleboard a local lake.

Mesa is 19km east of Phoenix airport.

ALTERNATIVES

Extension: Sedona

This touristy town north of Phoenix is known for its red-sandstone cliffs and its mystical auras. Whether or not you believe that the area channels the earth's energy, you can explore the landscape in state parks like Slide Rock or Red Rock. The town has good Mexican dining spots and gourmet Southwest restaurants.

Sedona is 185km north of Phoenix.

Diversion: The Grand Canyon

The Grand Canyon is one of Arizona's – indeed the world's – natural highlights. Carved by the Colorado River, its deep gorges show off layers of rock, striated with dusky colours. Stay in a lodge or campground near the South Rim; or make for the gorgeous, quieter North Rim, normally open only from May or June till October – or whenever the snows begin.

The South Rim is 360km north of Phoenix.

CANADA'S YUKON TERRITORY

Whitehorse – Dawson City

Explore Canada's highest peaks and vast icefields, travel back to the Gold Rush era and delve into indigenous cultures on this adventurous Yukon road trip.

FACT BOX

Carbon (kg per person): 168
Distance (km): 1350
Nights: 7-10
Budget: $$
When: Jun-Sep

❶ Whitehorse

In Yukon's capital, explore the past at MacBride Museum of Yukon History, hop aboard an historic stern-wheeler at SS Klondike National Historic Site and check out indigenous art at Kwanlin Dün Cultural Centre. Hike in scenic Miles Canyon, then soak in Takhini Hot Pools. Drive south along the Klondike Hwy towards Carcross to see Canada's smallest desert – yes, there's a desert in the Yukon. Tromp across the sand dunes, once the bottom of a glacial lake, then stop in the nearby town of Carcross to visit the studios of indigenous carvers and other artisans before returning to Whitehorse.

Drive 160km west from Whitehorse to Haines Junction, gateway to Kluane National Park.

❷ Kluane National Park

Think of Canada's tallest mountains and you probably think of the Canadian Rockies. But actually, 17 of the country's 20 highest peaks lie within the Yukon's Kluane National Park. Plan a wilderness trek or day-hike, raft the Alsek River, paddle Kathleen Lake or scout for over 100 species of birds. To end on a literal high, take a flight-seeing tour over immense icefields – keep your camera handy. Before leaving Haines Junction, check out the Da Kų Cultural Centre, where

ABOVE: DAWSON CITY

ABOVE: KLUANE NATIONAL PARK ICEFIELDS

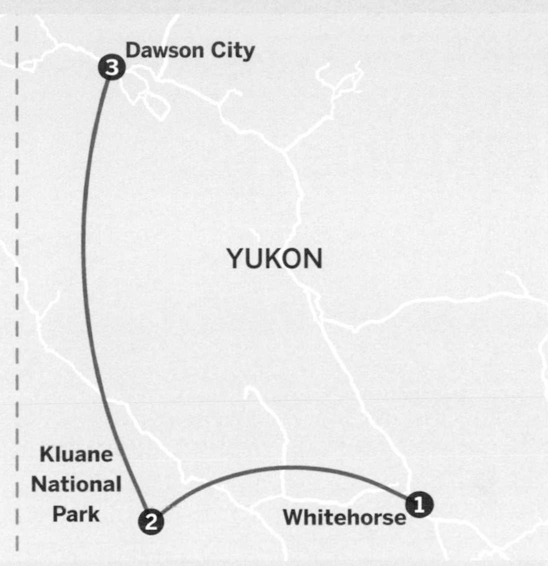

you can learn about the Indigenous Champagne and Aishihik people whose traditional lands span the surrounding region.

From Haines Junction, drive 660km north to Dawson City.

3 Dawson City

The 1898 Klondike Gold Rush comes alive in this frontier town's heritage buildings. Parks Canada staff in period costumes lead informative walking tours through Dawson's unpaved streets that take you back to the Gold Rush era. Cruise the river on a restored paddlewheeler, or hear stories of Gold Rusher turned author Jack London in his cabin, now a museum. Dawson sits on the traditional territory of the Tr'ondëk Hwëch'in First Nation; visit the Dänojà Zho Cultural Centre to learn more. If you dare, sample the town's most famous drink – the sourtoe cocktail, which contains the remains of a human toe. Late at night, when the sky is completely dark, head for Midnight Dome, a lookout point above Dawson City, to try to spot the aurora borealis – the northern lights – as they dance across the sky. This natural light show is most visible between August and April.

Return to Whitehorse (530km south) via Hwy 2, the Klondike Hwy.

ALTERNATIVES

Extension: Tombstone Territorial Park

Hike among the wildflowers and craggy mountains of this 2200-sq-km parkland within day-trip distance of Dawson City. The Dempster Hwy, which continues north to the Northwest Territories, bisects the park. Pack a picnic, and stop at the park's interpretive centre for advice about trails and other activities.

Tombstone Interpretive Centre is 110km northeast of Dawson City.

Diversion: Top of the World Highway

From Dawson, this scenic Alaska-bound day trip crosses the Yukon River via the *George Black* ferry to travel the isolated Top of the World Hwy (normally open May to September; check before leaving), winding through dramatic mountains to North America's northernmost border crossing. If it's open, continue on to lunch in Chicken, Alaska, before backtracking to Dawson.

It's 170km from Dawson to Chicken.

BIG SKIES, BIG PARKS AND BUSTLING CITIES IN TEXAS

Austin – Big Bend National Park

This great US road trip features the coolest city in Texas plus alternative culture, desert landscapes and one massive mountain-filled national park.

FACT BOX

Carbon (kg per person): 219
Distance (km): 1760
Nights: 10-12
Budget: $$
When: Oct-May

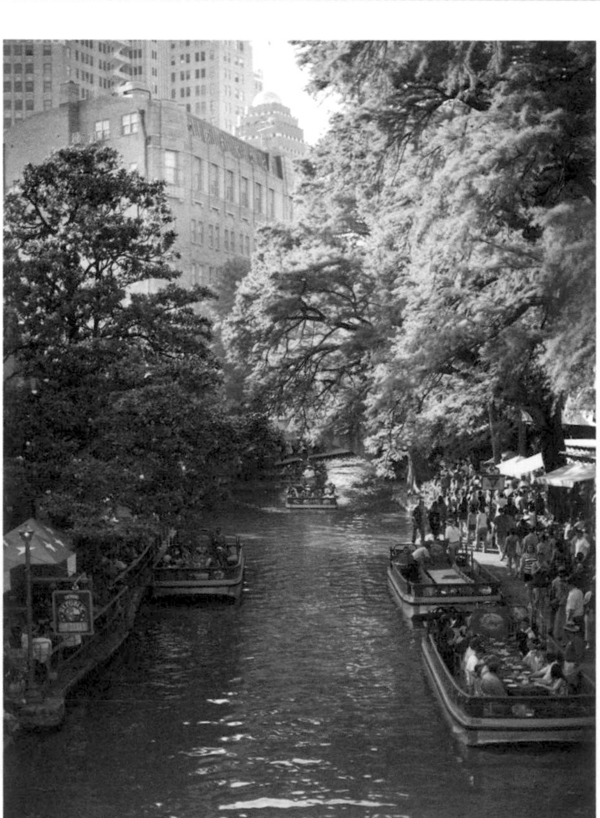

ABOVE: SAN ANTONIO RIVER WALK

❶ Austin

One of the brightest, coolest stars in Texas, Austin is famed for its live music scene. Any night of the week, you can catch rockabilly, blues or punk, and its music festivals (like Austin City Limits) are worth planning a trip around. For urban exploring, head to South Congress Ave, with its thrift stores and eye-catching boutiques, creative food trucks, and bars and eateries with tree-shaded patios. When the sun beats down, cool off with Amy's Ice Cream or a dip in ice-cold Barton Springs. As the sun sets, head to Congress Ave Bridge for the nightly bat swooping spectacle.

Drive 130km southwest to San Antonio.

❷ San Antonio

Despite its prodigious size, San Antonio has a walkable downtown that's home to museums, leafy parks and historic buildings, including the famous Alamo. You can tour the former Spanish mission turned battle site, then stroll the restaurant-lined banks of the scenic River Walk. Here you can hop onto an eco-friendly river shuttle and take in the views of pretty gardens and stone bridges en route to Pearl Brewery, a complex of shops, cafes, bars and food markets.

It's a long 650km drive west from San Antonio to Marfa.

ABOVE: BIG BEND NATIONAL PARK

❸ Marfa

Sitting pretty on the edge of a vast desert expanse, the tiny town of Marfa is home to a wild mix of cowboys and artists. The streets are dotted with galleries, imaginative restaurants and quirky lodging options (including El Cosmico, where you can bunk in retro-chic travel trailers). The star attraction is the Chinati Foundation, a former army post housing the world's largest permanent collection of minimalist art.

Continue driving 220km south.

❹ Big Bend National Park

In the heart of the Chihuahuan Desert, Big Bend may appear at first glance like a barren wasteland. But, in fact, amid its steep-walled canyons, fertile river corridor, desert scrubland and forested ridges, you'll find varied microclimates that support abundant wildlife – including more recorded bird species than in any other national park in the US. There's much to do here too, from hiking some of the 240km of trails to soaking in hot springs, horse riding and river rafting on the Rio Grande.

It's a 760km drive back to Austin and its airport.

ALTERNATIVES

Extension: Luckenbach

The cluster of Old-West-era buildings that make up this tiny settlement typify Texas Hill Country charm. Grab a Shiner Bock beer and listen to twanging guitar solos under a giant live oak tree. You can practice your two-step in the dancehall, or catch up on local gossip inside the old saloon, which also doubles as the general store and post office.

Luckenbach is a 116km drive west of Austin.

Diversion: Terlingua Ghost Town

An easy add-on to the Big Bend experience, Terlingua is a former mining boom town that went bust in the 1940s. Today you can explore the ruins of old adobe buildings and peek in an abandoned cemetery. By night, head to the Starlight Theatre, a roofless former cinema reborn as a restaurant and live-music spot.

From Big Bend, it's a 67km drive northwest to Terlingua.

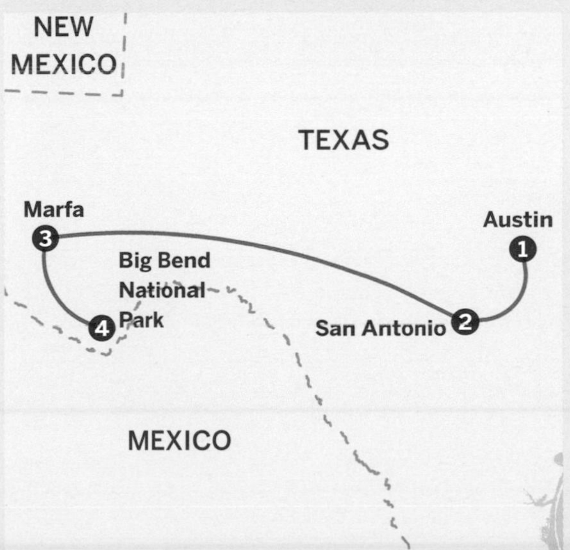

CREOLE AND CAJUN CULTURE IN LOUISIANA

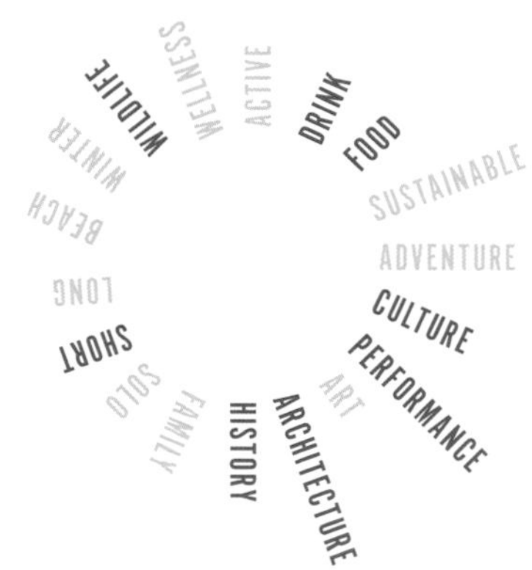

New Orleans – Lafayette

New Orleans' brassy beats and Cajun fiddles form the soundtrack to this romping US road trip that takes you from city to swamp.

FACT BOX

Carbon (kg per person): 57
Distance (km): 457
Nights: 5-7
Budget: $$
When: Oct-Apr

❶ New Orleans

Tucked along a scenic bend in the Mississippi, New Orleans is a city of brassy jazz celebrations in the street, Mardi Gras revelry and colorful Creole architecture. The food scene is equally decadent, with chargrilled oysters, crawfish étouffée and rich bowls of gumbo. All go nicely with a classic New Orleanian cocktail like the French 75 (champagne, gin, lime juice). By day, there's much to explore, from historic sights in the French Quarter to grand mansions in the Garden District, plus gorgeous parks like Audubon or City Park, easily reached on the city's vintage streetcar lines.

Drive 80km west, following the south bank of the Mississippi.

❷ Whitney Plantation

Grand plantation homes dot the banks of the Mississippi along the River Rd, though only one historic site, the Whitney, focuses exclusively on the lives of enslaved people. Take a tour and visit the original cabins in which the enslaved lived, and the first owner's 1790 house, learning about the lives of the 350 people held in bondage here. There are several powerful memorials, including life-size casts of enslaved children inside the Antioch Baptist Church.

Continue 160km west.

ABOVE: WHITNEY PLANTATION

ABOVE: KAYAKING LAKE MARTIN

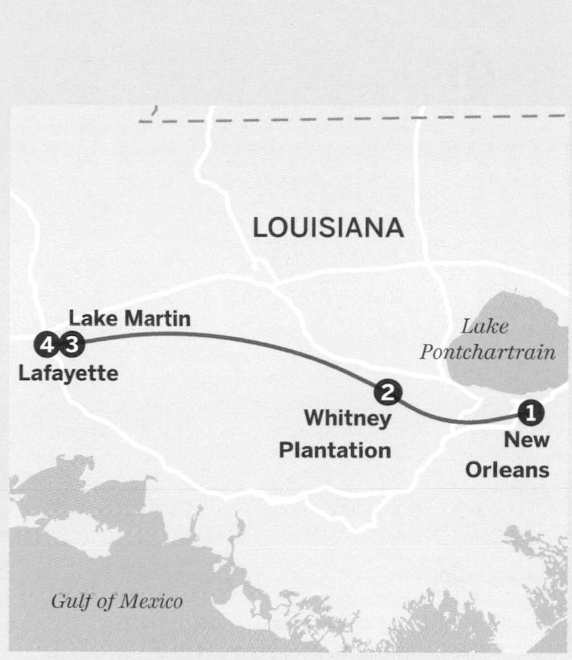

❸ Lake Martin

Amid the swamp-filled wilderness of Lake Martin, you can spy great blue herons, snowy egrets and roseate spoonbills flitting through the bald cypress trees, while alligators glide through the inky waters below. There's a 4km levee trail to walk, but the best way to experience the lake is to get out on the water. Cajun Country Swamp Tours run boat trips, which are led by knowledgeable guides who grew up amid this aquatic wonderland. For a more leisurely pace, Pack & Paddle offers kayaking tours.

Drive 17km west along the Prairie Hwy.

❹ Lafayette

The self-proclaimed capital of French Louisiana is a good place to soak up some Cajun culture. You can learn about Cajun history at Vermilionville, a recreated 19th-century village where bands perform in the barn on weekends. Afterwards, feast on crawfish étouffée, rice with smoked andouille sausage and other Cajun delicacies at famed Prejean's. In the evening, catch a concert at Blue Moon Saloon, which has zydeco bands working the backyard dance crowd most nights of the week.

Drive 200km back to New Orleans and its airport for onward connections.

ALTERNATIVES

Extension: St Francisville

You'll feel like you've stepped back in time at this tranquil town, with its abundance of original 18th- and 19th-century homes. Stroll the historic district, then visit the Oakley Plantation, where naturalist John James Audubon spent a few productive months in 1821 painting birds from the surrounding forest. In the evening, take a ghost tour through the Myrtles Plantation, one of America's most haunted homes.

St Francisville is 140km northeast of Lafayette.

Diversion: Mandeville

Easy-going Mandeville, on the north side of Lake Pontchartrain (Louisiana's largest), sports a lakefront promenade, elegant homes and outdoor restaurants and cafes. Cyclists can hop onto the Tammany Trace, a 50km-long former rail trail. There's also good wildlife viewing (birds, alligators) in nearby Fontainebleau State Park.

Mandeville is 56km across the causeway from New Orleans.

CRAGGY COASTS AND PRETTY TOWNS OF MAINE

Portland – Acadia National Park

Get your fill of rugged shorelines, outdoor activities and plenty of fresh-caught seafood on this oceanside road trip along the US coast in Maine.

FACT BOX

Carbon (kg per person): 66
Distance (km): 535
Nights: 7
Budget: $$$
When: May-Jun, Sep-Oct

❶ Portland

Perched on a peninsula fronting Casco Bay, Maine's biggest city, Portland, has a celebrated food scene, some excellent museums and abundant green spaces overlooking the brooding Atlantic. There's much to explore in the Old Port district, with its cobblestone streets and 19th-century brick buildings harbouring cafes, restaurants, shops and microbrew-minded bars. You can't leave town without walking or biking along the Eastern Promenade, taking in the views of rocky islets and sailboats gliding across Casco Bay. It's also worth visiting photogenic Portland Head Light, Maine's oldest functioning lighthouse.

Drive along Hwy 1 to Camden, some 130km northeast.

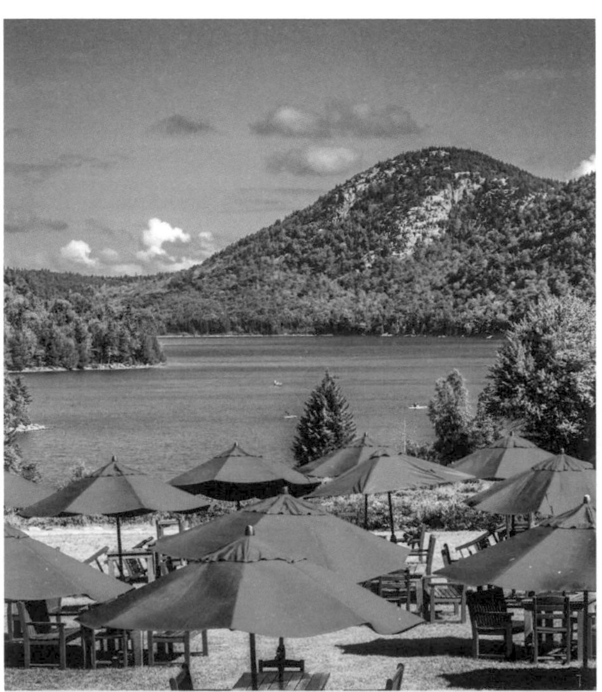
ABOVE: JORDAN POND HOUSE, ACADIA NATIONAL PARK

❷ Camden

The tiny seaside town of Camden boasts a picture-perfect harbour and gallery-dotted back streets that are a delight to wander. Home to the state's largest fleet of windjammers (multi-masted sailing ships), it's no surprise that Camden is a great place to arrange a sailing excursion. Just north of town, the Camden Hills State Park has nearly 50km of trails, including the short jaunt up Mt Battie, which provides wide-ranging views over island-dotted Penobscot Bay.

Drive a scenic 90km around Penobscot Bay to Castine.

ABOVE: PORTLAND, MAINE, AND ITS OLD PORT DISTRICT

3 Castine

One of Maine's prettiest villages, Castine has a handful of historic buildings and a pleasant waterfront anchoring the end of Main St. Though Castine is diminutive in size, it's worth not rushing straight through: make time for kayak trips, for walks through Witherle Woods and for relaxing with coffee and a good book while taking in the meditative pace of village life.

Back in the car, drive 75km east to Acadia National Park.

4 Acadia National Park

The mountains meet the sea in dramatic Acadia National Park, a wonderland of forest-fringed lakes, rugged coastline and craggy clifftops on glacier-formed Mt Desert Island. There's much to see and do: hiking dozens of trails; cycling 72km of carriage roads; or braving the chilly waters at Echo Lake or Sand Beach. When you need a break from the action, you can enjoy tea and popovers (a sweet version of the UK's Yorkshire pudding) at the historic Jordan Pond House, and get your fill of shopping and fine dining in upscale Bar Harbor.

It's 240km back to Portland and onward connections.

ALTERNATIVES

Extension: Deer Isle and Isle au Haut

Just off the beaten track, Deer Isle is a collection of tranquil islands joined by causeways. Highlights include kayaking trips, and the fantastical sculptures and homemade jams at Nervous Nellie's. In Stonington, feast on seafood or take the 45min boat trip to Isle au Haut for fabulous coastal trails, thick forests and wave-battered sea cliffs.

From Castine, it's 42km to Deer Isle and another 16km to Stonington.

Diversion: Southwest Harbor

The less crowded western side of Mt Desert Island is home to the pretty village of Southwest Harbor. From here, you can take a boat trip out to the Cranberry Islands or drive south to enjoy the views from scenic Bass Harbor Lighthouse. Then hit Thurston's Lobster Pound, a legendary spot for a dockside crustacean feast.

Southwest Harbor is next to Acadia National Park, on the western side of Somes Sound.

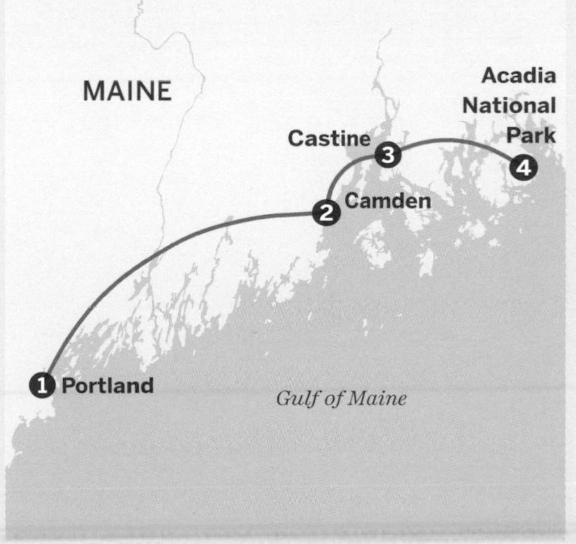

DRIVE THE POWDER HIGHWAY IN BRITISH COLUMBIA

Kicking Horse – Fernie

Blaze a Canadian trail on British Columbia's Powder Highway, enjoying spectacular scenery, world-class ski resorts and cool mountain towns along the way.

FACT BOX

Carbon (kg per person): 88
Distance (km): 710
Nights: 7-10
Budget: $$$
When: Dec-Apr

❶ Kicking Horse

Avid skiers and snowboarders visiting Kicking Horse for the first time might think they've died and gone to ski-resort heaven. The terrain here is exceptional, even by British Columbia's high standards; it's steep, too, though there are also some fun mellow pistes for beginners and improvers to cruise around on. But the real draw is the snow, which is so light, fluffy and frequent that the resort was allowed to trademark the strapline 'Champagne Powder Capital of Canada'. Head to the Eagle Eye restaurant up the mountain for panoramic views of the Canadian Rockies and visit nearby Golden for unpretentious cool-mountain-town vibes.

Drive a scenic 150km south.

❷ Panorama

Welcome to one of the best resorts in BC for families or mixed ability groups thanks to great beginner areas and some fantastic former heli-ski terrain for more advanced skiers and snowboarders. This is a resort where it's really easy to escape the crowds and soak up the wonderful nature and jagged mountain backdrops all around. The village is compact and easy to navigate, with outdoor hot-tubs and an ice rink for après-ski fun; the nearby lake town of Invermere is worth a visit too.

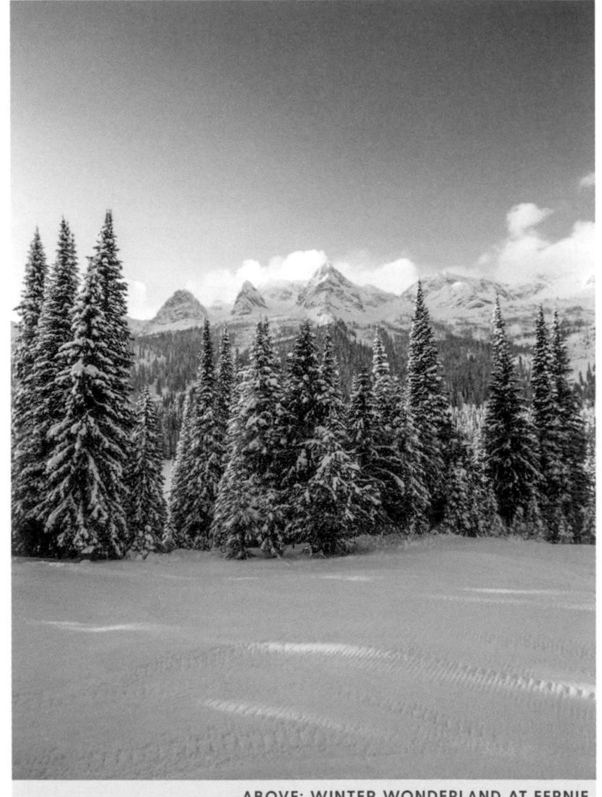

ABOVE: WINTER WONDERLAND AT FERNIE

ABOVE: OFF-PISTE SKIING, KICKING HORSE

Back in the car drive 140km south.

3 Kimberley

Kimberley is one of BC's smaller and less well-known resorts, but as a result it's popular with locals and has a friendly feel to it. It's a good mountain for beginners and intermediates to progress on, but it still has some decent technical terrain for experts to test their mettle. The town of Kimberley has a hip mountain feel – but also an unlikely Bavarian heritage, as seen with its cuckoo clock (the largest in Canada) and its woodcarving shop run by a former Austrian yodelling champion.

Continue 120km east.

4 Fernie

Fernie's long dreamy tree runs, fantastic bowls at altitude and excellent snow quality have given it a quasi-mythical status in skiing and snowboarding circles – the stunning landscapes help too. Stay in the historic centre for a taste of this former mining town's unique heritage. Fernie is also a good base for snowcat adventures into the backcountry.

Drive 300km north to Calgary Airport for onward connections.

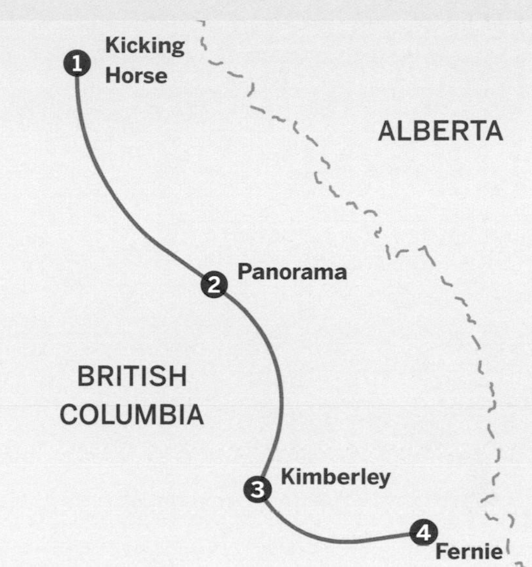

ALTERNATIVES

Extension: Radium Hot Springs

Soak tired limbs in the rejuvenating waters of Radium Hot Springs, with a backdrop of Sinclair Canyon in the Kootenay National Park. Enjoy the contrast of the naturally heated large outdoor mineral pool and the freezing mountain air all around. There's also a cooler pool for lane swimming and the option of yoga classes and spa treatments on site.

Radium Hot Springs is 36km northeast of Panorama

Diversion: Whitewater

It's an out-of-the-way choice but Whitewater is well worth the trip if you're looking for a super-authentic BC mountain experience, thanks to its strong snow record – notably of dry fluffy powder – and its old-school locals' scene. Stay at nearby Nelson, an atmospheric frontier town with antique buildings and a buzzing bar scene. Whitewater is also a good base for ski touring or splitboarding.

Whitewater is 250km west of Kimberley.

SURFING AND SKIING ACROSS CALIFORNIA

Los Angeles – San Francisco

Drive the ultimate US surf-and-snow road trip, starting on the coast in LA, heading inland for the snow, and finishing in the waves around San Francisco.

FACT BOX

Carbon (kg per person): 128
Distance (km): 1030
Nights: 7-10
Budget: $$$
When: Spring

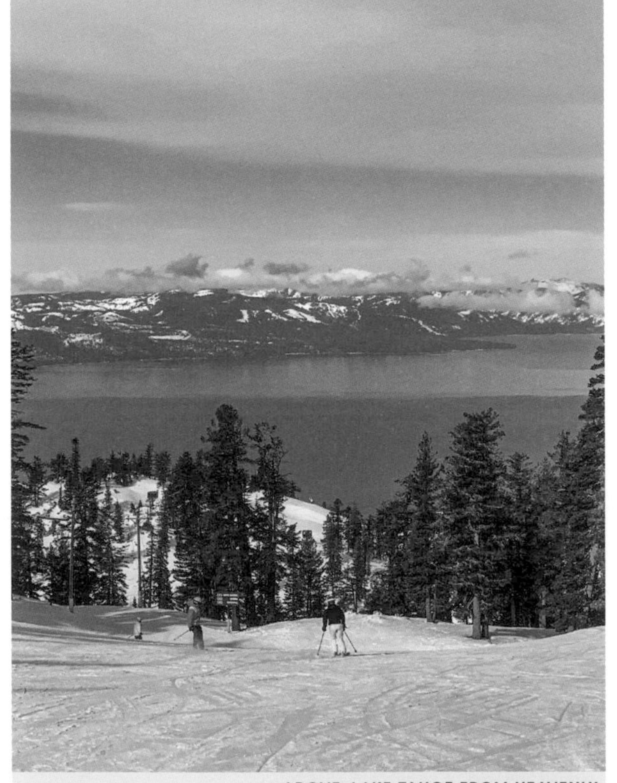

ABOVE: LAKE TAHOE FROM HEAVENLY

❶ Los Angeles

Surfing may have been born in ancient Polynesia, but it came of age on the beaches in and around Los Angeles. Huntington Beach and Malibu are two of the most celebrated and remain consistently good surf spots to savour that history and ride some epic waves (both have plenty of rental places). Huntington Beach is also home to the International Surfing Museum. Be sure to visit Venice Beach, another top surf spot that's also famous for its oceanfront skatepark and its role in shaping modern skateboarding in the 1970s, as chronicled in the film *Dogtown and Z-Boys*.

Leaving the LA coast, drive 500km north into the Sierra Nevada.

❷ Mammoth Mountain

The proximity of Mammoth to LA and its influential surf and skate scene means this is no ordinary ski resort. For a start, you'll find just as many snowboarders as skiers, if not more. Mammoth also has a big focus on freestyle snowboarding and skiing – its snow parks and halfpipes are revered the world over and bring a hip, youthful crowd to the resort. But the regular skiing and snowboarding here is fantastic too, with pistes for all levels, some great tree runs and breathtaking views over the Sierra Nevada mountains.

ABOVE: LA SURF SPOT HUNTINGTON BEACH

Continue 230km north, past eerie, geologically interesting Mono Lake.

3 Heavenly

There are 15 ski resorts around Lake Tahoe, but none has views of the lake that dazzle quite like those at Heavenly. The resort, which straddles California and Nevada and averages 300 days of sunshine a year, is vast, with a mix of snowboard terrain from gentle groomers to steep lines and runs through forests of ancient pines.

Leave the mountains behind and drive 300km west to San Francisco.

4 San Francisco

Often wild, windy and foggy, Ocean Beach at San Francisco's western edge is a very different kind of Californian surf experience from LA's – but that's part of its charm. It certainly doesn't deter the hardcore surfers, and it gives this part of town a more relaxed, less tech-money and touristy feel compared to the rest of the city. It's not a spot for beginners though, who should instead head to Linda Mar or Half Moon Bay (20km and 40km south respectively).

SFO Airport has onward national and international connections.

ALTERNATIVES

Extension: Santa Cruz

Santa Cruz is a laidback Pacific beach town with a pier and an atmospheric seafront amusement park (definitely ride the vintage wooden roller coaster). It's also a haven for surfers thanks to the quality and variety of its breaks. It was here that Jack O'Neill invented the wetsuit in the 1950s, changing the sport forever by opening up new cold-water surf spots like Santa Cruz.

Santa Cruz is 125km south of San Francisco.

Diversion: Sausalito

The best way to reach pretty Sausalito, across the bay from San Francisco, is by bike. Set off from SF's waterfront (bike rental shops can be found here), enjoy an awe-inspiring cycle across the Golden Gate Bridge and descend into Sausalito's mix of colourful floating homes and art galleries before getting the ferry back to SF.

Sausalito is around 13km from San Francisco.

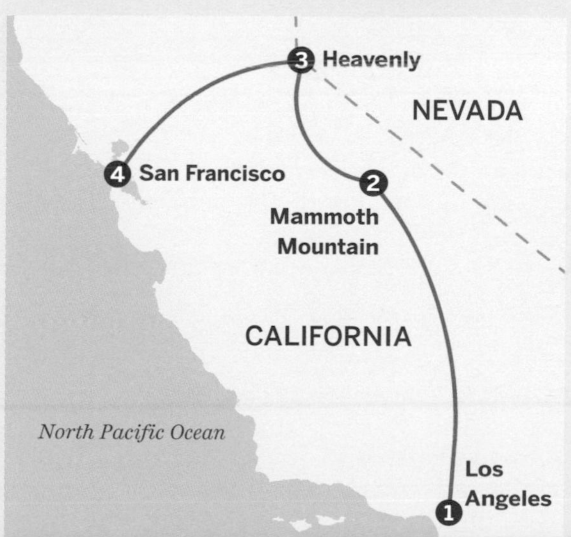

OUTDOOR LIVING ON THE OLYMPIC PENINSULA

Seattle – Aberdeen

Explore seaside towns and natural wonders on a US car and ferry trip round the Olympic Peninsula, one of the Pacific Northwest's most accessible outdoor havens.

FACT BOX

Carbon (kg per person): 77
Distance (km): 614
Nights: 4
Budget: ££
When: May-Oct

❶ Seattle

Before delving into nature, begin with some urban fun in Seattle. Visit the recently renovated Space Needle and inspect the art in the Olympic Sculpture Park, then find lunch at famous Pike Place Market and stroll some of the city's picturesque neighbourhoods.

Take the ferry to Bremerton (every 60-90min; 1hr); then drive 83km north to Port Townsend.

❷ Port Townsend

Port Townsend includes an historic downtown on the water and tall cliffs atop which historic homes peer out over Puget Sound. History buffs will love Fort Worden State Park outside town where you can learn about military operations in the region, while adventurous travellers will enjoy kayaking or a wildlife cruise in the town's namesake bay. Either way, keep your eyes peeled for the resident orcas that are often seen offshore here.

Continue 76km west to Port Angeles.

❸ Port Angeles

There's a fine view from Port Angeles across the Salish Sea toward British Columbia's Victoria Island, and the area's inland adventures are just as appealing. Head south into the mountains of Olympic National Park – Hurricane Ridge is a popular summer destination with a visitor centre and plenty of hiking. After working those muscles, relax at Sol Duc Hot Springs Resort,

ABOVE: OLYMPIC NATIONAL PARK

ABOVE: PORT TOWNSEND MARINA

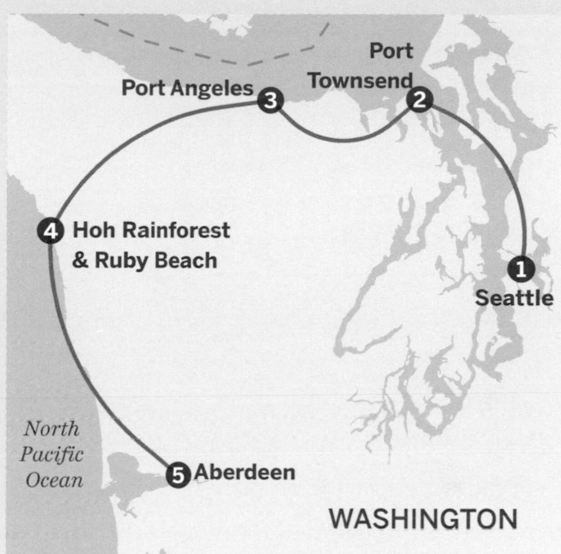

which welcomes day visitors and overnight guests to enjoy mineral springs fed by the mountains you've just conquered.

Drive 141km to Hoh Rainforest.

4 Hoh Rainforest and Ruby Beach

Along the Pacific Coast, there are several waypoints worth stopping at. Hoh Rainforest is one of the wettest places you can visit, receiving an average of 3.55m of rain per year – the moss-covered trees combine with the dense leafy canopy overhead to create an otherworldly experience. Ruby Beach, 52km west, offers the chance to explore the wild coastline, with sea stacks, driftwood and piles of sea foam even on calm days.

Continue south on US-101 for 130km.

5 Aberdeen

You'll pass through it anyway en route back to Bremerton, but Aberdeen is worth a short stop to see Nirvana lead singer Kurt Cobain's childhood home. There's a small sign near the house and a memorial a short walk away.

Head 132km east across the peninsula back to Bremerton for the return ferry to Seattle and onward connections.

ALTERNATIVES

Extension: Sequim and Dungeness Spit

Stretch your legs by exploring the community of Sequim and the surrounding area. During summer, the fields here are famous for their lavender (there's a lavender-themed festival during July). Just north is Dungeness Spit, the longest natural sand-spit in the US. You can walk out to the Dungeness National Wildlife Refuge and Dungeness Lighthouse.

Sequim is 50km west of Port Townsend, on the way to Port Angeles.

Diversion: Victoria

A quick ferry trip gets you to Canada and British Columbia's pretty capital, Victoria. Craigdarroch Castle is a great example of the city's Victorian architectural style, while the Royal BC Museum gives an excellent overview of the region, including local indigenous culture. Round off a visit with afternoon tea at the Fairmont Empress.

Ferries run between Port Angeles and Victoria (2 daily; 1hr 30min).

FOLLOW THE GEORGIA O'KEEFFE TRAIL IN NEW MEXICO

Santa Fe – Ghost Ranch

Drive through northern New Mexico's multicoloured mesas, canyons and deserts, tracing the ethereal beauty of the US state dubbed the 'Far Away' by Georgia O'Keeffe.

FACT BOX

Carbon (kg per person): 29
Distance (km): 233
Nights: 6-10
Budget: $$
When: year-round

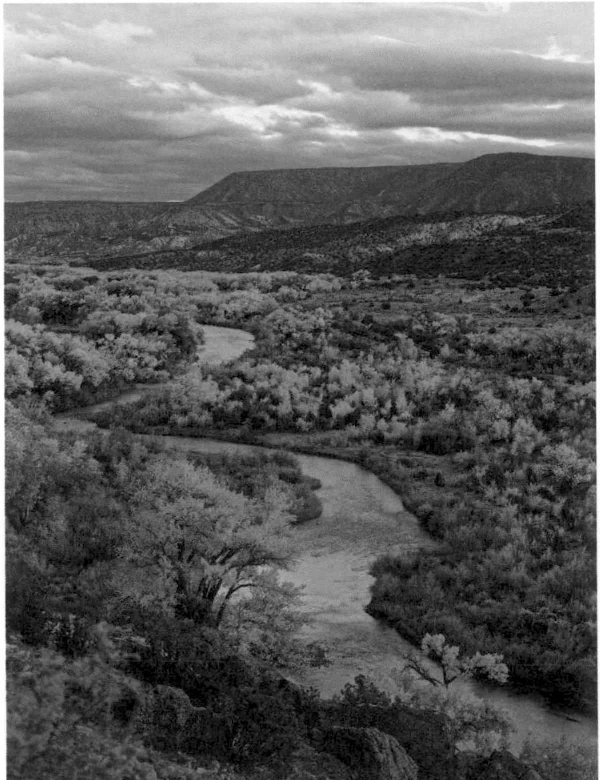

ABOVE: RIO CHAMA VALLEY

❶ Santa Fe

The Georgia O'Keeffe Museum is a must for putting the artist's New Mexico life into context. It's also just one highlight of Santa Fe, a picturesque Spanish colonial-era town filled with adobe architecture and leafy streets. Kick back with an O'Keeffe-themed supper at Eloisa, whose dishes are served in ways that reference her work (one course arrives on a cow's skull).

Drive 76km northwest of Santa Fe to Abiquiú.

❷ Abiquiú and the Rio Chama Valley

Book ahead to tour the Georgia O'Keeffe Home and Studio on the outskirts of Abiquiú. Housed in what was the artist's winter home, it inspired more than 20 of her paintings, many of them centred on the cottonwood trees and snaking river of the Rio Chama Valley which stretches northwest from Abiquiú and offers some lovely walks. 'Gas up', as O'Keeffe used to say, and try a green chili burger at Bode's General Store in Abiquiú before getting back on the road.

Continue 30km west to Cerro Pedernal.

❸ Cerro Pedernal and Abiquiú Lake

A 3006m plateau peak overlooking the reservoir known as Abiquiú Lake, Cerro Pedernal (Spanish for

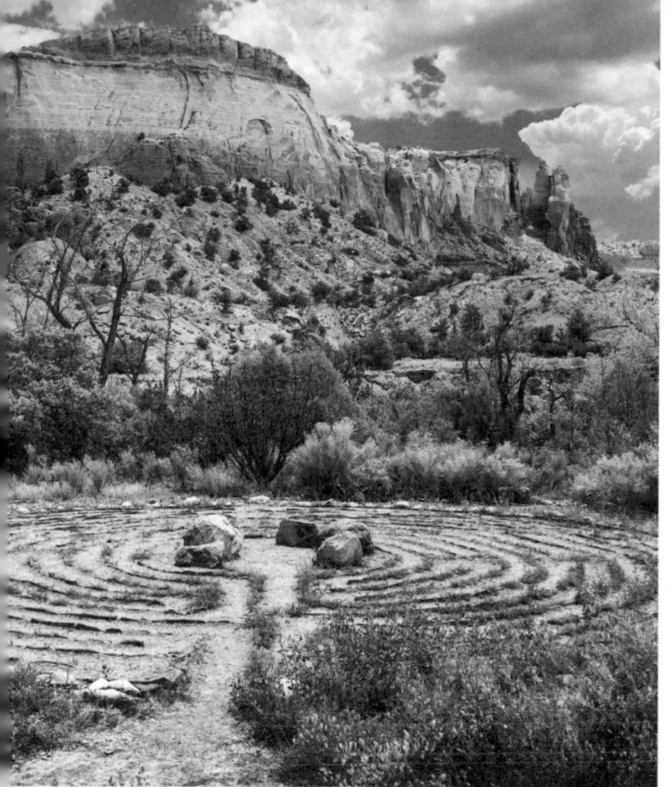

ABOVE: CHARTRES LABYRINTH, GHOST RANCH

'flint hill') beguiled O'Keeffe – and will surely have the same effect on you. She described it as her private mountain, saying 'It belongs to me. God told me if I painted it enough, I could have it.' Fittingly, her ashes were scattered from its flat peak after her death in 1986 at the age of 98.

Drive 30km north to Ghost Ranch.

4 Ghost Ranch

Located northwest of Abiquiú, on a road bounded by multicoloured natural attractions, lies Ghost Ranch, the dude ranch discovered by O'Keeffe in 1929. Booking ahead for one of the bus and walking tours of the sites O'Keeffe painted nearby is essential. Don't miss the eerie rock formations found in Plaza Blanca, characterised by rock chimneys, white cliffs and hoodoos. O'Keeffe called it her White Place, and it was one of her favourite spots to paint. Many visitors come here for the hiking trails, the most famous of which is the 5km round-trip climb to Chimney Rock. Pick up a trail guide at the welcome centre.

It's 97km back to Santa Fe and onward connections.

ALTERNATIVES

Extension: Taos Pueblo and Taos

Taos Pueblo, New Mexico's most extraordinary and most beautiful Native American site, was a favourite location for O'Keeffe and her friend Ansel Adams, who photographed the Unesco World Heritage Site on numerous trips. In Taos itself, you can stay in a room that O'Keeffe slept in at the Mabel Dodge Luhan House.

Taos Pueblo is 120km north of Santa Fe.

Diversion: White Sands National Monument

For a memorable, lengthy, add-on, head south of Santa Fe to the astonishing White Sands – O'Keeffe fans may think of her *Sky with Flat White Cloud* when faced with its vast white gypsum dunes. Follow the scenic loop drive – sunrise and sunset are the most photogenic times to be here – and get out of the car for some sandy hiking.

White Sands is 460km south of Santa Fe via Albuquerque.

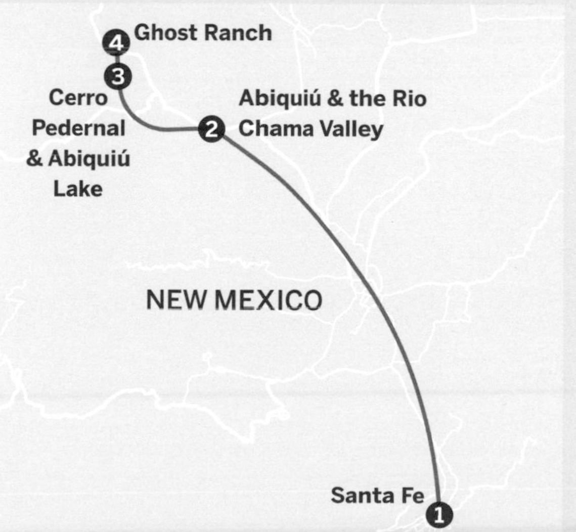

A TEXAN FIVE-WAYS ROAD TRIP

Houston – Marfa

This near-1400km driving trip makes a fine introduction to the music, food, art, history and natural attractions of the US' hugely underestimated Lone Star State.

FACT BOX

Carbon (kg per person): 172
Distance (km): 1378
Nights: 15-21
Budget: $$
When: year-round

❶ Houston

The biggest city in Texas, Houston has lots of earthly and heavenly delights to entertain visitors. The unmissable Space Center Houston has information on past, present and future space exploration. Similarly impressive is the fine art at the Menil Collection, the awe-inspiring Rothko Chapel and James Turrell's Twilight Epiphany Skyspace at Rice University's Moody Center for the Arts, as well as the architecturally arresting Buffalo Bayou Park Cistern, a 1920s decommissioned underground reservoir.

Drive 260km west to Austin.

❷ Austin

Austin is Texas at its coolest. Spend nights in iconic music venues like the Continental Club and atmospheric spots like the Driskill Bar, and days exploring the Blanton Museum of Art, the State Capitol and Contemporary Austin's sculpture garden on the banks of Laguna Gloria. In between, take a bat-spotting cruise on Lady Bird Lake.

Continue driving 127km southwest to San Antonio.

❸ San Antonio

Take in a Texan institution at the San Antonio Stock Show and Rodeo. Then stroll the pretty cypress-lined River Walk to the San Antonio Museum of Art (SAMA) and the Pearl, a converted brewery that's now home to some of the state's best eateries. Spanish colonial history comes to life in the 18th-century Mission San José, the largest of the city's five historic Missions, and

ABOVE: SXSW MUSIC FESTIVAL, AUSTIN

ABOVE: LADY BIRD LAKE, AUSTIN

at the storied Alamo, site of one of the USA's most iconic sieges.

Drive 114km north.

4 Fredericksburg

Quaint Fredericksburg, whose long main street is filled with cute shops, wineries and ice-cream parlours, is a good place to enjoy a wander and stock up on local wine and a cowboy hat and boots for your next stop, the arid land and vast skies of western Texas.

Make sure the tank is full for the 565km drive to Marfa.

5 Marfa

Set in a landscape film fans might recognise from *Giant*, *There Will Be Blood* and *No Country for Old Men*, big Texas country and big art really come together in tiny, arty Marfa, where Donald Judd turned a huge former army base on the outskirts of town into the vast and unforgettable Chinati Foundation art space. Marfa itself is a coolly creative town filled with galleries, restaurants, giant murals featuring scenes of James Dean from *Giant* and lots of design stores.

Drive 312km west to El Paso for onward connections from its airport.

ALTERNATIVES

Extension: Lyndon B Johnson National Historical Park

Set in the heart of Texan wine country amid rocky hills planted with vines, yucca, cactus, juniper and live oak, the home of Texas schoolteacher-turned-US president Lyndon B Johnson and his much-loved first lady is a fascinating look at the man and his life.

The site is 27km east of Fredericksburg.

Diversion: Big Bend Ranch State Park and National Park

The vast desertscapes of these two parks promise a region of dramatic canyons and vertiginous overlooks, cut through by the bright green ribbon of the Rio Grande. Add in the characterful mining town of Terlingua, huge inky-black night skies and an otherworldly sense of tranquillity and you have an unforgettable side-trip.

Terlingua, between the two parks, is 180km south of Marfa.

EXPLORING WISCONSIN'S DOOR COUNTY

Sturgeon Bay – Bailey's Harbor

Strike out on an easy US road trip from Green Bay to Door County, Wisconsin's Lake Michigan playground: expect lots of local cherries and sweeping water views.

FACT BOX

Carbon (kg per person): 51
Distance (km): 241
Nights: 2-3
Budget: ££
When: year-round (summer for fresh cherries)

❶ Sturgeon Bay

Sturgeon Bay is widely considered the entry point to Door County. Located on a shipping corridor that cuts across the Door Peninsula, this is the largest community in the county and has quickly become a hub for less-industrial types including artists and restaurateurs. While here be sure to visit the Door County Maritime Museum to see its displays on the Great Lakes, and admire the historic Michigan Street Bridge.

ABOVE: SISTER BAY HARBOUR

Drive 45km along the western side of Door County Peninsula, passing through Egg Harbor and Fish Creek.

❷ Ephraim

Historic Ephraim is nestled on a quiet bay where you can find some of the first settlements in Door County alongside art galleries and homestyle diners. One main attraction beyond the panoramic waterfront is Peninsula State Park, where towering limestone cliffs reveal the Niagara Escarpment, carved by glaciers over millions of years. Time-travel more by enjoying a famous 'fish boil' at one of Ephraim's restaurants, a nod to the region's Scandinavian heritage brought here by immigrants; be sure to end the meal with local cherry pie.

Continue 7km northeast.

❸ Sister Bay

It takes less than ten minutes to drive from Ephraim to Sister Bay, but the two feel worlds apart and

ABOVE: STURGEON BAY LIGHTHOUSE

offer very different experiences for visitors. Ephraim's picturesque waterfront architecture is replaced with more modern structures in Sister Bay and a focus on food. Stop off at Fred & Fuzzy's Waterfront Bar & Grill for their famous Door County cherry margarita, or dine and drink at Al Johnson's Swedish Restaurant where rooftop goats are the main attraction. Burn off the calories with a stroll between Sister Bay Beach and the nearby marina.

Drive 15km south Door County's east coast and Bailey's Harbor.

4 Bailey's Harbor

Leave Green Bay's calm waters behind and brave the Lake Michigan side of Door County in Bailey's Harbor. This lakefront community is known for its natural wonders, from the Ridges Sanctuary where you can spot birds and spy lighthouses to getting out on the lake yourself with some kayaking to 'sea' caves. When you're ready to refuel after a day of adventure, take your pick from the many dining opportunities with views of the water.

Drive 107km southwest to Green Bay for onward connections.

ALTERNATIVES

Extension: Washington Island

Book the ferry from Northport, at the northern end of the Door Peninsula, to Washington Island and cross Death's Door – the waterway which gives the county its name. Stroll through lavender fields, look for sea glass on Schoolhouse Beach, or explore the even more remote, neighboring Rock Island State Park.

Take the ferry to Washington Island from Northport (5-6 daily in summer).

Diversion: Plum Island

Take a cruise to Plum Island from the town of Gills Rock on the Door Peninsula's north shore. Head out on a hike, admiring the lighthouses and navigation markers that make this small island significant. Ask your captain to stop at the Grape Shot Shipwreck off the coast; this well-preserved Civil War era ship sits just below the water's surface.

A number of private operators offer Plum Island cruises, primarily from Gills Rock.

THE ULTIMATE ALASKA ITINERARY

Anchorage – Seward

America's Last Frontier beckons with wide-open spaces, wildlife and unique cultural experiences. Enjoy the best of what Alaska has to offer by plane, train and automobile.

FACT BOX

Carbon (kg per person): 164
Distance (km): 1557
Nights: 5
Budget: £££
When: Jun-Aug

❶ Anchorage

Anchorage serves as the base from which to head out into the rest of Southcentral and Interior Alaska. The state's biggest city also has plenty of things to do before you set off. Start by renting bicycles from any of several downtown retailers and exploring the Tony Knowles Coastal Trail, which offers views over Cook Inlet – and the chance to encounter moose. If you're lucky, you might spot 6190m Denali, the highest peak in the US, on the northern horizon. In the afternoon, explore the historic district of Downtown Anchorage, including a stop at the Anchorage Museum where you can admire a Smithsonian-affiliated collection of Alaska Native crafts and artefacts.

Take the morning Alaska Railroad train from Anchorage to Denali (daily in summer; 8hr).

❷ Denali National Park

Among the breathtaking sights of Alaska, Denali is the crown jewel. Sprawling Denali National Park, where the eponymous peak is found, covers some six million acres and is only accessible by modified school bus provided by the National Park Service. Book the eight-hour Tundra Wilderness Tour for the best opportunity to spot wildlife in the park and hopefully glimpse the namesake mountain, which is visible only 30% of the time.

ABOVE: MOOSE, DENALI NATIONAL PARK

ABOVE: VISIT KENAI FJORDS NATIONAL PARK FROM SEWARD

Back on the train, head north from Denali to Fairbanks (daily in summer; 4hr).

3 Fairbanks

Fairbanks is a winter wonderland for aurora chasers; during the summer months, it's perfect for exploring the outdoors and learning Alaskan history. Whether you opt for a day of hiking and hot springs or hop from the Museum of the North to the Morris Thompson Cultural Center, you can experience a lot of Interior Alaska in a short time here.

Take a flight from Fairbanks back to Anchorage (1hr) and pick up a rental car for the 200km drive south to Seward.

4 Seward

Take your time driving the Seward Hwy, as the route along the shoreline of the body of water known as Turnagain Arm offers opportunities for whale-spotting. In Seward itself, book a full-day cruise into Kenai Fjords National Park for more wildlife viewing opportunities. You'll also get up close to the glaciers in the park – listen out for the 'white thunder' of these massive ice giants.

Head back to Anchorage for onward connections.

ALTERNATIVES

Extension: Girdwood

En route to Seward, explore small town of Girdwood, nestled in a mountainous valley and host to art festivals in the summer. Ride the Alyeska aerial tram up the mountain for a moderate hike or to dine at the resort's restaurant with jaw-dropping views. There are more delicious restaurants, shops and galleries on the valley floor.

Girdwood is 63km from Anchorage on the Seward Hwy.

Diversion: Eagle River

North of Anchorage, the community of Eagle River allows you to explore the Chugach National Forest. From Eagle River Nature Center at the back of the valley, you can opt for an easy or more strenuous hike depending on your time and fitness. Afterwards, refuel at an Eagle River dining spot, such as the Matanuska Brewing Company, with food and local craft beer.

Eagle River is 25km north of Anchorage.

MEXICO'S RIVIERA MAYA ROAD TRIP

Cancún – Tulum

Take your time travelling along Mexico's most popular coastline, stopping at beaches and Maya ruins and looking out for wildlife along the way.

FACT BOX

Carbon (kg per person): 33
Distance (km): 266
Nights: 4-5
Budget: $$
When: Nov-Jul

ABOVE: GREEN TURTLE, AKUMAL BAY

❶ Cancún

This hard-partying resort city isn't for everyone, but those looking for a convivial, get-away-from-it-all attitude and lots of sun will find Cancún ideal. There's a wealth of hotels and excellent restaurants to choose from (some with Michelin stars), plus arguably some of Mexico's most gorgeous beaches on which to flop. More authentic local culture is best experienced in Cancún Centro, inland from the beach resorts. There's also a surprisingly historical side to the city too, with the excellent Museo Maya de Cancún containing 400 artefacts relating to the ancient Maya civilisation that once flourished here. The adjoining San Miguelito archaeological site is also well worth checking out.

Pick up your rental car and drive 68km south along the coast to Playa del Carmen.

❷ Playa del Carmen

Playa del Carmen is an international melting pot with yogic flair. Wellness-minded restaurants and juice bars sit next to old-school *taquerías*, and nightlife hangouts range from swanky hotel rooftops with views to avant-garde arthouse venues. Playa's beaches can get a little crowded, but they also have a sense of community often missing from guarded resort beaches, and the reefs here are excellent for spotting marine life.

ABOVE: TULUM MAYA RUINS

Continue 39km south.

❸ Akumal

As well as being a convenient stop between Playa del Carmen and Tulum, Akumal is worth a visit for the turtles that frequent its shores. Don a pair of fins and snorkel and get into the shallow waters for a chance to spot some shelled reptiles and a vibrant reef. Just along the shoreline you'll find Yal-Ku, an inland lagoon where a cenote (natural pool) empties fresh water into the sea.
Drive 29km south.

❹ Tulum

Renowned for its bohemian-chic Hotel Zone, Tulum is an Instagrammer favourite for good reason. Rent a bungalow by the beach and gaze at the blue water and white sand. The town's namesake Maya ruins, sitting romantically atop a dramatic cliffside looking out over the Caribbean, reveal the area's rich Maya past. The city itself has a fun personality and offers several budget-friendly digs alongside its upmarket options. The Reserva de la Biosfera Sian Ka'an, 10km south, has guided walks through forest that's home to monkeys and over 300 bird species.
Make the return trip back to Cancún (130km) for onward connections.

ALTERNATIVES

Extension: Cobá

Cobá is home to the area's highest pyramid. Climb (carefully) to the top, wander the other ruins, then rent a bike and pedal out to its three cenotes, off-the-beaten-track swimming holes where you won't have to battle other tourists for a dip. Don't miss Tankach-Ha, a cenote so deep that it has a jumping platform inside.
Cobá is 48km northwest of Tulum.

Diversion: Valladolid

Located in the heart of the Yucatán, Valladolid is a good hub for anyone looking to see more of the region beyond the coast. Characterised by its pastel colonial architecture, the town sits close to several cenotes as well as the iconic Chichén Itzá, the largest Maya city on the peninsula (go early to avoid the crowds), and the quieter ruins of Ek Balam.
Valladolid is 150km southwest of Cancún.

MYSTICAL MAYA TOUR THROUGH MEXICO

Palenque – Lagos de Montebello

Immerse yourself in the world's best Maya sites, nestled in southern Mexico's pine-forest highlands and sultry jungles, where travel conditions can make your trip an adventure.

FACT BOX

Carbon (kg per person): 132
Distance (km): 730
Nights: 5-6
Budget: $-$$
When: Nov-May

❶ Palenque

The archaeological site of Palenque (accessed from a town of the same name) is one of the top destinations in Chiapas and one of the best examples of Maya architecture in all of Mexico. The dense jungle covering these hills forms an evocative backdrop to the area. Hundreds of ruined buildings are spread over 15 sq km, but only a fairly compact central area has been excavated. Spend an entire day wandering through the site, taking time to explore the awesome Templo de las Inscripciones, El Palacio and Templo del Sol. Don't miss the museum situated outside the site.

Regular minibuses ply the 9km between Palenque town and the archaeological zone. Back in Palenque, the easiest way to undertake the rest of this trip is by organising a car and driver, giving you an unrushed, stress-free experience. Next stop, Bonampak, is 145km southeast.

❷ Bonampak

A dense jungle setting kept Bonampak, one of Chiapas' most outstanding archaeological sites, hidden from the outside world until 1946. Beyond its beautiful location, its highlights include vivid frescoes that bring the Maya world to life.

Continue 44km to Frontera Corozal on the Guatemala border, where you can arrange a motorboat trip to Yaxchilán (1hr).

ABOVE: SCARLET MACAWS, EJIDO REFORMA AGRARIA

ABOVE: TEMPLO DE LAS INSCRIPCIONES, PALENQUE

❸ Yaxchilán

Jungle shrouded Yaxchilán, one of the region's most important Classic Maya cities with ornamented facades and roofcombs, has an impressive setting above a horseshoe loop along the Río Usumacinta.

Stay in Frontera Corozal before heading south 113km to Ejido Reforma Agraria.

❹ Ejido Reforma Agraria

Home to the endangered *guacamaya* (scarlet macaw), this tiny community has a welcoming ecolodge, Las Guacamayas, and you can take tours of macaw nesting sites. The Montes Azules biosphere is on the opposite bank of the river and has plenty of howler monkeys and birds.

Continue west in the car for 155km.

❺ Lagos de Montebello

Spend a day exploring the beautiful Montebello lakes – there are over 50 – that are nestled in pine and oak forest. You can spend the night in the village of Tziscao.

Drive 150km to San Cristóbal de las Casas or 208km to Tuxtla Gutiérrez, both of which have good onward connections.

ALTERNATIVES

Extension: Misol-Ha and Agua Azul Waterfalls

These spectacular water attractions – the 35m jungle waterfall of Misol-Ha and the thundering cascades of Agua Azul – are surrounded by verdant jungle. It's a case of wandering and enjoying the nature here; outside wet season you can swim at Misol-Ha but don't be tempted at Agua Azul (the current is deceptively fast).

The easiest way to visit the falls is on an organised day trip from Palenque.

Diversion: San Cristóbal de las Casas

Set in a gorgeous highland valley surrounded by pine forest, this colonial city has fabulous markets, historic churches and cutting-edge restaurants. Don't miss the indigenous fabrics at Centro de Textiles del Mundo Maya, or the Museo de Sergio Arturo Castro Martínez, displaying traditional costumes and gifts of gratitude from villagers to the museum owner.

The city's facilities and transport links make it the ideal place to end your trip.

MEXICO CITY PYRAMIDS AND CULTURE

Mexico City – Teotihuacán

From the vibrant capital take trips to climb pre-Hispanic pyramids, feel the spiritual vibes of Tepotzlán and traverse the centre of the world's largest pyramid in Cholula.

FACT BOX

Carbon (kg per person): 14
Distance (km): 512
Nights: 4
Budget: $$
When: year-round

❶ Mexico City

Mexico's monumental capital is a pleasant surprise for visitors. Stroll leafy Roma and Condesa neighbourhoods, lingering at cafes and top-quality restaurants, and visit the many museums – standouts are the Museo Nacional de Antropología, for insights into pre-European civilisations, and Templo Mayor, once the centre of the Aztec empire. For more modern Mexican culture head to Museo Frida Kahlo, where the artist's house is maintained just as she left it.

Take a Pullman de Morelos coach from Taxqueña station to Tepoztlán (every 40min; 1hr 15min).

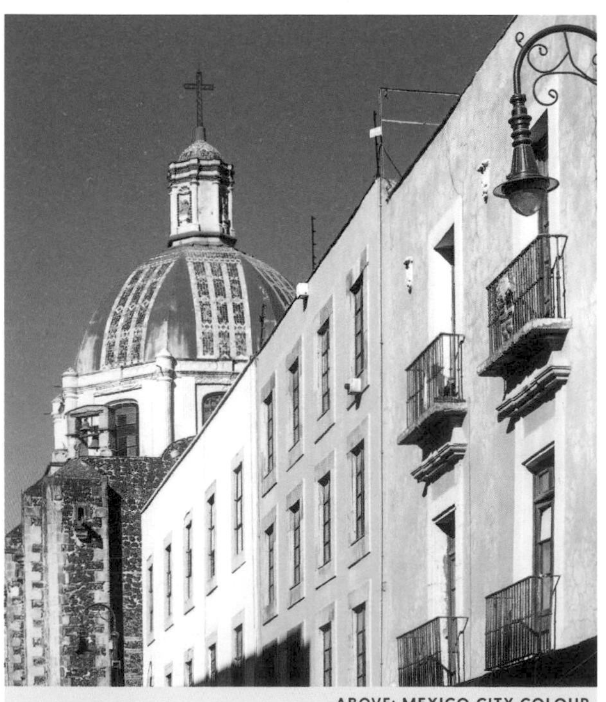

ABOVE: MEXICO CITY COLOUR

❷ Tepoztlán

Tepoztlán's colourful market dishes out pre-Hispanic and vegetarian food, hotels offer *ayahuasca* ceremonies (for herbally induced enlightenment) and shops tinkle with crystals. But the main appeal of the town is El Tepozteco, the pyramid rising above its cobblestoned streets. After a steep hike, the top of the pyramid rewards the effort with views across the town and valley. El Tepozteco is dedicated to Tepoztēcatl, the god of *pulque* (fermented agave alcohol); quench a post-hike thirst by sampling it one of the many bars.

Return to Mexico City. From TAPO bus terminal, head to CAPU terminal in Puebla (frequent; 2hr) and then take a bus or taxi to Cholula.

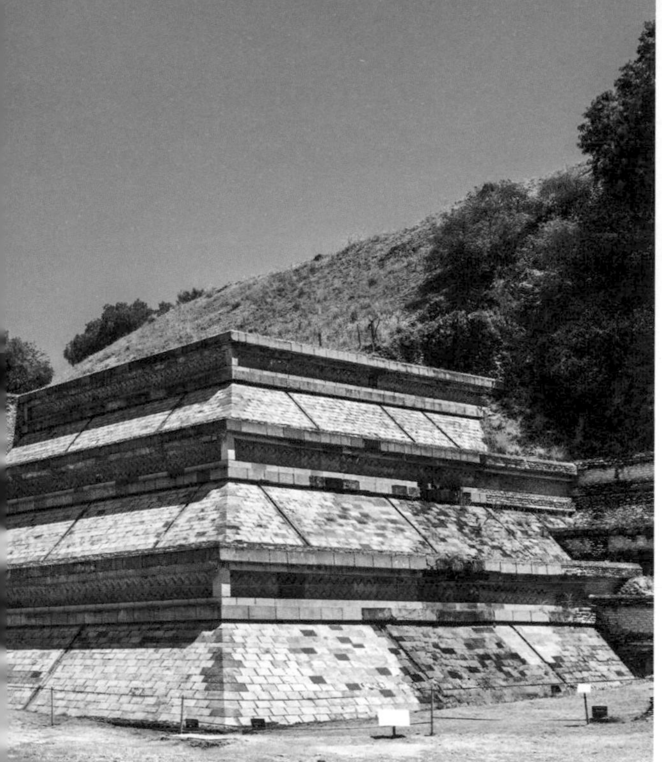

ABOVE: PIRÁMIDE TEPANAPA, CHOLULA

❸ Cholula

The world's largest pyramid by volume, Pirámide Tepanapa is deceptive, looking more like a hill these days. But this mighty construction has ancient manmade origins. It's actually a series of pyramids built on top of each other, and its monumental size is best appreciated by walking the 800m tunnel through the centre. Outside are the remains of stone altars and murals. Climb to the church at the summit for panoramic views and a further sense of the pyramid's former magnificence.

Back in Mexico City arrange a minivan tour to Teotihuacán. They leave every morning, have bilingual guides and last 6-10hr.

❹ Teotihuacán

Teotihuacán is one of Mexico's most imposing and mysterious sites. Not much is known about the civilisation that created this giant ancient city but their legacy, including two huge pyramids, still wows visitors. Both the Pirámide del Sol (Pyramid of the Sun, the third largest in the world) and the Pirámide de la Luna (Pyramid of the Moon) can be climbed for bird's-eye views of the sprawling metropolis below, containing fascinating temple ruins with plumed-serpent carvings and murals.

Head back to Mexico City and its onward international connections.

ALTERNATIVES

Extension: Xochimilco

The Spanish built Mexico City over a lake and its waterways, and some are still visible in Xochimilco. Festive gondola-like *trajinera* boats glide through tree-lined canals, floating by *chinampas*, artificial islands where crops have been grown since Aztec times. For incongruous horror, drift by Island of the Dolls, where discarded toys dangle from trees.

Take the light rail from Taxqueña station to Xochimilco (frequent; 35min).

Diversion: Puebla

Stopping in the historic centre of Puebla on the way to Cholula is as culturally rewarding as any pyramid. The cobbled streets put on a show of colourful churches, gorgeous *talavera* (ornate tiles) architecture and restaurants serving *mole poblano* dishes (Puebla's signature chocolate- and spice-infused sauce). Daring eaters should try buttery *escamole* (ant larvae).

Puebla's CAPU bus terminal has frequent services from Mexico City.

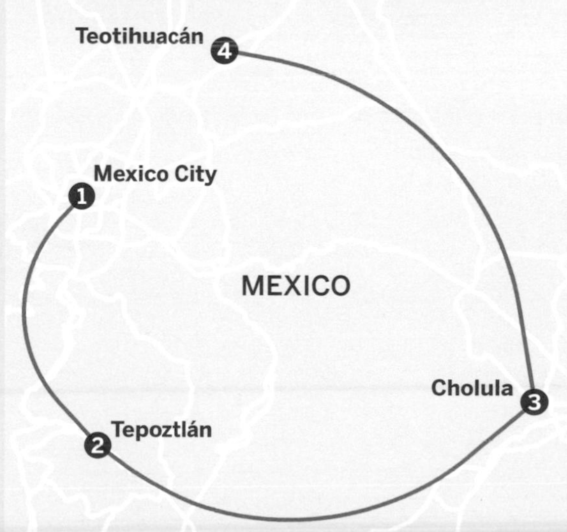

MEXICO'S COPPER CANYON RAILWAY

Los Mochis – Creel

All aboard the spectacular Copper Canyon Railway, rolling through the Barrancas del Cobre past clifftop vistas, pine-clad mountains and picturesque colonial towns.

FACT BOX

Carbon (kg per person): 9
Distance (km): 430
Nights: 6
Budget: $$
When: Sep-Oct

❶ Los Mochis

The sprawling lowland city of Los Mochis doesn't get much love from travellers, but it's worth spending a day here before boarding the famous Copper Canyon Railway, aka El Chepe. Breathe in the verdant greenery of the Benjamin Francis Johnston Botanical Garden, then take a taxi out to Playa El Maviri for a seafood feast and a stroll along the beach. You can also arrange scenic boat tours around Bahia de Ohuira from nearby Topolobampo.

Both the luxury El Chepe Express and the bare-bones regional train head to El Fuerte (3 weekly each; 2hr 30min).

❷ El Fuerte

Founded in 1564, the peaceful town of El Fuerte oozes historical character and makes a delightful overnight while riding the rails. Aside from wandering the colonial-era streets, you can take a boat trip along the Río Fuerte, spying kingfishers, flycatchers and herons, before stopping for a look at 2000-year-old petroglyphs. You can also arrange visits to indigenous Mayo communities and witness the high-energy *danza del venado* (deer dance).

Continue on either the express or regional train to Divisadero (3 weekly each; 6hr).

ABOVE: DIVISADERO STREET FOOD

ABOVE: COPPER CANYON RAILWAY VIEWS

3 Divisadero

A train stop without a village, Divisadero is home to a stunning viewpoint overlooking the Copper Canyon that gives the train line its name. The adjoining food market serves up the world's best *gorditas* (maize cakes stuffed with goodies) cooked up before your eyes. There are excellent places to stay with clifftop views and a highly recommended adventure park nearby, Parque de Aventura Barrancas del Cobre, which has ziplines right over the canyon plus indigenous guides who lead walks in the area.

Continue on the train for the short hop to Creel (3 express/3 regional weekly; 1hr 20min).

4 Creel

It's easy to while away a few days in the sleepy town of Creel, where you can arrange a wide range of adventures. Saddle up for some horse riding, check out towering waterfalls, soak in hot springs, visit indigenous Rarámuri communities, browse local handicraft stores or just relax beside the pine-fringed waters of Arareko lake, some 9km south of town.

From Creel, retrace your journey back to Los Mochis (9hr), or continue on to Chihuahua by regional train (6hr) or bus (4hr 30min) for onward flights.

ALTERNATIVES

Extension: Chihuahua

Capital of Mexico's biggest state, Chihuahua is a fascinating city with a friendly vibe and a pleasant colonial centre. Pedestrian lanes lead off graceful central plazas past brightly painted storefronts, while museums include the Museo Histórico de la Revolución, former home of revolutionary Pancho Villa.

Frequent buses connect Creel with Chihuahua (4hr 30min); slower regional trains (6hr) run three times weekly.

Diversion: Batopilas

A charming town at the bottom of the Copper Canyon, the former silver-mining village of Batopilas is a sleepy place where you can unwind after days of exploring. The ride down into the canyon is nothing short of spectacular with twists, turns and a few heart-in-the-mouth vertical drops.

Minibuses make the trip down from Creel (daily Mon-Sat; 4hr).

EXPLORING MEXICO'S RIVIERA NAYARIT

Puerto Vallarta – Bucerías

Drive along Mexico's Pacific coast through the state of Nayarit for a peaceful getaway of whale watching, sunset spotting, gallery hopping and quiet relaxation.

FACT BOX

Carbon (kg per person): 39
Distance (km): 317
Nights: 4
Budget: $$
When: year-round

❶ Puerto Vallarta

Puerto Vallarta offers acclaimed beaches and boat trips out onto the Pacific with the chance (Dec-Mar) of spotting whales. It's also the place to indulge in some raucous nightlife, if that's your thing, as the rest of the stops on this trip are significantly more laidback.

Drive 40km northwest.

❷ Sayulita

The best-known of Nayarit's coastal towns, cute-as-a-button Sayulita is your first stop. Take photos beneath the signature *papel picado* flags strung across the town's streets, peruse the numerous shops (boho-chic reigns supreme) or pop into one of its many restaurants and bars for seafood-inflected fare and mezcal. Sayulita is also Nayarit's surfing centre – book a lesson and hit those waves.

Back in the car it's a quick 7km north to friendly San Francisco.

❸ San Francisco (aka San Pancho)

San Francisco, affectionately known as San Pancho (a nickname for Francisco) is a former fishing village turned arty, interesting town – check out fun shops like Mexicolate (to try all things cacao) or La Baba del Diablo (to sample house-made mezcal), or walk the leafy main drag down to the beach for seasonal whale watching. Be sure to check the local events calendar: if you're lucky, you can catch a show at

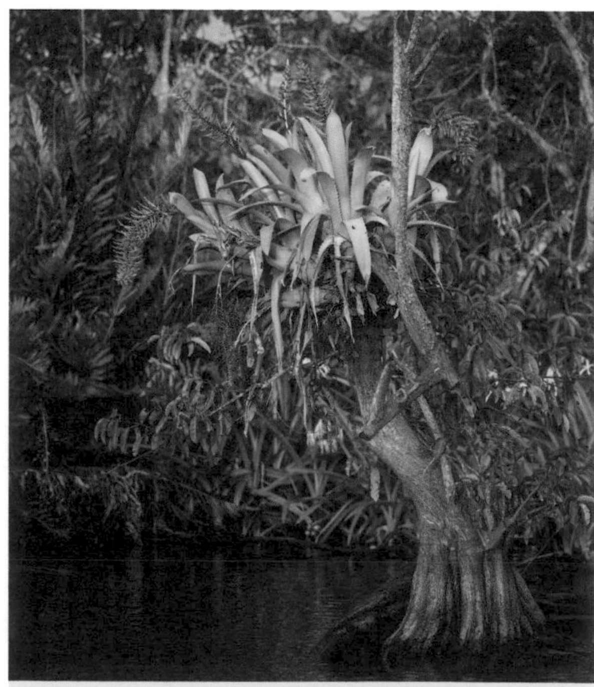

ABOVE: LA TOVARA NATIONAL PARK, NEAR SAN BLAS

ABOVE: PUERTO VALLARTA

to the local circus school, run by the former head of Cirque du Soleil.

Drive 113km up the coast.

4 San Blas

Perched on Nayarit's northern end, San Blas is the state's historic centre – it once served as an important port for the Spanish empire. Today it has a slower pace and is a particularly popular spot for birdwatchers. Nearby La Tovara National Park is home to hundreds of bird species, some endemic only to this area, and a boat trip through its mangroves yields excellent wildlife watching – you might spot some seriously large crocodiles.

Head 135km back down the coast.

5 Bucerías

You're on the home stretch but make one more stop in Bucerías. It's best-known for its sprawling flea market, but equally worth a look if you're in the mood to shop are its numerous art galleries and home decor stores; and from November to April, you can join one of the hosted art walks. Playa Bucerías is also a great place to catch the sunset.

Drive 22km back to Puerto Vallarta Airport for onward international connections.

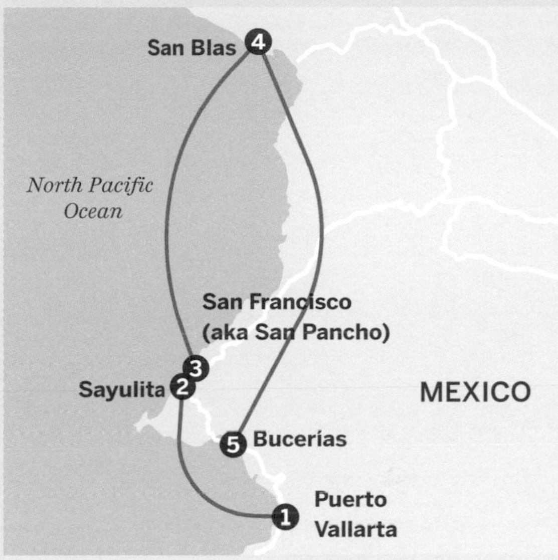

ALTERNATIVES

Extension: Lo de Marcos

If you're hankering for a peaceful seaside respite, turn off the highway and spend an afternoon in the small fishing village of Lo de Marcos. This tiny town is an antidote to the bustle of Sayulita and San Francisco – sit back, grab drink and enjoy its quiet stretch of sand and sea.

Lo de Marcos is 11km north of San Francisco.

Diversion: La Cruz de Huanacaxtle

If you're heading back to Bucerías on a Sunday, be sure to detour to La Cruz, home to Nayarit's best market. The numerous stalls feature everything from locally made clothing to indigenous art and jewellery to freshly made smoothies. The atmosphere is jovial, helped by live music, and the setting along the marina is wonderfully atmospheric. The market is cash only, so come prepared.

La Cruz is 4km west of Bucerías.

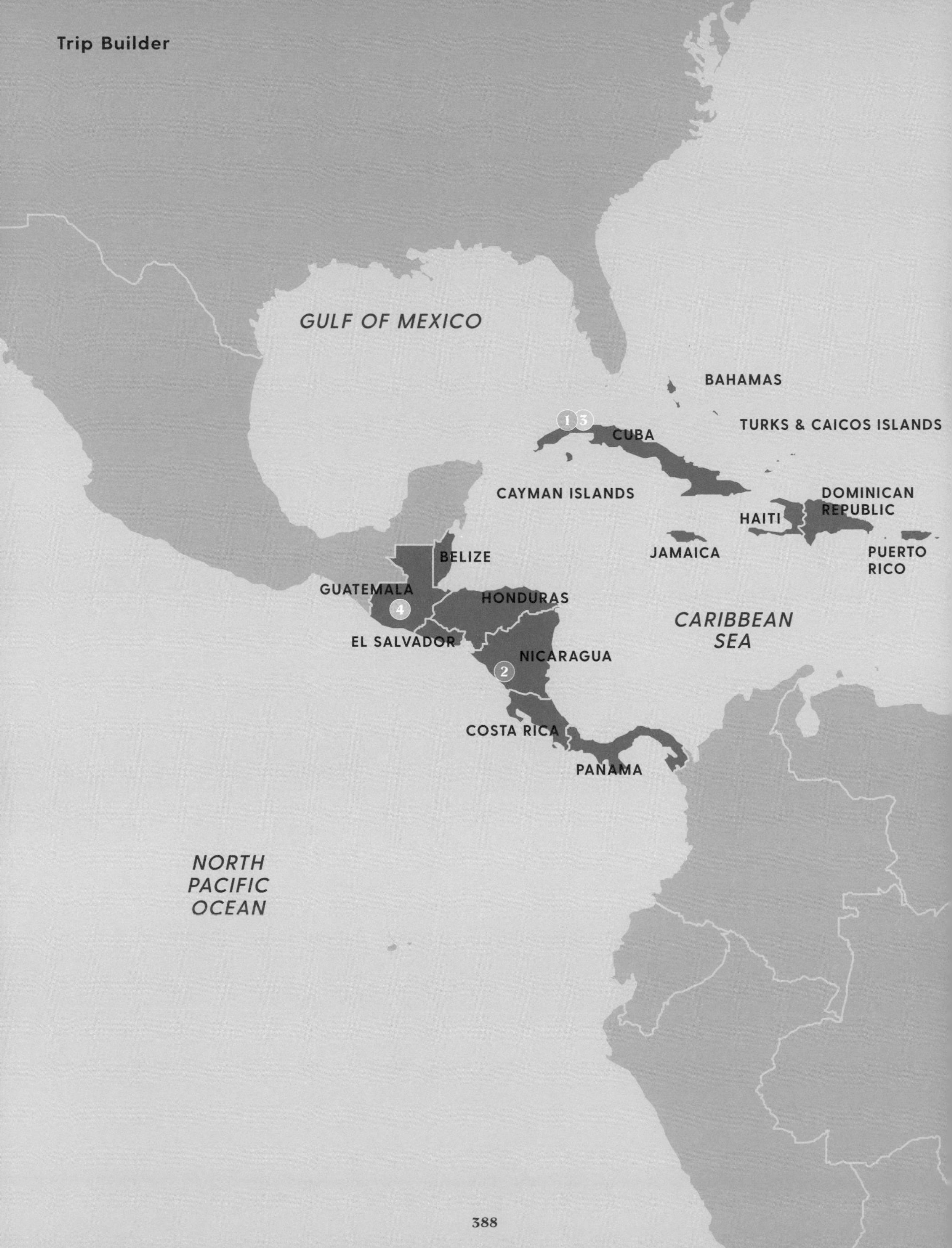

GULF OF MEXICO
BAHAMAS
TURKS & CAICOS ISLANDS
1
3
CUBA
CAYMAN ISLANDS
DOMINICAN REPUBLIC
HAITI
BELIZE
JAMAICA
PUERTO RICO
GUATEMALA
HONDURAS
4
CARIBBEAN SEA
EL SALVADOR
NICARAGUA
2
COSTA RICA
PANAMA
NORTH PACIFIC OCEAN

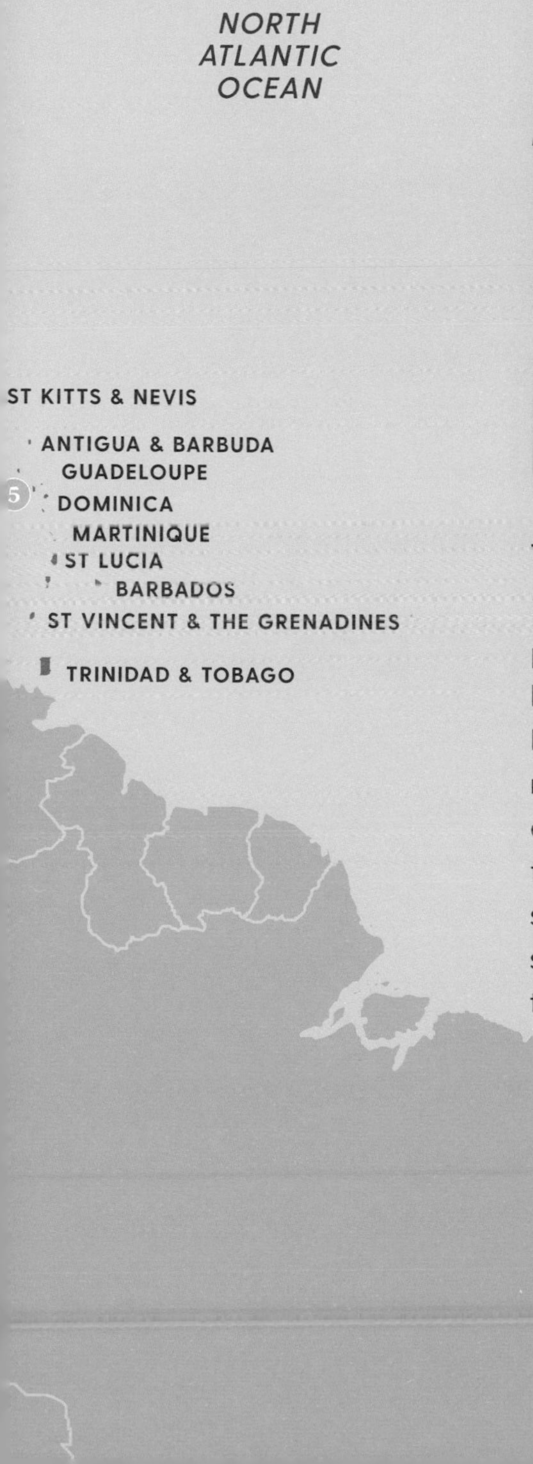

NORTH ATLANTIC OCEAN

ST KITTS & NEVIS
ANTIGUA & BARBUDA
GUADELOUPE
5 DOMINICA
MARTINIQUE
ST LUCIA
BARBADOS
ST VINCENT & THE GRENADINES
TRINIDAD & TOBAGO

CENTRAL AMERICA & THE CARIBBEAN

Yes, there are beaches, from the powder-sand, swaying palms and turquoise sea variety to rugged volcanic-black shores – but there's so much more to discover. Follow a revolutionary trail or tap in to the electrifying music scene in Cuba; uncover ancient Mayan pyramids and wildlife-stuffed national parks in Guatemala; kayak the backwaters in Nicaragua; dive into waterfall pools or snorkel the reefs on Dominica and St Lucia; and revel in some very Gallic charm in Guadeloupe and Martinique, toasting the sunset with a frosty rum cocktail in hand.

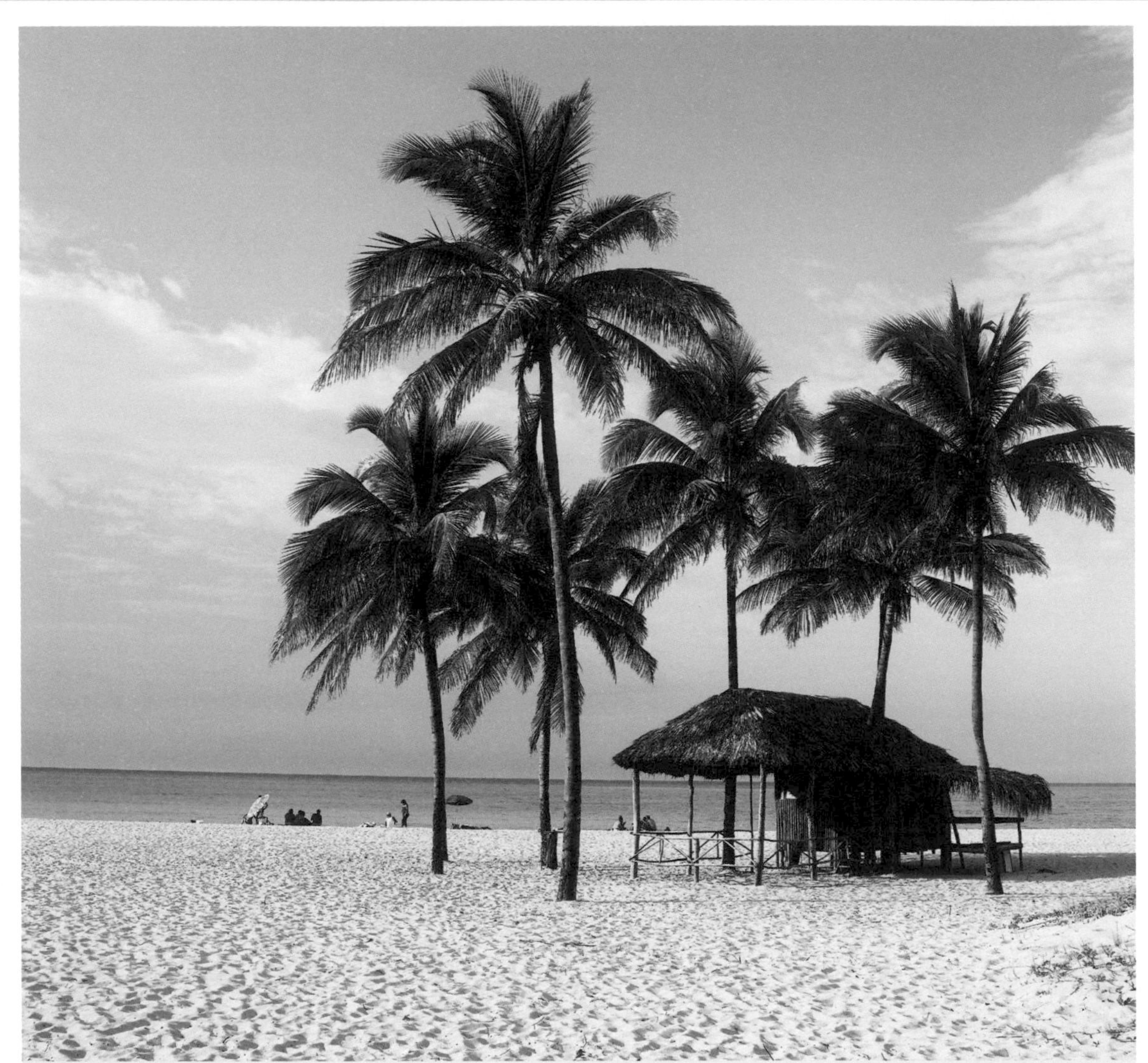

ABOVE: PLAYAS DEL ESTE, CUBA

CONTENTS

CLASSIC CUBAN ROAD TRIP

Havana – Playas de Este

A classic Cuba road trip means gorgeous Havana, Viñales tobacco region, pretty Trinidad and its enticing music, and downtime at palm-flecked eastern beaches.

FACT BOX

Carbon (kg per person): 53
Distance (km): 956
Nights: 12
Budget: ££
When: year-round but best Nov-Apr

❶ Havana

Seductive, sultry Havana is all about music, avant-garde art, elegant architecture, colourful classic cars and killer cocktails. Stroll the Old Quarter and Centro on foot, and then see upmarket El Vedado and Miramar in a classic car. Take in the museums, historic sights, new artisan shops and cafes by day; hop between bars and music haunts by night.

Book a classic car with driver from nostalgicarcuba.com and drive 185km west to Viñales.

❷ Viñales

Viñales is Cuba's adventure capital. Go for the plant-draped pudding-bowl mountains, horse riding, valley walks, morning mist, fresh fruit and juicy pork with *malanga* (taro) fritters. Minty mojitos at sundown are de rigeur – either sipped with your B&B host on the porch or on a terrace at one of the bars in town.

Organise round-trip transport to Cayo Jutías with agencies facing Viñales' plaza.

❸ Cayo Jutías and Vuelta Abajo tobacco plantations

A day trip of two parts begins with Cuba's premium cigar terroir at Finca Hector Luis Prieto in tobacco heartland, Vuelta Abajo. Tour the Nicotiana tabacum plantation and learn about the path from tiny seed to luxury smoke. Continue north to the coast and the scimitar of honeyed sand arcing into the Atlantic Ocean at Cayo Jutías. Swim in sapphire sea, order BBQ lobster for lunch and head to the sandy tip to spot starfish in the shallows before dragging yourself back to Viñales.

ABOVE: HAVANA STREET MUSICIAN

ABOVE: VIÑALES VALLEY

Take the Víazul bus (daily; 9hr 30min) or a collective (group), turn-up-and-go taxi (6hr) to Trinidad.

❹ Trinidad

The 19th-century sugar industry catapulted tiny Trinidad to stratospheric wealth. The 'Florence of the Caribbean' brims with paste mansions, churches and fancy interior decor built along cobblestone streets. Ramble the picturesque city by day but save some energy for dancing to live music by night.

Head back on the Víazul bus to Havana (daily; 6hr 30min). At Havana's bus station, look for a private or collective taxi or head to Parque Central for the hop-on hop-off HabanaBusTour to the Eastern Beaches (30min).

❺ Playas del Este

The 'Eastern Beaches' are Havana's local sun-and-sea playground. Pretty, palm-dotted Santa María beach is popular; Playa Mégano is quieter. Hire beach loungers, parasols and catamarans for sailing from nautical huts. Pack a picnic and grab a piña colada to have with it from makeshift bars on the sand.

Take the regular HabanaBusTour back to Havana (30min).

ALTERNATIVES

Extension: Las Terrazas

Lush mountain forests burbling with the sounds of rivers, waterfalls and birds surround eco retreat Las Terrazas. Hike forested trails, discover coffee plantation ruins, visit artists' studios, dine at pioneering vegetarian restaurant El Romero and walk the fragrant national orchid garden at nearby Soroa.

One Víazul bus a day heads from Havana to Las Terrazas (1hr); or detour en route to Viñales.

Diversion: Playa Caletón, Bay of Pigs

The sparkling ultramarine sea at the Bay of Pigs conceals a huge coral wall, shoals of rainbow-coloured fish and war wrecks from the ill-fated CIA-backed 1961 invasion. Go for the snorkelling, boho scene at Playa Caletón (with its B&Bs and beachfront bars) and the stellar birdwatching in neighbouring Zapata National Park.

One Víazul coach a day from Trinidad to Havana stops at Playa Larga for Caletón (3hr).

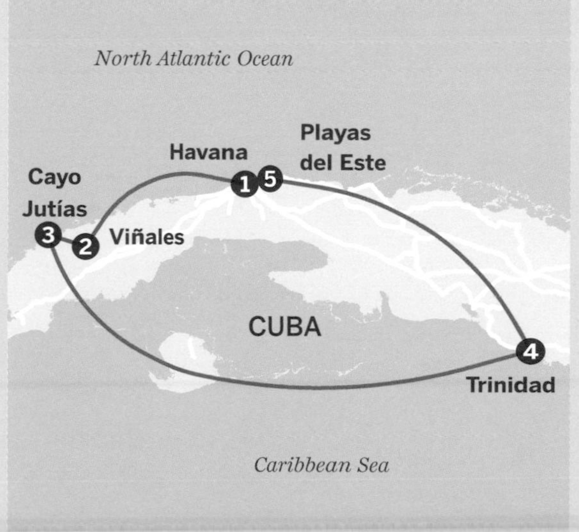

FIRST-TIME NICARAGUA FROM CITY TO COAST

Managua – San Juan del Sur

Explore an emerging Central American destination featuring compelling history, surprising natural scenery and the ultimate laidback Pacific surf town.

FACT BOX

Carbon (kg per person): 17
Distance (km): 502
Nights: 10-13
Budget: $$
When: Nov-May

❶ Managua

Big, bustling and chaotic, Nicaragua's capital is a compelling but confounding destination. Embrace the urban energy with a stroll along the waterfront Malecón, and take in stellar city views from nearby Tiscapa Lagoon.

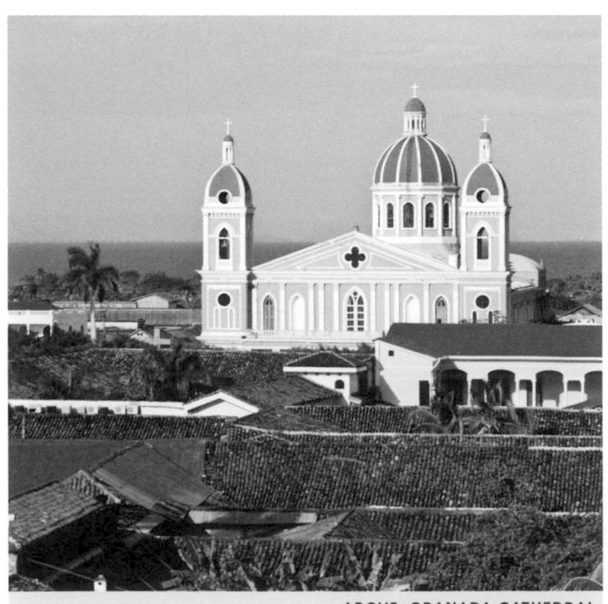

ABOVE: GRANADA CATHEDRAL

Take the bus from Managua's Bóer bus station to Léon (frequent; 2hr). Many travellers arrange direct shuttle transport to León upon arrival at Managua's international airport (2hr).

❷ León

The raffish university town of Léon was a stronghold for the Sandinista rebels during Nicaragua's civil war from 1978 to 1990. A brotherhood of former Sandinista fighters gathers daily to show visitors around Léon's Museo de la Revolución. Vistas from the nearby rooftop of Léon Cathedral are best at sunset, when local families gather in Parque Central.

Take a microbus to Granada (hourly; 2hr 30min).

❸ Granada

Founded in 1524, lakeside Granada is the oldest colonial city in the Americas and Nicaragua's architectural gem. When dusk approaches and shadows lengthen, food vendors set up to dispense street snacks around the town's cathedral. Negotiate with horse-and-carriage drivers to arrange a leisurely exploration of the colourful backstreets and lakefront.

Catch a morning bus from Granada to Rivas (1hr 30min), followed by a bus or taxi to San Jorge (20min), and then a ferry to Moyogalpa

ABOVE: KAYAKING LAGO NICARAGUA

on Isla de Ometepe (1hr).

❹ Isla de Ometepe

Punctuated by the twin peaks of Concepción and Maderas volcanoes, Isla Ometepe is the biggest island in Lago Nicaragua. Options for exploration include kayak trips around the Rio Istián wetland area, where graceful herons and egrets, caiman alligators and the occasional troop of howler monkeys all make an appearance.

Journey by ferry back to San Jorge, take a bus or taxi to Rivas, then catch a bus south to San Juan del Sur (hourly; 45min).

❺ San Juan del Sur

Tattoo shops and organic juice bars enliven the Pacific Coast surf town of San Juan del Sur, and colourful clapboard houses offer some of Nicaragua's best eating and drinking. Craft beers concocted by the San Juan del Sur Cerveceria are poured at the brewery's hip taproom, while local restaurants do clever things with lobster, prawns and swordfish.

Take the bus back to Rivas and catch another to Managua (every 30min; 2hr 30min). Direct shuttles to Managua's international airport (2hr 30min) are also available.

ALTERNATIVES

Diversion: Laguna de Apoyo

Escape Granada's heat and dust around Laguna de Apoyo, where comfortable lodges sit lakeside on the edge of a 23,000-year-old volcanic caldera. Stay overnight, or commandeer a kayak for an afternoon exploring quiet coves and concealed bays. Most lodges have simple restaurants and bars to recharge on beer, cocktails and ceviche.

Tour agencies in Granada offer lake-lodge shuttles.

Diversion: Around San Juan del Sur

Most San Juan del Sur accommodation options can arrange relaxed days out on catamarans, tacking lazily a few hundred metres offshore both north and south of town. Ask the skipper to pull in at Playa Rosa to the north – there's great swimming off the beach and a simple beachside shack selling grilled lobster tails smothered in garlic.

Book full-day catamaran sailing adventures in San Juan del Sur.

A JOURNEY THROUGH THE CUBAN REVOLUTION

Havana – Santiago de Cuba

Take in the full sweep of Cuba's revolution and independence wars in museums, memorials and mountaintop encampments on this cross-country bus trip.

FACT BOX

Carbon (kg per person): 29
Distance (km): 1005
Nights: 8-9
Budget: $$
When: Nov-Mar

❶ Havana

Anything can happen in Havana, from spontaneous music jams to Hemingway-style drinking binges. History buffs normally kick off at the Museo de la Revolución, in the former presidential palace, displaying a detailed if sometimes bombastic chronology of Cuban history with a strong emphasis on the Fidel Castro years.

Take the Víazul bus from Havana's main bus station to the Bay of Pigs (2 daily; 3hr 15min).

❷ Bay of Pigs

The scene of a botched US-backed invasion in 1961 when the Cold War almost got hot, the Bay of Pigs is more famous for its diving than its military exploits these days. Propaganda billboards line the coast road leading into the small village of Playa Girón, which maintains a small museum dedicated to the invasion; the building is guarded by a British Hawker Sea Fury used by the Cuban Air Force.

Take a Víazul bus to Cienfuegos (3 daily; 1hr 30min) and change there for Santa Clara (2 daily; 1hr 30min).

❸ Santa Clara

Home to Cuba's edgiest nightlife outside Havana and a veritable pilgrimage site for anyone with an interest in Che Guevara, Santa Clara's essential Che memorabilia includes his monumental mausoleum and Café Revolución, a museum-like drinking hole run by a Guevara-loving Spaniard.

Continue on the Víazul bus to Camagüey (4 daily; 4hr).

❹ Camagüey

In Cuba's third-largest city the revolutionary spirit centres on local son Ignacio Agramonte, a hero of the Independence War of 1868-78. It is also the heart of

ABOVE: DIVING THE BAY OF PIGS

ABOVE: OLD HAVANA

Catholic Cuba and harbours half-a-dozen churches with styles ranging from Gothic to Baroque.

Take the Víazul bus to Bayamo (4 daily; 3hr 30min).

5 Bayamo

Sedate contemporary Bayamo masks a much livelier past – this was the first city to be liberated during the Independence War. The tree-shaded main square includes a memorial dedicated to the Cuban national anthem, first played here in 1868, and the home/museum of 'father of the motherland' Carlos Manuel de Céspedes.

Víazul runs buses to Santiago de Cuba (4 daily; 2hr 15min).

6 Santiago de Cuba

The so-called 'city of revolutionaries' is anchored by Moncada Barracks, now a museum and school, where Fidel Castro attempted a failed coup in 1953. Other rebellious outposts include the Santa Ifigenia cemetery, location of an ornate mausoleum to Cuban national hero José Martí; and the museum/birthplace of Independence War general, Antonio Maceo.

There are daily flights between Santiago de Cuba and Havana.

ALTERNATIVES

Extension: La Plata

Fidel Castro's wartime HQ, high up in the cloud forests of the Sierra Maestra, has been left pretty much as it was in the late 1950s. Head to the tiny mountain village of Santo Domingo, overnight in the local hotel and hire a national park guide the next morning to part drive, part walk to the heavily camouflaged encampment. From here the rebels plotted their unlikely takeover of Cuba.

To get to Santo Domingo, take a taxi from Bayamo.

Diversion: Baracoa

Infused with a large dose of imperceptible magic, Baracoa is the country's quirkiest, zaniest and most unique outpost. Come here for distinctive spicy food; wild, rootsy music venues; unkempt palm-fringed beaches; and a profusion of deep-green foliage bursting with endemic species. And for the Baracoans, a law unto themselves.

Take a bus to this isolated town from Santiago de Cuba (daily; 4hr 45min).

CLASSIC GUATEMALA DISCOVERER

Antigua Guatemala – Tikal

From colonial cities travel to fiery volcanoes, forests filled with tropical birds, lush lakes and rivers, Maya craft markets and ancient jungle-dressed monuments.

FACT BOX

Carbon (kg per person): 43
Distance (km): 1571
Nights: 18
Budget: $
When: Oct-Apr

❶ Antigua Guatemala

Gracefully ruined by an 18th-century earthquake, dreamy Antigua is all churches, pastel mansions and gourmet spots. Hike to the crater of Pacaya, one of a trio of volcanoes around this gorgeous Spanish colonial city.

Minibuses depart daily for Panajachel (2hr).

❷ Lake Atitlán

From Panajachel, boats head across this volcano-ringed crater lake to traditional Maya town Santiago Atitlán (30min), bohemian San Pedro La Laguna (30min), eco-conscious San Marcos La Laguna (20min) and artisan hotspot San Juan La Laguna (40min).

Take a minibus to Quetzaltenango (regular; 2hr).

❸ Quetzaltenango

Unpolished, lively Quetzaltenango (known as Xela) is a base for volcanos, hot springs, colourful San Andrés Xecul church and Chicabal crater-lake (sacred to the Maya).

Minibuses depart frequently for Guatemala City (3hr).

❹ Guatemala City

The capital is no beauty, but its museums are an education: Ixchel for richly worked Maya textiles;

ABOVE: WHITE-NOSED COATIMUNDI, TIKAL

ABOVE: LAKE ATITLÁN

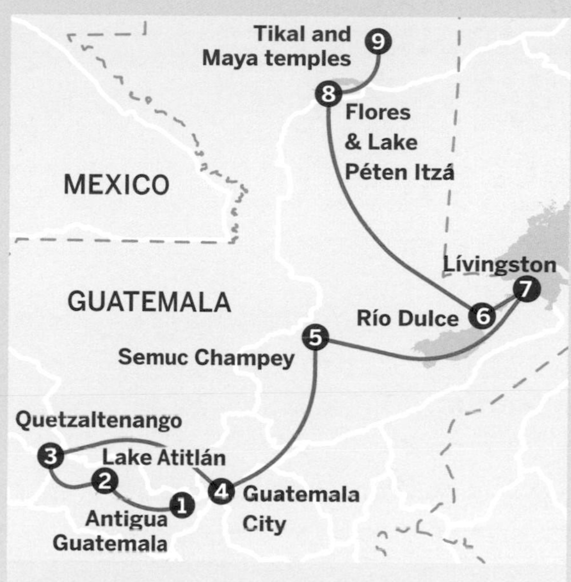

Archaeology for carved ancient stones; Popol Vuh for colonial and indigenous artefacts.

Take buses to Cobán (5hr), then to Lanquín (2hr 30min), then to Semuc Champey (30min).

5 Semuc Champey

The River Cahabón's turquoise waters spill over limestone terracing into enticing forest pools at Semuc Champey – perfect for swimming.

Take the Adrenlina Tours shuttle bus to Río Dulce (daily; 6hr).

6 Río Dulce

A humid, wild spot on Lake Izabal with lo-fi jungle accommodation, this is the place to board a slow boat down the teal-coloured river, journeying through a lush canyon to Guatemala's Caribbean coast.

Take one of the regular boats to Lívingston (1hr 10min).

7 Lívingston

Friendly and fun Lívingston is a palm-fringed Caribbean escape of waterfalls, slices of beach and music performed by the Arawak and African-rooted Garifuna locals.

Return to Río Dulce for buses to Flores (4hr).

ALTERNATIVES

Extension: Todos Santos Cuchumatán

This Central Highlands village is famous for its menfolks' dazzling purple and red-hued local dress. Come for the exquisite textiles, colourful bags, woven *huipil* (women's tunics) and fresh-air trekking. The 1 November horse racing and masked dancing festival is a drink-fuelled jamboree for the Mam-speaking Maya of this remote village.

Regular buses run here from Huehuetenango bus station, north of Quetzaltenango (2hr).

Diversion: Monterrico

Monterrico's Pacific Coast vibe is chilled. By day, read in a hammock under the tropical sun at the cluster of laidback hotels and hostels; boat through mangrove swamps spotting birds; and see endangered Olive Ridley and leatherback turtle babies released into the ocean. In the evening, sip cocktails, toes dug into warm sand, as the sun sets over the ocean.

Shuttles head here from Antigua Guatemala (2hr) and Guatemala City (3hr).

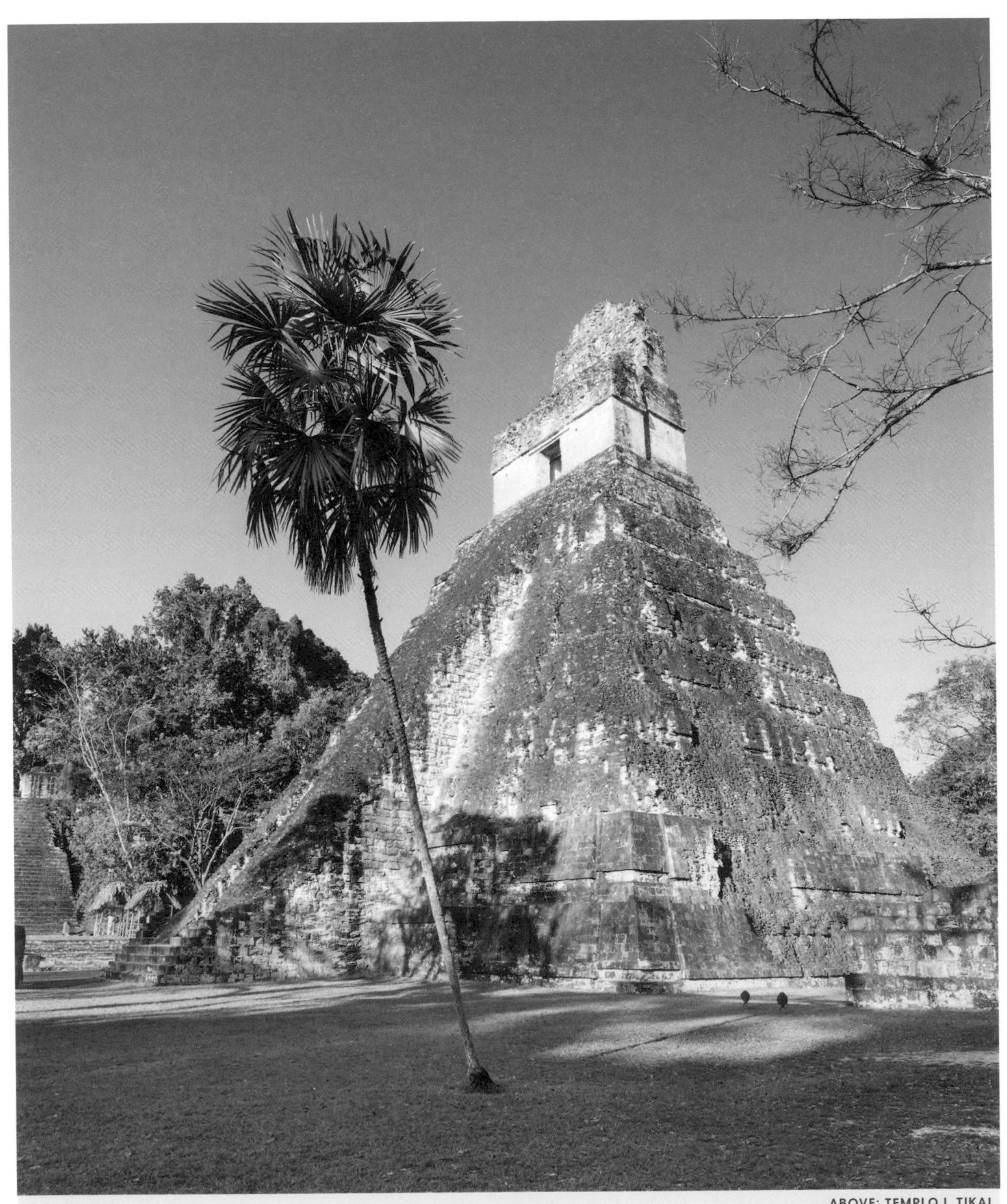

ABOVE: TEMPLO I, TIKAL

8 Flores & Lake Péten Itzá

Flores overlooks Lake Petén Itzá and is the island gateway to fantastic Maya monuments. Alternatively, slow-paced waterfront El Remate on the eastern lakeshore gives access to under-the-radar Maya site Yaxhá.

Regular minibuses run from Flores (via El Remate) to Tikal (1hr 15min).

9 Tikal and Maya temples

Maya powerhouse Tikal's ancient temples, pyramids and stelae are immersed in jungles inhabited by vocal howler monkeys and toucans. Further north, soak up Uaxactún's cosmic energy touring the largest known astronomy complex in the Maya world.

Take a minibus back to Flores and its international airport/buses.

ABOVE: *HUILPI* TUNICS FOR SALE, CENTRAL HIGHLANDS

ABOVE: ANTIGUA GUATEMALA

CARIBBEAN COMBO FROM GUADELOUPE TO ST LUCIA

Guadeloupe – Saint Lucia

No need for a private yacht: sail between the loveliest of the Lesser Antilles by local ferry, combining French-Caribbean flair, fine sands and untamed tropics.

FACT BOX

Carbon (kg per person): 6
Distance (km): 300
Nights: 14-21
Budget: $$
When: Dec-Apr

❶ Guadeloupe

Why stop at one Caribbean isle when you can combine four? Start with Guadeloupe, where you receive a warm 'bienvenue' to the Caribbean at its French-est. Guadeloupe's two main, linked islands – Grande-Terre and Basse-Terre – have different vibes: less developed is Basse-Terre, most of which is forest-cloaked national park dominated by La Soufrière volcano (1467m); this is the place for hikes and bikes. For context on the whole Caribbean, visit the Mémorial ACTe museum in Pointe-à-Pitre, Grande-Terre, exploring the brutal history of slavery in the region.

Infrequent L'Express ferries link Point-à-Pitre and Roseau (around 4 weekly; 2hr 30min).

ABOVE: PURPLE-THROATED CARIB HUMMINGBIRD, GUADELOUPE

❷ Dominica

Boats dock at Dominica's little capital Roseau, where you can wander the Old Market (once used to trade enslaved people) and charming French Quarter. But the 'Nature Isle' is more about exploring the lush interior or heading offshore for aquatic adventures. Journey inland to Morne Trois Pitons National Park to hike to its waterfalls and swim through its gorges; then head out to sea for great diving (try Scotts Head Marine Reserve) and whale spotting year-round in the warm Caribbean.

Board the L'Express ferry to Fort-de-France (schedules vary seasonally, check locally; 2hr).

❸ Martinique

Take another ferry to another dose of France – though for all its euros and Gallic chic, the island

ABOVE: SOUFRIÈRE BAY AND THE PITONS, ST LUCIA

of Martinique retains deep Creole roots. The great beaches here run from white-sand Grand Anse to black-sand Anse Noire. The drinking is even better – this is the Caribbean's rum capital and Martinique's *rhum agricole* has its own *appellation d'origine contrôlé*. Take a distillery tour (Habitation Clément is a good option) and browse the bottles at Fort-de-France's covered market.

Board another L'Express ferry for the trip to Saint Lucia's Castries (check seasonal schedule; 1hr 30min).

4 Saint Lucia

Lush, volcanically rumpled St Lucia has plenty of attractions to charm visitors: high-end retreats for honeymooners along with cheap cabins, buses and adventure tours for backpackers too. All travellers should make their way to the famous Pitons, two iconic, pointy volcanic plugs – hike up or sail below the two peaks and trek the Tet Paul Trail for its Pitons and sea views. If you still have some energy, try mountain-biking through former plantations or learning to make your own chocolate.

George FL Charles Airport, near Castries, connects to other Caribbean islands; international flights leave from Hewanorra Airport, in the south.

ALTERNATIVES

Extension: Marie-Galante, Guadeloupe

Round, flat Marie-Galante island is fringed by powdery sand – ideal for flopping. Allow a couple of days. When you're bored of beaches, tickle your tastebuds at the island's historic rum distilleries (Domaine de Bellevue has a lovely old windmill) and visit the Habitation Murat eco museum.

Ferries run to Grand-Bourg on Marie-Galante from Pointe-à-Pitre (daily; 1hr).

Diversion: Les Saintes, Guadeloupe

This archipelago is a charming side-trip. Only two of the islands are inhabited; main hub is Terre-de-Haut, where classy bistros and jaw-dropping beaches sit below verdant mountains. Hike to Fort Napoléon, swim to the islet off beautiful Baie de Pompierre and hop over to even quieter Terre-de-Bas for hiking and artisan crafts.

Ferries run to Terre-de-Haut from Trois-Rivières (daily) and Pointe-à-Pitre (weekly).

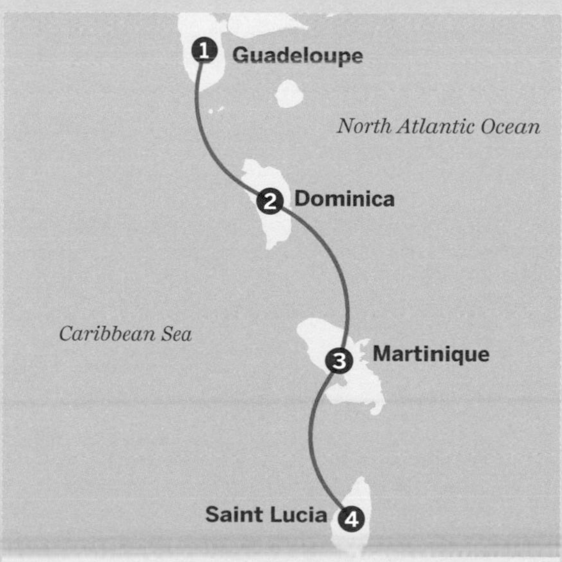

Trip Builder
NORTH ATLANTIC OCEAN
VENEZUELA
GUYANA
11
SURINAME
FRENCH GUIANA
COLOMBIA
ECUADOR
8
BRAZIL
6
3
PERU
BOLIVIA
9
10
PARAGUAY
CHILE
4
1
2
URUGUAY
5
SOUTH PACIFIC OCEAN
ARGENTINA
7
SOUTH ATLANTIC OCEAN

SOUTH AMERICA

With snow-capped Andean peaks and towering sand dunes, jungle-thronged waterways and teeming wetlands, South America offers bountiful natural allure, from stargazing in Chile's otherworldly Atacama Desert to meeting the wildlife and a very large waterfall in Guyana. Then there's following the wine trail in Argentina or Uruguay, exploring uncrowded ancient sites in Peru and hitting the beach on Colombia's Caribbean coast.

CONTENTS

ABOVE: RIO DE JANEIRO'S COPACABANA BEACH, BRAZIL

ROAD-TRIP ARGENTINA'S WONDERFUL WINE COUNTRY

Mendoza – Uco Valley

Between the snow-capped Andes and the Argentine grasslands lies South America's primo wine region. Come for the Malbec, stay for the dazzling food and landscapes.

FACT BOX

Carbon (kg per person): 26
Distance (km): 207
Nights: 7
Budget: $$
When: Mar–Jun, Sep–Nov

❶ Mendoza

The city at the heart of Argentina's most famous wine-growing region is one of the country's largest. It's also noted for its youthful population (there's a major university here), busy pedestrianised streets and elegant Parque General San Martín, a sprawling urban park dotted with lakes and laced with wooded paths. With several good hotels and restaurants, it's also a convenient place to base yourself while exploring the surrounding region.

Drive 12km south to Maipú.

❷ Maipú

Sometimes overlooked by visitors in their hurry to get to the region's most famous wineries, the town of Maipú, just outside Mendoza proper, is a worthwhile stop to soak up some historical context. Learn more about the region's wine-growing past at the venerable Wine and Grape Harvest Museum, housed in an Art Nouveau villa framed by olive groves and vineyards. Then stop off at the Wine Museum in Bodega La Rural, one of the oldest wineries in the area, to taste Rutini wines and view exhibits of antique tools used in the harvesting of grapes and production of wine.

Continue 13km south.

ABOVE: UCO VALLEY WINERY

ABOVE: MENDOZA PROVINCE VINEYARD

❸ Luján de Cuyo

Driving away from Mendoza city along legendary Ruta 40, you'll reach the famed vineyards of Luján de Cuyo. It was in this terroir, in a valley along the banks of the Mendoza River, that Malbec – a grape brought from France – first flourished, slowly developing into a varietal that caught the attention of wine enthusiasts from far and wide. A number of Mendoza's classic wineries are located here, and are open to tour by reservation. Go on a tasting excursion to a handful in the same day, either by car or bicycle: wine routes connect key destinations along the way.

Drive 72km south towards Uco Valley.

❹ Uco Valley

The Uco Valley is considered Argentina's most exciting up-and-coming wine region. The high-altitude landscape edges up against the Andes so you'll enjoy sweeping views of the towering mountain range from tasting rooms and vineyards alike. It's a wonderful place for a down-to-earth outdoor lunch at a rustic winery like Bodega La Azul, or for a visit to Siete Fuegos, Francis Mallmann's restaurant at the upscale Vines Resort & Spa.

Head back the 110km to Mendoza for onward connections.

ALTERNATIVES

Day trip: Potrerillos Dam

With an extra day in the Mendoza area, road-trip to Potrerillos Dam in the foothills of the Andes. Fed by the whitewater current of the Mendoza River and surrounded by mountains, the bright blue reservoir is a gorgeous venue for a picnic – bring a bottle from one of the local wineries. Nearby you'll also find the pre-Hispanic settlement of Uspallata. Look out for condors overhead.

Potrerillos is 45km west of downtown Mendoza.

Day trip: Horseback riding

Many outdoor outfitters and travel agencies offer horseback excursions around Mendoza. Some cross the Mendoza River, others take you along the vineyards. But the experience of galloping in the Andean foothills, where wild horses run, is unforgettable, especially if the trip stops at a traditional *estancia* (ranch).

Most outings start and end in the city and include transportation.

DISCOVER PABLO NERUDA'S CHILE

Santiago – Isla Negra

Be inspired on this ode-yssey in the footsteps of Chile's best-known poet, Pablo Neruda, connecting hilly capital Santiago, a storied port and the crashing Pacific Ocean.

FACT BOX

Carbon (kg per person): 9
Distance (km): 297
Nights: 4-6
Budget: $
When: Dec-Apr

ABOVE: MEMORIAL AT NERUDA'S HOME, ISLA NEGRA

❶ Santiago

Chile is known as the *pais de poetas* (country of poets) in no small part thanks to Pablo Neruda, one of the nation's two Nobel Prize-winning poets. Neruda's verses champion the country's mountains, wine and, most vividly, its coastline. But begin your circuit of Chile inland, in the capital Santiago. Head into the city's magnificent hills to the summit of Cerro San Cristóbal, topped by an immense statue of the Virgin Mary. Its outlook encompasses Andean peaks and the metropolis below and is enough to move you to poetry – as it did when Neruda lived here. His Bellavista-neighbourhood house, La Chascona, is open to the public, displaying objects showing the fabulous quirks of Neruda's personality.

Take a bus to Valparaíso (hourly; 1hr 45min).

❷ Valparaíso

'I love Valparaíso,' wrote Neruda. 'Queen of all the world's coasts, true headquarters of waves and ships'. Plummeting down hillsides so sheer its streets often become steps, or are connected only by *ascensores* (lifts), this jumbled, street-art-bedaubed port is huge fun to explore. The house Neruda built here, La Sebastiana, crowned with a ship's funnel, juts out of upper Valparaíso like a vessel above the ocean. The building commands sublime city views and its

ABOVE: VALPARAÍSO

interior decoration shows the debonair side of a poet achieving international fame, replete with eccentric memorabilia from foreign travels.

Take another bus to Algarrobo (hourly; 1hr).

3 Algarrobo

There is no Neruda connection in Algarrobo, one of Chile's principal seaside resorts, but Neruda loved the sea and this is a great place to enjoy it. There are glorious stretches of Pacific-washed sand, watersports and, just north in San Alfonso del Mar, one of the world's largest swimming pools.

Board a bus to Isla Negra (hourly; 10min).

4 Isla Negra

At peaceful Isla Negra, Neruda could live on the very edge of the ocean that inspired more of his words than almost anything else. His residence here is full of ship's figureheads, shells and sea charts, and inspired much of his best work. 'Look for me here,' the poet wrote 'between the stone and the ocean, in the light storming in the foam.' Poetic tour done, wander Isla Negra's tempestuous shoreline or visit one of the tempting seafood restaurants.

Take a bus back to Santiago for onward connections (hourly; 1hr 45min).

ALTERNATIVES

Extension: Viña del Mar

Valparaíso's coastal neighbour is about as different from Neruda's port city as can be. Along with wide boulevards and a sand-rimmed bay, Viña del Mar offers absorbing museums, acclaimed festivals and beautiful parks like Parque Quinta Vergara to back up a vibrant beach scene and nightlife. Don't miss Castillo Wulff, an early 20th-century mock castle-gallery leaning over the sea.

Valparaíso Metro connects with Viña Del Mar (4 hourly; 10min)

Diversion: Casablanca

This town is the hub of Valle de Casablanca wine region. Despite being closer to the equator than almost all other renowned wine-producing areas, cool Pacific fogs make vineyards not only possible here, but even excel at producing Chardonnay, Sauvignon Blanc and Pinot Noir. Sample vintages at Casas del Bosque. Neruda loved Chilean wine: see whether you agree.

The town is between Isla Negra and Santiago, connected to both by bus (hourly; 1hr).

AWAY FROM THE CROWDS IN PERU

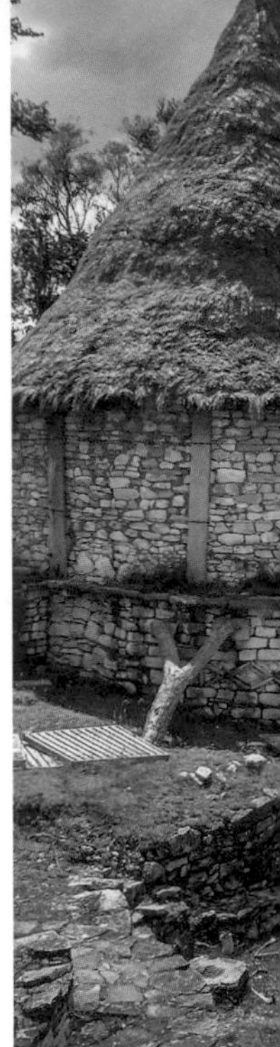

Cajamarca – Tarapoto

Take a spectacular descent from the Andean mountains to the Amazon jungle in northern Peru, stopping off at indigenous sites and pre-Columbian ruins en route.

FACT BOX

Carbon (kg per person): 21
Distance (km): 704
Nights: 7
Budget: $
When: Jun-Aug

❶ Cajamarca

Cajamarca was founded in the 1320s and was the site of a great Inca-Spanish confrontation in 1532 that would direct the future path of South America. You can feel the history in its only remaining pre-conquest building, El Cuarto del Rescate, and sample its beauty in a medley of impressive churches, including the squat but ornate cathedral and the sprawling Belén religious complex.

Take the bus to Celendín (4 daily; 2hr 30min).

❷ Celendín

A sleepy small town known for its straw hats, Celendín is little more than an overnight stop on the scarily spectacular route between Cajamarca and Leimebamba, a journey through plummeting canyons, ghostly cloud forests and patchwork-green hillsides on a narrow, vertiginous road.

Bus company Virgin del Carmen has buses to Leimebamba (2 daily; 7hr).

❸ Leimebamba

Essential for a basic understanding of the Chachapoya, the people who inhabited this region in pre-colonial times, tiny Leimebamba is dominated by an excellent museum built to house 200 15th-century mummies unearthed at the nearby Laguna de los Cóndores in 1996.

Take the bus following the Utcubamba River Valley to Chachapoyas (5 daily; 2hr).

❹ Chachapoyas

Anchored by a huge main square, this well-equipped whitewashed town is fast becoming the cultural and adventure nexus of northern Peru. Several pre-Inca ruins and burial sites are reachable on day trips, including Kuélap and less developed Yalape. Equally

ABOVE: CHACHAPOYAS

ABOVE: KUÉLAP

epic are the nearby waterfalls, ideal for the adrenalin sport of canyoning; a local company can organise guided excursions.

Bus company Turismo Selva will get you to Moyobamba (5 daily; 6hr).

5 Moyobamba

Little-visited Moyobamba is worth an overnight for its flora and hot springs. Tourist authorities, working together with local communities, have developed a number of ecological activities in the area including hiking, river-boat trips and appreciation of the town's emblematic orchids.

Take a Turismo Selva bus to Tarapoto (4 daily; 2hr).

6 Tarapoto

On the edge of the Amazon, this city of buzzing motorcycles, honking traffic and sticky humid air can feel hot and frenetic. More reason to decamp to its more pleasant surroundings, a tropical melange of waterfalls, lakes and artisan villages; local travel companies can organise day trips. Walkable from the city centre is Alto Shilcayo, a small slice of protected jungle populated by monkeys, colourful birds and five rarely visited waterfalls.

Tarapoto has an airport with daily flights to Lima.

ALTERNATIVES

Extension: Kuélap

No one comes to these parts without visiting Kuélap, former ceremonial centre of the mysterious Chachapoya people. Coined the 'Machu Picchu of the north' and recently made more accessible via a cable car, the hilltop ruins with their 20m-high walls sit 3000m above sea-level on a flat-topped misty mountain.

Take a bus from Chachapoyas to the village of Nuevo Tingo (1hr), where the cable car begins its ascent.

Diversion: Gocta

Ride a crowded minibus for 45 minutes out of Chachapoyas before getting off at an inconspicuous curve in the road called Cocahuayaco. Thus begins your day-hike to jungly Gocta, one of the world's tallest waterfalls – barely mapped until the early 2000s and still bizarrely under-visited. The circular hike to upper and lower falls is 15km.

You'll need to hail a local motorbike taxi to escort you to/from the start/finish points.

ALONG COLOMBIA'S CARIBBEAN COAST

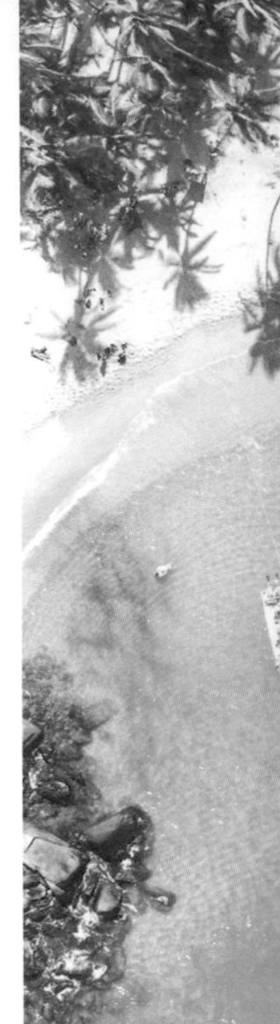

Cartagena – Guajira

Enjoy a Colombian Caribbean bus trip, from a Unesco-listed city to a stark desert inhabited by the indigenous Wayuu people via a protected national park.

FACT BOX

Carbon (kg per person): 63
Distance (km): 574
Nights: 6
Budget: $$
When: Nov-Apr

❶ Cartagena

One of the historical heavyweights of Latin America, Cartagena is a city rich in colonial architecture, including the continent's largest Spanish-built fort and an 11km ring of defensive walls. Immerse yourself in the Old Town's splendour for a couple of days, enjoying hushed courtyards, elegant balconies and atmospheric hotels.

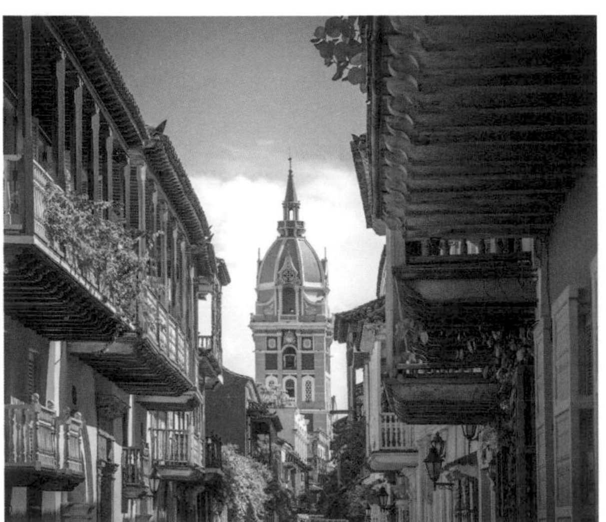

ABOVE: CARTAGENA OLD TOWN

Book a minibus to take you along the coast to Santa Marta (4hr): some will pick up and drop off from your accommodation.

❷ Santa Marta

Though not as handsome as its coastal cousin Cartagena, Santa Marta can lay claim to being the more senior city – indeed it is the oldest European-founded settlement in South America. It's best known for a boisterous street vibe and thoroughly reasonable hotels that act as good bases for the surrounding rural sights, including Minca village and Ciudad Perdida.

Regular buses link Santa Marta with the entrance of Tayrona National Park (every 30min; 1hr).

❸ Tayrona National Park

Pocket-sized but ecologically priceless, Tayrona is where the lush rainforest comes down to kiss the cerulean waters of the Caribbean. A hot but shady hike from near the El Zaino entrance gate leads to several walk-in-only beaches headlined by Cabo San Juan del Guía, where sunbathing crowds can't spoil the joy of the paradisical setting.

From the park's El Zaino entrance hail a small

ABOVE: TAYRONA NATIONAL PARK

bus to Palomino, where you can change for a bus to Riohacha (every 30min; 1hr 30min).

❹ Riohacha

By the time you reach Riohacha, the palms have given way to cacti and the jungle has dissolved into semi-desert. This rough and ready seaside town sits at the entrance to the Guajira Peninsula, for which it acts as an unfussy base camp. There's a long stroll-worthy pier, some good cheap eats and a couple of deluxe hostels.

The only practical way into the Guajira Peninsula is by jeep. Book day- or overnight trips with one of the many travel agencies in Riohacha.

❺ Guajira

This arid peninsula flush up against the Venezuelan border has long been the home to the unconquerable Wayuu people, whose culture still flourishes. Hire a jeep and driver to tackle the rough off-piste roads that lead to the top-of-the-continent paradise of Punta Gallinas, where desert dunes collide with crashing waves.

Jeeps will drop you back in Riohacha, from where buses head back to Cartagena (hourly; 7hr).

ALTERNATIVES

Extension: Ciudad Perdida

Colombia's 'lost city' lay covered and forgotten until its rediscovery in the 1970s. The remote walk-in-only site is located deep in the Sierra Nevada and was built between the 11th and 14th centuries by the indigenous Tayrona people. The moderately difficult guided hike takes between four and six days including time spent at the ruins. Both ruins and landscapes are magnificent.

Numerous travel agencies in Santa Marta can organise the trip.

Diversion: Palomino

A laidback beach village favoured by the affluent offspring of 1980s backpackers, Palomino is low-rise and agreeably sleepy. Accommodation oscillates between rustic beach shacks and hippy-chic eco-resorts. Mountains, long-time home to the Tayrona people, glimmer inland while the beach is a mix of fishers and hammock-hoarding travellers. River tubing is a fun local activity.

Buses between Santa Marta and Riohacha stop at Palomino.

WINELANDS TO WETLANDS IN URUGUAY

Montevideo – Laguna de Rocha

From exotic Montevideo, road-trip to little-known vineyards, eat at traditional *asado* barbecues, and discover spectacular Atlantic beach resorts and lagoon wetlands.

FACT BOX

Carbon (kg per person): 60
Distance (km): 484
Nights: 5
Budget: $$
When: Oct-Jun

❶ Montevideo

Uruguay's capital is a pleasant surprise – small, safe and walkable (though the 24km scenic Rambla along the River Plate is best enjoyed by bike). Begin at Plaza Independencia, whose towering 1920's Palacio Salvo resembles both the Chrysler Building and Gaudi's Palau Güell. The iconic Old Town is an intriguing maze of Art Deco and colonial mansions, *bodegas* and tango bars, with the venerable Mercado del Puerto the place for juicy grilled steaks. For a perfect slice of local life, plunge into the teeming Tristan Narvaja Sunday flea market.

Rent a car and drive 50km north to Canelones.

❷ Canelones

The vineyards of the pampas and fertile hills surrounding Canelones are the historic centre of Uruguay's wine industry. Nearly all of the 80 wineries here offer not just tastings but gourmet restaurants and vineyard picnics. The innovative Bodega Bouza has a tasting room filled with vintage cars; Establecimiento Juanico experiments with orange wine aged in ceramic amphorae; and Vinos de Lucca is a rock and roll garage cantina. Overnight options run from B&B cottages to luxury hotel spas.

Drive 74km southeast to Atlántida.

❸ Atlántida

At the charmingly old-fashioned seaside resort of Atlántida, the muddy-brown River Plate meets the

ABOVE: PLAZA INDEPENDENCIA, MONTEVIDEO

ABOVE: PUNTA DEL ESTE BEACH

deep blue waters of the Atlantic: the perfect spot for a beachside barbecue of freshly caught prawns. Just outside town you can get a feel for the Italian immigration that marks Uruguay's national identity with a visit to either Bodega Pablo Fallabrino winery or nearby Bodega Bracco Bosca. Fallabrino uses Italian grapes like Arneis in homage to his heritage; Bracco Bosca serves prosciutto, olives and salami with delicious Tannat and Moscatel vintages.
Continue 160km east, following the coast to Laguna de Rocha.

4 East to Laguna de Rocha

Driving along the breathtaking Atlantic Coast, stop at glamorous Punta del Este, where South America meets Miami. Adjoining is La Barra, famed for surfing, gourmet restaurants and nightclubs, followed by the latest hotspot, José Ignacio, a fishing village turned luxury bolthole. Bodega Oceánica José Ignacio produces wine and olive oil and exhibits giant avant-garde sculptures, while the scenic vine-covered hills surrounding Bodega Garzón's state-of-the-art winery are known as Uruguay's Tuscany. Base yourself in funky Garzón village to explore Laguna de Rocha's Unesco Biosphere wetlands on foot or by kayak.
Drive west 200km back to Montevideo.

ALTERNATIVES

Extension: Colonia del Sacramento

Uruguay's earliest settlement is a romantic enclave of tropical beaches and colonial monuments, with Buenos Aires an hour's ferry ride across the Rio de la Plata – a sunset river cruise is unforgettable. Gaucho horsemen traverse the pampas grasslands around the town, also home to welcoming Colonia vineyards like Alcamén de la Capilla, tended over five generations by the same Italian family.
Colonia is 180km west of Montevideo.

Diversion: Fray Bentos

An exciting bus ride through endless cattle country brings you to the mythical town of Fray Bentos, a global byword for corned beef and meat pies. Food to feed the world – and the armies of both World Wars – was produced here in an immense 1863 industrial factory, abandoned but preserved by Unesco as a World Heritage site.
A direct bus runs from Montevideo's Tres Cruces terminal to Fray Bentos (3 daily; 4hr 20min).

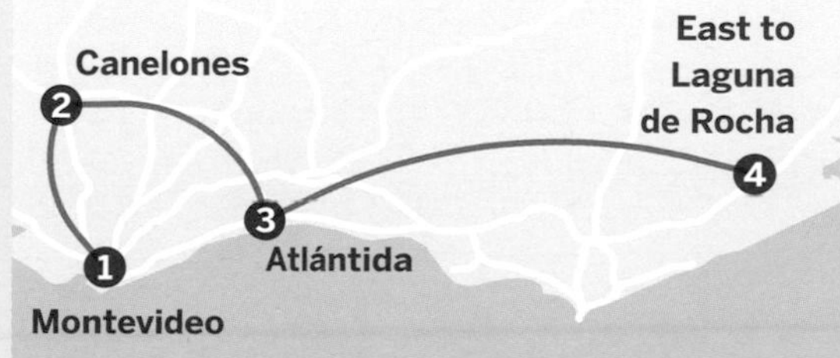

FIND DUNES AND FINE FOOD IN BRAZIL'S NORTHEAST

Natal – Jericoacoara

Hop on beach buggies and 4WDs for an epic coastal adventure amid the protected dunes, powdery sands and sapphire seas of Brazil's sunbaked northeast.

FACT BOX

Carbon (kg per person): 29
Distance (km): 235
Nights: 4
Budget: $$
When: year-round

❶ Natal

Rio Grande do Norte's capital is a welcoming seaside city that acts mostly as a regional travel hub. Spend a fun night in Ponta Negra, mingling with fellow travellers and ogling Natal's signature sand dunes – a taster of what's to come.

Several private operators such as Natal Vans offer day trips to Genipabu, 10km north of Natal.

❷ Genipabu

This outstanding dunescape features mountains of sand with harrowing names like Wall of Death and Vertical Descent. Your day-trip dune adventure here comes *com emoção* or *sem emoção* (with or without excitement) – choose wisely.

Back in Natal, take an Expresso Cabral bus to São Miguel de Gostoso from Rodoviária Nova bus station (3-6 daily; 2-3hr).

❸ São Miguel de Gostoso

Gostoso is a beach-peppered paradise that's also a haven for wind sports. Spend a few days relaxing in this far-flung coastal village, undiscovered thus far by package tourism.

Arrange a beach buggy or 4WD to make the 2hr 15min trip along the coast. to Galos/Galinhos.

❹ Galos/Galinhos

The isolated twin fishing villages of Galos and Galinhos are set between a brilliant blue inlet and the ocean. Time slows down here among colourful beach shacks, astounding sunsets and breeze-swept beaches. They're difficult to reach – which is part of the charm.

Westward public transport is nearly non-existent, so arrange a buggy or 4WD for the 4-5hr ride along the coast to Canoa Quebrada (often split into two days).

ABOVE: CANOA QUEBRADA

ABOVE: GENIPABU

5 Canoa Quebrada

Fiery sunsets bounce off craggy, pink-hued sandstone cliffs in Canoa Quebrada, one of Ceará state's coastal beauties. The landscape here – beautiful beaches and towering dunes complemented by Carnauba palms and coconut trees – is mesmerizing.

Take a São Benedito bus to Fortaleza (4-6 daily; 3hr 15min).

6 Fortaleza

Excellent dining, big-city infrastructure and several wonderful urban breaches (most notably Praia do Futuro) greet travellers in Ceará's bustling capital. If you need a mid-trip dose of metropolis, this is the place to get it.

To get to Jericoacoara take a shared van or a bus/open-sided 4WD combos operated by Fretcar (5 daily; 5hr).

7 Jericoacoara

Streets forged of sand characterise remote 'Jeri,' where otherworldly dunes, blue lagoons and scenic beaches hide a relaxed, once quiet fishing village turned dining, windsport and nightlife heaven. Stay a day. Or a week.

Head back to Fortaleza and its international airport for onward connections.

ALTERNATIVES

Extension: Fernando de Noronha

No other destination in Brazil (South America, really) evokes visions of bliss more than Fernando de Noronha island, 325km out to sea from Natal as the crow flies. It's a fiercely protected national marine park, spinner dolphin/sea turtle sanctuary and world-class diving destination. It's also home to Brazil's top three beaches, all wonderfully uncrowded thanks to controlled entry.

Flights run from Natal to the island (daily; 1hr 15min).

Diversion: Parque Nacional dos Lençóis Maranhenses

You may be all duned-out, but Maranhão state's Parque Nacional dos Lençóis Maranhenses, nearly 400km west of Jeri, is next-level wow. An amazing 1550 sq km of protected expanses of white dunes (*lençóis* means 'sheets' in Portuguese) broken up by crystal-blue lagoons, it's simply spectacular.

To get to the park from Jeri, arrange private transport to Barreirinhas via Camocim.

EXPLORE ARGENTINIAN AND CHILEAN PATAGONIA

Puerto Madryn – Puerto Natales

Vast, windswept Patagonia is a very special place. See whales and wide-open spaces, icy glaciers and cloud-shrouded mountains on this unforgettable two-country road trip.

FACT BOX

Carbon (kg per person): 226
Distance (km): 1820
Nights: 14
Budget: $$
When: Oct-Mar

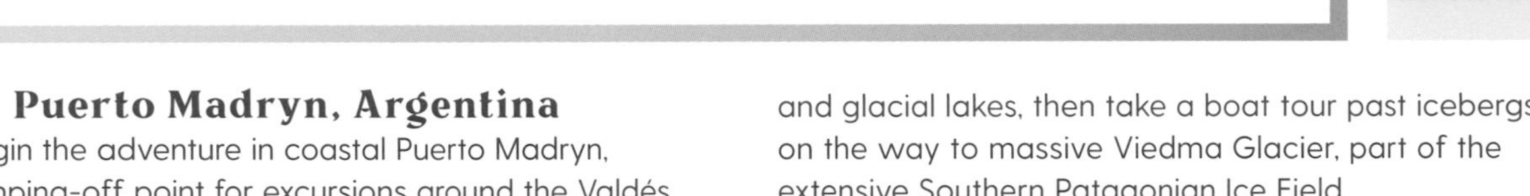

❶ Puerto Madryn, Argentina

Begin the adventure in coastal Puerto Madryn, jumping-off point for excursions around the Valdés Peninsula, where southern right whales swim offshore from June through December – put on a life jacket and board a boat for an up-close view of them breaching and playing in the chilly water. Outside whale season, tour the Magellan penguin colony in Punta Tombo and wander the region's wonderful beaches.

Drive 870km south along Ruta Nacional 3.

❷ Puerto San Julián, Argentina

Break your journey in the sleepy port town of Puerto San Julián, considered the cradle of Patagonian history. The Portuguese explorer Ferdinand Magellan arrived in its natural harbour in 1520, spending the winter here before going on to sail the strait that now bears his name. With its dramatic cliffs and colourful cottages, the town is a picturesque stop on the trip.

Drive 465km west.

❸ El Chaltén, Argentina

A magnet for outdoor adventurers, the village of El Chaltén offers easy access to the natural wonderland of Parque Nacional Los Glaciares. Hike to the Condor Lookout for breathtaking views over snowy mountains and glacial lakes, then take a boat tour past icebergs on the way to massive Viedma Glacier, part of the extensive Southern Patagonian Ice Field.

Back in the car, drive an incredibly scenic 210km south.

ABOVE: PUNTA TOMBO PENGUINS, ARGENTINA

ABOVE: PARQUE NACIONAL LOS GLACIARES, ARGENTINA

❹ El Calafate, Argentina

El Calafate is named after a native berry – the story goes that once you eat it, you're guaranteed to come back to Patagonia, and a very pleasant way to get that guarantee is with a berry-infused pisco sour. Afterwards, plan your adventure to nearby Perito Moreno glacier. The gargantuan natural attraction is a wonder to behold from the viewing platforms, especially when big chunks of ice break off and splash into the water below.

Drive 275km south, crossing the border into Chile.

❺ Puerto Natales, Chile

The last stop on the itinerary is arguably the most special, location-wise. Puerto Natales is close to the world-class Parque Nacional Torres del Paine and the town is a great place in which to base yourself before and after hiking the famous W Circuit, a trail that passes through pristine forests and glacial valleys. Make sure you're properly outfitted before heading into the park – as the local saying goes, you'll experience all four seasons in a single day in Patagonia.

Catch a flight from Puerto Natales to Chile's capital Santiago for international connections.

ALTERNATIVES

Extension: Gaiman and Trelew

In 1995, Diana, Princess of Wales came to Argentinian Patagonia to visit communities settled by the Welsh in the late 19th century. She took tea in one of Gaiman's traditional teahouses, still wonderfully quaint places to visit today. In nearby Trelew (where all the street signs are in Welsh), there's a museum devoted to the legacy of Welsh immigrants to the region.

Gaiman is 80km southwest of Puerto Madryn.

Diversion: Comodoro Rivadavia

You'll have considerable distances to cover in Patagonia, meaning breaking up the journey is advisable. One option on the first leg of the trip is Comodoro Rivadavia: there's not much charm to the industrial city, but it's a practical spot for an overnight stay on the long trip between Puerto Madryn and Puerto San Julián.

Comodoro Rivadavia is 440km south of Puerto Madryn.

AVENUE OF THE VOLCANOES, ECUADOR

Quito – Cuenca

Follow Ecuador's Andean spine while taking in snow-capped volcanoes, high-altitude national parks and vibrant colonial centres on an action-filled adventure.

FACT BOX

Carbon (kg per person): 19
Distance (km): 650
Nights: 10
Budget: $$
When: Jun-Sep

❶ Quito

History lurks around every corner in Quito's stunning old town, a Unesco World Heritage site full of gilded 16th-century sanctuaries, atmospheric monasteries and palm-fringed plazas that draw a mesmerising cross-section of Ecuadorean society. After taking in the Spanish-Colonial architecture, fast-forward a few centuries to the evocative works created by the artist Guayasamín in his haunting Capilla del Hombre (Chapel of Man). Evening brings verve to the capital, with live music in La Ronda, plus multi-ethnic dining and late-night revelry in the buzzing Mariscal district.

To reach the Parque Nacional Cotopaxi it's best to book a private transfer (2hr 30min) with one of the haciendas there.

❷ Parque Nacional Cotopaxi

The perfectly conical peak of Volcán Cotopaxi (5897m) rises above the surrounding *paramo* (Andean grasslands), where herds of llama scamper beneath soaring Andean condors. Experienced climbers can attempt the two-day climb to the volcano's summit; and everyone can enjoy hikes, visit alpine lakes, go horse riding and spot myriad Andean birds. Afterwards, you can catch dramatic sunsets and warm up by the fire at one of the splendid haciendas in the area. El Porvenir is a great base for activities, while 400-year-old La Cienega is pure class.

Get a lift back to the national park entrance on the Pan-American Hwy; from there catch a south-bound bus to Ambato (frequent; 3hr), and change onto a bus to Baños (hourly; 1hr).

ABOVE: SWING AT CASA DE ARBOL VIEWPOINT, BAÑOS

ABOVE: *VACQUERO* CATTLE-HERDERS, PARQUE NACIONAL COTOPAXI

3 Baños

Tucked between the Andes and the Amazon, the highland town of Baños is the gateway to several active adventures. You can go rafting, hike a waterfall circuit, arrange jungle excursions or undertake the white-knuckle mountain-bike descent to the town of Puyo on the edge of the Amazon. Afterwards, soak away the day's aches and pains in hot springs just outside of town.

Take a bus heading to Ambato (hourly; 1hr), from where you can transfer to an onward bus bound for Cuenca (hourly; 7hr).

4 Cuenca

Dating from the 16th century, Cuenca is an historic jewel of a city with lofty steeples, geranium-filled balconies and buzzing markets where age-old craft traditions still flourish. If you've overdosed on colonial churches, instead spend the day learning about Ecuador's diverse indigenous cultures at the absorbing Museo Pumapungo; there's also browsing for the world's finest *toquilla* straw headwear (aka Panama hats) or strolling the banks of the Río Tomebamba.

Daily flights connect Cuenca with Quito (1hr), for onward connections.

ALTERNATIVES

Extension: Otavalo

On Saturdays, the indigenous town of Otavalo transforms into one vast market. You'll find traditional crafts, clothing, rugs, folk art, straw hats and jewellery, along with food stalls selling belly-filling classics like roast suckling pig served with *mote* (hominy) and *chicha* (a corn or yucca drink). A smaller version of the market runs daily.

Frequent buses make the trip from Quito (3hr).

Diversion: Parque Nacional Cajas

This park, just 30km west of Cuenca, is a spectacular expanse of *páramo*, mist-covered lakes and small forests of high-altitude polylepis trees. On day-hikes, take in the otherworldly scenery while keeping eyes peeled for llamas, alpacas, deer, foxes and over 150 species of bird.

From Cuenca, Guyquail-bound buses pass the Laguna Llaviucu entrance. Outfitters in Cuenca offer guided tours.

A SUNKISSED ODYSSEY THROUGH SOUTH-EAST BRAZIL

Rio de Janeiro – São Paulo

Brazil's southeast packs in tropical islands, endless beaches, colonial treasures and two of South America's most dynamic cities.

FACT BOX

Carbon (kg per person): 19
Distance (km): 690
Nights: 7-10
Budget: $$
When: Oct-Nov, Mar-Apr

❶ Rio de Janeiro

Planted between golden beaches and forest-covered mountains, the *cidade maravilhosa* (marvellous city) is one of the world's most captivating destinations. Days are spent basking on the sands in Ipanema, admiring the views from clifftop perches like Pão de Açúcar (Sugarloaf Mountain) and exploring the colonial lanes of the city centre. The nights are no less alluring – you can join revellers at music-filled samba clubs in Lapa, raise an ice-cold *chope* (draft beer) in Leblon or linger over creative cocktails at a hilltop bar in Santa Teresa.

ABOVE: PARATY

Take one of the frequent Costa Verde buses to Conceição de Jacareí (2hr 30min), from where island-bound ferries depart regularly to Vila do Abrãao (30min), Ilha Grande's main town.

❷ Ilha Grande

The vehicle-free island of Ilha Grande boasts tropical rainforest and pristine beaches lapped by aquamarine seas. Signposted trails lead through the lush jungle to photogenic shorelines like Praia Lopes Mendes, rated one of Brazil's prettiest beaches. You can also take a guided 6km hike up to the summit of Pico do Papagaio (982m), go diving or take a boat trip to more difficult to reach parts of the island.

Take a boat to Angra dos Reis (every 2hr; 1hr), and then a bus to Paraty (hourly; 2hr).

❸ Paraty

Backed by verdant peaks, the picturesque town

ABOVE: RIO DE JANEIRO

of Paraty has undeniable appeal. The cobblestone streets are lined with handsomely painted colonial buildings, including a handful of 18th-century churches, and the nearby beaches and islands are nothing short of spectacular. There's much to do in the area too, including kayaking, horse riding, mountain biking and walks in Atlantic rainforest. It's also a fine place to stay put for a few days – Paraty has atmospheric guesthouses, excellent restaurants and plenty of music-filled bars.

Take a Reunidas Paulista bus to São Paulo (6 daily; 7hr 30min).

4 São Paulo

Brazil's biggest, most cosmopolitan city is a skyscraper-studded behemoth packed with intrigue. Get the lay of the land from the 26th floor of the Farol Santander building, stop for lunch at the Belle Époque Mercado Municipal, then take in the nation's best collection of Brazilian art at the neoclassical Pinacoteca do Estado. For urban exploring, take a stroll through the Japanese enclave of Liberdade, the chic Jardins district or the bar-lined Vila Madalena neighbourhood.

São Paulo's GRU Airport has numerous international flights.

ALTERNATIVES

Extension: Ouro Preto

The nexus of Brazil's gold-mining boom in the early 1700s, Ouro Preto has a storybook colonial centre, with a tangle of narrow, hilly streets dotted with architectural treasures. You can admire the sculpted masterpieces of Aleijadinho, discover 18th-century African tribal king turned folk hero Chico-Rei and gaze upon opulent gilded churches.

Buses run to Ouro Preto from São Paulo (11hr) and Rio (8hr).

Diversion: Petrópolis

A mountain retreat with a decidedly European flavour, Petrópolis is where Dom Pedro II's imperial court spent the summer when Rio's temperatures soared. Its centre is awash with manicured parks, canals and old-fashioned street lamps. Don't miss the Museu Imperial, which served as the royal palace in the 19th century; and grand mansions like the Casa da Ipiranga.

Única Fácil buses run from Rio to Petrópolis (every 30min; 1hr 30min).

FROM DESERT TO BEACH IN NORTHERN CHILE

San Pedro de Atacama – Arica

Extraterrestrial-like valleys, ghost towns and breathtaking coastlines set the stage for big adventures on this bus trip around northern Chile.

FACT BOX

Carbon (kg per person): 25
Distance (km): 850
Nights: 7-10
Budget: $$
When: Mar-May, Sep-Nov

❶ San Pedro de Atacama

Perched at dizzying 2438m, this high desert village lies near an astonishing range of natural wonders. You can spend the day sandboarding down towering dunes at the Valle de la Muerte (Death Valley, known as Mars Valley due to its otherworldly appearance), soak in salt lakes while flamingoes strut in the distance and learn about indigenous cosmology on an evening stargazing excursion. Also available are sunrise trips to the world's highest geyser field, pre-Colombian ruins hidden in the desert and fabulous sunsets from the aptly named Valle de la Luna (Valley of the Moon).

Take the overnight bus from San Pedro de Atacama to Iquique (daily; 9hr 30min).

❷ Iquique

Set in a golden crescent of coastline, Iquique is one of Chile's premier beach destinations, boasting excellent surfing as well as other adventure sports (paragliding, sandboarding and climbing). Georgian-style architecture and old-fashioned wooden sidewalks hark back to the mining boom days of the 19th century.

Frequent buses as well as *colectivos* (shared taxis) run inland from Iquique to Humberstone (1hr).

❸ Humberstone

Ghost towns litter the desert in this corner of Chile, though none is more impressive than Humberstone. Once home to over 3000 workers and their families,

ABOVE: IQUIQUE

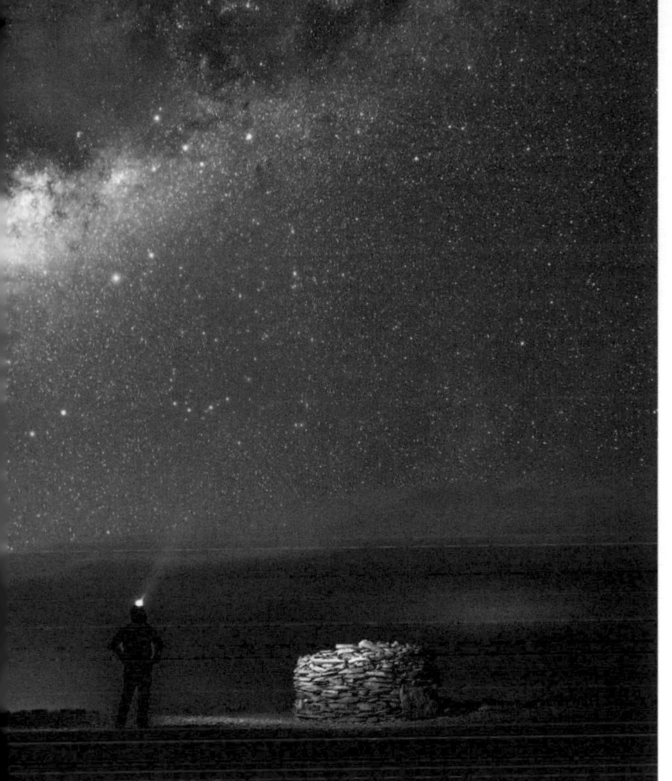
ABOVE: STARGAZING AT SAN PEDRO DE ATACAMA

the former nitrate mining town has dozens of buildings in various states of decay, some restored and safe to enter. You can roam the abandoned lanes, cross the weed-sprouting central plaza and peak into eerily deserted school rooms, markets, a hotel and a grand theatre that's rumoured to be haunted.

Head back to Iquique where several bus companies operate services north to Arica (daily; 4hr 30min).

❹ Arica

The delightful seaside town of Arica makes a great final leg on your north Chilean circuit. Take time to explore the charming pedestrian-friendly centre, loll on the beach and hike some memorable walks along the rocky shoreline spying sea lions, *chungungo* (marine otter) and *piqueros* (boobies). Looming above town, the Morro de Arica involves a short but steep climb with fine views over the city as the reward. After snapping some pics, stroll back to the waterfront for dinner in one of Arica's many outstanding seafood restaurants. With time, head 12km from town to the Azapa Valley, home to the world's oldest mummies.

From Arica, there are several flights daily to Santiago for onward international connections.

ALTERNATIVES

Extension: Salar de Uyuni, Bolivia

From San Pedro de Atacama, join a 4WD tour to the surreal landscapes across the border in Bolivia. You'll see colourful altiplano lakes, Dali-esque rock formations, volcanoes, flamingos and, most famously of all, the blindingly white Uyuni salt flat. Standard trips take three to four days and include food and accommodation; bring drinks, snacks, warm clothes and a sleeping bag.

Cordillera Traveller and other San Pedro companies offer trips.

Diversion: Parque Nacional Lauca

For a taste of the Andes, head 160km northeast of Arica to the stunning Parque Nacional Lauca. Spread across 1380 sq km of altiplano, this Unesco Biosphere Reserve is home to snow-capped volcanoes, sparkling lakes and remote hot springs. Get acclimatised and arrange guides in the lofty Aymara village of Putre (elevation 3530m).

One daily bus goes to Putre from Arica (2hr 30min).

GUYANA ADVENTURE, FROM WILDLIFE TO WATERFALL

Georgetown – Waikin Ranch

Travel into Guyana's jungle heart to discover why this is South America's most thrilling new adventure destination – especially if you opt to explore it by bus and boat.

FACT BOX

Carbon (kg per person): 18
Distance (km): 650
Nights: 8
Budget: $$
When: Sep-Apr

ABOVE: GUIANAN SQUIRREL MONKEY

❶ Georgetown

You can be forgiven for thinking you're in the West Indies when you arrive in Guyana's English-speaking, Caribbean-flavoured capital. Allow at least a day to take in Georgetown's vibrant markets and dilapidated colonial architecture, and enjoy a Creole feast at Backyard Cafe, a local culinary institution, before heading into Guyana's jungle heart. Most travellers opt to fly into the nation's interior on a tour, but if you time your visit right (outside the May-August rainy season), consider saving some money (and carbon) by travelling overland by bus, car and boat.

Minibuses leave Georgetown at 6.30pm bound for the border town of Lethem in the south. It takes 12hrs (if you're lucky) to cover the 350km to ATTA Rainforest Lodge.

❷ ATTA Rainforest Lodge

The wildlife-rich jungles and savannahs of the Rupununi region in Guyana's southwest are the star attractions for most travellers. Just off the dirt road passing for the country's main north-south highway, ATTA is a sensible first stop. Spend a couple of nights here enjoying the jungle setting, the great food and wildlife-spotting from the canopy walkway. Some travellers opt to tack on a night at nearby Surama

ABOVE: KAIETEUR FALLS

© Jonathan Gregson | Lonely Planet; Tim Snell | 500px

Ecolodge, one of several Amerindian community-run lodges in the region, before continuing the trip south.
Arrange a vehicle (and/or boat) transfer to Caiman House, which takes about half a day.

❸ Caiman House

Rather than jumping back on the Georgetown-Lethem bus towards Waikin Ranch, prearrange a transfer to Caiman House, a rustic riverside ecolodge in a remote village that doubles as a black caiman research station. During the day you can head out on wildlife viewing safaris by 4WD or canoe. Once the sun sets, join researchers on caiman-tagging expeditions.
Arrange a vehicle transfer to Waikin Ranch, which takes several hours.

❹ Waikin Ranch

In the savannah region of the Rupununi, a few kilometres off the highway, Waikin Ranch is a perfect spot to decompress from the jungle before heading back to Georgetown. Embark on an anteater-spotting safari by horseback, swim in the oasis-like pond and dine on paddock-to-plate produce.
Arrange a vehicle transfer to Lethem (28km/30min) to fly or bus back to Georgetown. The bus journey takes 16hrs on a good run.

ALTERNATIVES

Extension: Kaieteur Falls

Unless you're up for a multi-day hike, the only way to visit the world's highest single-drop waterfall is by small plane. Take a day trip from Georgetown, or include it as a stop on a tour of Guyana with the likes of Wilderness Explorers. Upon arrival, a local guide will escort you to breathtaking viewpoints.
Scheduled flights between Georgetown and Kaieteur Falls depart at 1pm and return by 5pm (1hr).

Diversion: Essequibo River

Baganara Island Resort is a relaxed retreat on its own island in the mighty Essequibo River. Sign up for a boat tour to spot sloths, toucans and more tropical wildlife, or kick back on the river beach with a cocktail at sunset.
Take a taxi from Georgetown to Parika (1hr) then a boat to Bartica (1hr), where the resort boat will pick you up.

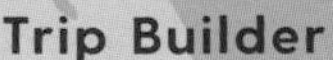

INDIAN OCEAN

AUSTRALIA

9 22 5 6 14

OCEANIA

Oceania is an outdoorsy dream: you can canoe New Zealand's Whanganui River, marvel at the technicolour Great Barrier Reef, sail the aquamarine waters of the Whitsundays and revel in glorious beaches on the Cook Islands and Fiji's Viti Levu. Hop in a car to road-trip Australia's Red Centre and Queensland rainforest, or break out the board to surf the Gold Coast. Culture vultures can lose days in world-class museums, and gourmands can explore winelands, farmers markets and top-class dining scenes The only problem is choosing where to start.

OCEANIA
PAPUA NEW GUINEA
SOLOMON ISLANDS
NORTH PACIFIC OCEAN
AMERICAN SAMOA
VANUATU
11
21
FIJI
12
NEW CALEDONIA
TONGA
18
COOK ISLANDS
1
8
13
19
2
3
20
17
16
NEW ZEALAND
10
7
TASMANIA
4
SOUTH PACIFIC OCEAN
15

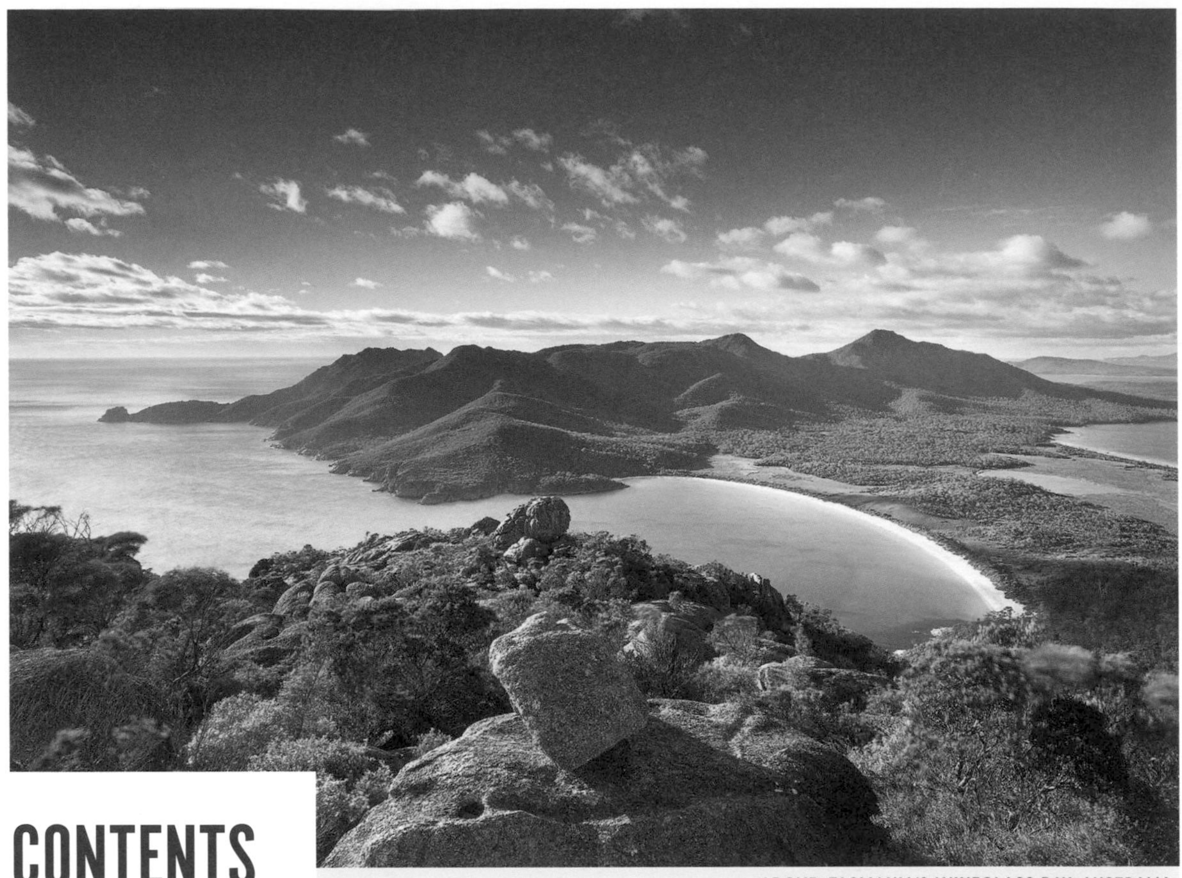

CONTENTS

ABOVE: TASMANIA'S WINEGLASS BAY, AUSTRALIA

ABOVE: CAPE OTWAY KOALA, AUSTRALIA

POLYNESIAN ADVENTURES ON THE COOK ISLANDS

Rarotonga – 'Atiu

Traditional Polynesian culture and tropical scenery combine on this super South Pacific adventure spanning three very different islands.

FACT BOX

Carbon (kg per person): 111
Distance (km): 700
Nights: 5-7
Budget: $$$
When: Apr-Nov

❶ Rarotonga

Combine gentle activity and tropical relaxation on the Cooks' most populous island: music and dance feature at the Highland Paradise Cultural Centre; windsurfing, paddleboarding and kayaking are options on sheltered Muri Lagoon; and food stalls and local flavours at the Punanga Nui Market are the biggest game in town every Saturday morning. The following day, attend a local church service for warm island hospitality and some of the South Pacific's most beautiful harmonies. Options to explore Rarotonga's forest-clad backroads include drive-yourself off-road buggies and exciting 4WD excursions into the island's rugged hinterland. For post-activity refuelling, sandwiches crammed with fresh tuna or mahi mahi are an essential purchase from The Mooring Fish Cafe, while the over-the-water bar at Trader Jacks is one of Polynesia's best spots for an afternoon beer or cocktail.

✈ Fly from Rarotonga to Aitutaki (daily; 45min).

❷ Aitutaki

Journey north across the wild blue expanses of the Pacific Ocean to Aitutaki's perfect lagoon. For sapphire-hued views across the compact fish hook-shaped atoll, negotiate the 30-minute hike to the top of Maungapu (124m), Aitutaki's highest peak. The sheltered waters of the lagoon are perfect for

ABOVE: COOK ISLANDS MARINE LIFE

ABOVE: AITUTAKI

snorkelling and learning to dive – more experienced divers can look forward to superb visibility amid a fascinating underwater seascape including drop-offs, wall dives and exciting cave systems. Game fishing opportunities include going after wahoo, giant trevally and Aitutaki's super-fast bonefish. Alternatively, glass-bottom boat trips and lagoon cruises offer a relaxed option for all travellers wanting to experience what's going on under the turquoise waters. Boat trips visit the islet of Akaiami, where trans-Pacific flying boats used to refuel in the 1950s.

Take a flight from Aitutaki to 'Atiu (3 weekly Apr-Nov; 45min).

3 'Atiu

Continue your Cook Islands adventure on rocky and remote 'Atiu. The compact island is honeycombed with limestone caverns, including Anatakitaki Cave, the only known home of the rare *kopeka* ('Atium swiftlet). Above ground, avian attractions in 'Atiu's tropical forests include the endangered *kakerori* (Rarotongan flycatcher). The perfect way to complete an active day on 'Atiu is chatting with locals at a *tumunu* (traditional 'bush-beer' club).

Fly back to Rarotonga for onward connections (5 flights per week, Apr-Nov; 45 min).

ALTERNATIVES

Day trip: Cross-Island Track

The soaring 413m peak of Te Rua Manga (The Needle) is the highlight of Rarotonga's Cross-Island Track. Tackling the trail takes most travellers three to four hours. Some sections are steep and the rocky path weaves between tangles of tree roots. Cool off at the finish on the south coast with a dip at Wigmore's Waterfall.

For the trailhead, take a local bus to Avatiu Stream car park on Rarotonga's north coast.

Day trip: Storytellers Eco Cycle Tours

Guided cycling or walking experiences with Storytellers explore the byways and backroads, with Rarotonga's history, environment and culture all explained in detail along the way. Highlights include learning about the island's whale conservation programme and exploring Ata Metua, an historic inland route between the villages.

Storytellers provide transfer services from island accommodation.

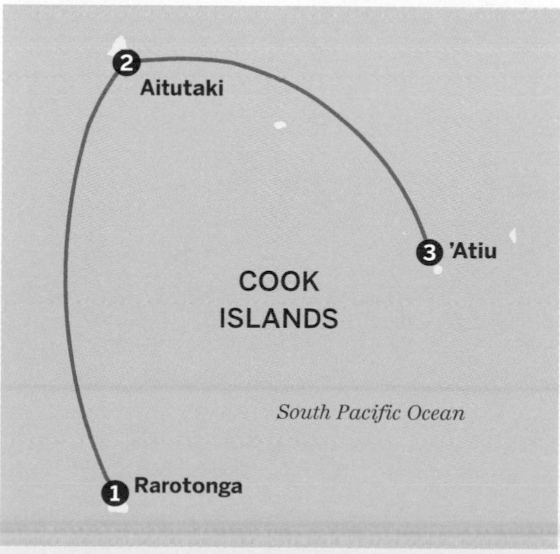

EAST COAST OZ DISCOVERER

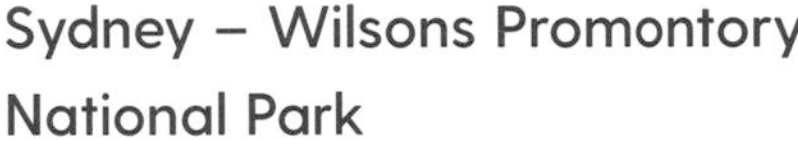

Sydney – Wilsons Promontory National Park

Travel Australia's east coast from Sydney to Melbourne on a spectacular journey combining food, wine, wildlife and history.

FACT BOX

Carbon (kg per person): 153
Distance (km): 1237
Nights: 7-10
Budget: $$$
When: Sep-Nov

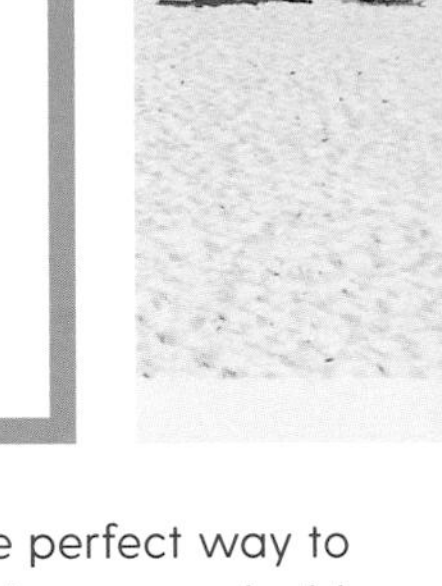

❶ Sydney

Begin your Australian coastal adventure in one of the world's great harbour cities. Climb the famous bridge for brilliant views of the Sydney Opera House. Catch the ferry from Circular Quay to Manly. And eat and drink in leafy Surry Hills, raffish Newtown or beachside Bondi. Usually dotted with food trucks, Marrickville's craft brewery trail is the perfect way to spend a weekend afternoon. A popular way to build up a thirst is to tackle the stunning Bronte to Coogee coastal walk, a lovely clifftop stroll extending for 6km.

Drive 220km south, taking in Royal National Park's coastal beauty – especially spectacular Sea Cliff Bridge – en route.

ABOVE: SYDNEY HARBOUR

❷ Mollymook

Those in search of top Australian seafood come to Mollymook. Rick Stein at Bannisters combines great views with dishes showcasing local scallops and snapper, while it's a short drive for beer and briny-fresh bivalves at the Ulladulla Oyster Bar. For excellent local wine, adjourn to Cupitt's Estate – produced from local grapes, their Sauvignon Blanc and Semillon are both highly regarded.

Continue on the coastal road via Bermagui and Tathra for 250km.

❸ Eden

Whales were hunted in Eden's Twofold Bay from the late 18th century to 1930, but now the focus is on protecting the giant marine mammals that return to these sheltered waters from September to November. Visit for the annual Eden Whale Festival in late

ABOVE: BONDI BEACH, SYDNEY

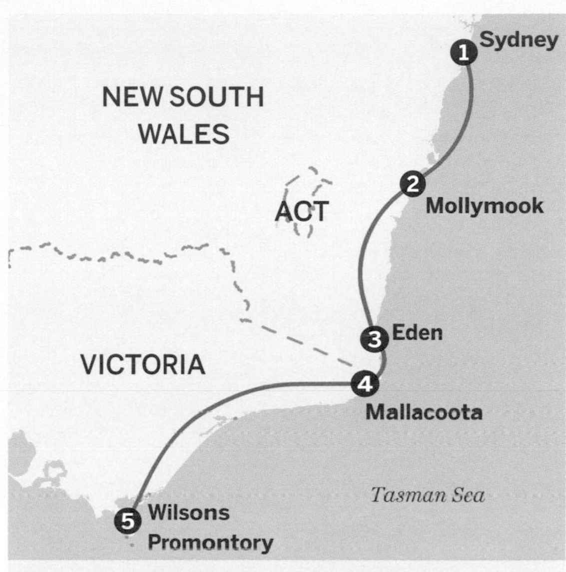

October, or take a cruise to see migrating humpbacks and southern right whales.

Drive 85km, crossing the border from New South Wales to Victoria.

4 Mallacoota

The easternmost town in Victoria is surrounded by the arcing surf beaches and sprawling sand dunes of Croajingolong National Park. Make the journey to Gabo Island to see seals, whales and dolphins, and consider staying overnight in the island's lighthouse accommodation. The Mallacoota Bunker Museum is a fascinating reminder of this remote area's high-security communications facility in WWII.

Follow Victoria's coast west for 472km.

5 Wilsons Promontory

Colloquially known as 'The Prom', Wilsons Promontory is one of Australia's most popular national parks. Explore more than 80km of walking tracks, get more adventurous on a guided canoeing, abseiling or paddleboarding trip, and view abundant wildlife, including kangaroos and wombats.

Finish your trip with the 210km drive northwest to Melbourne.

ALTERNATIVES

Day trip: Jervis Bay

The town of Huskisson is the main hub for exploring expansive Jervis Bay. Wander along the beaches south of 'Husky', reputed to have the world's whitest sand, or join a kayaking or paddleboarding tour. There's also whale- and dolphin-watching available from September to November and excellent bushwalking on Booderee National Park's coastal peninsula.

From the main highway town of Nowra, it's a 24km drive to Huskisson.

Day trip: Ben Boyd National Park

South of Eden, history and natural beauty combine at Ben Boyd National Park. Explore heritage architecture including the Green Cape Lighthouse and Boyd's Tower, a former whaling lookout. You could consider a longer detour by completing the Light to Light Walk, a 30km coastal track linking both towers via forest-clad beaches and brick-red rocky shores.

It's a 33km drive from Eden to Ben Boyd National Park.

HIGH COUNTRY VICTORIA EXPLORER

Beechworth – Omeo

Explore this surprising subalpine Australian region on a road trip combining iconic history, food and wine, and gentle two-wheel activity.

FACT BOX

Carbon (kg per person): 35
Distance (km): 279
Nights: 5-7
Budget: $$$
When: Oct-Apr

❶ Beechworth

Gold-mining history and the life and times of Australia's most infamous bushranger, Ned Kelly, come together amid Beechworth's National Trust-listed townscape. Join a walking tour around the honey-coloured granite buildings to learn about Kelly's adventures in northeast Victoria, and spend time exploring historic Beechworth Gaol. Dating from 1857, the town's Burke Museum is one of Australia's oldest and has displays on local history. Highlights of Beechworth's eating and drinking scene include Japanese-inspired fine dining at Provenance and craft beer at Bridge Road Brewers. Around 20km north of town, indigenous cave paintings can be seen the Yeddonba Aboriginal Cultural Site.

Drive 74km southwest to Whitfield in the heart of King Valley.

❷ King Valley

Originally settled by tobacco farmers from Italy, King Valley is now an emerging winemaking area. Sample refreshing Prosecco from Dal Zotto or try lesser-known Sicilian varietals like Grecanico and Nero d'Avola at Politini. Spend a night at Whitfield's cosy Mountain View Hotel after teaming their Italian-inspired menu with local craft beers from King River Brewing. Popular classes in cooking Italian cuisine are offered at Pizzini Wines.

Head 95km east to Bright.

ABOVE: BRIDGE ROAD BREWERS, BEECHWORTH

ABOVE: AUTUMN COLOURS IN THE TOWN OF BRIGHT

❸ Bright

Cosmopolitan cafes, the popular Bright Brewery and a convenient location on the Murray River to the Mountains Rail Trail make the tree-lined streets of Bright a popular weekend escape for Melburnians. Rent a bike and set out on the trail's scenic 6km riverside spur to Porepunkah. E-bikes make it an easy ride for everyone. On the way, cafes and a kids' playground make it a good option for families. Abseiling, kayaking and underground-river caving are available for adventurous travellers during spring and summer. On weekend afternoons, an essential detour is to nearby Wandiligong for live music at the Wandi Pub.

Negotiate the winding and wildly scenic Great Alpine Rd 110km southeast to Omeo.

❹ Omeo

After possibly pausing in Mt Hotham or Dinner Plain for some alpine mountain biking, you arrive in historic Omeo, on the edge of the Snowy Mountains and a good base for exploring nearby Snowy River National Park. Keep an eye out for the rare brush-tailed rock-wallaby – this is one of its last natural habitats.

From Omeo it's a 430km drive west to Melbourne and its international airport.

ALTERNATIVES

Day trip: Yackandandah

A one-street town of well-preserved 19th-century buildings, super-compact Yackandandah is one of Australia's loveliest small towns. Travellers can join weekend tours exploring an 1857 gold mine, and the local information office has maps for the nearby Yackandandah Gorge Scenic Walk. A gourmet pie from Yack's very own Gum Tree Pies is definitely recommended.

Yackandandah is a 22km drive east of Beechworth.

Day trip: Mt Buffalo

Looming above Bright and Porepunkah, Mt Buffalo National Park is a low-key, family-focused winter sports resort. During autumn and spring the access road is a popular challenge for road cyclists, and the mountain's granite outcrops, waterfalls, wildflowers and wildlife all combine to make it an excellent destination for bushwalking and mountain biking.

The winding drive from Bright to the summit of Mt Buffalo takes around 1hr.

SOUTHERN ALPS LOOP IN NEW ZEALAND'S SOUTH ISLAND

Christchurch – Lake Tekapo
Begin with one of the world's great train journeys before continuing on an independent exploration of the alpine highlights of New Zealand's South Island.

FACT BOX
Carbon (kg per person): 211
Distance (km): 1340
Nights: 8-12
Budget: $$$
When: Oct-Apr

❶ Christchurch

Exciting urban regeneration following a 2011 earthquake complements genteel 19th-century architectural heritage in the South Island's biggest city. Formal New Zealand art is showcased at the Christchurch Art Gallery, while street art lines the lanes of the emerging SALT District. Go punting or kayaking on the Avon River through leafy Hagley Park, and experience Christchurch's increasingly cosmopolitan diversity by taking your pick from the global array of food outlets at the Riverside Market or visiting Saturday morning's Riccarton Farmers Market.
Take the wonderfully scenic TranzAlpine train to Greymouth (daily; 5hr), through river valleys and over mountain passes. Rent a car in Greymouth and drive 175km south to Franz Josef Glacier.

❷ Franz Josef and Fox glaciers

Two extraordinary glaciers, the Franz Josef and the Fox, glide their way down towards the west coast just 26km apart. Get up close to these blue-ice, forest-surrounded wonders on short walks, explore their higher reaches on heli-hikes or take to the air for sky-high views.
Enjoy more marvellous views on the 350km drive south to Queenstown.

❸ Queenstown

Active adventures, stunning scenery and a cosmopolitan eating and drinking scene all merge in New Zealand's most popular alpine resort. Take a bungee jump above a river, cruise the vineyards of the Gibbston Valley or ascend on the Skyline Gondola

ABOVE: SKYLINE GONDOLA, QUEENSTOWN

ABOVE: LAKE TEKAPO LUPINS

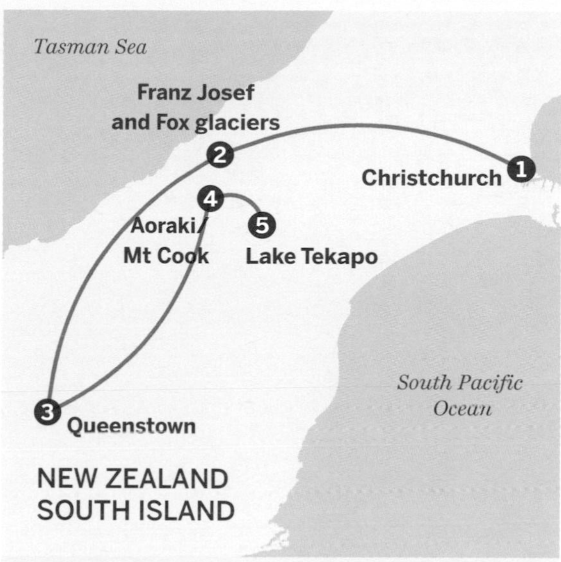

for lake and Southern Alps vistas. Options to explore Lake Wakatipu include jetboat rides and the historic steam-powered TSS *Earnslaw*.

Continue 260km north to Aoraki/Mt Cook.

4 Aoraki/Mt Cook

In the shadow of New Zealand's highest mountain, Aoraki/Mt Cook (3724m), you can embark on scenic day-hikes through alpine valleys to glacial lakes. Get out onto Lake Tasman at the foot of the Tasman Glacier by kayak or in an inflatable zodiac, and fly by helicopter to land on expansive ice-flows high in the mountains.

Drive 100km back along the western shore of Lake Pukaki to Lake Tekapo.

5 Lake Tekapo

At the heart of Mackenzie Country, turquoise-coloured Lake Tekapo is a brilliant spot to observe the southern hemisphere's night sky from atop nearby Mt John. After stargazing from the International Dark Sky Reserve, relax with lake and mountain views at Tekapo's hot springs and admire the superb lake vistas from the Church of the Good Shepherd.

Complete your Southern Alps circuit by driving 225km back to Christchurch.

ALTERNATIVES

Extension: Punakaiki

Detour north to the spectacular Pancake Rocks at Punakaiki. Created by a natural weathering process called stylobedding, towers of limestone rocks have been transformed into formations resembling stacks of giant pancakes. Visit around high tide for the most impressive natural display, as the Tasman Sea surges into rocky caverns and explodes showers of salty spray through coastal blowholes.

Punakaiki is 44km north of Greymouth.

Diversion: Arrowtown

Established during an 1860's boom, more than 60 well-preserved heritage buildings along Arrowtown's main street reflect the region's gold-mining heyday. Uncover history at the local museum and former Chinese miners' settlement before hiring a bike to negotiate the scenic Arrow Bridges River Ride (12km). End the day with local craft beers at Lake & Wood Brew Co.

Arrowtown is just 20km from Queenstown.

SOUTHWEST AUSTRALIA DISCOVERER

Perth – Esperance

Enjoy Perth's big-city buzz then take in history, fine wine and food, and spectacular national parks on a journey along one of Australia's most underrated coastlines.

FACT BOX

Carbon (kg per person): 140
Distance (km): 1125
Nights: 7-10
Budget: $$$
When: Sep-Mar

❶ Perth

Anchored by the Swan River and the Indian Ocean, Australia's most isolated state capital is a vibrant and dynamic destination. Photograph city views from sprawling King's Park before exploring Northbridge's Western Australian Museum, revitalised and reopened in late 2020 after a four-year redevelopment. Nearby, the Art Gallery of Western Australia has a well-regarded collection of Aboriginal art. Leederville and Mount Lawley are packed with good cafes, bars and restaurants, while oceanside Cottesloe or Scarborough are both ideal for sunset drinks.

Drive 270km south to Margaret River.

❷ Margaret River

One of the Australia's premium wine-growing areas, especially crafting Chardonnay and Bordeaux-style red wines, Margaret River is also one of Western Australia's most spectacular coastal regions. Negotiate the clifftop Cape to Cape Walk – either the full 140km, five- to six-day adventure or on shorter sampler walks – or, from September to December, ride the waves on a whale-watching adventure. Vineyard restaurants range from informal bistros to stylish fine-dining.

Continue 375km east to Albany on the coastal route, stopping at the giant tingle trees around Walpole and the spectacular Elephant Rocks coastal pools near Denmark.

ABOVE: LUCKY BAY KANGAROO, ESPERANCE

ABOVE: MARGARET RIVER VINEYARD

❸ Albany
Albany is one of Western Australia's oldest cities, and its heritage is best seen in the historic precinct along Stirling Terrace. Visit the National Anzac Centre, an excellent museum remembering the Australian and New Zealand soldiers who departed Albany by naval convoy to fight in WWI. The region's dramatic seacapes can be best appreciated at the spectacular Natural Bridge in nearby Torndirrup National Park.
Drive 480km east along the coast to Esperance.

❹ Esperance
You might be tired after a road trip of more than 1000km, but Esperance's sublime location makes it definitely worth the journey. Surrounded by aquamarine waters, this coastal town is a relaxed base for exploring some of Western Australia's best national parks. Drive the 40km Great Ocean Drive near town, just a taster of the natural brilliance further east. Kangaroos are occasionally seen on the crystalline white-sand beaches of Lucky Bay, while the area's best sea views are from atop Frenchman Peak, reachable via a steep 3km, two-hour loop trail in Cape Le Grand National Park.
Take a flight back to Perth to finish the trip (3 daily; 90min).

ALTERNATIVES

Extension: Cape Leeuwin
Hang on to your hat at wild Cape Leeuwin, the often windy meeting point of the Indian and Southern oceans, and the most southwesterly point of the Australian continent. Don't leave without signing up for a guided tour of the Cape Leeuwin lighthouse – Western Australia's tallest, it offers unsurprisingly superb views of the wild coastline.
Take the scenic 60km route via Hamelin Bay from Margaret River to Cape Leeuwin.

Diversion: Fremantle
Wander Fremantle's Victorian townscape, packed with cafes, bars and design shops, before visiting the maritime galleries of the Western Australian Museum's 'Freo' branch. An extended excursion can also include catching a ferry to nearby Rottnest Island, for sandy beaches, bike adventures and quokka selfies with the Island's much-loved marsupials.
Frequent trains travel from Perth Station to Fremantle (45min).

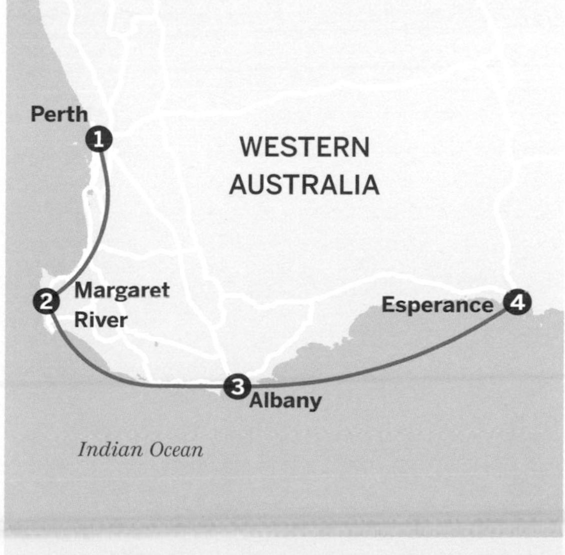

EXPLORE AUSTRALIA'S EYRE PENINSULA

Port Lincoln – Streaky Bay

The vast, triangular Eyre Peninsula is South Australia's big sky country – not to mention big distances, big national parks, big seafood dinners and big sharks.

FACT BOX

Carbon (kg per person): 78
Distance (km): 627
Nights: 5-7
Budget: $$
When: May-Oct

1 Port Lincoln

Port Lincoln (a 50min flight from Adelaide) is the biggest town on South Australia's gargantuan Eyre Peninsula. Fronting onto broad Boston Bay, the 'Tuna Capital of the World' is undeniably appealing: the local fishing fleet brings in the big bucks, which translates to some great eateries and accommodation, buzzy pubs and an upbeat waterfront vibe. Tours to swim with sea lions and sharks are almost obligatory, while just south of town, Lincoln National Park offers sandy solitude. Port Lincoln's sea-salty maritime museum is a great rainy-day detour.

Drive 47km west to Coffin Bay.

2 Coffin Bay

Slurp some oysters in ominous-sounding Coffin Bay (it was actually named in 1802 by British explorer Matthew Flinders after his buddy Sir Isaac Coffin). Tour a working oyster farm or order a slippery dozen at the local oyster bar (best consumed with a cold Long Beach Lager, brewed in Port Lincoln). The rolling dunes in nearby Coffin Bay National Park are overrun with wildlife: kangaroos, emus, ospreys and fat goannas. See what you can spy on a hike or sea-kayaking adventure.

Drive 32km north to Coulta, then 12km west down a dirt road to Greenly Beach.

ABOVE: COFFIN BAY OYSTERS

ABOVE: GREENLY BEACH ROCKPOOL

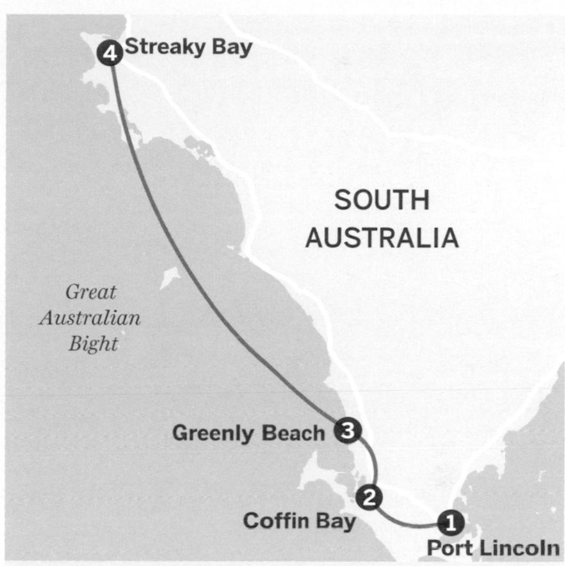

3 Greenly Beach

Beach picnic, anyone? A short hop north of Coffin Bay, Greenly Beach offers gnarly surf and a much-photographed aquamarine rock pool at its northern end (fabulous for a swim). If you feel like exploring further, the hike up Mt Greenly (305m) is short and sharp.

Back in the car, drive 241km northwest to Streaky Bay.

4 Streaky Bay

Beyond a string of snoozy fishing towns is middle-sized Streaky Bay. Signposted just south of town, Murphy's Haystacks are a globular congregation of 'inselbergs' - colourful, weather-sculpted granite outcrops, an estimated 1500 million years old. No less impressive is the 5m-long great white shark inside Streaky Bay's petrol station – a replica of the real thing, caught here in 1990. Regain your composure at the pub's breezy terrace, or take the 38km loop drive around Cape Bauer to discover wild beaches, reefs, blowholes, crumbling limestone cliffs and the endless ocean grinding into shore – a good dose of wild stuff before you track back to Port Lincoln.

Drive 295km back to Port Lincoln and fly back to Adelaide.

ALTERNATIVES

Extension: Port Augusta

Explore the Eyre Peninsula's east coast town of Port Augusta, aka the 'Crossroads of Australia', from where highways roll west into WA, north to Darwin, south to Adelaide and east to Sydney. Don't miss the excellent Australian Arid Lands Botanic Garden (ever seen a Sturt's Desert Pea?) and a walking tour of the Old Town and waterfront, where kids backflip off jetties into Spencer Gulf.

Port Augusta is 345km northeast of Port Lincoln; 304km northwest of Adelaide.

Diversion: Head of Bight

In the region between May and October? From Streaky Bay, continue west to the raffish oyster town of Ceduna and Head of Bight. From the clifftop viewing platforms here (literally at the head of the Great Australian Bight), gaze down and spy colossal southern right whales cavorting just offshore. These leviathans migrate from Antarctica to breed here every year.

Ceduna is 111km from Streaky Bay, with Head of Bight a further 285km west.

EAST-COAST TASMANIA FROM CITY TO CITY

Launceston – Bruny Island
Develop a taste for Tasmania's urban delights in Hobart and Launceston via an east coast road-trip taking in bushwalking and swimming in Freycinet National Park.

FACT BOX

Carbon (kg per person): 54
Distance (km): 434
Nights: 7
Budget: $$
When: Nov–Mar

❶ Launceston

Kick off your Tasmanian adventures in hilly Launceston, an underrated little city with a cache of well-preserved architecture and a simmeringly tasty food scene. Craggy Cataract Gorge and the cool-climate Tamar River Wine Region are on Lonnie's back doorstep, meaning you can both launch into a bushwalk and sip your way through some winery cellar doors.

Drive 175km south to Freycinet National Park via Campbell Town and Coles Bay.

❷ Freycinet National Park

As most of Tasmania's rain falls on its west coast, by the time the clouds reach the east coast, and Freycinet National Park, they're empty. Which means you can soak up the sun here, maybe on the steep climb up the pink-granite Hazards mountains to gaze down on famous Wineglass Bay – a glorious goblet of white sand and gin-clear water. Better yet, continue down to the beach and cool off in the waves (visit early and you'll have it to yourself). Refuel afterwards in Coles Bay's restaurants, or duck into Freycinet Marine Farm for some super-fresh local seafood.

Continue 193km south to Hobart.

❸ Hobart

Australia's second-oldest city (behind Sydney) has been doing things a little differently since 1804. At

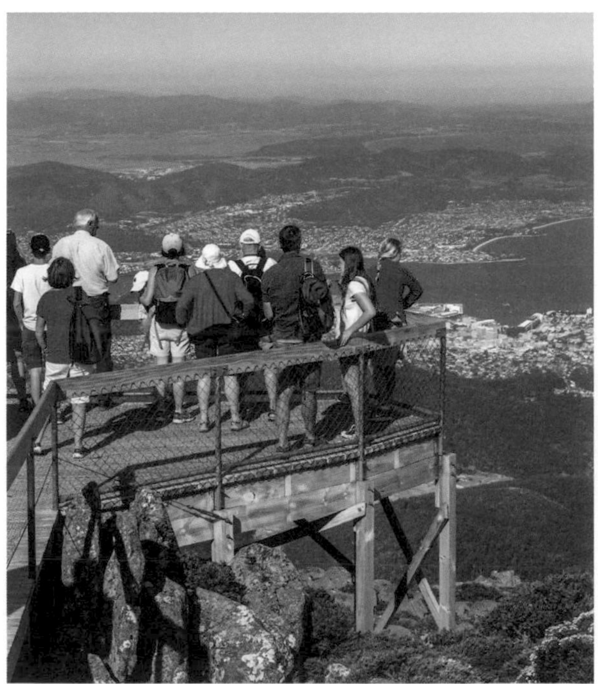

ABOVE: HOBART FROM KUNANYI/MT WELLINGTON

ABOVE: BRUNY ISLAND

the foot of kunanyi/Mt Wellington on the slate-grey Derwent River, it's a gorgeous place. Spend a day exploring historic Battery Point, sipping beers at Cascade Brewery (the country's oldest), eating and drinking on the waterfront or chugging upriver to see spectacular MONA, the Museum of Old and New Art. With challenging, entertaining and downright weird exhibits, markets and festivals, MONA has been Hobart's twisted cultural highlight for over a decade now. A mountain-bike ride from the summit of kunanyi/Mt Wellington (1271m) down to the waterfront is another Hobart must-do.

Drive 33km south to Kettering and catch the ferry across the D'Entrecasteaux Channel (15min).

❹ Bruny Island

An island, off an island, off an island. It may be pushing the theme to the limit, but Bruny is worth it. The food scene here is a distillation of Tasmania's best offerings: handmade cheeses, oysters, berries, whisky, honey, craft beer and wines from Australia's most southerly vineyard. Spy penguins and seals on a boat cruise, head off on a bushwalk in South Bruny National Park, or hit the (chilly) surf at Cloudy Bay.

Head back to Hobart and onward connections from its airport.

ALTERNATIVES

Extension: Derby

Former home of the world's most prosperous tin mine, Derby was on a fast track to ghost-town status before mountain biking took off in the surrounding hills around 2015. The 'Blue Derby' MTB network now comprises 125km of world-class trails for all skill levels. Don't expect to find any accommodation here (or even a sandwich) when there's a tournament on.

Derby is 95km northeast of Launceston.

Diversion: Port Arthur Historic Site

On the isolated Tasman Peninsula, Tasmania's #1 tourist attraction is a compelling mix of gorgeous coastal scenery, architectural heritage and human tragedy. Between 1830 and 1877, 12,500 convicts did brutal prison time here – a grim legacy compounded by a 1996 shooting massacre. The site's history is explored on poignant tours.

Port Arthur is 100km southeast of Hobart.

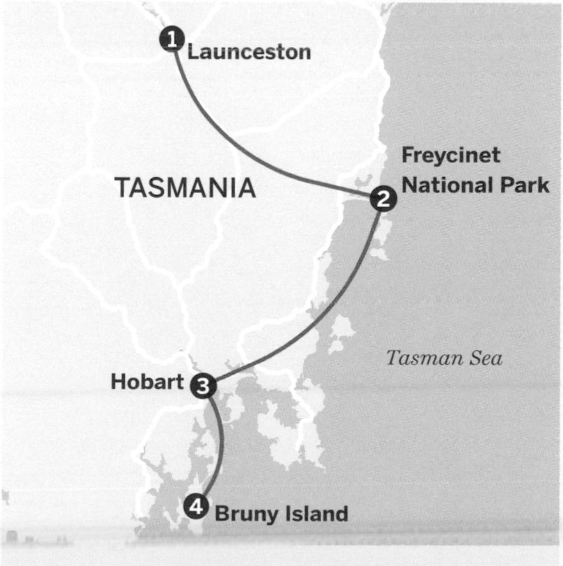

AUSTRALIA'S OUTBACK ON THE BIRDSVILLE TRACK

Brisbane – Adelaide

Drive from Brisbane to Adelaide on an exhilarating, isolated trip following back-roads and the 4WD Birdsville Track between one-kangaroo towns and Outback pubs.

FACT BOX

Carbon (kg per person): 388
Distance (km): 3119
Nights: 10-14
Budget: $$$
When: May-Sep

❶ Brisbane

One of Australia's most dramatic off-road drives, the Birdsville Track traces the footsteps of Aussie pioneers, but you'll have to get there first from Brisbane, Queensland's upbeat capital. Pause for some big-city comforts, then fuel up your 4WD and unleash your inner explorer.

Drive 513km west to St George via Toowoomba.

❷ St George

Heading inland, the landscape becomes ever emptier and the outposts more scattered and wind-whipped. Toowoomba is a virtual metropolis, Goondiwindi a bustling crossroads, and St George a mere blip on the map, but the latter is good for breaking up the journey with a bed and a brew.

Continue 285km west to Cunnamulla.

ABOVE: SOUTH BANK PARKLANDS' STREETS BEACH, BRISBANE

❸ Cunnamulla

The road runs arrow-straight through the kangaroo-coloured scrub to Cunnamulla – a friendly outpost of 1200 souls, with pubs, a public pool and fine sunsets over the Warrego River.

Drive 683km northwest to Windorah.

❹ Windorah

It's another full day of desert driving to reach Windorah, the next overnight stop, close to the spot where explorers Burke & Wills met their unfortunate end, exhausted and starving on their return trip after traversing Australia south to north.

Drive 532km west to Birdsville.

❺ Birdsville

The start of the Birdsville Track is something of an Aussie legend thanks to its blood-red sand dunes and the Birdsville Cup horse race. A stubby (small bottle of beer) and steak at the Birdsville Hotel is pretty much

ABOVE: OUTBACK DUNES

mandatory before you check your water and fuel supplies and set off into the emptiest country yet.
Head 315km south to Mungerannie.

6 Mungerannie Hotel

As one-horse towns go, Mungerannie doesn't even have a horse, or a town, but it does have the Mungerannie Hotel, a precious oasis providing fuel, vittles and accommodation.
It's 202km south to Marree.

7 Marree

The Birdsville Track rolls on, dusty and desolate, to Marree, with the inevitable bush pub and a 4km-long figure of an Aboriginal hunter etched into the desert. Continuing to Adelaide, switch to the gentler Oodnadatta Track, passing ruined telegraph stations, hot springs and some stunning country campsites.
Drive 589km south to Adelaide.

8 Adelaide

After the quiet of the Outback, Adelaide will wrap you in a warm embrace of city life. Eat well, sleep well, then hit the Barossa wine country or the beach.
Adelaide airport has national and international connections.

ALTERNATIVES

Extension: Barossa Valley

Put all those Outback pub beers and charred steaks out of your mind. Just an hour's drive north of Adelaide, the Barossa Valley is arguably Australia's most epicurean wine region, studded with elegant B&Bs, gourmet restaurants, farmers markets and wineries offering tastings and tours. Stay in Angaston and rent a bike for gentle, wobbly days of wine sampling.
Angaston is 89km northeast of Adelaide.

Diversion: Kangaroo Island

If you've had enough of empty landscapes, beat a trail to Kangaroo Island where seals, dolphins, penguins and seabirds throng the coast, and kangaroos, koalas, echidnas, possums and countless other critters frolic inland. Come by ferry from Cape Jervis and camp to keep down your overheads while you track down some of Australia's signature wildlife.
Drive 200km south of Adelaide and take the 45min ferry to the island.

INDIGENOUS AUSTRALIA ON THE DAMPIER PENINSULA

Broome – Cygnet Bay Pearl Farm
From Broome, head north through the Dampier Peninsula to the spectacularly beautiful Cape Leveque via some culturally rich and fascinating indigenous communities.

FACT BOX
Carbon (kg per person): 58
Distance (km): 469
Nights: 4-5
Budget: $$$
When: May-Aug

❶ Broome

Plunge into the turquoise waters of Broome's Cable Beach, the region's most famous landmark. Explore the red pindan cliffs in search of dinosaur footprints, and learn about the town's pearling history on a walking tour. Then, when you've had your fill, throw yourself into the indigenous cultures of the Dampier Peninsula along the Cape Leveque road.
Beagle Bay is 147km north of Broome. You can drive on your own or, for convenience, join a multi-day tour led by local operators. Tours may stop at one or two of the following indigenous communities en route. If you're going it alone, note that while the main road is sealed, the access roads to the communities are not – you'll need a 4WD.

❷ Beagle Bay

Beagle Bay's community showcases the beautiful church built by two German Pallottine priests between 1915 and 1918. Confined to the community during WWI, the priests busied themselves by constructing this unusual building, using mother-of-pearl, cowrie and trochus shells to embellish the nave, altar and Stations of the Cross.
Lombadina is 82km northwest.

❸ Lombadina

Lombadina boasts a remarkable church with a paperbark ceiling, and a small arts and crafts centre.

ABOVE: BEAGLE BAY CHURCH

ABOVE: KOOLJAMAN CLIFFS, CAPE LEVEQUE

Various Aboriginal-led tours – from mudcrabbing and whale-watching to kayaking and 4WD adventures to 20,000-year-old dinosaur footprints – can be booked through the Lombadina website.

Cape Leveque is 17km further north.

❹ Cape Leveque

Red, remote and at the tip of the Dampier Peninsula, Cape Leveque is high on the list of must-visits for many travellers to this section of Australia. Deserted beaches, birdlife and incredible nature tours with local Bardi Jawi people are highlights here. Spend the night in a solar-powered wilderness camp operated by the excellent, Aboriginal-run business Kooljaman.

The pearl farm is 15km southeast of Cape Leveque.

❺ Cygnet Bay Pearl Farm

Located on Dampier Peninsula at the head of the Buccaneer Archipelago (a series of more than 800 islands) is Australia's oldest pearl farm, Cygnet Bay. Learn about all things oysters-and-pearls on an interesting tour, followed by a boat trip – spot snubfin dolphins, dugongs and turtles on the way. Pitch a tent in a bush site or nestle into a luxury safari tent or cabin.

Head back to Cape Leveque and finish the trip with the 208km drive back to Broome.

ALTERNATIVES

Extension: Whale Song Cafe

For a real off-the-beaten-track adventure (that is if you don't think you're already well off the beaten track) head to Munget on stunning Pender Bay and the eco-friendly Whale Song Cafe. The heart of the community, the cafe serves up the peninsula's best coffee and decent snacks. You can bed down for the night at its cliff-top campsite.

From Beagle Bay, the cafe is 30km northwest.

Diversion: Middle Lagoon

The empty, if remote, beaches of Middle Lagoon are perfect for swimming, snorkelling, fishing and lazing around with a book. Bring your binoculars to spot birds, turtles, dugongs and passing whales.

Middle Lagoon is around 30km northwest off Cape Leveque Rd (but it can take several hours to get there, depending on road conditions).

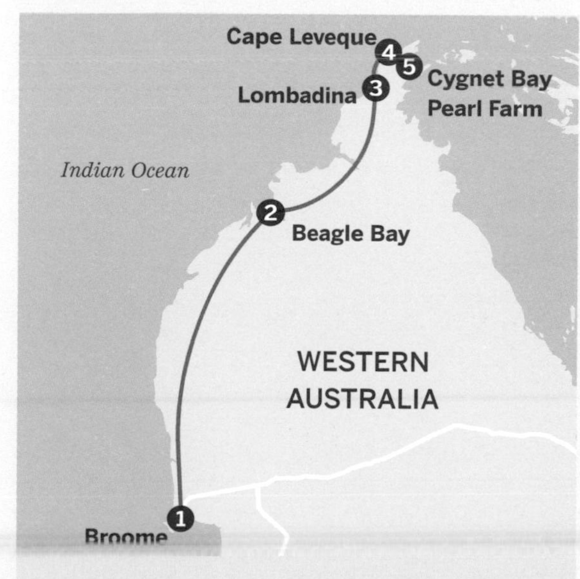

ROAD-TRIP TASMANIA'S EPIC NORTHWEST

Devonport – Lake St Clair

Cradle Mountain is Tasmania's unofficial emblem, but the state's northwest also delivers quirky cities, pretty towns, native wildlife and fabulously far-flung places.

FACT BOX

Carbon (kg per person): 86
Distance (km): 693
Nights: 7
Budget: $$
When: Dec–Apr

❶ Devonport

Getting to Tasmania from mainland Australia is half the fun: the overnight ferry from Melbourne arrives at Devonport at dawn. Roll down the off-ramp and get started: stroll out to the red-and-white lighthouse on Mersey Bluff, then dive into some local nautical heritage at the Bass Strait Maritime Centre.

Drive 50km west along the coast, taking a look at the Big Penguin at Penguin on the way.

❷ Burnie

Burnie is an industrial city that's reinventing itself as a creative hub. Swing into the Burnie Regional Museum for some local history, and the Makers' Workshop to see how the city's future is looking. There's more penguin action at the Little Penguin Observation Centre and superb Tasmanian whisky to sip at Hellyers Road Distillery.

Continue your drive 79km west.

❸ Stanley

You can't miss Stanley. An impossibly pretty little fishing village, soaked in history and inextricably bound to the sea, it sits at the base of the Nut, a colossal volcanic formation towering over the coast. Hike (or take the chairlift) up the Nut, then check out Highfield Historic Site, a perfectly preserved homestead dating back to 1835.

Drive 184km southeast to Cradle Mountain.

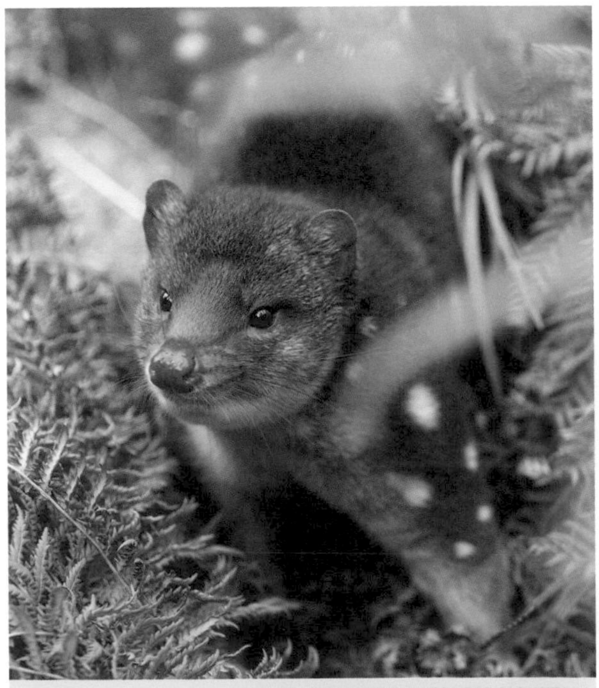

ABOVE: TIGER QUOLL, CRADLE MOUNTAIN

ABOVE: CRADLE MOUNTAIN

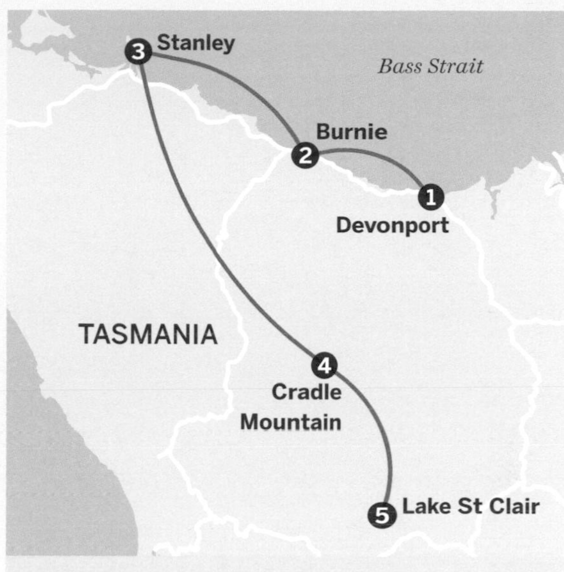

❹ Cradle Mountain

You've seen the photos, now see it for yourself. The alpine trails around Cradle Mountain can get seriously busy in summer, but head beyond the Dove Lake circuit and it's not difficult to find solitude – you might end up with just this beloved peak to keep you company. Native wildlife is a big part of any Cradle Mountain visit: look for wombats, Tasmanian devils, pademelons, tiger quolls, wallabies and sleeping tiger snakes coiled in the sun. The famous five-day Overland Track starts here, too (parks pass required – book early).

Drive 200km south to Lake St Clair via Queenstown.

❺ Lake St Clair

At the southern end of Cradle Mountain-Lake St Clair National Park is Lake St Clair itself – plummeting 200m into a dark glacial crevasse, it's Australia's deepest lake and well worth the journey. There are some terrific walks around the heavily forested lakeshore, and no shortage of serene spots for a picnic by the mirror-flat waters.

Head 180km back to Devonport (or about the same distance to Hobart).

ALTERNATIVES

Extension: Marrawah

From Stanley, keep trucking west and you'll hit remote Marrawah, said to have the cleanest air on the planet. The Roaring Forties winds blast in from the Southern Ocean here, generating huge swells. The Marrawah West Coast Classic surf comp has been running since the mid-1970s; the ancient petroglyphs in the nearby Preminghana Indigenous Protected Area have been here a lot longer.

Marrawah is 60km southwest of Stanley.

Diversion: Strahan and Queenstown

Strahan is an essential west coast destination: a photogenic fishing town with some top accommodation, eating options and myriad tours onto vast Macquarie Harbour (six times the size of Sydney's) and into the rainforest along the Gordon River. Ride the West Coast Wilderness Railway to nearby Queenstown, an unpretentious mining town with a gritty past.

Strahan is 137km southwest of Cradle Mountain.

AUSTRALIAN REEFS AND RAINFOREST IN QUEENSLAND

Cairns – Daintree and Cape Tribulation
Embark on a tour of Far North Queensland's World Heritage-listed reefs, rainforest and refined beach breaks, driving from Cairns to the Daintree.

FACT BOX
Carbon (kg per person): 47
Distance (km): 350
Nights: 14
Budget: $$$
When: year-round

❶ Cairns

Most visitors use Cairns as a base to rush off to the Great Barrier Reef or the Daintree Rainforest, but it's worth spending some time in town first. Stroll the 3km boardwalk and take a dip in the dazzling artificial saltwater lagoon, gawk at marine life at the aquarium and get a birds-eye view of the rainforest on one of the world's longest gondolas.

Cairns has plenty of operators offering day trips out to the Great Barrier Reef.

❷ Great Barrier Reef

This World Heritage-listed underwater wonderland is a top-of-the-list destination for many a marine lover. The Great Barrier Reef stretches over 2000km off the coast of Queensland and is home to an incredible ecosystem and a vast array of marine life including fish, rays, dugongs, humpback whales, dolphins and endangered sea turtles. Experience a small section of it on a diving or snorkelling trip from Cairns.

Back in Cairns, drive 30km north to Palm Cove.

❸ Palm Cove

This gorgeous coastal village is the sophisticated darling of Far North Queensland and, thanks to a more placid atmosphere than Cairns or next stop Port Douglas, is perfect for spa days and date nights. The melaleuca-lined promenade is made for romantic balmy evening strolls, after which you can take your pick of contemporary restaurants and luxury digs.

Drive 42km north following the coast.

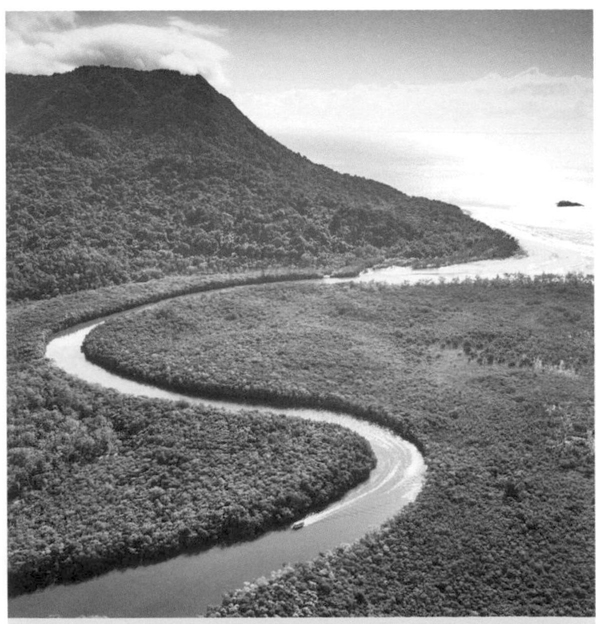

ABOVE: DAINTREE RIVERMOUTH AT CAPE TRIBULATION

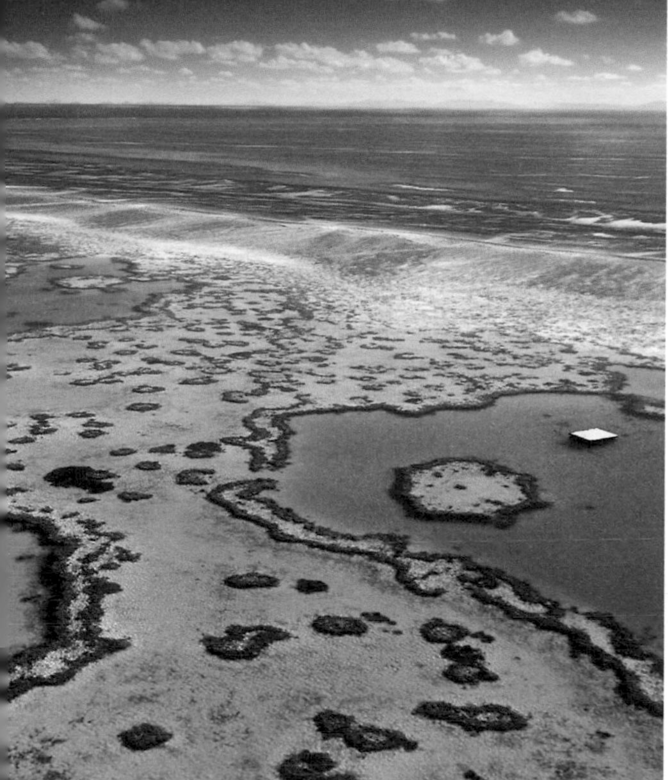
ABOVE: A FRACTION OF THE GREAT BARRIER REEF

❹ Port Douglas

Port Douglas is equal parts luxury and laidback and serves as an alternative base to Cairns for exploring the Daintree and the Reef. It caters to cashed-up, loved-up couples as well as young, active families looking for fun, from the yacht-filled marina and flash resorts to the native animal sanctuary and watersports galore. Join a tour to Cooya Beach (Kuyu Kuyu), the traditional fishing lands of the Kuku Yalanji people. A Kubirri Warra guide will teach you about indigenous culture and traditional plant uses.

Head 69km north to the Daintree Rainforest.

❺ Daintree and Cape Tribulation

The unspoiled ancient rainforest of the Daintree and Cape Tribulation is simply sublime, a place where forest-blanketed mountains spill down to deserted golden beaches. A 15-min ferry takes cars across the Daintree River to this remote, World Heritage-listed tropical oasis filled with wildlife (including plenty of crocs) and biodiversity.

Finish the trip with a leisurely 126km drive back to Cairns to pick up national and international onward connections.

ALTERNATIVES

Extension: Kuranda

An arts-focused retreat hidden in the rainforest, Kuranda is a fantastic day trip from Cairns via a scenic railway ride. Check out the local arts and crafts scene at the Kuranda Original Rainforest Market and drop by the beloved Kuranda Hotel for a pub meal of kangaroo pot-pie or locally caught barramundi on the broad deck.

Daily trains to Kuranda depart at 8.30am and 9.30am from Cairns Central (2hr), returning at 2pm and 3.30pm.

Diversion: Mossman Gorge

Before you reach the Daintree Rainforest, stop off for a splash at this swimming-hole sanctuary with emerald-green water surrounded by giant boulders under a canopy of rainforest. It's part of the traditional lands of the Kuku Yalanji people and is a magical spot.

The gorge is about 22km inland from Port Douglas via the Captain Cook Hwy.

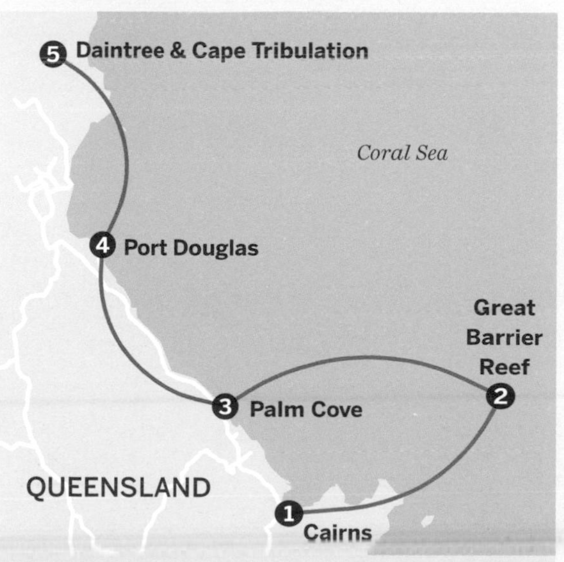

SAILING AROUND THE WHITSUNDAY ISLANDS

Airlie Beach – Hamilton Island

Charter a boat and set sail for Australia's tropical-paradise islands and some of the best beaches on the planet. Drop anchor to snorkel and swim along the way.

FACT BOX

Carbon (kg per person): 1
Distance (km): 62
Nights: 7-10
Budget: $$$
When: Jun-Aug

❶ Airlie Beach

Airlie Beach, on the mainland Queensland coast, is the gateway to the wonderful Whitsunday Islands. It's here you can organise your bareboat charter (no sailing licence or experience required) before embracing the backpacker party scene or keeping it low-key and stocking up on essentials for your sailing trip. The town has an appealing croc-and-jellyfish-free swimming lagoon, yachts bobbing on the water and laidback beer gardens to enjoy before it's anchors away.

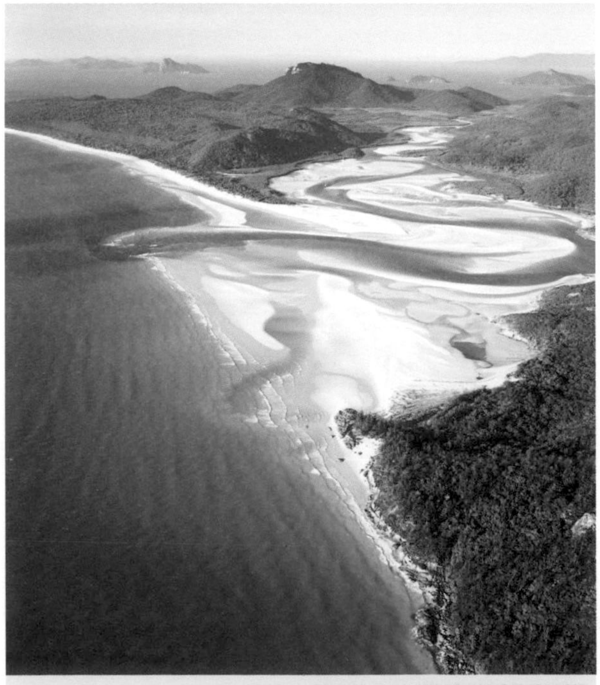
ABOVE: HILL INLET, HOOK ISLAND

It's around 2hr sailing time from Airlie Beach to Hook Island's Nara Inlet anchorage.

❷ Hook Island

Set sail on the sky-blue waters for the first stop on the Whitsundays' second-largest island, Hook Island. Anchor at Nara Inlet and hike to the indigenous rock-art sites here and at Hill Inlet, with informative signage on the Indigenous Ngara people. Overnight here, then the next day head to the north of the island for the delightful beach at Butterfly Bay (yes, home to the fluttering beauties). Spend the day swimming, snorkelling and soaking up the sun.

From Nara Inlet anchorage, set sail southeast for a 2-3hr journey to Whitsunday Island's Cid Harbour.

❸ Whitsunday Island

The namesake of the Whitsunday Island group,

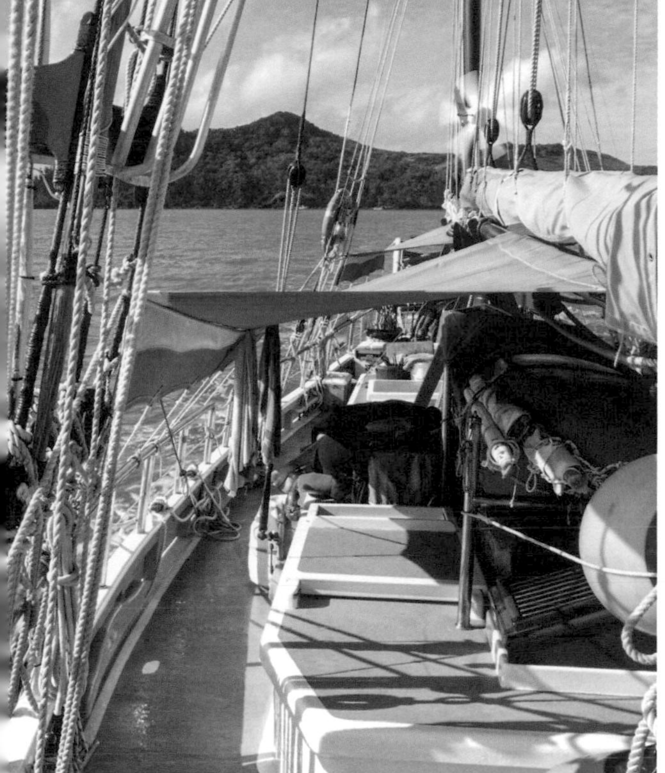

ABOVE: SAILING OFF WHITEHAVEN BEACH, WHITSUNDAY ISLAND

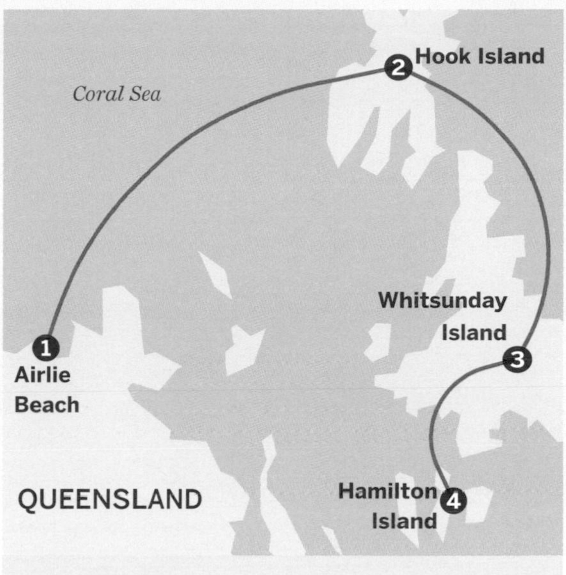

this is the largest of the chain and is home to the spectacular, 7km long, world famous, blindingly white-sand Whitehaven Beach. Cruise to Whitehaven, drop anchor and disembark for a challenging but rewarding trail up to Whitsunday Peak where the reward is dazzling views over this tropical idyll. Cool off with a swim, laze on the sand or get under the water for fantastic snorkelling.

Continue sailing southwest from Whitehaven Beach for around 2-3hr to get to Hamilton Island's marina.

4 Hamilton Island

Home to the largest resort in the Whitsundays, Hamilton Island is a great spot to restock any essentials and just chill out with day-use of the eponymous resort's facilities. Dine at one of the waterfront restaurants, strike out on the bushwalking trails or get out on the water in a kayak, windsurf or SUP. If that sounds too energetic, sunbathe on the sands at Casteye Beach or Escape Beach (the latter accessed by a two-hour hike).

The last stretch of the trip is back to the mainland to drop anchor for the last time and return the boat at Airlie Beach.

ALTERNATIVES

Extension: Langford Island

Located opposite the exclusive Hayman Island (only accessible to resort guests), tiny Langford makes for a great stop: it's reasonably protected from the winds and features a lovely long slice of sand-spit. It's suitable for beginner snorkellers as you can snorkel straight off the beach and the waters are teeming with fish and other marine life.

Sail from Airlie Beach to moor at Langford Island before heading to Hook Island.

Diversion: Chalkies Beach

Across the water from Whitehaven Beach, on the western side of Haslewood Island, Chalkies Beach boasts the same soft-white silica sands of Whitehaven – but, being off the beaten track, you might just have it all to yourself. Good snorkelling is on offer at mid- to low tide.

You can moor your boat at Chalkies Beach.

SUN, SURF AND SUBTROPICAL HINTERLAND ON THE GOLD COAST

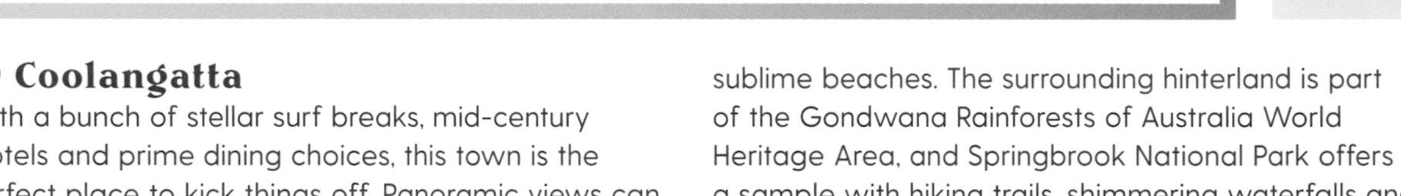

Coolangatta – Lamington National Park
Glam it up in Australia's Gold Coast beach towns, wining and dining and taking to the water before recuperating in the rainforest.

FACT BOX
Carbon (kg per person): 23
Distance (km): 187
Nights: 7
Budget: $$
When: year-round

❶ Coolangatta
With a bunch of stellar surf breaks, mid-century motels and prime dining choices, this town is the perfect place to kick things off. Panoramic views can be had at Point Danger; surfers can take to the waves at Greenmount, Kirra or experts-only Snapper Rocks; and a sundowner on the deck of the Rainbow Bay Surf Life Saving Club, overlooking a curve of glorious shoreline, is a must.
Drive 13km north along the coast.

❷ Burleigh Heads
If you're looking for one of the Gold Coast's best dining scenes and a taste of nature before hitting the hinterland, 'The Heads' is the place. The pine-tree-backed beach is lovely, kids enjoy spotting native animals at the David Fleay Wildlife Park, and the Jellurgal Aboriginal Cultural Centre runs tours to learn about the original inhabitants of the land, the Yugambeh people.
Head inland and southwest for around 37km.

❸ Springbrook National Park
Despite its reputation for tacky glitz and fame as a partying paradise for school-leavers, the Gold Coast is also a place of pristine beauty, and not just in its sublime beaches. The surrounding hinterland is part of the Gondwana Rainforests of Australia World Heritage Area, and Springbrook National Park offers a sample with hiking trails, shimmering waterfalls and eye-popping lookouts.
Drive 40km northeast, back towards the coast.

ABOVE: SURFING BURLEIGH HEADS

ABOVE: SURFERS PARADISE

❹ Surfers Paradise

The butt of many a Gold Coast joke, Surfers has a deserved reputation as Australia's 'Sin City'. Despite that, there's no denying that the wide, pearly stretch of sand and glittering Pacific is simply stunning, and there are enough good coffee spots and boutique hotels in amongst the mainstream to keep everyone happy. Family-friendly activities are plentiful and even the skyscrapers look pretty in the evening glow from the observation deck at Q1 – once the world's tallest residential building at 322m.

Drive 46km southwest via Nerang to the Binna Burra section of Lamington National Park.

❺ Lamington National Park

The landscape of Lamington National Park is known to the Yugambeh people as Woonoongoora, meaning 'quiet and timeless', and that's exactly how it feels – and exactly what you need after Surfers – when you're deep in its rainforest. The park is split into two accessible sections, Binna Burra to the east and Green Mountains (O'Reilly's) to the west; both parts are a haven for birdwatchers and hikers.

Return to Surfers Paradise for onward connections.

ALTERNATIVES

Extension: Theme Parks

The Gold Coast's theme parks are a major draw. Bare those white knuckles on Dreamworld's rollercoasters and rides; splash around on a hot summer day at Wet 'n' Wild waterpark; or try to hold down your lunch on the Batwing Spaceshot at Warner Bros Movie World.

The theme parks are 20-25km northwest of Surfers Paradise.

Diversion: South Stradbroke Island

If you fancy a bit more nature, cruise over to car-free, 21km-long South Stradbroke and stay at Couran Cove, the island's family-friendly, dated sole resort, or opt for rustic camping. The beach is Robinson-Crusoe-esque (though a bit, er, sharky) and the native wildlife ranges from snakes to wallabies.

Ferries to Couran Cove depart from Hope Harbour Marina, around 30km north of Surfers Paradise.

WINING AND DINING IN ADELAIDE AND AROUND

Adelaide – McLaren Vale

Sample Adelaide's food and culture, then head for the hills to drive through beautiful countryside and taste some of Australia's best wine (designated driver required).

FACT BOX

Carbon (kg per person): 27
Distance (km): 215
Nights: 5-7
Budget: $$
When: year-round

❶ Adelaide

South Australia's capital is a big city with country charm, and its breezy, relaxed attitude flows from its church-cluttered centre to its sandy beaches – cooling off in the sea on one of the latter is a welcome respite on Adelaide's famously hot summer days. Dubbed the 'Festival Capital of Australia', the city's calendar is jammed with high-calibre events complementing a thriving, year-round arts and food scene. Amble around the aromatic Central Market (one of the country's best food markets), check out big-name Australian art at the grand Art Gallery of South Australia, jump on a tram to the beach at Glenelg and brush up on your wine knowledge at the National Wine Centre before heading out to the wine regions.

ABOVE: CENTRAL MARKET, ADELAIDE

Drive roughly 70km northeast of Adelaide to the Barossa Valley.

❷ Barossa Valley

The world-famous wine region of the Barossa Valley is a pilgrimage spot for oenophiles, with over 170 wineries run by long-established families, some into their sixth generation, to choose between. The Barossa is best known for its Shiraz and Riesling and you can sample them at any number of cellar doors, from the legendary Penfolds and well-respected Peter Lehmann to the delightful Rockford Wines and the bluestone beauty of a building at Seppeltsfield.

ABOVE: MCLAREN VALE VINEYARDS

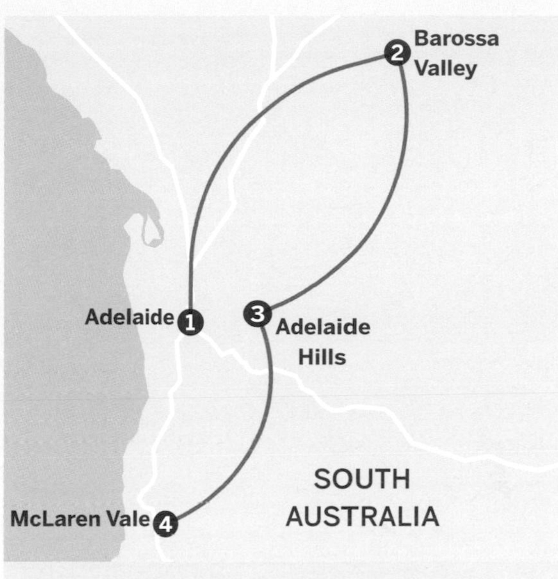

Take the scenic, 60ish-km route to the Adelaide Hills via Mt Pleasant, Birdwood and Woodside.

3 Adelaide Hills

The hills are alive with the sound of more wine imbibing, along with top-notch local produce, European flair and brilliant landscapes. The Adelaide Hills are perfect for bouncing from town to town – kitschy Hahndorf, bustling Mt Barker and the pretty village of Stirling – stopping along the way at craft-beer breweries and cool-climate wineries, and at the Mt Loft Summit for spectacular views.

Drive 45km southwest.

4 McLaren Vale

Tear your eyes away from the bucolic setting of the McLaren Vale wine region to brief yourself on the winery map (available from the visitor information centre), before embarking on a tour of as many as you can fit in. Swish mouthfuls of Shiraz at charming cellar doors, linger over piled-high food platters and dine at fantastic degustation restaurants before reluctantly ending your trip.

From McLaren Vale it's a 40km drive north back to Adelaide.

ALTERNATIVES

Extension: Clare Valley

Once you've sampled what the Barossa has to offer, continue north to another of South Australia's premier wine destinations, Clare Valley. The stars of the show are the Riesling and the wonderful countryside vistas – rent a bike and hit the Riesling Trail to get a taste of both.

Clare Valley is roughly 100km north of the Barossa.

Diversion: Willunga

Quaint Willunga village is a great alternative to McLaren Vale winery-hopping: visit on a Saturday to buy organic produce at the Willunga Farmers Market, then head down High St, lined with heritage buildings housing eating options, pubs and craft-beer breweries. For a memorable meal with Fleurieu Peninsula views, head west to Port Willunga's Star of Greece restaurant.

Follow Main Rd from McLaren Vale south for about 7km to Willunga.

QUIRKY NEW ZEALAND FROM ŌAMARU TO DUNEDIN

Ōamaru – Dunedin

From steampunk-crazy Ōamaru to street-art-splattered Dunedin, this coastal drive takes in one of NZ's best restaurants and a scattering of marbles from a giant's game.

FACT BOX

Carbon (kg per person): 14
Distance (km): 116
Nights: 2
Budget: $$$
When: Sep-Feb

1 Ōamaru

Ōamaru is off-beat small-town New Zealand at its very best – this place has made the most of its ramshackle Victorian warehouses by embracing steampunk. Wander the streets to meet artists, inventors, antiquarians and bohemians of all stripes, boldly celebrating the past and the future with an ethos of 'tomorrow as it used to be'. Plus there are penguins. Every night, up to 250 adorable blue penguins (the world's smallest penguin species) waddle ashore to their nests in town. You can watch the parade from special stands, and even peer inside their nesting boxes.

Drive south for 40km.

2 Moeraki

The name Moeraki means 'a place to sleep by day', which should give you some clue as to the pace of life here. Where this tiny fishing village does spring into action, though, is with top quality food – it's home of one of the South Island's best and most popular seafood restaurants, Fleur's Place. There's an appealingly ramshackle look about the tin-and-timber hut, while inside the menu includes exquisitely fresh shellfish, tender blue cod and other recently landed ocean bounty; book ahead. Afterwards take

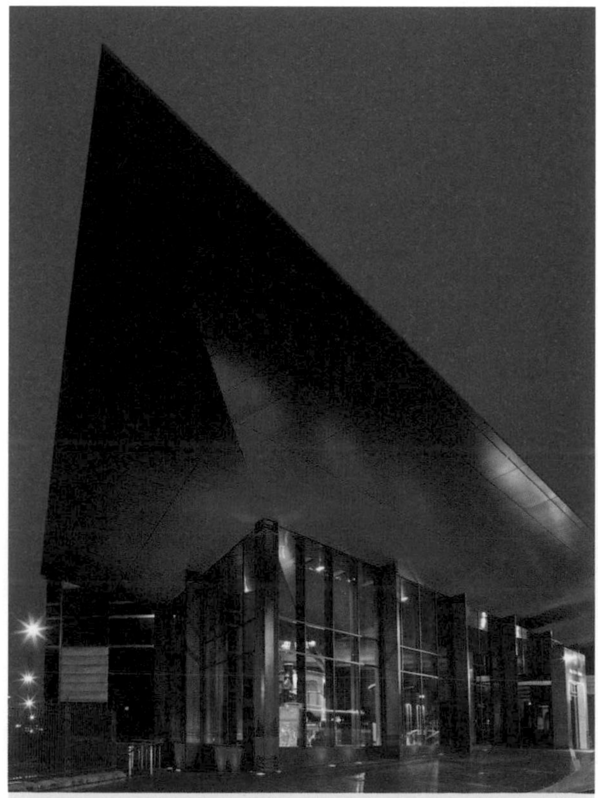

ABOVE: TOITŪ OTAGO SETTLERS MUSEUM, DUNEDIN

ABOVE: MOERAKI BOULDERS

a stroll on the beach to see the strange Moeraki Boulders, looking for all the world like a giant's game of seashore marbles.

Head south for 76km.

3 Dunedin

This southern city is full of hidden surprises. A Unesco City of Literature and home to the acclaimed Toitū Otago Settlers Museum, it's also firmly on the international street-art map – artists from around the world compete for the best wall space in town and there are works by a United Nations of artists. Many of them are drawn to the city's blossoming warehouse precinct, which is home to funky burger joints and vegan cafes. You'll find it to the south of The Octagon. Dunedin also boasts a vibrant live music scene, supported by the students of New Zealand's oldest university. And the city has the world's steepest street, Baldwin St, with a 34.8% gradient. In 2019, the Guinness World Record title was temporarily lost to Ffordd Pen Llech, in Harlech, Wales, but reinstated in 2020, after Dunedin surveyor Toby Stoff made a formal appeal.

Dunedin's airport has flights to major cities across New Zealand.

ALTERNATIVES

Extension: Otago Peninsula

This picturesque idyll of rolling hills, secluded bays and clifftop vistas harbours abundant wildlife: penguins, fur seals, sea lions and, most impressivly, albatrosses. They're on the peninsula year-round, but nest here from December to March, with parents taking turns to guard the young and hunt. See them via a Royal Albatross Centre tour.

The tip of the peninsula is a scenic 30km drive from Dunedin.

Diversion: Riverstone Kitchen

Award-winning chef Bevan Smith's Riverstone Kitchen combines oddball extravagance with sophisticated dining. Much of the produce comes from the on-site gardens (you can go for a stroll around them). And what's that looming over the complex? A moated castle of course, the brainchild of Bevan's mum Dot Smith, the self-described queen of the castle.

Riverstone Kitchen is 20km north of Ōamaru.

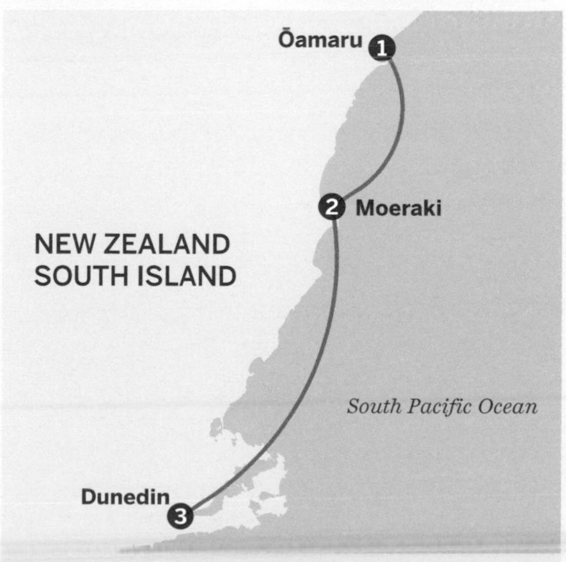

CANOE THE WHANGANUI RIVER IN NEW ZEALAND

Tongariro National Park – Pipiriki

Paddle the Great Walk that's not a walk along the Whanganui, the North Island river that's a legal living entity, passing through lush landscapes and Māori settlements.

FACT BOX

Carbon (kg per person): 8
Distance (km): 188
Nights: 5-6
Budget: €€
When: Oct-Apr

❶ Tongariro National Park

Tongariro is New Zealand's oldest national park and the North Island's volcanic hotspot, dominated by the Ruapehu, Ngauruhoe and Tongariro volcanoes. It's among the country's most popular national parks too, known for winter skiing and the Tongariro Alpine Crossing, often dubbed the 'world's best one-day trek'. Gaze at the Emerald and Blue lakes from trails beneath the grumbling cones, or explore by bike, stand-up paddleboard or kayak.

Take a bus from the national park to Taumarunui (daily; 35min).

ABOVE: THE WHANGANUI RIVER NEAR TAUMARUNUI

❷ Taumarunui

Settle into your canoe for the first leg of the Whanganui Journey at Ngahuinga (Cherry Grove) in Taumarunui, drifting past farmland then steep valleys forested with rata and rimu trees, and tackling a couple of rapids to keep things fresh.

Canoeing the 22km first leg to Ohinapane takes 3-5hr.

❸ Ohinapane

Vegetation becomes denser as you continue downstream, keeping an eye – and ear – out for *tuna* (eels), *kererū* (wood pigeon) and the noisy, bright *tui* bird. A short side-trip up the Ohura tributary rewards with views of the namesake falls.

The 35km paddle to Whakahoro takes 6-8hr.

❹ Whakahoro

As you enter Whanganui National Park proper, the valley becomes a sheer-sided gorge pocked with pole-holes made by Māori boatmen. The implacable flow carries you around the Kirikiriroa Peninsula,

ABOVE: EMERALD AND BLUE LAKES, TONGARIRO NATIONAL PARK

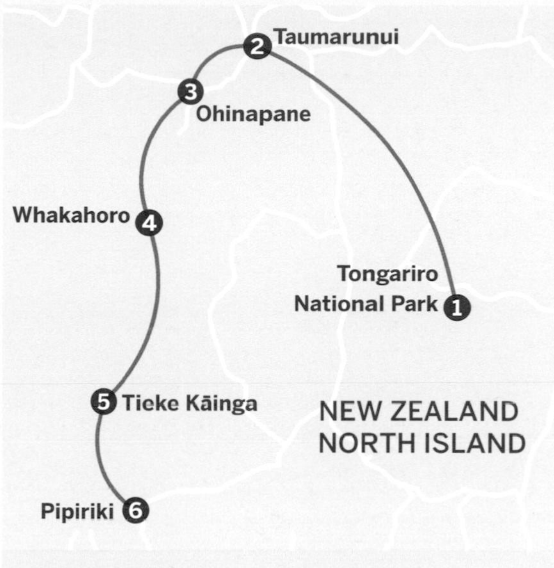

through Tarepokiore (whirlpool) rapids and past moss-dripping Tamatea's Cave to John Coull Hut and campsite, its grassy banks star-dusted with glow-worms. Spend the night before continuing on to Tieke Kāinga.

It's 37.5km to John Coull (7-9hr), then 29km (7-9hr) to Tieke Kāinga.

5 Tieke Kāinga

Highlight of this remote, verdant stretch comes at its endpoint – Tieke Kāinga, a living Māori community with a hut and campsite alongside a traditional *marae* (meeting ground). As well as admiring the intricately carved *pou whenua* (carved wood pole), you'll likely be greeted with a *pōwhiri* (welcoming ceremony).

Paddle 21.5km (4-6hr) to Pipiriki.

6 Pipiriki

Swoop past a series of caves and eel-traps, the confluence with the Manganui o te Ao River – look out for *whio* (blue duck) – and through Ngaporo and Autapu rapids to reach the haul-out endpoint at Pipiriki.

If not on an organised tour, pre-book transfers to Ohakune, Whanganui or other transport hubs.

ALTERNATIVES

Extension: Bridge to Nowhere

From Mangapurua Landing, between John Coull and Tieke Kāinga, a trail winds to one of the most atmospheric sites in Whanganui National Park – the Bridge to Nowhere. Built during an abortive post-WWI settlement project and surrounded by lush forest, it lies on the 35.5km Mangapurua Track, a great one-day cycle or two-day hike.

Several operators offer jetboat tours to Mangapurua Landing from Pipiriki.

Diversion: Mail run to Whanganui

The Whanganui River snakes south to meet the coast at its namesake town, once a major hub for river traffic, its streets lined with elegant Victorian and Edwardian architecture. The best way to reach this arty burg is via the mail bus that serves rural communities, meeting locals and enjoying views across the Whanganui Valley as you go.

Book the 2hr mail run from Pipiriki to Whanganui in advance.

ROAD-TRIP THROUGH NEW ZEALAND'S NORTH ISLAND

Auckland – Wellington

Drive through the steamy centre of the North Island, from New Zealand's biggest city to its buzzing capital, with volcanoes, beaches and museums along the way.

FACT BOX

Carbon (kg per person): 87
Distance (km): 700
Nights: 5-7
Budget: $$
When: year-round

❶ Auckland

Volcanoes loom large on this itinerary, and the multicultural city of Auckland, NZ's largest, kicks it off with a bang, being built on 50 of them. There are also two harbours, beaches, islands, wineries, rainforest, an excellent museum and art gallery and a sophisticated dining scene to enjoy before departing on your road trip.

Pick up your rental car and drive south of Auckland for 230km.

❷ Rotorua

Sulphur-scented Rotorua sits at the heart of the country's most active geothermal zone. Steam rises from drains, mud boils by the lakeside and geysers shoot into the air beside historic Māori villages – Māori culture is highly visible here and there are plenty of opportunities to engage with it. And everybody enjoys the chance to soak in the naturally heated lakeside hot pools at the end of a long drive.

Back on the road head 80km south.

❸ Taupō

The vast lake at the centre of the North Island is actually a hugely powerful volcano. Not that you'd know it from the tranquil scene today – although, if you're in the mood for a relaxing dip, hot springs still conveniently bubble up at several locations. Like Rotorua, this little waterside town has developed a reputation as an adventure hub, offering everything from bungee jumping to ski diving.

ABOVE: BUNGEE JUMPING, TAUPŌ

ABOVE: CHAMPAGNE POOL, ROTORUA LAKES

Drive 320km south, skirting the lake and continuing through the dramatic landscape of the North Island Volcanic Plateau.

4 Kāpiti Coast

With good train connections, the stretch of coast north of Wellington has become part of the city's commuter belt. The long, uncrowded beaches are the big attraction and there are some cute cafes scattered around the main settlements: Ōtaki, Waikanae, Paraparaumu and Paekākāriki. If you've got time, walk the Paekākāriki Escarpment Track and take a boat to the Kāpiti Island nature reserve.

The final drive of the trip is 70km to Wellington.

5 Wellington

The nation's capital squeezes a lot into its compact centre, creating a buzz that's totally disproportionate to its size. Spend your time exploring its world-class museums – most notably Te Papa – as well as its galleries, bars, cafes and restaurants. If it's not blowing a gale, take a walk along the harbour or in the surrounding ranges.

Wellington has an international airport with global connections. If you need to head back to Auckland, the quickest way is to fly.

ALTERNATIVES

Extension: Mt Maunganui

The beautiful Bay of Plenty is one of NZ's summertime hotspots due to its glorious sandy beaches, and the lively little surf town of Mt Maunganui is a popular spot. Stretching along a narrow peninsula and terminating in a 232m extinct volcano, it offers good swimming, hot pools and cool bars.

Mt Maunganui is an easy detour off the Auckland to Rotorua leg of the trip, 210km from the former.

Diversion: Tongariro National Park

Dominating the landscape of this national park are three active volcanoes: Tongariro (1967m), Ngauruhoe (2287m) and Ruapehu (2797m). You may recognise Ngauruhoe from its starring role as Mt Doom in the *Lord of the Rings* movies; Ruapehu is the North Island's premier ski destination. Excellent hikes include the famous Tongariro Alpine Crossing.

The national park is 90km south of Taupō.

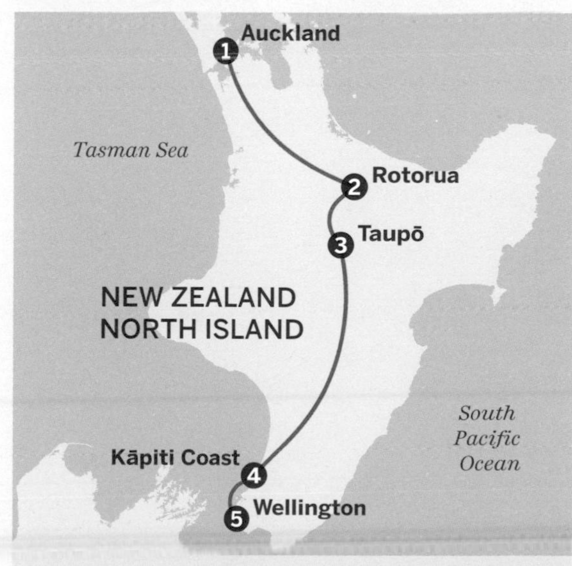

AUSTRALIA'S GREAT OCEAN ROAD

Torquay – The Twelve Apostles
Take one of Australia's most iconic road trips along the Great Ocean Road in southwest Victoria, enjoying surf beaches, wildlife and iconic views.

FACT BOX

Carbon (kg per person): 53
Distance (km): 429
Nights: 3
Budget: $$
When: year-round

❶ Torquay

As the official start of the Great Ocean Road and the spiritual home to Australian surfing, Torquay's the perfect spot to wax down those boards and start your surfin' safari. Whether you're a novice learning to stand or a pro looking for monster waves at nearby Bells Beach, here it's all about getting out on the water for a paddle. In between sets don't miss the town's surf museum and youthful vibe with hip coastal diners, microbreweries and coffee roasters.

Drive the Great Ocean Road (aka the B100) 20km southwest.

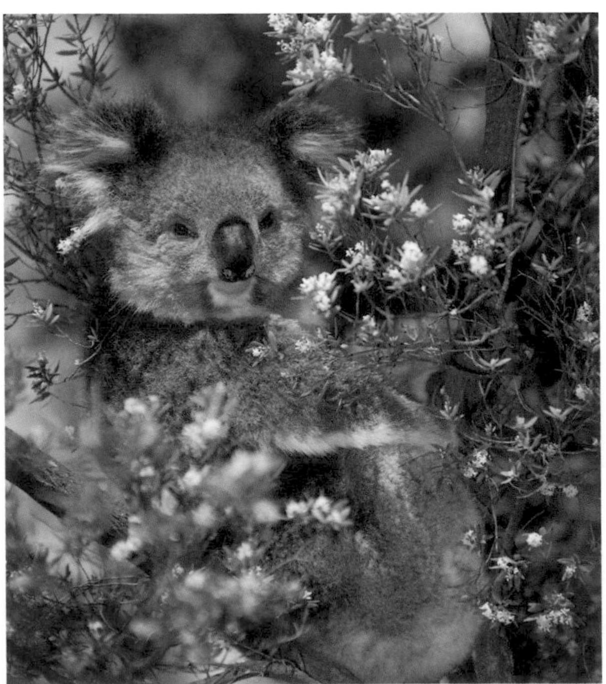

ABOVE: KOALA, CAPE OTWAY

❷ Angelsea

Get a taste for old-school Aussie-style summer holidays at this nostalgic coastal town known for its relaxed atmosphere and beautiful calm beaches. This is the place to eat fish and chips, get out on the river and walk along dramatic coastal cliffs and heathland in search of wild orchids. Animal lovers should make a beeline to the town's golf course to find a celebrated resident mob of kangaroos.

Back in the car continue 28km southwest.

❸ Lorne

A favourite of holidaymakers for generations, Lorne remains one of the Great Ocean Road's most popular seaside towns. Its endearing sweep of white sand and gorgeous backdrop of eucalyptus forest charm

ABOVE: THE TWELVE APOSTLES

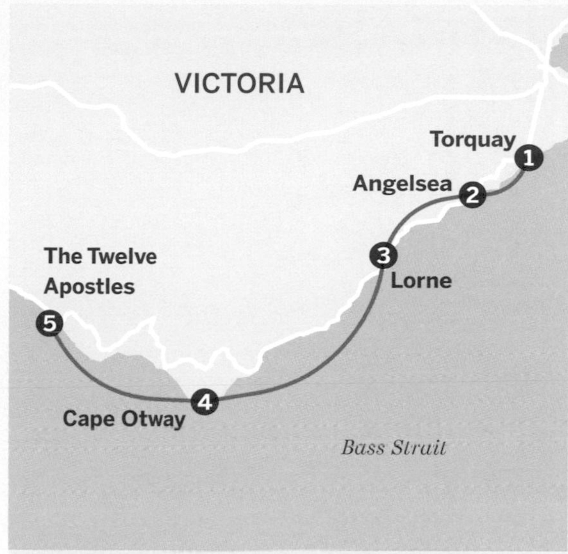

all who visit. Once you're done lazing by the beach, be sure to head inland to visit the area's stunning waterfalls.

Drive 78km southwest to Cape Otway.

4 Cape Otway

One of the more dramatic legs along the route is the section to Cape Otway, where you head inland and find yourself immersed in a shady canopy of dense rainforest. Look for koalas feeding in the trees as you head to one of Australia's oldest lighthouses (1848) at Cape Otway itself – scan for whales, read up on the region's shipwreck past and learn about the Indigenous Gadubanud people.

Continue on the Great Ocean Road 73km west.

5 The Twelve Apostles

Offering a fitting finale to this spectacular drive are this series of rock formations that stand majestically out in the churning ocean. Sunset is the best time to visit, offering a mesmerising sight as the limestone stacks are bathed in a soft golden glow among a sky of swirling pinks and reds.

Drive to Melbourne – either inland (230km) or back along the coast (270km) – for onward international connections.

ALTERNATIVES

Extension: Birregurra

Hidden away on the fringes of the lush Otways rainforest, tiny Birregurra has firmly stamped itself on the map as a food hotspot. Here you'll find Brae, one of Australia's finest restaurants (and regular fixture on the World's Top 100 list) where you can treat yourself to creative, contemporary cuisine showcasing indigenous-inspired dishes using locally sourced ingredients.

Birregurra is 40km inland from Lorne.

Diversion: Bellarine Peninsula

Take an eastern turn off the Great Ocean Road between Melbourne and Torquay to the Bellarine Peninsula, a string of superb beaches and coastal hamlets. After a swim at Ocean Grove, step back in time among the grand heritage buildings of Queenscliff before hitting the wineries, known for quality Pinot Noirs and Chardonnays.

Queenscliff is 100km south of Melbourne, 40km east of Torquay.

AUSTRALIA'S BREATHTAKING NSW NORTH COAST

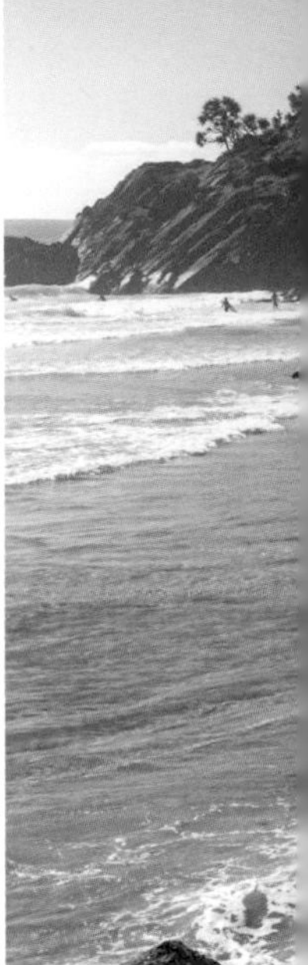

Byron Bay – Cabarita Beach

Experience brilliant beaches, bohemian vibes, forests and fine food on a driving foray from Byron Bay into the hinterland and up the north coast.

FACT BOX

Carbon (kg per person): 16
Distance (km): 131
Nights: 8-10
Budget: $$$
When: year-round

❶ Byron Bay

Byron Bay became a counter-culture haven in the '60s and '70s and the town's still clinging on to its boho credentials. But for each shoeless backpacker, surfer and ageing hippy, you've now got fat-walleted property developers, Instagram influencers pouting at smartphones and celebs and young families on a beach break. While some cry that paradise is lost, Byron's incredible sands, chilled attitude, natural rugged beauty and exciting dining scene are stellar draws that keep them all coming back. Start with breakfast at a hip cafe then head off on a hike up to the lighthouse for spectacular views. Lunch at any of the restaurants crammed into the town's centre before hitting the beach. The evening brings cocktails and epic sunsets followed by dinner and perhaps some live music at the local pub.

Drive 15km inland to Bangalow.

❷ Bangalow and Newrybar

If Byron's traffic-clogged streets are driving you mad, head into the nearby verdant hinterland for some respite, where the alternative community spirit continues in the delightful towns of Bangalow and Newrybar, 6km apart. Here, organic produce reigns supreme at excellent dining spots and community markets.

Continue 21km north.

❸ Mullumbimby

Mullumbimby is a laidback inland town with palm-lined streets. Drop by for a morning coffee made with locally roasted beans, stock up at the health

ABOVE: BYRON BAY'S HEADLAND

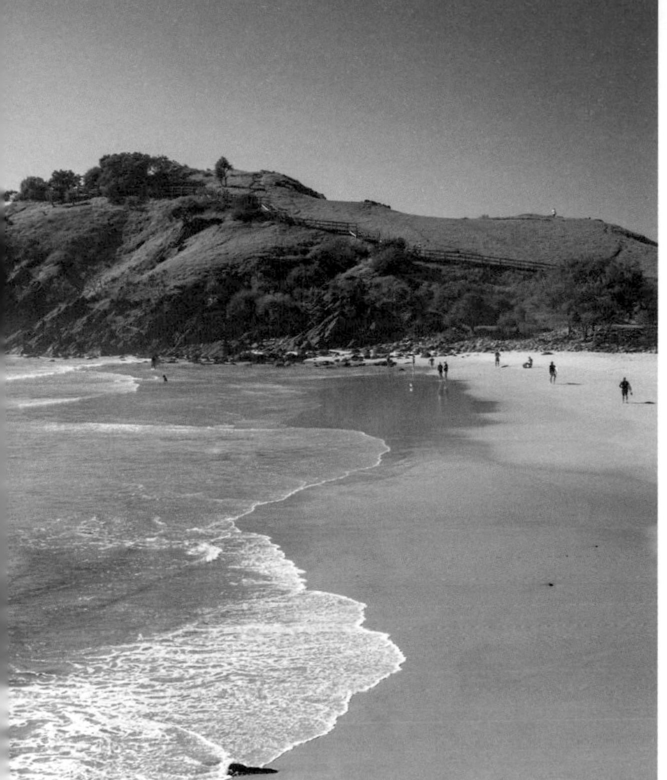
ABOVE: CABARITA BEACH

stores, relax with arvo beers at the pub, then ease into a balmy evening with dinner and drinks. Join the locals on Fridays at the excellent farmers market, set beneath shade-giving fig trees.

Drive 10km east back to the coast.

4 Brunswick Heads

The seaside town of Brunswick Heads offers a pared-down, more sedate alternative to Byron. You still get a long stretch of surf beach, alongside sensational eating options for a town of this size (including Fleet, a fine-dining star), a beloved 1940s pub with epic beer garden, and a fabulous mid-20th-century community arts hub, the Brunswick Picture House.

Head 35km up the coast.

5 Cabarita Beach

Nestled in between Byron Bay and over-the-Queensland-border Gold Coast, Cabarita Beach brings a bit of glam to the Aussie beach holiday. Stay at the Hamptons-esque Halcyon House boutique hotel and dine at its acclaimed restaurant, Paper Daisy. The bone-white sandy beach here was voted Australia's best in 2020 by Tourism Australia.

It's 50km south back to Byron Bay.

ALTERNATIVES

Extension: Nimbin

The original counter-culture capital of Australia is the inland town of Nimbin, an easy day trip from Byron. Remnants of its psychedelic '70s peace-and-love vibe still linger and the aroma of weed wafts in the air. Keep an open mind, chat to friendly alternative-minded locals, enjoy a meal at the pub and pick up some organic local honey or vegan ice-cream to take home.

Nimbin is 65km west of Byron Bay.

Diversion: Nightcap National Park

From Byron it's well worth making the journey to Nightcap National Park, traditional land of the Widjabul people and part of the Gondwana Rainforests World Heritage Area, for a spot of hiking, picnicking and admiring the magnificent waterfalls.

The park is 63km west of Byron Bay; day trips can also take in Nimbin, just 33km from Nightcaps.

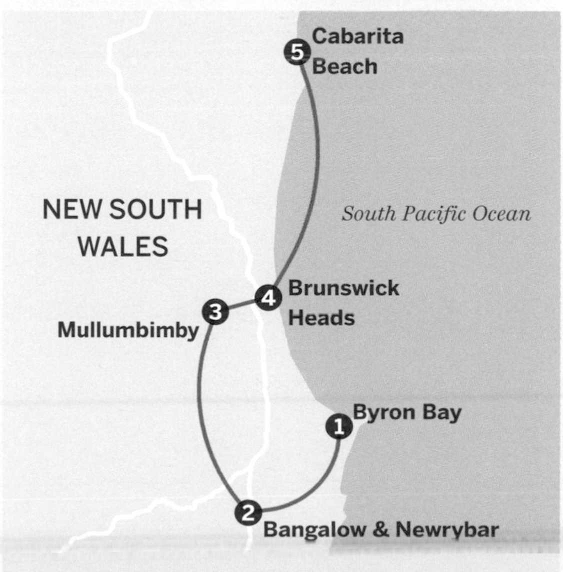

STRIKE THROUGH THE GOLDFIELDS OF VICTORIA

Ballarat – Kyneton

Take a road trip northwest of Melbourne to meander through Victoria's attractive countryside in search of Australia's gold-rush-era splendours.

FACT BOX

Carbon (kg per person): 34
Distance (km): 275
Nights: 5
Budget: $$
When: year-round

❶ Ballarat

Start your trip at the grandest and most famous of Victoria's goldfield towns, Ballarat, which was at the forefront of the 1850s gold rush. Here its legacy remains clear with a city centre comprising nothing but majestic heritage buildings from yesteryear – many converted into elegant hotels and restaurants.

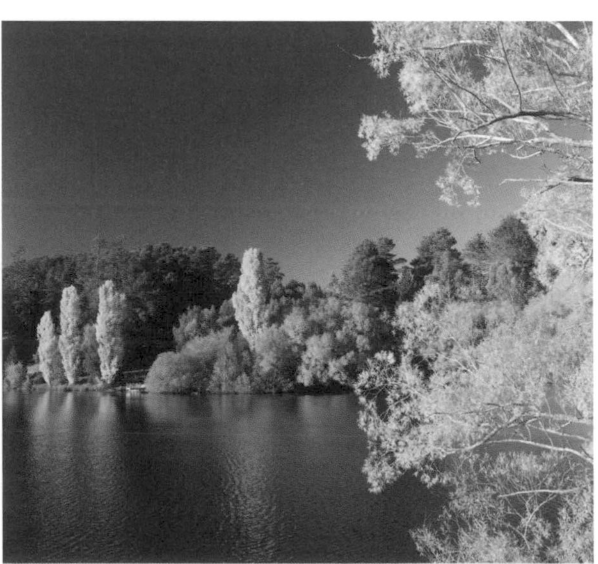

ABOVE: LAKE DAYLESFORD

As well as its blockbuster family-friendly attraction, Sovereign Hill (a re-creation of a gold rush town), Ballarat has a wonderful art gallery, a scenic lake, many historical sights and a burgeoning food and drink scene.

Drive 45km northwest to Daylesford.

❷ Daylesford

This attractive spa town is popular among Melburnians who come here to indulge in a weekend of wellness and relaxation. Daylesford is known for its curative natural spring waters: take a soak in a heated spa, sample its sparkling waters and sign up for endless options of massages and treatments. Enjoy its cafe culture and local wines by an open fire and be sure to explore the immediate region, home to picturesque small towns where you'll find cider houses, acclaimed chefs offering menus of local produce, and some fantastic bushwalks.

Continue 38km north.

❸ Castlemaine

Historical Castlemaine offers a fascinating interplay between blue-collar country charm and arty inner-city pretension. At its centre is a stunning theatre, recently restored to its original glory and featuring a

ABOVE: SOVEREIGN HILL, BALLARAT

calendar of quality live music, food and events. The regional gallery is another must-see, with works by acclaimed Australian artists.

Back in the car, head 38km north to Bendigo.

4 Bendigo

Rivalling Ballarat in the grandness of its architecture, Bendigo was also built on gold and is home to no shortage of ornate 19th-century buildings. It boasts a thriving arts scene, world-class art gallery and respected theatre in a converted, once-terrifying prison. Bendigo is a city packed with family-friendly attractions, and its dining has Melbourne food fans taking notice.

Drive 65km south.

5 Kyneton

Small-town Kyneton has led the way in regional Victoria's meteoric rise as a culinary destination, establishing itself with a quarter-mile of gastronomic delight centred along Piper St. Here you'll find a stretch of nothing but quality restaurants, gastropubs, cafes, bars and a gin distillery, all set up by talented proprietors who know their stuff.

It's an 89km drive to Melbourne and onward connections.

ALTERNATIVES

Extension: Hanging Rock

An ancient volcanic rock of haunting beauty and mystery, Hanging Rock is famed as the spooky setting for the Joan Lindsay novel *Picnic at Hanging Rock*. Set atop a bush-covered mountain, it's worth taking the moderate walk to its summit for sweeping views of the Macedon Ranges. Pack a picnic (and a bottle from the nearby winery) to enjoy on the lawns among grazing kangaroos.

Hanging Rock is 21km southeast of Kyneton.

Diversion: Heathcote Wine Region

Known for its Shiraz and other full-bodied reds, Heathcote has earned its well-deserved reputation as an epicurean escape with some 60 wineries to visit. Most have cellar doors to taste the goods, along with stops at breweries and bistros offering paddock-to-plate cuisine.

Heathcote's wine region is around a 45km drive southeast of Bendigo.

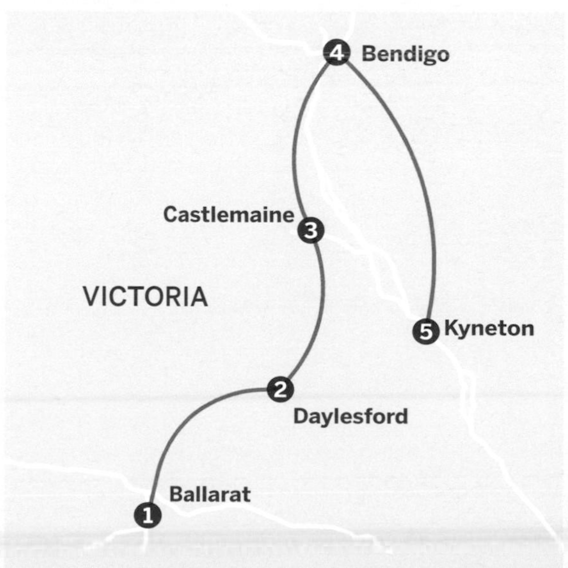

A FAMILY-FRIENDLY FIJI ADVENTURE

Denarau Island – Pacific Harbour

Fiji's biggest island, Viti Levu ticks all the family-friendly holiday boxes: sunshine, blue seas, adventurous activities, a welcoming population and kids' clubs galore.

FACT BOX

Carbon (kg per person): 5
Distance (km): 200
Nights: 10-12
Budget: $$
When: May-Oct

❶ Denarau Island

Only 6km from Nadi – Viti Levu's main transport hub and home to Fiji's international airport – diminutive Denarau heaves with child-friendly resorts, appealing shops and attractions such as Big Bula Waterpark. What it lacks in traditional Fijian authenticity, it more than makes up for with whizz-bang family fun: all Denarau resorts boast sparkling pools, live kid-centric entertainment and activity programmes, along with eating options designed to entice even the fussiest family member.

The *Yasawa Flyer* high-speed catamaran transports visitors to all island resorts in the Yasawas and many in the Mamanucas. It departs every morning from Denarau Marina; trip times vary from 30min to 5hr, depending on your destination.

❷ Mamanuca and Yasawa islands

The 50-plus islands making up these volcanic archipelagos are the stuff holiday dreams are made of: white sands; dazzling seas; genuine hospitality. Islands in the Mamanucas such as Treasure, Castaway and Malolo Lailai are famously family-oriented destinations, while in the further-flung Yasawas, resorts including Octopus, Paradise Cove and Blue Lagoon are renowned for their kids' clubs.

The *Yasawa Flyer* catamaran returns to Denarau Marina every afternoon. Back on the main island, take a bus to Sigatoka (several daily; 1hr 30min) from Nadi's bus station (the *Flyer* arrives at the marina at 5.45pm and the last Sigatoka-bound bus departs between 6.20pm and 6.50pm, so it's best to grab a taxi for the 13min drive to the bus station).

ABOVE: BULL SHARKS OFF PACIFIC HARBOUR

ABOVE: MAMANUCA ISLANDS

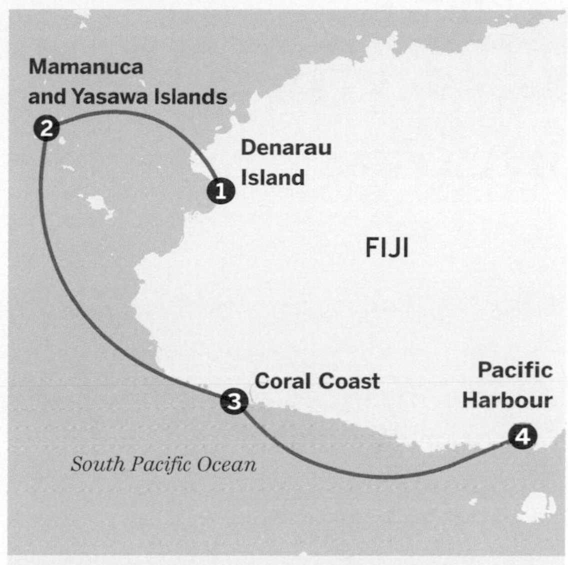

❸ Coral Coast

The Coral Coast – a stretch of shoreline that's one of the most scenic on Viti Levu – is named after the wide fringing reef that keeps swells here to a minimum. Sigatoka, the region's largest town, is a great base for reaching the area's many attractions and activities, such as sandboarding down the Sigatoka Sand Dunes, the Kula Wild Adventure Park, and the rustic Coral Coast Scenic Railway, plus traditional bamboo rafting, jet-boating and village visits.

Take the bus to Pacific Harbour (several daily; 2hr).

❹ Pacific Harbour

Though Pacific Harbour is billed as Fiji's 'Adventure Capital' (activities include diving with bull sharks and ziplining), it's also a great jumping-off point for several cultural and off-the-beaten-track experiences that the whole family will likely enjoy. Check out the firewalking shows and mock battles at the Arts Village, and don't miss a 4WD tour into the steamy, impossibly green Namosi Highlands.

Take a bus back to Nadi (frequent; 3hr 30min) for onward connections from the airport.

ALTERNATIVES

Extension: Sabeto Hot Springs

The mud pool at the foot of Mt Batilamu (known as the 'Sleeping Giant') makes for a deliciously dirty day trip. Kids will have a gloopy, grubby good time wallowing about in the reputedly therapeutic mud; parents will be grateful for the neighbouring sulphuric hot springs that wash it all off at the end.

The springs are about 30min by taxi from Denarau.

Diversion: Nananu-i-Ra

This small, hilly, low-key island off Viti Levu's northern tip offers watersports, tropical fish-spotting off the jetty, easy (but rewarding) hiking and back-to-basics accommodation.

From Nadi, catch a bus to Ba (frequent; 1hr 30min) then another to Ellington Wharf (frequent; 3hr). Arrange boat transfer (15min) with your island accommodation in advance. Return to the Coral Coast via Ba, then a Sigatoka bound bus from Nadi.

RAVISHING RED CENTRE LOOP

Alice Springs – Yulara

Soak up some of Australia's most breathtaking landscapes on a road trip through the nation's desert heart. A 4WD is ideal for this adventure.

FACT BOX

Carbon (kg per person): 141
Distance (km): 1129
Nights: 6
Budget: $$
When: May-Oct

❶ Alice Springs

Allow at least a day to explore the vibrant desert town of Alice Springs, which is home to some of Australia's top Aboriginal art galleries including the Araluen Arts Centre. 'Alice' also has excellent mountain-biking trails near historic Telegraph Station, the birthplace of the township, where you can rent bikes and gear; look out for dingoes and rock wallabies.

Drive 132km west to Ormiston Gorge Campground.

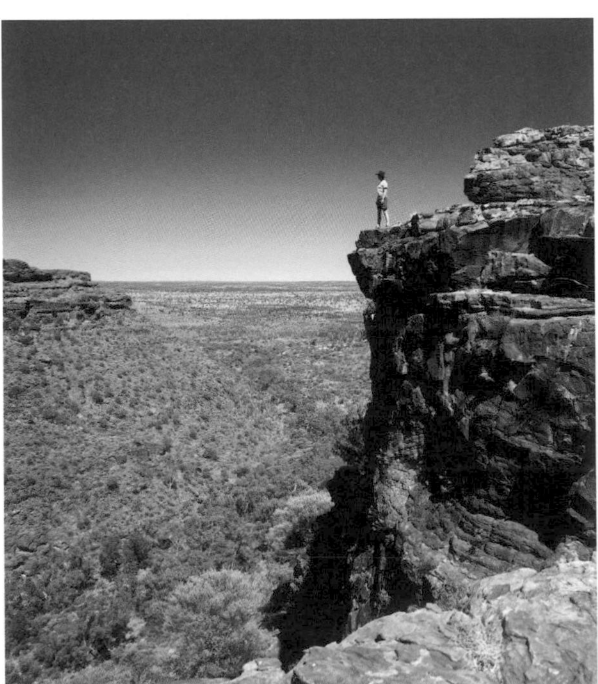

ABOVE: KING'S CANYON, WATARRKA NATIONAL PARK

❷ Tjoritja (West MacDonnell Ranges)

Tjoritja West MacDonnell National Park stretches for 161km west of Alice. Along its length are spectacular natural formations, with unmissable sights including Simpsons Gap, Ochre Pits, Standley Chasm and the permanent waterholes at Ellery Creek and Ormiston Gorge, perfect for a refreshing dip. Camping in the national park is currently your only accommodation option.

Continue 245km southwest to Kings Canyon via the Mereenie Loop (part of the Loop is unsealed so 2WD road-trippers need to double back to Alice and drive to Kings Canyon via the Stuart and Lasseter hwys).

❸ Kings Canyon

If travelling by 4WD, check out the huge Tnorala (Gosse Bluff) crater en route to Petermann, your base

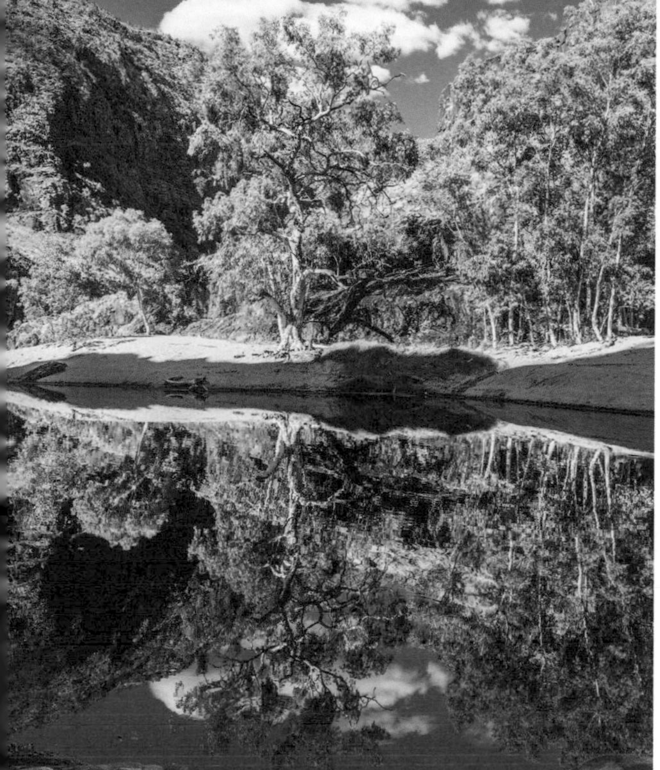

ABOVE: ORMISTON GORGE, TJORITJA WEST MACDONNELL NP

for exploring the rugged Watarrka National Park and its geological showpiece, Kings Canyon. Book a glamping tent at Kings Canyon Resort, and wake early the next day to tackle the 6km Kings Canyon Rim Walk (allow three to four hours) before heading to Yulara, Uluru-Kata Tjuta National Park's tourism hub.
Continue 305km southwest from Kings Canyon to Yulara.

4 Yulara

Drop off luggage in your accommodation and head straight out to one of the viewing platforms to witness mighty Uluru change colour at sunset (watch for kangaroos on the road when driving at night). Allow at least two days to explore Uluru-Kata Tjuta National Park, starting with the free ranger-guided Mala Walk conducted at the base of the park's sacred sandstone monolith. Make sure to hike into the 'many heads' of Kata Tjuta, the rock formations to Uluru's west.
To head back to Alice (a 447km drive from Yulara) and return the rental car, take the Lasseter and Stuart hwys, being sure to stop at the Mt Conner Lookout (to see the rock known as 'Fool-uru') and Outback roadhouses such as Curtin Springs Station and Erldunda Roadhouse en route.

ALTERNATIVES

Extension: Hermannsburg

Instead of heading straight from the Tnorala crater to Kings Canyon, detour east on Larapinta Dr to Hermannsburg, where you can explore the historic buildings of the former Lutheran mission and see works by local Aboriginal artists. Another 50km southeast, take a rock-art tour and buy Aboriginal pottery at the mission's former outpost, Wallace Rockhole.
Hermannsburg is 62km southeast of the Tnorala (Gosse Bluff) crater.

Diversion: Arltunga Historical Reserve

A worthwhile day trip from Alice is the Arltunga Historical Reserve. Born out of a gold rush in 1887, Arltunga was officially central Australia's first town and once supported a population of 300. Learn more at the visitor centre before taking a walk through the reserve to see relics of its past, from mines to former government buildings.
The reserve is 114km northeast of Alice Springs.

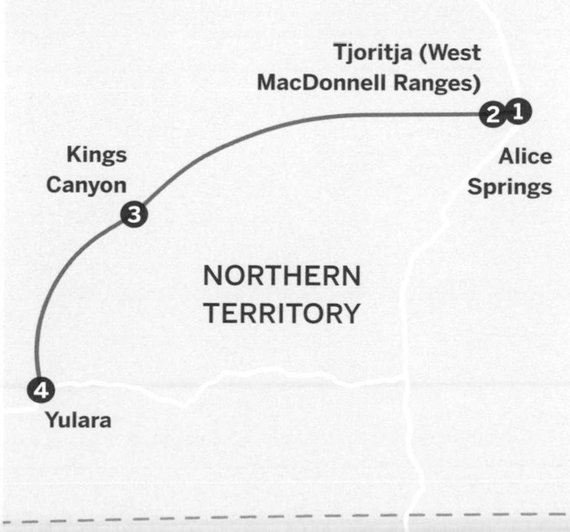

Trip Builder

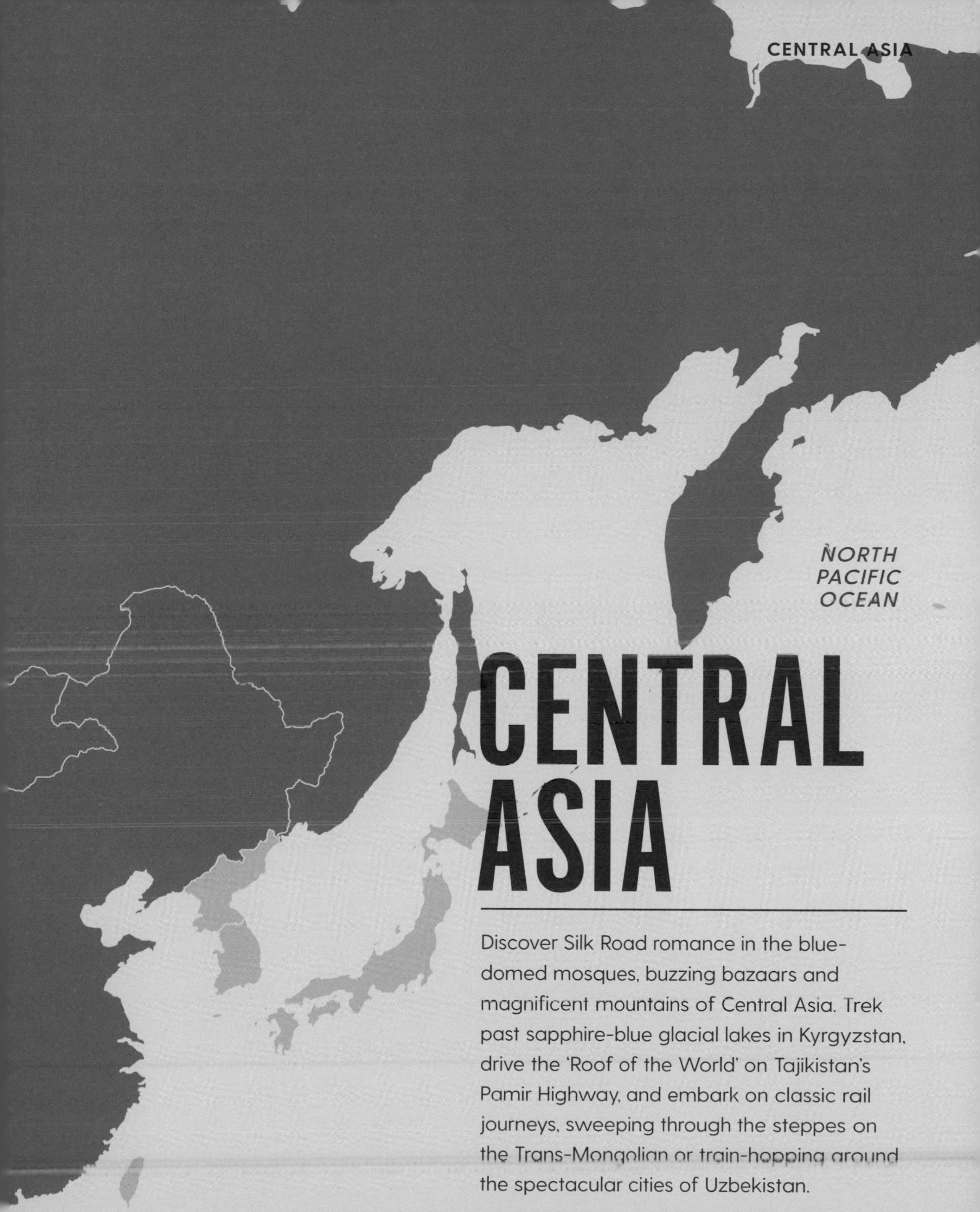

NORTH
PACIFIC
OCEAN

CENTRAL ASIA

Discover Silk Road romance in the blue-domed mosques, buzzing bazaars and magnificent mountains of Central Asia. Trek past sapphire-blue glacial lakes in Kyrgyzstan, drive the 'Roof of the World' on Tajikistan's Pamir Highway, and embark on classic rail journeys, sweeping through the steppes on the Trans-Mongolian or train-hopping around the spectacular cities of Uzbekistan.

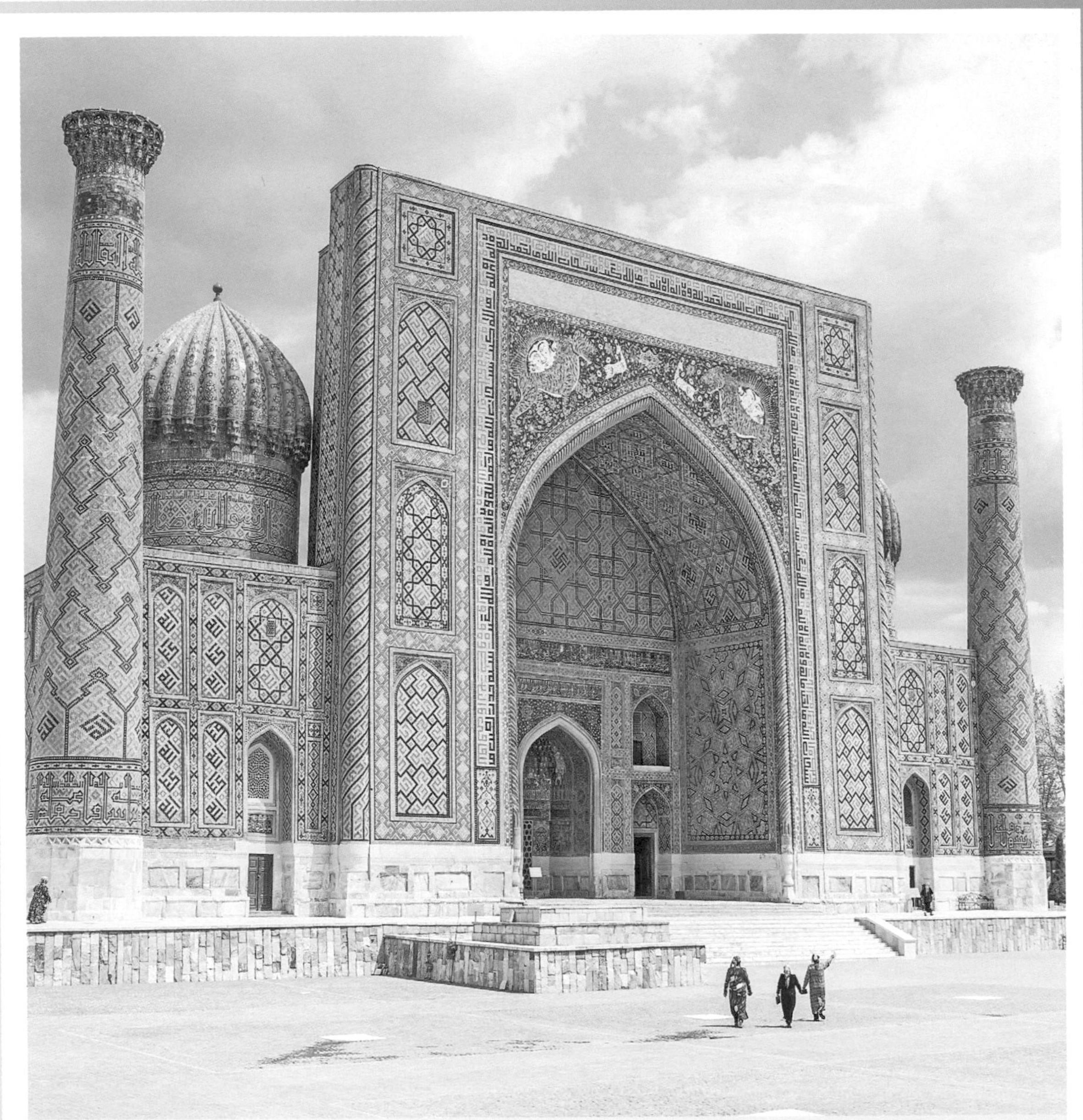

ABOVE: SAMARKAND'S REGISTAN, UZBEKISTAN

CONTENTS

THE PAMIR HIGHWAY, TAJIKISTAN

Dushanbe – Murgab

Follow Central Asia's most scenic mountain road on a trip through remote Badakhshan to the Pamir Plateau and on to Osh in Kyrgyzstan.

FACT BOX

Carbon (kg per person): 214
Distance (km): 1217
Nights: 8
Budget: $$
When: Jun-Oct

❶ Dushanbe

Tajikistan's capital is Central Asia's most scenic. Admire the pastel-coloured neoclassical facades of Rudaki St, stare at the world's largest teahouse and see Central Asia's biggest Buddha statue at the National Antiquities Museum, before recovering from the midday heat over a pot of tea and kebabs at the Persian-style Chaykhona Rokhat.

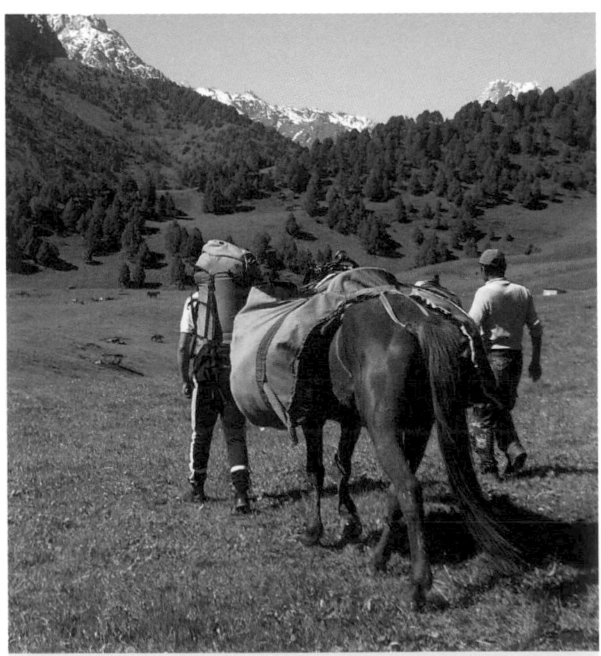

ABOVE: PAMIR MOUNTAINS TREKKING NEAR OSH

The daily morning flight to Khorog is weather-dependent so be prepared for an extra day in Dushanbe.

❷ Khorog

Having reached Khorog after a spectacular (if somewhat unnerving) flight, you are now firmly in the heart of the dramatic, rugged and beautiful western Pamir Mountains. The streets of Badakhshan's only town are framed by caramel-coloured mountains and tall poplar trees. Spend the day chatting with Khorog's Ismaili Muslim Wakhi residents, stocking up on supplies in the bazaar (you won't find any real shops until Osh) and confirming your onward transport for the next day.

From here to Osh you'll need to hire a vehicle and a driver in advance through a travel agency. The first drive is 240km southeast to Langar in the Wakhan Valley.

❸ Wakhan Valley

The Wakhan is one of the most beautiful and little-

ABOVE: YAMCHUN FORT, WAKHAN VALLEY

known corners of Asia, so take your time here. Wakhi farmers harvest wheat by hand at the side of the road, with Afghanistan and the towering snow-capped peaks of the Hindu Kush looming a stone's throw to the south. The road meanders past Silk Road forts, relaxing hot springs and even the remains of an ancient Buddhist stupa, all the while following Marco Polo's route towards China. Homestay hospitality and hundreds of petroglyphs make Langar an obvious place to overnight.

Continue 230km to Murgab.

4 Murgab

From Langar, the Pamir Hwy climbs quickly onto the Pamir Plateau, a rolling, barren and treeless landscape. It sits above 4000m – hence its nickname 'Roof of the World' – and is dotted with only the very occasional tomb or hardy yurt camp. Arrange to overnight in a Kyrgyz yurt just off-route or at a simple guesthouse in the end-of-the-world settlement of Murgab. North of here, over the 4655m Ak-Baital Pass, you'll pass the mesmerisingly blue high-altitude lake of Kara-Kul, before entering Kyrgyzstan over the Kyzyl-Art pass.

Drive 410km to reach Osh and its onward connections.

ALTERNATIVES

Extension: The Alay Valley

Before crossing the final pass into Fergana Valley to end the trip at Osh, detour west into the wide, high and remote Alay Valley, sandwiched between the snowbound Pamir-Alay and Trans-Alay mountain ranges. Homestays in Sary-Mogul village or yurt-stays at the foot of hulking 7134m Lenin Peak allow memorable hiking to climbers' base camps and high alpine lakes.

Make this detour with your hired vehicle and driver.

Diversion: Osh

After the rigours of your overland trip, it's worth resting a couple of days in the Silk Road city of Osh in Kyrgyzstan. Riotous bazaars, outdoor beer gardens, a Soviet-era Lenin statue and a medieval Islamic pilgrimage site overlooking the city will keep you busy. You can even visit the world's only three-storey yurt.

Osh's airport has international connections.

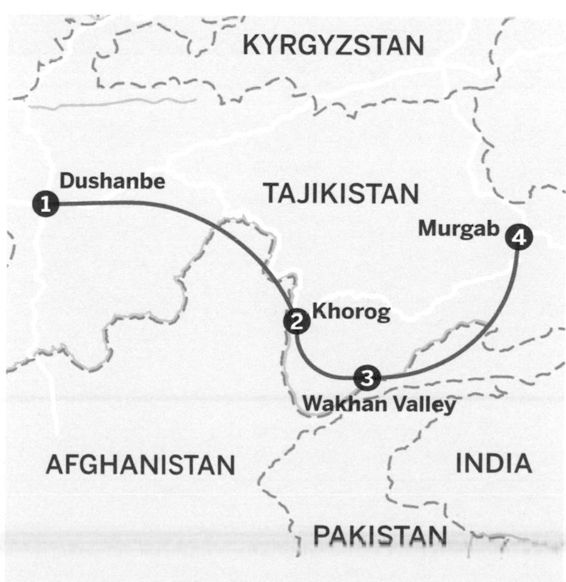

UZBEKISTAN'S STORIED SILK ROAD

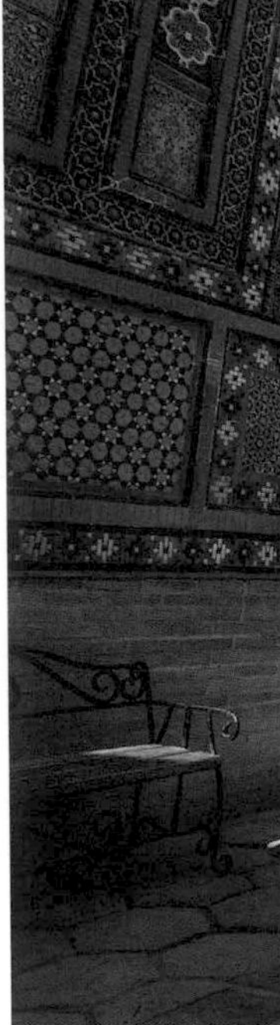

Tashkent – Khiva

Travel the Silk Road by rail, overnighting at Central Asia's most impressive and historic caravan stops along the way to discover the region's fascinating past.

FACT BOX

Carbon (kg per person): 40
Distance (km): 2025
Nights: 12
Budget: $$
When: Apr–Jun, Sep & Oct

❶ Tashkent

Uzbekistan's Soviet-built capital is worth a day or two for its museums on history, fine arts and applied arts, the country's three best. Ride the spectacular Soviet-era metro (photos only allowed since 2018) and then admire the world's oldest Quran in the Khast Imam complex. Tashkent has easily the best bars and restaurants in the country, so make the most of them while here.

Take a high-speed train to Samarkand (3-4 daily; 2hr 15min).

❷ Samarkand

Samarkand is above all the city of Timur (Tamerlane), the 14th-century Central Asia empire builder. The city's focus is very much on his spectacular and unsurpassed Timurid-era buildings: the intricate tilework of the Shah-i Zinda street of tombs; the tomb of Timur himself at the Gur-e-Amir; the epic if overambitious Bibi-Khanym Mosque, which began to crumble during its construction, partially collapsed after an 1897 earthquake and was rebuilt from the 1970s onwards. Even more spectacular is the Registan, a sublime ensemble of public square framed by three beautifully decorated medieval medressas. If time allows, fit in a visit to Ulugh Beg's 15th-century observatory and the city's traditional paper-making workshop.

Continue from Samarkand on the fast train to Bukhara (2 daily; 90min).

ABOVE: KHIVA MARKET

ABOVE: SAMARKAND

❸ Bukhara

The medieval emirate of Bukhara is Central Asia's most interesting historic town, so factor in some extra time here. The massive Ark citadel reveals the opulent life of the Bukharan emirs, and you can track down the ghosts of the Great Game (19th-century British-Russian rivalry in Central Asia) in its nearby 'Bug Pit' prison. The city boasts dozens of dramatic Islamic monuments, notably the Kalon Minaret and Ismail Samani Mausoleum, and its backstreets are the perfect place for a wander before a well-earned pot of tea at the Lyabi-Hauz pond, Bukhara's cultural heart.

Take the train to Khiva (3 weekly; 5hr).

❹ Khiva

A full day is just enough to visit Khiva's cluster of architectural gems, most of which are enclosed within the medieval city wall. The khan's two residences (the Kukhna Ark and Tosh-Hovli Palace) are fun to explore, and there are mosques, medressas and sufi shrines at every turn, most featuring exquisite tilework and small museums. At sunset head to the top of the Islom Hoja Minaret for remarkable views over the mud-brick town.

An overnight train trundles back to Tashkent (16hr) and its international airport.

ALTERNATIVES

Extension: The Desert Citadels

The best day trip from Khiva is a loop around Karakalpakstan to see some of its two-dozen ancient desert forts, palaces and Zoroastrian sites. Top of the list should be the 2000-year-old former capital of Toprak Qala and the three forts of Ayaz Qala, which feature a yurt camp if you want to overnight.

Renting a car with driver is the only option for getting here.

Diversion: Shakhrisabz

For more of Timur's epic architecture head over the mountains from Samarkand to the warlord's home town of Shakhrisabz. Recent renovations have sapped the town of its charm, but the scale of Timur's ruined Ak Seray Palace still impresses. The 14th-century Dorus Siyadat mausoleum holds the resting place of his favourite son, and the crypt he planned for himself.

Get here via shared or private taxi, or rent a car and driver through a travel agency.

TREKKING ACROSS KYRGYZSTAN

Jyrgalan Valley – Jeti-Ögüz Resort
Follow the high ridge of Kyrgyzstan's Terskey Ala-Too range, from ecotourism hub Jyrgalan to stunning Ala-Köl Lake and the Soviet-built Jeti-Ögüz health resort.

FACT BOX
Carbon (kg per person): 0
Distance (km): 109
Nights: 7
Budget: $
When: Jun-Oct

❶ Jyrgalan Valley

In its eponymous valley, Jyrgalan village is a focus for ecotourism, a great base for day-hikes, and the trailhead for numerous multi-day treks (as well as the place to buy topographic maps). Explore the local lakes and waterfalls, then follow the verdant Jyrgalan river valley to the top of Terim-Tor Bulak. Camp beneath a ridge of rugged peaks.

ABOVE: JYRGALAN VALLEY

From Jyrgalan village it's 28km hike through the Terim-Tor and Boz-Uchuk passes to Boz-Uchuk lakes.

❷ Boz-Uchuk lakes

Tucked atop small plateaus at the head of the valley, Boz-Uchuk's three lakes perfectly reflect the remote nature and untamed terrain that define Kyrgyzstan. This wide-open pastureland cut by a glacier-fed stream narrows as it climbs to a rock tumble beneath a jagged ridge, dotted throughout by semi-nomadic families' yurt camps and livestock herds.

Head southwest for 30km from Boz-Uchuk over a series of three passes to Altyn Arashan.

❸ Altyn Arashan

Midway through your week-long trek, the soaking pools at Altyn Arashan (Golden Spring) are a welcome relief for walk-weary muscles. Play it like the locals do – stay in the hot springs until you hit your heat limit, then hop into the chilly river water just a few steps away.

Follow the well-trafficked 12km path to Ala-Köl.

ABOVE: ALA-KÖL LAKE

❹ Ala-Köl Lake

This is the iconic Kyrgyzstan trekking destination. Climbing the final scree slope to reach 3907m Ala-Köl Pass will take your breath away, but it's the view from the top that will truly leave you breathless. The 3km length of Ala-Köl Lake stretches out 500m below, fed by the blindingly-bright Takyr-Tor Glacier that hangs just above and backed by a row of 5000m-plus peaks beyond. The first glimpse is a moment that defines not just this trek, but the entire trip to Kyrgyzstan.

Exit Ala-Kol to the southwest over Panorama Pass, continuing onwards to Telety Pass and descending to Jeti-Ögüz, a total of 39km.

❺ Jeti-Ögüz Resort

After over 100km of trekking, a little old-school R&R is just the thing. In Jeti-Ögüz you can soak in the Soviet-era vibes alongside your radon bath, go for a healing electric mud therapy or just stick with the more traditional post-hike massage.

Minibuses depart throughout the day from the resort's main gate to the city of Karakol, a major tourist town and transport hub.

ALTERNATIVES

Extension: Turnaly Kol

Before leaving Jyrgalan, spend a day making short hikes near the village. The 13km loop that climbs to Turnaly Kol (Crane Lake) is an easy half-day option – and doable by horse as well if you're saving your own legs for the main trek.

Turnaly Kol is due north of Jyrgalan village – climb the obvious ridge northeast of town to drop down to the lake, and then loop back through the small forest to the west.

Diversion: Chong Kyzyl-Suu Valley

If 100km isn't enough, extend the trek 22km more by continuing west across the tough 3890m Archa-Tor Pass, atop which you'll see a handful of small lakes. There's lots of wildlife in this area – your chances of spotting wolves and Marco Polo sheep are high.

Hired transport can reach the Kyzyl-Suu Meteorological Station trailhead, otherwise it's 22km further to public transport in Kyzyl-Suu village.

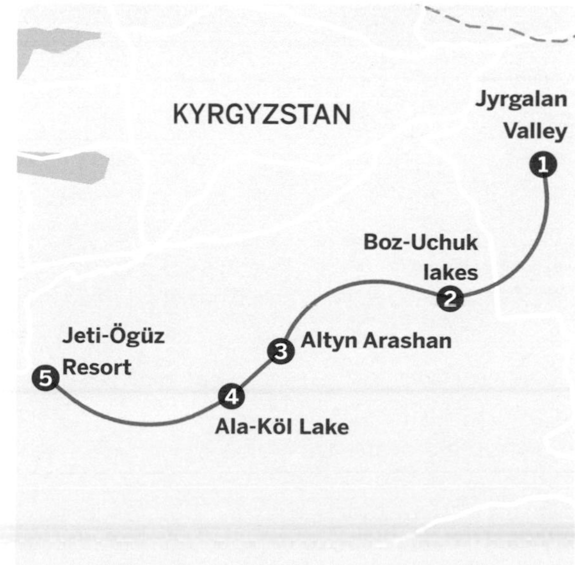

ALL ABOARD THE TRANS-MONGOLIAN

Moscow, Russia – Beijing, China

One of the world's great train trips, this epic journey hurtles you through Russia, Mongolia and China on a cross-Asia adventure with plenty of highlights.

FACT BOX

Carbon (kg per person): 299
Distance (km): 7621
Nights: 6-10
Budget: $$
When: year-round

❶ Moscow, Russia

The beginning of this iconic journey starts in Russia's glorious capital. Famed for the Kremlin and Red Square, beyond Moscow's formidable history is a cosmopolitan city with lively bars, cutting-edge dining, fantastic art, gorgeous parks and a friendly new generation of switched-on locals. It's the perfect place to enjoy some urban adventure before the long train trip ahead.

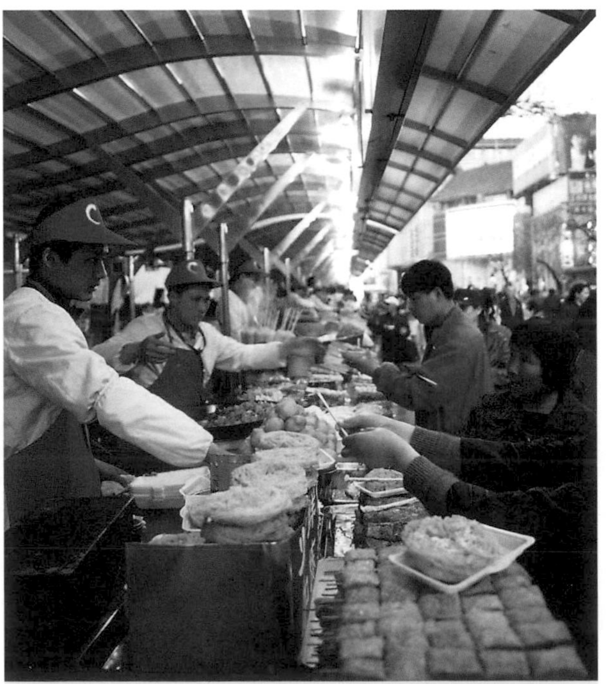
ABOVE: DONGHUAMEN MARKET, BEIJING

Take the train from Moscow to Irkutsk (7 daily; 3-5 days), a popular Siberian stop for breaking up the journey.

❷ Irkutsk and Lake Baikal, Russia

Take your seat and settle in with a hefty Dostoevsky or Tolstoy novel for several days of quintessential slow travel, rattling through endless birch forests, bucolic farmland and past *dachas* (Russian country houses) as you bisect the heart of Siberia. Disembark at historic Irkutsk to enjoy its grand 19th-century architecture and epicurean offerings, before reboarding the Trans-Mongolian to continue the journey. And just when you think the scenery is becoming monotonous, the dramatic snow-capped mountains appear, soon followed by ocean-sized Lake Baikal, its waters lapping mere metres from the train window.

Take the train from Irkutsk to Ulaanbaatar (daily; 27hr).

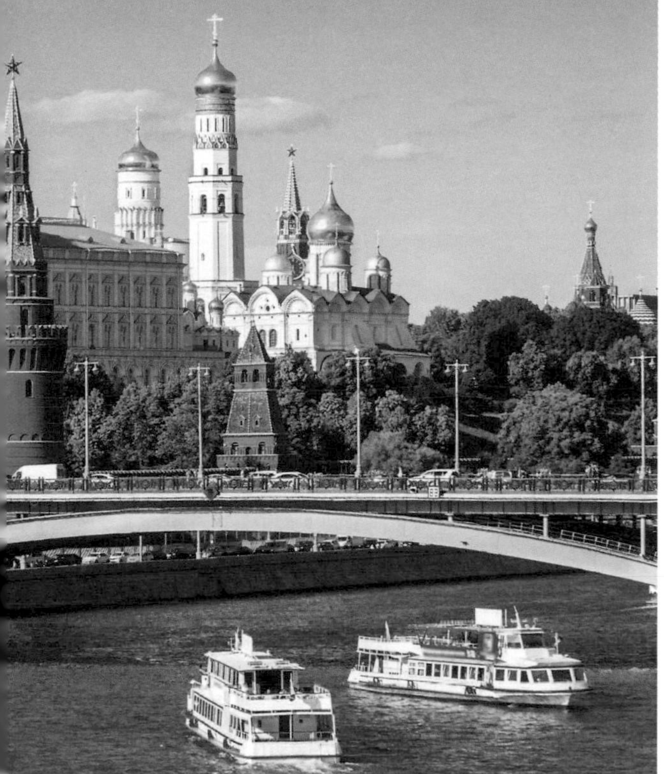

ABOVE: THE KREMLIN AND RIVERFRONT, MOSCOW

❸ Ulaanbaatar, Mongolia

One of the highlights for those coming from Russia is the moment, upon waking, when you realise you've just arrived in Mongolia. It's a dramatic shift in landscape as you take your first glimpse of its vast, undulating steppe wilderness. Most passengers disembark at capital Ulaanbaatar to explore Buddhist temples, museums and Soviet-era sights, then moving on to its countryside for a night in a traditional *ger* (yurt).

The last leg of the Trans-Mongolian is the train from Ulaanbaatar to Beijing (daily; 28hr).

❹ Beijing, China

The final section of this epic journey has you passing through the extraordinary Gobi Desert. You'll then cross the border into China, which announces itself with beautiful mountains covered in pine forest. Pull into Beijing and set about exploring the Chinese capital's blockbuster sights. Get lost in the Forbidden City. Stroll the Imperial Palace. Be amazed by the size of Tiananmen Square. And take a trip out to see the Great Wall. Then round off your travels with a night out in Beijing's labyrinthine network of *hutongs* (streets), filled with buzzing bars and restaurants.

Beijing Airport has flights to dozens of international destinations.

ALTERNATIVES

Extension: Vladivostok, Russia

Rather than heading across the border into Mongolia, you can continue through Russia to the country's far east and the port of Vladivostok, the final stop along the Trans-Siberian train route. It's a city offering plenty of charm, culture and lively nightlife.

From Irkutsk take the train to Vladivostok (3 daily; 72hr), which has an international airport offering flights across the world.

Diversion: Harbin, China

One for those who prefer the less-travelled road (or rail in this case), the Trans-Manchurian railway also starts out from Moscow and journeys across Siberia, then skirts around Mongolia and into China via the Russian border. Disembark at frontier town Harbin, famous for its ice festival (December to February), before continuing on to end the trip in Beijing.

From Harbin there are many trains on to Beijing (8 daily; 10-16hr).

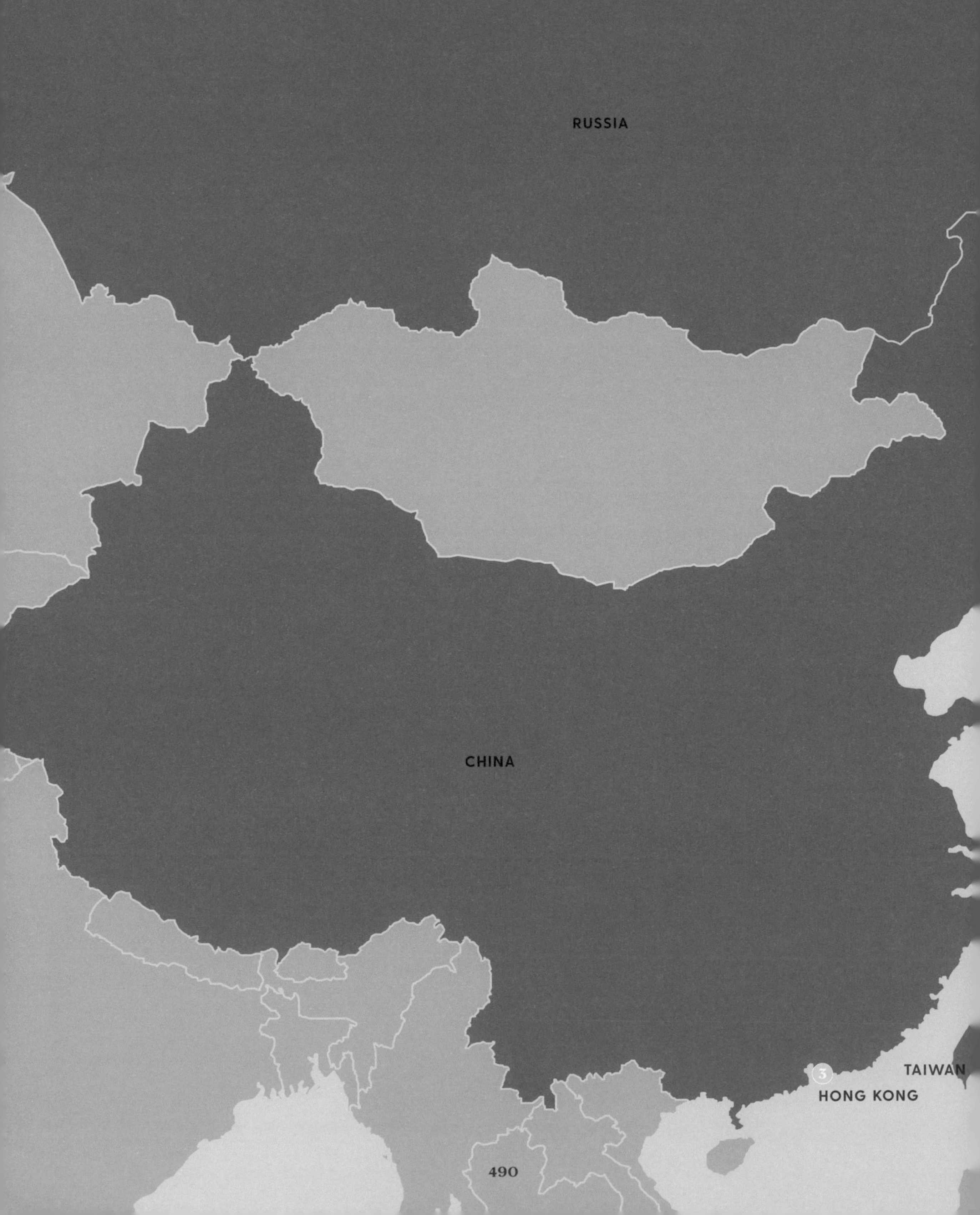

RUSSIA

CHINA

3

TAIWAN

HONG KONG

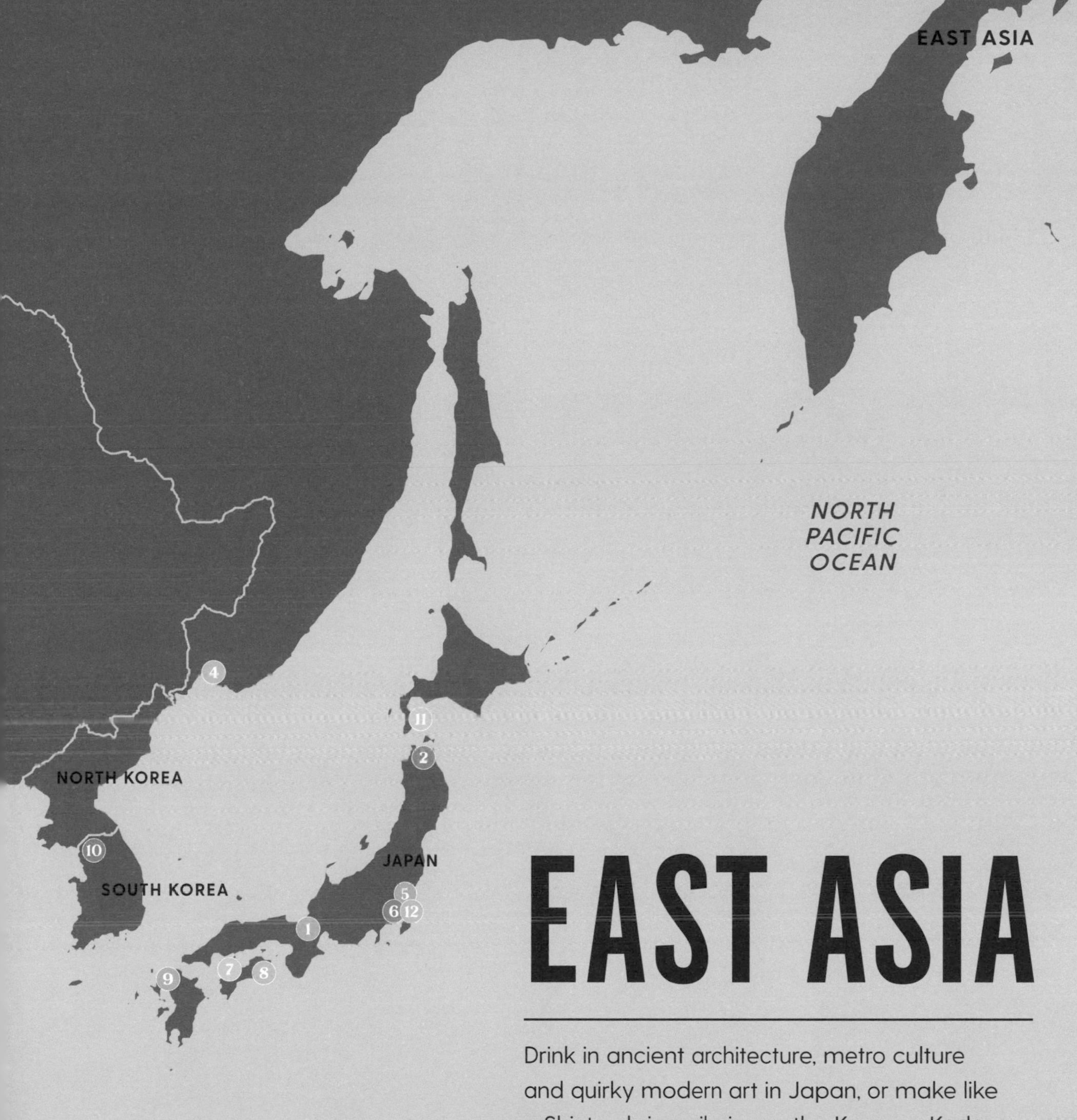

EAST ASIA

Drink in ancient architecture, metro culture and quirky modern art in Japan, or make like a Shintu-shrine pilgrim on the Kumana Kodo, become a *sakura*-spotter on the cherry-blossom trail and ski the slopes in Hokkaidō. Take a gastronomic tour of Hong Kong, island-hop from Seoul to South Korea's Jeju Island for urban bling and beachtime, and ride the rails through Russia's stunning far east.

ABOVE: TOKYO'S SENSŌ-JI TEMPLE, JAPAN

CONTENTS

HOTEL GRACERY
TOHO CINEMAS
IMAX
歌舞伎町に
ウシジマ社長、降臨!!
10.22
KABUKI町
GODZILLA ROAD
MARUHAN
SHIDAX
BIC DRUG
BIC CAMERA x SHIDAX
747 カラオケランド
マルハン新宿東宝ビル店
Booth
ネットカフェ & カプセル
8F ~ 5F
鳥貴族
TORIKIZOKU
虎の子
4F
山内農場
5F
焼肉屋 さかい
4F
お好み焼き
2階
とらそば
1F
新宿 かっぱ
B1
海の家
大衆酒場
トラノコ
歩行者専用

KUMANO KODO PILGRIMAGE THROUGH JAPAN

Tanabe – Kumano Nachi Taisha

One of earth's great pilgrimage routes (now reachable by local buses and hiking) takes in mountains, forests, hot-spring villages and three of Japan's most sacred Shinto shrines.

FACT BOX

Carbon (kg per person): 4
Distance (km): 128
Nights: 3
Budget: $
When: Mar-Jun or Sep-Dec

❶ Tanabe

Arriving by train from Osaka or Kyoto, pick up trail snacks at local markets, and travel tips (bus timetables, hiking routes, lodging and luggage assistance) at the tourist information centre. Once on the road, buses between each stop run several times a day; same-day porterage saves hassles, and inns offer dinner and breakfast.

Take the bus to Kawayu Onsen (2hr).

❷ Kawayu Onsen

Overnight in this secluded hot-spring hamlet for access to its very special river; thermally heated waters flow into it from underground and guests construct soaking pools from smooth river stones.

Take another bus to Hosshinmon-oji (30min).

ABOVE: KAWAYU ONSEN

❸ Kumano Hongu Taisha

From the Hosshinmon-oji bus stop, a popular 7km hike leads to Hongu Taisha. This is the first of Kumano's three main shrines, situated on a ridge up a long, stone staircase, through a cedar and cypress forest. Nearby, Japan's largest *torii* (ceremonial Shinto gate) towers over rice paddies.

Take the bus to Yunomine Onsen (15min).

❹ Yunomine Onsen

A rushing river runs through this hot-spring village that has welcomed visitors for 1800 years. A stay in one of its home-style lodgings feels like a step back in time, as does the tiny bathhouse next to a Buddhist temple. Eat dinner at your inn, or visit one of the shops selling ingredients to steam over natural vents by the river.

Take the bus to Michi-no-eki Kumano-gawa (40min).

❺ Kumano Hayatama Taisha

Over the centuries, thousands of pilgrims have floated

ABOVE: TEMPLE-KEEPERS AT KUMANO HONGU TAISHA

down the mountain-ringed river, Kumano-gawa; these days, the ride by flat-bottomed boat takes ninety minutes. As you approach the river mouth, alight near the bright vermillion Kumano Hayatama Taisha, second of the grand shrines. Enjoy a picnic lunch at the nearby hillside shrine, Kamikura-jinja, where the rope around the giant cliffside boulder Gotobiki-iwa signifies its status as mythical landing site of the gods.

Take another bus to Kii-Katsuura Station (35min).

6 Katsuura

This seaside village makes for a handy overnight stop thanks to its several hot-spring inns, some with coastal views. It's also a fishing port, so expect to enjoy some of Japan's freshest catch for your dinner.

Continue on a bus to Daimon-zaka (20min).

7 Kumano Nachi Taisha

Reached by a dramatic descent on a 2km, tree-lined staircase, the third of Kumano's main shrines is the most spectacular, with a postcard-worthy pagoda framing Japan's tallest waterfall, 133m-high Nachi-no-taki.

Bus back to Kii-Katsuura Station and from there return by train to Ōsaka or Kyoto.

ALTERNATIVES

Extension: Kōya-san

On the north-central Kii Peninsula, between the Kumano region and Osaka, this mountaintop monastery has been a centre of Buddhist Shingon worship since the early 9th century; non-believers will also find inspiration in its temples, monuments and natural beauty. Dozens of *shukubō* (temple lodgings) offer overnight stays, vegan meals and spiritual encounters.

Cable car and express train connect Kōya-san with Osaka (1hr 30min).

Diversion: Ise-Jingū (Ise Grand Shrine)

Between Kumano and Nagoya, Shinto's most sacred site covers two sprawling forests. Near the train station, Gekū (outer shrine) honours the god of agriculture; 6km away, Naikū (inner shrine) salutes Amaterasu-Ōmikami, goddess-guardian of the Japanese nation. Both sites have historic shopping and dining districts.

Take a Kintetsu Line express train to Ise-Jingū from Nagoya (1hr 15min), Osaka (1hr 45min) or Kyoto (2hr 15min).

JAPANESE GOURMET DELIGHTS IN TOHOKU

Aomori – Niigata

Journey by train – from bullet to local rattler – through one of Japan's least-visited but most enchanting regions, tasting local specialities along the way.

FACT BOX

Carbon (kg per person): 15
Distance (km): 461
Nights: 6
Budget: $$
When: May-Oct

❶ Aomori

Tohoku occupies the northeast region of Japan's main island, Honshu, an area far from the standard tourist route but one that's a hotbed of gastronomic delight. Begin this journey in Aomori, on the north coast, a pleasant place known for the hugely popular annual summer Nebuta floats festival (as well as a museum dedicated to it), plus the Furukawa Fish Market.

Take the train south to Hirosaki (frequent; 35min).

❷ Hirosaki

Welcome to apple country. Hirosaki is known for all things apple – apple pie, apple cider, apple blossom – and it's worth seeking out the local apple-based specialities. This is also a famously historical city, former seat of Tsugaru shogunate, home to famous Hirosaki Castle and several impressive temples and pagodas.

Take the Ou Main Line Limited Express to Akita (4 daily; 2hr).

❸ Akita

There are three things you have to try in Akita, on Japan's west coast: Inaniwa noodles, a type of hand-stretched and silky udon noodle often served with soy- or sesame-based dipping sauces; *kiritanpo*, a tube of rice-meal that's roasted over hot coals and smothered in miso; and sake, for which the Akita region is famous.

Waddle back to the train station and take an Inaho Line train to Sakata (3 daily; 1hr 30min).

ABOVE: HIROSAKI CASTLE

ABOVE: TSURUOKA'S SHOJIN RYORI CUISINE

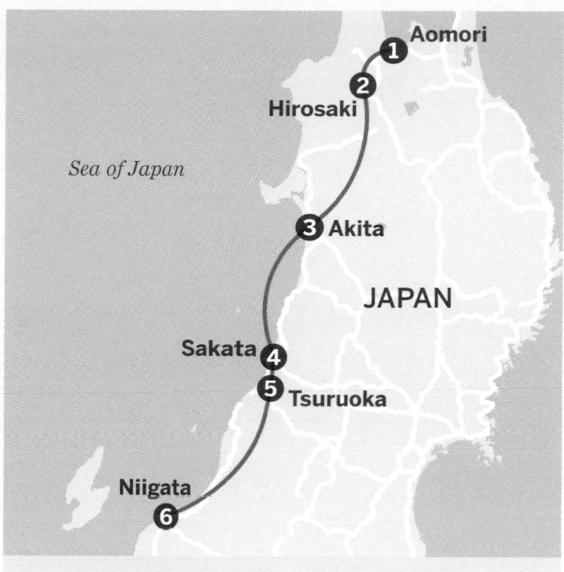

4 Sakata

You're in Sakata for one reason, and one reason only: ramen. Sakata ramen is cooked up with handmade noodles soaked in a clear broth made with flying fish, dried anchovies and bonito. And it's exceptionally tasty. Grab a bowl and then call into the Sankyo Rice Storehouse, a warehouse turned museum.

Take the Uetsu Line train south to Tsuruoka (7 daily; 30min).

5 Tsuruoka

As a Unesco City of Gastronomy, Tsuruoka knows its cuisine. Try sushi and sashimi that uses fresh local produce; feast on sesame tofu; and dine on Shojin Ryori, the vegan cuisine of the local Buddhist monks.

Take the Inaho Line to Niigata (4 daily; 2hr).

6 Niigata

Coastal Niigata is known for two items whose quality go hand in hand: rice and sake. Some claim the rice from this area is Japan's finest, and the sake that's made from that rice is up there with the country's best. Enjoy both products at a local *izakaya* (drink and snack bar).

Take the Shinkansen bullet train to Tokyo (hourly; 2hr) for onward connections.

ALTERNATIVES

Extension: Three Mountains Of Dewa

Also known as Dewa-Sanzan, these three peaks near Tsuruoka – Mt Haguro, Mt Gassan and Mt Yudono – are holy to the Shinto religion. The highlight here is Mt Haguro, where there's a five-storey pagoda that's been declared a national treasure; and Hagurosan Saikan, a traditional restaurant serving vegan Shojin Ryori cuisine in the middle of a forest.

Take the bus (5 daily; 1hr) from Tsuruoka Station to Mt Haguro.

Diversion: Sukayu Onsen

Ever fancied jumping into a steaming hot bath with 1000 of your closest, naked friends? Perhaps not, but you should still book a stay at Sukayu Onsen, a gorgeous hamlet in the mountains south of Aomori. The onsen here is known for its 'sen-nin-buro', or thousand-person-bath, a slight exaggeration for a large, mixed-gender, natural-springwater public bath revered for its healing properties.

Take the bus (frequent; 50min) from Aomori Station to Sukayu Onsen.

FOLLOW THE HONG KONG FOOD TRAIL

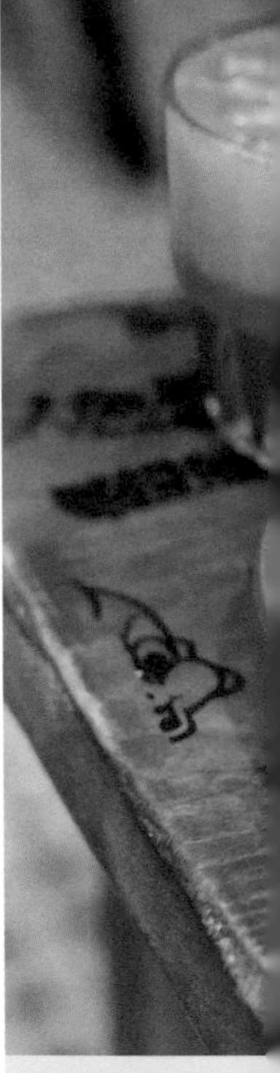

Hong Kong Island – Lantau Island

Discover the exotic Asian and global gourmet world of Hong Kong's outdoor markets, street food and fine dining by foot, tram, subway and boat.

FACT BOX

Carbon (kg per person): 5
Distance (km): 47
Nights: 4
Budget: $
When: year-round

❶ Hong Kong Island

The heart of Hong Kong are the Central, Wan Chai and Causeway Bay neighbourhoods where the food scene bubbles with contrasts, from the exceptional Italian Otto e Mezzo and Asian-molecular Bo Innovation (both 3-star Michelin) to renowned hole-in-the-wall eateries like Mak's Noodle, serving beef soup and wonton noodles at budget prices. Ride the vintage tram – known affectionately as the Ding-Ding – and every stop reveals a pulsating wet market, medicinal shop, traditional tea salon, noodle maker or soy sauce fermenter.

Take the Star Ferry across Victoria Harbour to Tsim Sha Tsui in Kowloon (frequent; 10min).

❷ Kowloon

Kowloon begins at the glittering waterfront and shopping malls of Tsim Sha Tsui, where T'ang Court serves exquisite Cantonese cuisine and the smoky stalls of the Ladies Market are a street-food paradise, cooking everything from grilled octopus and beef balls to dim sum. Take the efficient MTR subway to Sai Kung, in the northeast New Territories, where fishermen moor sampans and sell their catch, then order chili crabs, pepper clams and sea urchins at Chuen Kee Seafood.

Cross back to Central on the Star Ferry and pick up the Lamma Ferry (every 20min; 1hr).

ABOVE: PO LIN MONASTERY BUDDHA, LANTAU ISLAND

ABOVE: LUNCH, HONG KONG-STYLE

3 Lamma Island

Lamma is a secluded green haven where local fisherfolk live alongside a cosmopolitan mix of international expats. There are dozens of seafood restaurants serving the freshest seafood, from lobster to abalone, but for a quiet romantic meal, trek over the hills to the picturesque port of Yung Shue Wan for a waterside feast of razor clams, steamed grouper and mantis prawns at Lamcombe Seafood Restaurant. Its speciality steamed scallops are smothered with transparent vermicelli and finely-diced ginger.

Take the ferry back to Hong Kong Island and switch to the ferry for Lantau (frequent; 1hr 30min).

4 Lantau Island

A Hong Kong highlight is hiking up to Lantau's iconic giant Buddha statue at Po Lin Monastery, refuelling in the restaurant at the top with its delicious vegetarian food prepared by monks. Alternatively, family-friendly Discovery Bay offers sophisticated international dining, a beach and marina. For a change from food, especially if you're with children, visit Lantau's prime attraction, Hong Kong Disneyland.

Hong Kong's international airport, adjoining Lantau, has global onward connections.

ALTERNATIVES

Extension: Macau

Visitors flock to this former Portuguese colony for two very different reasons. Macau's more recent appeal lies in the giant casinos of its 'Las Vegas of the East'. The less flashy temptation is the island's Unesco-protected, centuries-old cultural heritage of Portuguese and Chinese churches, mansions and temples.

High-speed ferries from Hong Kong Island to Macau leave Central's Sheung Wan pier (every 30min; 55min).

Diversion: Dragon's Back Trail

Behind its facade of futuristic skyscrapers, Hong Kong conceals some wild mountain and coastal countryside, perfect for trekking. The daunting MacLehose Trail is for the seriously fit, but everyone can enjoy the gentle trail along Dragon's Back mountain ridge, down to the sea at Shek O Beach.

From Central MTR station, take the Island Line to Shau Kei Wan, then bus 9 to To Tei Wan (1hr 15min).

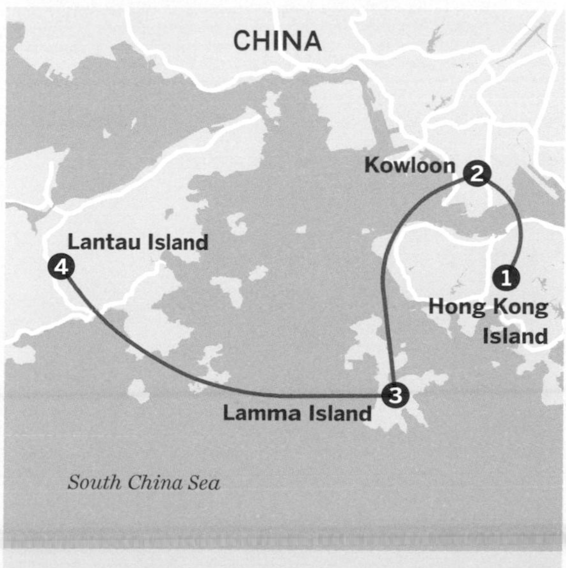

HIGHLIGHTS OF THE RUSSIAN FAR EAST

Vladivostok – Birobidzhan

Visit remote communities and take in cinematic landscapes on a rail journey across Russia's eastern edge.

FACT BOX

Carbon (kg per person): 174
Distance (km): 4224
Nights: 28
Budget: $$
When: Apr-Oct

❶ Vladivostok

The Russian Far East's most dynamic and cosmopolitan city, Vladivostok seduces with its hillside setting overlooking the Pacific coast. A naval base since the late 19th century and the terminus for the Trans-Siberian Railway, it offers a striking mix of historic and contemporary architecture, interesting museums and an excellent dining scene. Don't miss visiting the Zarya Centre for Contemporary Art, housed in a former clothing factory; or taking the one-minute funicular ride up to a viewpoint providing a glorious panorama of the giant suspension bridges hopping across Golden Horn Bay to Russky Island.

Take the train to Khabarovsk (8 daily; 11-15hr).

❷ Khabarovsk

A lovely park – with its own beach – along the Amur River is the trump card of this pleasant, historic city. Your eye will also be caught by impressive turn-of-the-20th-century architecture, such as the handsome building containing the fine Regional Museum.

Get on the Baikal-Amur Mainline (BAM) train and head to Komsomolsk-na-Amure (2 daily; 10hr).

❸ Komsomolsk-na-Amure

Some 400km north of Khabarovsk, and also on the Amur River, this Soviet-era industrial town sports colourful socialist-realist murals and Stalin-era neoclassical buildings.

Continue on the BAM train to Tynda (daily; 38hr).

❹ Tynda

As former HQ for the BAM train project, this workaday town's main attraction is, unsurprisingly, the BAM

ABOVE: TRINITY CO-CATHEDRAL, TYNDA

ABOVE: KHABAROVSK RAILWAY STATION

Museum, with exhibits about the construction of the railway as well as artefacts from the native Evenk people. Tynda's striking railway station resembles a sci-fi film set.

On odd numbered days there's a direct train to the next stop, Blagoveshchensk (16hr).

5 Blagoveshchensk

Facing the Chinese city of Heihe across the Amur River, this modest border town is another Far East backwater. Take a leisurely wander to find dozens of handsome tsarist-era buildings, including the Amur Regional Museum.

Take the train to Birobidzhan (daily; 10hr 45min).

6 Birobidzhan

Welcome to the capital of the so-called Jewish Autonomous Region, a Stalin-era creation and still the world's only officially Jewish territory outside Israel. Clues to the town's religious heritage include a giant menorah at the train station, a fascinating museum on the town's early settlers, and the Jewish culture centre Freud.

Take the train back to Khabarovsk (6 daily; 2hr 30min) from where there are onward flight connections.

ALTERNATIVES

Extension: Yakutsk

The capital of the Sakha Republic, built on permafrost, is remote even by Far East standards. Most visitors take a cruise to the Lena Pillars, a 35-million-year-old geological marvel. Also worth a visit is Permafrost Kingdom, an underground gallery of ice sculptures.

An overland trip is possible by rail and road, but the fastest way here is to fly from either Khabarovsk (2hr 45min) or Vladivostok (3hr 15min).

Diversion: Kamchatka

Snow-capped volcanos, bubbling geysers, salmon-packed rivers and vast reindeer herds tended by native peoples are the attractions of this peninsula of fire and ice at Russia's furthest eastern edge, accessed via Petropavlovsk-Kamchatsky. The pretty village Esso, a 10-hour drive north, is near the trails of the well-managed Bystrinsky Nature Park.

You can fly here from either Khabarovsk (3hr) or Vladivostok (3hr 30min).

AROUND TOKYO, MT FUJI AND THE FIVE LAKES

Tokyo – Mt Fuji

Start in Japan's mind-blowing capital, Tokyo, before working your way by train and bus to iconic Mt Fuji and its fabulous Five Lakes region.

FACT BOX

Carbon (kg per person): 11
Distance (km): 343
Nights: 10
Budget: $$
When: Mar-Oct (early Jul-early Sep for climbing Fuji)

❶ Tokyo

Spend a few nights soaking up the sights and sounds of Japan's capital city. People-watch in Harajuku; join the throngs at the famous pedestrian crossing in Shibuya; sample the nightlife in Shinjuku; and get a feel for ancient Japan in atmospheric Asakusa district.

Take the JR Yokosuka line train to Kamakura (every 15min; 55min).

❷ Kamakura

While just under an hour from the chaos of Tokyo, Kamakura feels light years away with its laidback feel, organic eating options, seaside setting and multitude of brilliant temples. The town is most famous for its landmark Amida Buddha, an imposing bronze statue dating back to 1252.

Take the JR Yokosuka line to Ofuna (every 10-15min), transfer to the Tōkaidō line to Odawara (every 20min), then take the Hakone Tozan Railway to Hakone-Yumoto Station (every 10min); total journey time is 1hr 15min.

❸ Hakone

Centred on Lake Ashino, Hakone is a popular escape from Tokyo. Admire the iconic view of Mt Fuji with shrine gates rising from the lake in the foreground; spend hours at the Open-Air Museum with works by Rodin, Miró and Picasso; take in the mountain scenery; stay in a charming *ryokan* (traditional inn); and soak in a steaming onsen (hot-spring bath).

Take a bus from Hakone-Yumoto Station to Gotemba, transferring at Sengokuhara (every 20min; 50min). From Gotemba the bus runs to Kawaguchi-ko (every 20min; 45min)

ABOVE: SHIBUYA CROSSING, TOKYO

ABOVE: MT FUJI FROM ASHI-NO-KO CRATER LAKE

❹ The Fuji Five Lakes

There's no prize for guessing what the star of the Fuji Five Lakes region is, but the lake towns set around the iconic volcano – Yamanaka-ko, Kawaguchi-ko, Sai-ko, Shōji-ko and Motosu-ko – are worth visiting in their own right. Get your camera ready to capture Mt Fuji from a number of vantage points reflected in the water, and lace up your walking boots for picturesque hiking in the great outdoors.

Take the bus from Kawaguchi-ko to the Fifth Station on Mt Fuji (1hr). The bus runs roughly every 90min from around 9am to 3pm during climbing season.

❺ Mt Fuji

Unesco World Heritage-listed Mt Fuji is the symbol of Japan. At 3776m high, it's not a mountain to be taken lightly so if you decide to climb it, be prepared. It's a gruelling slog and the conditions can change in an instant, but you are rewarded with stunning scenery and bragging rights. The most popular route is the Yoshida Trail (allow around six hours to reach the summit). Official climbing season is from early July to early September.

Take the direct bus from the Fifth Station on Mt Fuji back to Tokyo for onward connections (hourly; 2hr).

ALTERNATIVES

Extension: Yokohama

It's worth making a day or overnight trip to the cosmopolitan waterfront city of Yokohama. It offers a booming arts, dining and nightlife scene, and is particularly enticing for beer lovers as the city is crammed with microbreweries. There are also plenty of family-friendly attractions and a lively Chinatown.

JR line trains connect Tokyo Station with Yokohama (every 10min; 25-30min).

Diversion: Izu Peninsula

Extend your trip by a few days and head to the Izu Peninsula, just 100km southwest of Tokyo and a fabulous place to slow things down. Set your sights on some rugged coastline and beaches; stay in delightful seaside towns such as Itō, Shimoda, Ō-shima and Shirahama; and bathe in open-air onsen or hit the surf.

Trains run to the main towns from Tokyo and buses service the peninsula but having your own wheels is best.

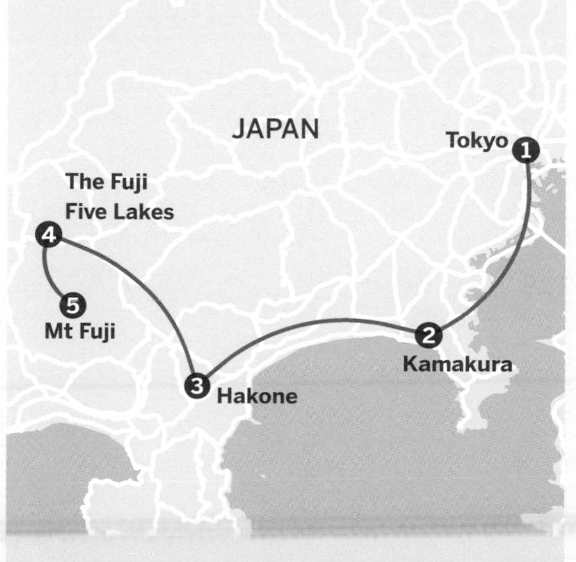

FIRST-TIME JAPAN BY BULLET TRAIN

Tokyo – Hiroshima

Shoot from the chaos of Tokyo to the beauty of Kyoto and on to thought-provoking Hiroshima on a Shinkansen (bullet train) introductory tour of Japan.

FACT BOX

Carbon (kg per person): 32
Distance (km): 1600
Nights: 8
Budget: $$$
When: year-round

❶ Tokyo

Godzilla-sized neon billboards, cloud-nudging skyscrapers, robots in restaurants and sumo wrestlers battling it out – tick off a list of Japanese clichés on even a short stay in the country's megalopolis capital.

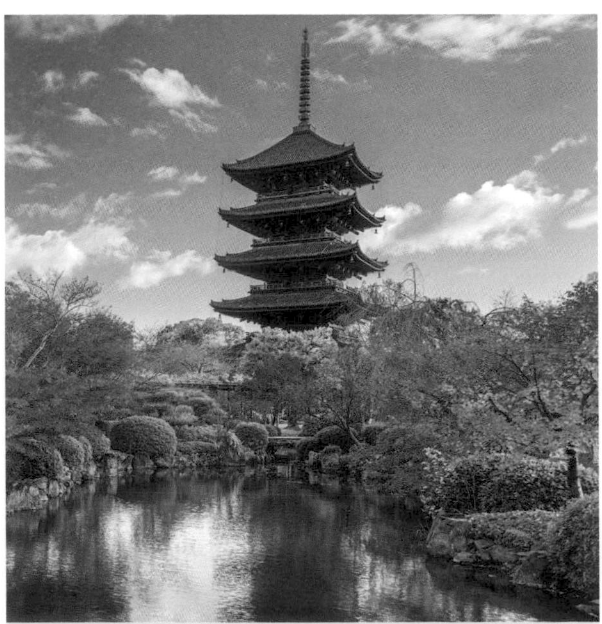

ABOVE: TŌ-JI PAGODA, KYOTO

In just a few days you can sample Michelin-starred restaurants, bar-hop around Shinjuku and belt out a tune at karaoke, visit hip art galleries and centuries-old temples, stroll through vast parks and spend hours at Tetris-like department stores filled with futuristic gadgets.

Take the Shinkansen train (the Nozomi is the fastest) from Tokyo Station to Kyoto Station on the Tōkaidō line (every 10-20min; 2hr 30min).

❷ Kyoto

Swap Tokyo's whirl of feverish activity for a more tranquil few days in the country's cultural heart. Wander the serene gardens and admire the masterful architecture of some 2000 atmospheric temples, hope for a glimpse of elegant *geiko* (the Kyoto name for geisha) shuffling down backstreets, dine in traditional *kaiseki* (haute cuisine) restaurants and sip matcha tea in ancient teahouses before hitting the famed Nishiki food market.

Take a Kintetsu line train from Kyoto Station to Nara Station (every 10min; 35min).

❸ Nara

The charming town of Nara, Japan's first capital, is a rewarding day trip from Kyoto. The main sights

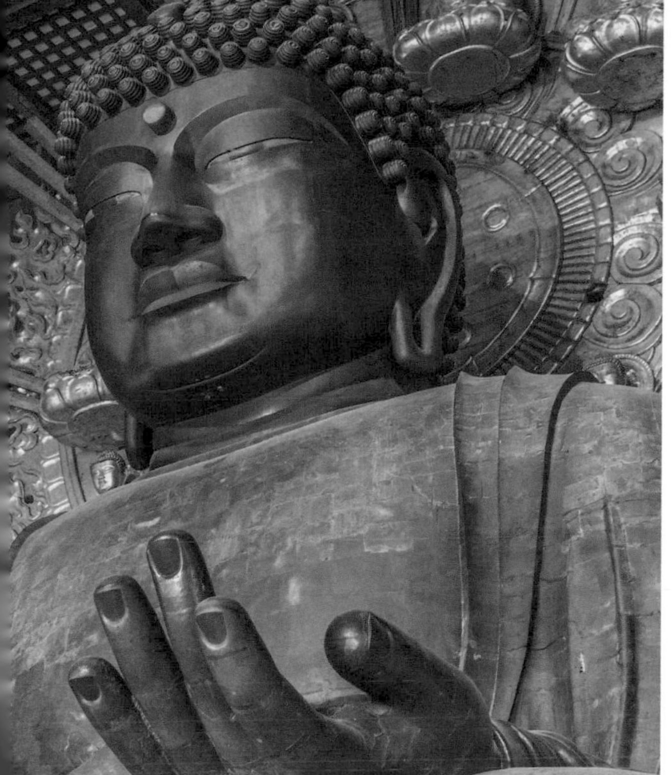

ABOVE: DAIBUTSU (GREAT BUDDHA), NARA

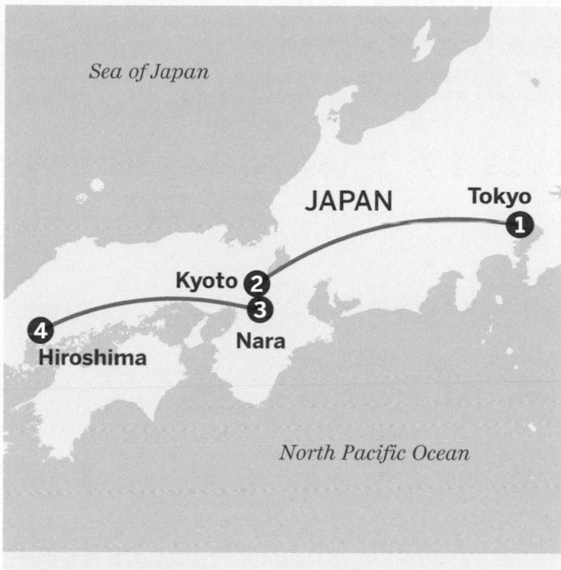

are located in and around its large park, Nara-kōen, famous for its resident deer. Highlights include the imposing 16m-high bronze statue of Daibutsu (Great Buddha) and the Nara National Museum, a must-see for its incredible collection of Buddhist art and artefacts.

Take the Kintestu line back to Kyoto then board another Shinkansen to Hiroshima Station on the JR San-yō line (every 20min; 1hr 30min).

4 Hiroshima

Forever ingrained in the world's memory as the site of the first atomic-bomb attack, Hiroshima draws crowds from around the globe who come here to reflect on the city's tragic past. Sights such as the Peace Memorial Park, the Hiroshima Peace Memorial Museum and Atomic Bomb Dome serve as confronting reminders of the destruction caused, and as moving messages of peace for the future. While the city is undoubtedly scarred by history, it's also a modern, cosmopolitan place with an excellent dining scene (be sure to try the delicious local speciality, Hiroshima-style *okonomiyaki* – a type of thick savoury pancake).

Round the trip off on the JR San-yō line Shinkansen train back to Tokyo (every 20min; 4hr).

ALTERNATIVES

Extension: Himeji

En route from Kyoto to Hiroshima make a stop at Himeji, home to one of Japan's most beautiful and best-preserved feudal castles. Nicknamed the White Egret Castle owing to its white-plaster facade, it's a stunning sight and there are superb city views from its top floor.

Himeji is a stop on the JR Shinkansen San-yō train line between Kyoto and Hiroshima.

Diversion: Miyajima

Instead of rushing back to Tokyo, extend your trip on this tiny, World Heritage-listed island. Miyajima's 16m-high vermillion *torii* (shrine gates) are regarded as one of the three best views in the country – at high tide they are immersed in the water, giving the appearance of magically floating.

Tram 2 from Hiroshima Station arrives at the Miyajima-guchi ferry terminal, from where the 10min ferry trip to the island runs every 15min.

ON THE CHERRY BLOSSOM TRAIL IN JAPAN

Matsuyama – Tokyo

Travel Japan tracking the elusive *sakura* (cherry blossom) on its journey from the south to the north of the country, and discover the best of the viewing spots.

FACT BOX

Carbon (kg per person): 37
Distance (km): 980
Nights: 10
Budget: $$$
When: mid-March to early April

❶ Matsuyama

Matsuyama, on the island of Shikoku, is an elegant city and home to the legendary Dogo Onsen, a 19th-century hot-spring bath. It also draws visitors for its impressive castle balanced on the top of Mt Katsuyama in the town's centre, the grounds of which make for a particularly beautiful spot in spring, when its hundreds of *sakura* (cherry-blossom trees) come into bloom.

Take the hydrofoil ferry to Hiroshima (1hr 15min).

ABOVE: YOSHINO'S *SAKURA*-COVERED HILLS

❷ Hiroshima

Among the tragedy of Hiroshima, there is much beauty. This is a sophisticated, attractive and friendly city, despite the sadness and horror of its history as the site of the world's first atomic-bomb attack. Along with the war memorial sights such as the Atomic Bomb Dome and the Peace Memorial Museum, there is a clutch of art galleries, great restaurants and leafy spaces, including Hijiyama-kōen. This park is just outside the city centre and a locals' favourite spot for cherry-blossom viewing.

Take the Shinkansen bullet train to Kyoto (hourly; 1hr 20min).

❸ Kyoto

It's no surprise that one of the most enchanting cities in Japan is a premier cherry-blossom viewing destination. Tourists crowd Kyoto's landmarks and parks during the season to get a glimpse of the delicate blossoms bursting from the trees lining canals, temples and gardens. Popular spots include

ABOVE: MEGURO RIVER, TOKYO

Maruyama-kōen, the Philosopher's Path and Kyoto Imperial Palace Park.

Take a Kintetsu Yoshino line train from Kintetsu Kyoto Station to Yoshino (every 30min; 1hr 45min).

❹ Yoshino

It might be a remote mountain village at the northern end of the Kii Mountain Range, but Yoshino is the most famous spot to view cherry blossoms in all Japan. So, of course, you won't have it to yourself. The mountains here are blanketed in pale-pink blooms on over 30,000 trees – the best place to see it in all its natural glory is from the *hitome-senbon* ('1000 trees in a glance') viewpoint.

Take the Kintetsu line back to Kyoto and transfer to a Shinkansen train to Tokyo (every 20min; 2-3hr).

❺ Tokyo

Sakura spectators rush to Tokyo's top cherry-blossom sights to celebrate with sake and reflect on the fleeting nature of life. *Hanami* (cherry-blossom parties) take place in the city's parks, particularly Yoyogi-kōen, Ueno-kōen and Shinjuku-gyōen, and on the paths lining the Meguro River.

Tokyo's two international airports connect to destinations worldwide.

ALTERNATIVES

Extension: Hirosaki

To continue on the cherry-blossom trail beyond Tokyo, head to the Aomori prefecture town of Hirosaki on the northern tip of Honshu, which sits in the shadow of Mt Iwaki. The huge parkland, Hirosaki-kōen, has the well-preserved Hirosaki Castle at its centre, and is home to over 5000 cherry blossom trees.

Take the Shinkansen train to Shin-Aomori (hourly; 3hr 30min) then a local train to Hirosaki (hourly; 35min).

Diversion: Osaka

For a truly local *hanami* experience, there's no better place than Osaka. Osakans are known for their spirited, less conservative manner, out in full force when the city is in cherry blossom party mode. Join them at the raucous get-togethers, knocking back sake and snacking on street food, that fill the surrounding parkland of the iconic Osaka Castle.

Osaka is easily accessible by train from Kyoto (every 5-10min; 30min).

JAPAN'S SACRED AND SCENIC SHIKOKU ISLAND

Tokushima – Takamatsu

Take a road trip around the southern Japanese island of Shikoku, passing sacred temples, castles, hot springs and emerald-green forests on the way.

FACT BOX

Carbon (kg per person): 69
Distance (km): 554
Nights: 10-12
Budget: $$
When: year-round

❶ Tokushima

Tokushima sits in the shadow of impressive Mt Bizan and is carved in half by the Shinmachi River. It's the starting point for Shikoku's ancient 88 Sacred Temples pilgrimage and there are a number of nearby examples to check out, as well as the Tokushima Modern Art Museum and a gondola that will whizz you up Mt Bizan for city views.

Drive 120km drive southwest to Muroto-misaki.

❷ Muroto-misaki

This spectacular cape juts into the Pacific Ocean. A path leads to Temple 24, Hotsumisaki-ji, sitting on a hilltop and one of the most important on the pilgrimage trail: it's believed to be the place where the monk Kobo Daishi – founder of the Shingon Buddhist sect – achieved enlightenment.

From the cape, drive 80km northwest.

❸ Kōchi

Kōchi makes a great base for investigating the surrounding area with easy day trips to beaches, caves and mountain hikes. Admire the architecture of Kochi-jō, one of the few Japanese castles to have survived with its keep intact, and time your visit for a Sunday to coincide with the excellent market on the main road leading to the castle.

Drive 110km northwest via scenic Nakatsu Gorge to Matsuyama.

❹ Matsuyama

Matsuyama is the island's largest city, situated in a lush river basin with its famous castle perched on a hill in the middle of town and trams trundling along

ABOVE: DOGO ONSEN, MATSUYAMA

ABOVE: RITSURIN-KŌEN, TAKAMATSU

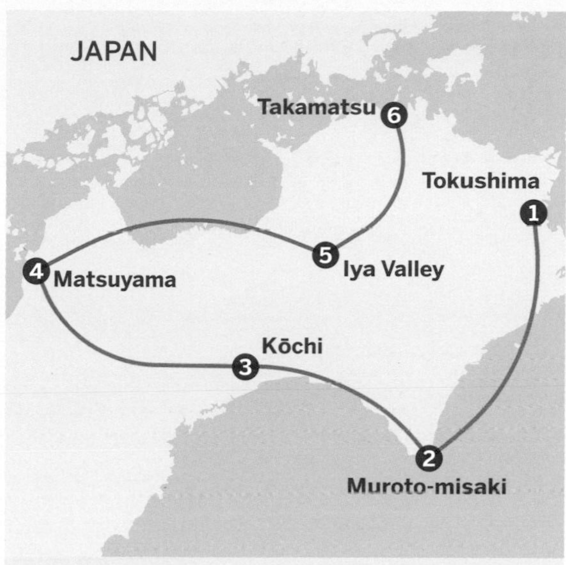

the streets. But the real drawcard is its legendary 19th-century Dogo Onsen (hot spring bath), with a reputation for the curative properties of its waters.
Head 160km east and inland, taking cliff-hugging Rte 32 to Iya Valley.

5 Iya Valley

Roads ribbon around dense mountain forests, free-flowing rivers and vertigo-inducing gorges in the magical Iya Valley. It's the perfect spot for hiking or kayaking, or for joining the tourists who flock here to lay eyes on the famous vine bridges and to hop from onsen to onsen.
Drive 84km north.

6 Takamatsu

As well as being the transport hub of the island, Takamatsu is also a lively town with plenty to enjoy. Don't miss one of Japan's most beautiful gardens, Ritsurin-kōen, woven with ponds, teahouses and bridges, and be sure to slurp back a bowl of the local favourite, *Sanuki-udon*, in one of the town's noodle restaurants.
Trains run to Okayama where you can connect to a Shinkansen bullet train to major cities including Kyoto, Nagoya and Tokyo.

ALTERNATIVES

Extension: Ashizuri-misaki

If you've got time, rather than heading to Matsuyama after Kōchi, take the longer but more scenic route along the coast and stopping at the dramatic cape, Ashizuri-misaki. It's famous for Temple 38, Kongōfuku-ji, where you can wander around the gardens and ponds in the grounds. There are a couple of hotels overlooking the sea in which to overnight.
The cape is a 150km drive south of Kōchi, then 200km north to Matsuyama.

Diversion: Naoshima

The small, tranquil island of Naoshima on the Seto Inland Sea is an art-lover's paradise and the location of the Benesse Art Site (an art project on the island). Contemporary works dot the island in museums and as installations and outdoor sculptures – the most iconic being the *Yellow Pumpkin* by Yayoi Kusama, perched at the end of the pier.
Ferries run between Takamatsu and Naoshima (up to 5 per day; 50min).

ISLAND HIGHLIGHTS IN JAPAN'S KYŪSHŪ

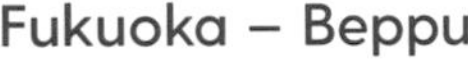

Fukuoka – Beppu

Take a bus and train trip to soak up the history and soak in the hot springs on this highlights tour of Kyūshū, Japan's southernmost main island.

FACT BOX

Carbon (kg per person): 36
Distance (km): 1032
Nights: 8-10
Budget: $$
When: Mar-Oct

❶ Fukuoka

The gateway to Kyūshū, Fukuoka comprises two former towns, Fukuoka and Hakata, which merged in 1889. There's plenty to do here, from exploring the excellent Asian Art Museum and wandering the hip Daimyō district to strolling the traditional gardens and savouring a bowl of the local speciality, Hakata ramen.

Take the JR line train from Hakata Station to Sasebo Station (hourly; 2hr), then the direct bus to Hirado (1hr 30min).

ABOVE: THE SENGAN-EN GARDENS NEAR KAGOSHIMA

❷ Hirado

During the early Edo Period (1603–1868), Hirado was the trading post of the Dutch East India Company. Today, it's a delightfully secluded beach escape with historic streets, an imposing castle, Christian churches (reminders of the hidden Christians who lived here), museums and fresh-off-the-boat seafood.

Take the bus back to Sasebo, then a JR express bus to Nagasaki (every 30-60min; total trip 3hr).

❸ Nagasaki

Despite its tragic history as the site of the second atomic-bomb dropping, Nagasaki is a spirited and friendly city with a charm that belies the horror of what it endured. Reflect on its past at the war memorials and museums, then discover what else the city has to offer – the historic site of the former Dutch trading post Dejima, shrines, temples, churches, beautifully landscaped gardens and an enticing culinary scene (try Nagasaki's speciality noodle dish, *champon*).

Take the JR Limited Express train to Hakata in Fukuoka (at least hourly; 1hr 30min), then the

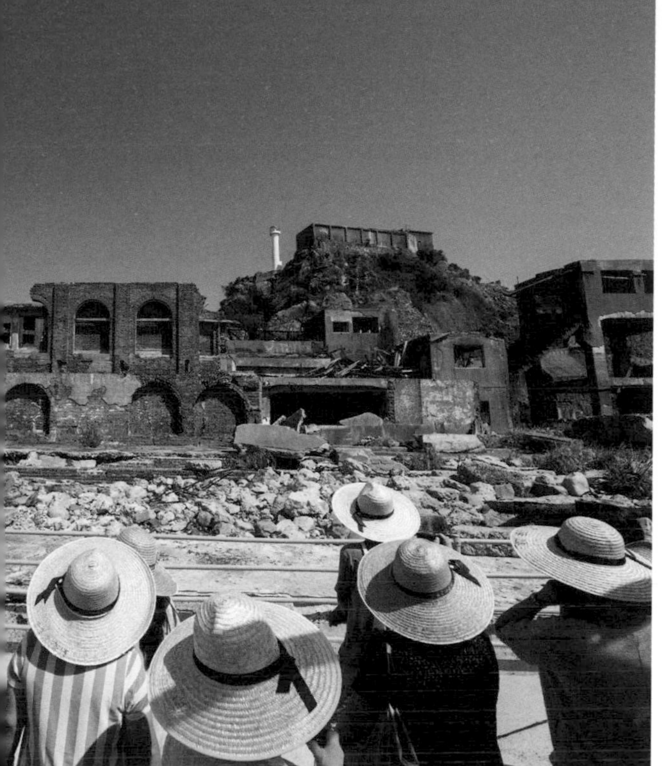
ABOVE: HASHIMA, AKA GUNKANJIMA (BATTLESHIP ISLAND)

Shinkansen to Kagoshima-Chuo (every 30min; 1hr 30min-2hr).

4 Kagoshima

Kagoshima has year-round sunshine – though umbrellas are sometimes needed to protect from the ash coming from active volcano, Sakurajima, across the bay. If you like onsen (hot-spring baths), this is the city for you, with over 50 to choose from. Don't miss a meander through the exquisite Sengan-en gardens and a bar-hop around the atmospheric Meizanbori neighbourhood.

Take the Shinkansen from Kagoshima-Chuo Station to Kokura (every 20-40min; 1hr 15min), then a JR Sonic train to Beppu (every 30min; 1hr 20min).

5 Beppu

If you haven't had enough onsen action yet, you're in luck; Beppu is famous for its eight distinct hot-spring areas with waters ranging from pleasantly warm to near scalding. Once your skin has wrinkled like a prune, head out and hit the *izakaya* (Japanese pubs) – Beppu has a vibrant nightlife scene thanks to its student population.

You can fly to Tokyo and Osaka for onward connections from nearby Oita Airport.

ALTERNATIVES

Extension: Hashima

Nicknamed Gunkanjima (Battleship Island) due its resemblance to a warship, the Unesco World Heritage island of Hashima is an eerie, apocalyptic day trip from Nagasaki. This abandoned coal mine featured in the Bond film, *Skyfall*, as the villain's lair and it was once the most populated place on Earth before becoming a ghost island.

Visit Hashima on a guided tour; cruises run from Nagasaki (Apr-Oct; 3hr).

Diversion: Kurokawa Onsen

Consistently ranked as one of the country's top onsen – no easy feat – Kurokawa is definitely worth a detour. This picturesque hot-spring village is peaceful, charming and surrounded by nature, with open-air riverside baths. Daytrippers can access a few onsen in town, but a night in a *ryokan* (traditional Japanese inn) offers the full experience.

The village can be reached by bus from Hakata Station in Fukuoka (2 daily; 3hr).

SOUTH KOREA DISCOVERER: SEOUL AND JEJU ISLAND

Seoul – Seogwipo

Start with the dazzle of the country's metropolis, then fly to Jeju to unwind, beach-, volcano- and waterfall-hopping across the island on local buses.

FACT BOX

Carbon (kg per person): 73
Distance (km): 582
Nights: 5-8
Budget: $$$
When: Apr-May & Sep-Nov

❶ Seoul

South Korea's ultramodern capital is multiple worlds in one. Stroll the Cheonggyecheon to futuristic Dongdaemun Design Plaza, finishing with kimchi *mandu* (spicy dumplings) at Gwangjang Market. Beauty products, K-pop, hip bars and the Gangnam high life sit alongside traditional Korean temples and royal palaces, yet it rarely feels overwhelming.

Fly to Jeju Airport on Jeju island (every 20min; 1hr 10min). Buy a rechargeable T-Money card (from airports and convenience stores) to use on transport across Korea, including around Jeju Island.

❷ Jeju City and around

Jeju City is a convenient base for exploring the north of Jeju Island by local bus. A traditional market and plenty of restaurants with seafood and Korean black pork are concentrated in the city. East Jeju spans quiet beaches to the Manjanggul Cave, forged from lava; west Jeju has tea fields; and south of the city sits Hallasan National Park and Jeju Loveland, a sex-themed sculpture park – more laugh-out-loud funny than raunchy.

Bus 201 from Jeju City follows the east coast to Seongsan Ilchul-bong (every 30min; 2hr).

❸ Seongsan Ilchul-bong

Climbing to the rim of Seongsan Ilchul-bong, an extinct volcano, is a Jeju island highlight – and

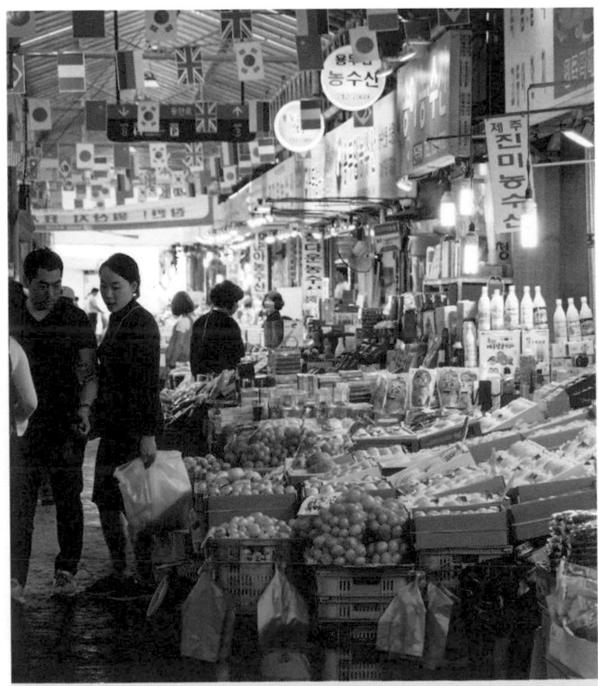

ABOVE: DONGMUN MARKET, JEJU CITY

ABOVE: JEJU ISLAND OLLE TRAIL

especially popular at dawn. The stairs are steep but only take twenty minutes to climb, with a peer across the vast overgrown crater at the end. The green surroundings are filled with other walks that are gentle on the legs but still provide knockout sea views. To get closer to the volcano, speedboats skim the water to the other side of the tuff cone. You can also witness the local *haenyeo* women divers catching seafood (1.30pm and 3pm).

Continue south on bus 201 to Seogwipo (every 30min; 2hr).

4 Seogwipo

The island's relaxed coastal spirit abounds in Seogwipo, the quieter of its two cities. Lee Jung Seob Art Gallery and friendly restaurants make up its centre. For an engaging couple of days venture out for cliffside views of Oedolgae and Cheonjiyeon Waterfall – Seogwipo's natural wonders are easy to reach following Olle trails (Jeju's signposted walking tracks).

Take bus 600 to Jeju Airport (frequent; 2hr) from where there are direct flights to Seoul (every 20min; 1hr 10min) and onward international destinations.

ALTERNATIVES

Extension: Udo

An easy visit from Seongsan Ilchul-bong, Udo is a mini island off the east coast of Jeju. Rent a fat-tyre bike, two-seater scooter or take the electric shuttle buses around the island. There are ruggedly beautiful caves, white-sand beaches and a smudge of blue waters dipping in and out of view. Finish with local peanut ice cream.

Ferries to Udo (frequent; 15min) leave from Seongsan Port near Seongsan Ilchul-bong.

Diversion: Busan by ferry

Take a leisurely 12hr ferry to Busan, South Korea's second city, where the pace of life is more relaxed than in Seoul and the seafood street snacks are big enough to constitute a meal. Port-side markets and nearby mountains to hike are other Busan highlights.

MS Ferry sail three days a week between Jeju Island and Busan.

SAVOUR A WINTER WONDERLAND IN HOKKAIDŌ

Hakodate – Sapporo

Take a winter trip to Japan's main northern island to enjoy top-grade skiing, onsen-dipping and ice-sculpture festivals, the most famous of which is Sapporo's Yuki Matsuri.

FACT BOX

Carbon (kg per person): 14
Distance (km): 335.3
Nights: 14
Budget: $$$
When: Jan-Feb

❶ Hakodate

Start in Hakodate, Hokkaidō's southernmost port and a straightforward Shinkansen (bullet train) ride from Tokyo. This historic city was one of the first locations in Japan to open to foreign traders in the mid-19th century. These European and US merchants, along with rich locals, built fancy wooden homes and elaborate churches on Hakodate's steep hillsides – many have been preserved in the Motomachi area. Other draws are the lively *asa-ichi* (morning market), famous for its fresh seafood; riding the cable car for the views from atop 334m Hakodate-yama; and the excellent exhibitions on the indigenous Ainu culture at the Hakodate City Museum of the Northern Peoples.

Travelling by train from Hakodate to the Niseko ski resorts, you have to change lines at Oshamambe. Depending on where you're staying in Niseko, alight at Niseko, Hirafu or Kutchan station (3-4hr).

ABOVE: DARUMA HONTEN RESTAURANT, SAPPORO

❷ Niseko

Asia's top ski destination has legendary powder snow throughout winter – up to 15m of the white stuff. You could easily spend a week skiing the scores of runs connecting the four resorts dotted on the eastern side of the mountain, Niseko Annupuri. There's a sophisticated international atmosphere in the various villages around the slopes, with Hirafu having the most accommodation and après-ski options.

Back on the train head to Otaru (frequent; 1hr 30min).

ABOVE: WINTER SPORTS IN NISEKO

❸ Otaru

This attractive, touristy port grew rich at the turn of the 20th century on the back of herring fishing and Hokkaidō's modern development. Handsome, heritage-listed buildings, often designed in Western styles, have been repurposed into hotels, shops, restaurants and bars, many sitting along the port's historic canal.

Continue on the train to Sapporo (frequent; 30-45min).

❹ Sapporo

Hokkaidō's prefectural capital offers everything you'd expect from a dynamic Japanese city. The Yuki Matsuri ice sculpture festival, held at several locations across the city in early February, is the main draw, but there are also historic buildings and institutions, including the beer museum in the original Sapporo Beer Brewery. If the winter chill gets too much there are also good onsen in the area where you can sink into wonderfully warm water.

New Chitose Airport, southeast of Sapporo by train (36min), has many flights to Tokyo along with international destinations.

ALTERNATIVES

Extension: Noboribetsu

A wooden boardwalk threads past bubbling, belching fumaroles in Noboribetsu's Jigkoku-dani (Hell Valley) – welcome to one of Hokkaidō's top onsen resorts. Soak in the hot-spring baths here and use the town as a base for visiting nearby attractions such as the U-popo-y Ainu Museum in Shiraoi, dedicated to the island's original inhabitants.

Frequent trains run to Noboribetsu from Hakodate (2hr 30min) and Sapporo (1hr 30min).

Diversion: Tōya-ko

At the southwestern side of Shikotsu-Tōya National Park, Tōya-ko is a picturesque caldera lake with an island in the middle. Nearby is highly active Usu-zan and its 'parasite volcano' Shōwa Shin-zan, which began growing out of the side of its parent in 1943. A cable car rises to a viewing platform overlooking Usu-zan's crater and provides stunning vistas of the surroundings.

Direct trains run from Sapporo (1hr 45min).

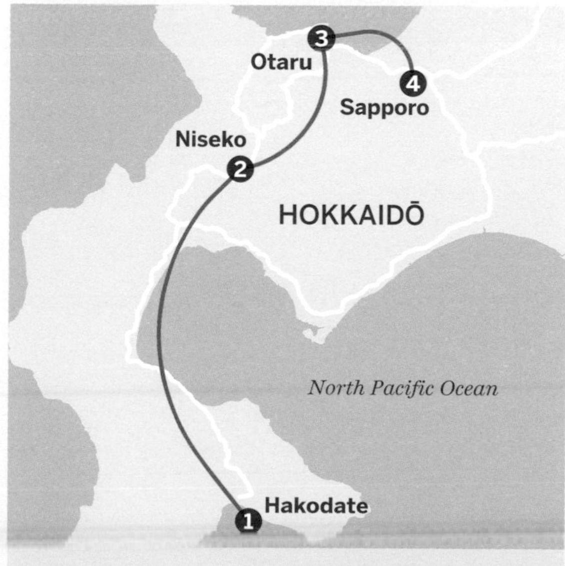

DISCOVER ANCIENT AND MODERN ART IN JAPAN

Tokyo – Naoshima Island

Discover centuries of exquisite, unique and idiosyncratic art and architecture on this train trip through some of Japan's most arresting cities and natural landscapes.

FACT BOX

Carbon (kg per person): 30
Distance (km): 730
Nights: 14
Budget: $$$
When: year-round

❶ Tokyo

Digital art collective teamLab is behind many of the most talked-about art installations of modern times, and teamLab Borderless – its permanent museum in Tokyo, housing 60 interactive works – is absolutely unmissable. Another must-see is the museum dedicated to the exuberant work of Yayoi Kusama, a remarkable space where even the toilets and roof feature the queen of maximalism's polka dots and pumpkins. And for a final flourish of wonder, Shinjuku district's Ghibli Museum, designed by the studio's founder and chief animator Hayao Miyazaki, is a magical experience for anime fans of all ages.

Take a train from Shinjuku Station to Yamanashi (every 30min; 1hr 30min).

❷ Itchiku Kubota Museum, Yamanashi

A museum with views of Mount Fiji is a win-win, particularly when that museum is filled with the exquisite work of Itchiku Kubota, a textile artist unlike any other. Having resurrected the centuries-old dyeing technique *tsujigahana* in the 20th century, his work is now displayed to breathtaking effect in a lofty wood-filled space that's in total harmony with its natural surroundings. Complement the museum visit with a soak in an idyllic onsen, trekking through local forests and some mountain-climbing.

Take a train to Kyoto (hourly; 4-5hr).

❸ Kyoto

The sumptuous National Museum of Modern Art in Kyoto has to top the list of the city's many cultural

ABOVE: CALLIGRAPHY BRUSHES, KYOTO

ABOVE: SHINJUKU'S GHIBLI MUSEUM, TOKYO

attractions, with a Japanese ceramics collection that's the best in the country don't miss the numerous works by Kawai Kanjiro. By contrast, the diminutive (one-roomed) Kyoto Ukiyo-e Museum is filled to the brim with woodblock prints by some of the country's best exponents, including Katsushika Hokusai, whose iconic *Great Wave off Kanagawa* is permanently displayed here.

Board the Hikari train to Okayama (hourly; 1hr 30min) and change for a train to Uno (50min), from where frequent ferries run to Naoshima (20min).

❹ Naoshima Island

This island in the Seto Inland Sea is a treat for lovers of contemporary art and architecture. Much of its allure lies in the contents and form of Tadao Ando's Benesse House Museum, but the Art House Project, in which traditional buildings have been turned into spaces for individual artists such as James Turrell and Hiroshi Sugimoto, is equally compelling – as are the island-wide installations.

Return to Okayama and take a bullet train from there back to Tokyo for onward international connections.

ALTERNATIVES

Extension: Nara

Everyone comes to Nara for the hugely impressive Daibutsu (Great Buddha), but make some time to properly explore the rest of the Buddhist art and architecture in this alluring 8th-century town. Highlights include the grand Kasuga Taisha Shinto shrine, the Isui-en Garden & Neiraku Art Museum, and the Nara National Museum.

Trains run from Kyoto to Nara (every 30min; 40-50min).

Diversion: Miho Museum, Koka

In an IM Pei building housing Japanese, Middle Eastern, Chinese and South Asian art and artefacts collected by its founder, Koyama Mihoko, Miho is so at one with the nature that you could easily miss it. Its location, half buried in a mountain near Shigaraki village, is a knockout too.

Take the JR Tokaidō Line from Kyoto Station to Ishiyama (frequent; 15min), then Teisan bus 150 (50min).

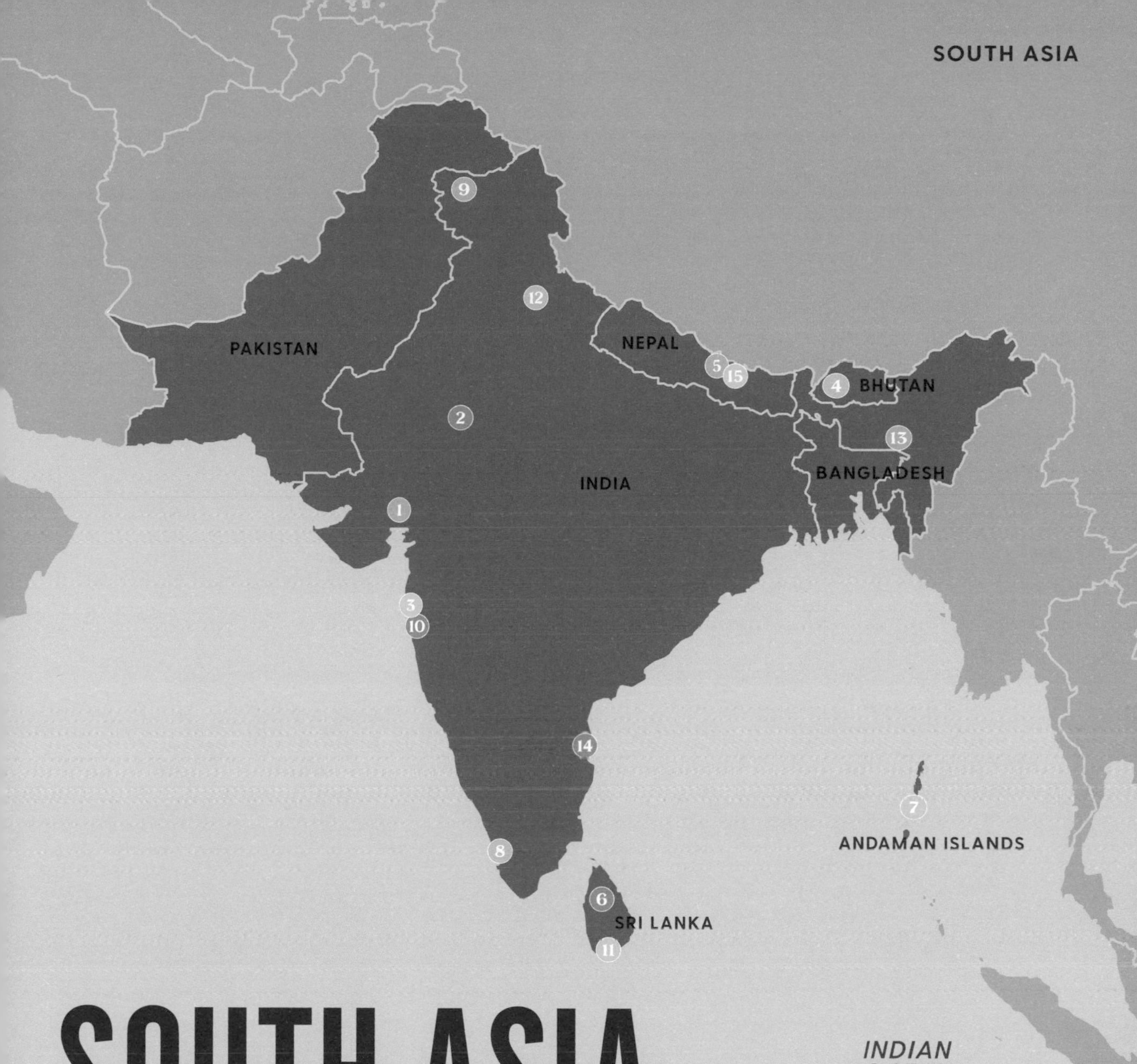

SOUTH ASIA

Temples and tigers and train trips and treks: South Asia is a sensory overload, whether you're heading into the Himalayas from Nepal or touring the sumptuous royal cities of India's Rajasthan. But there's plenty of scope for downtime, too, from gentle backwater boat trips in Kerala to road-tripping Bhutan or riding the rails in Sri Lanka's emerald-swathed Hill Country.

ABOVE: TIBETAN BUDDHIST MONKS IN BOUDHA, NEPAL

CONTENTS

LIONS, TEXTILES AND TEMPLES IN GUJARAT

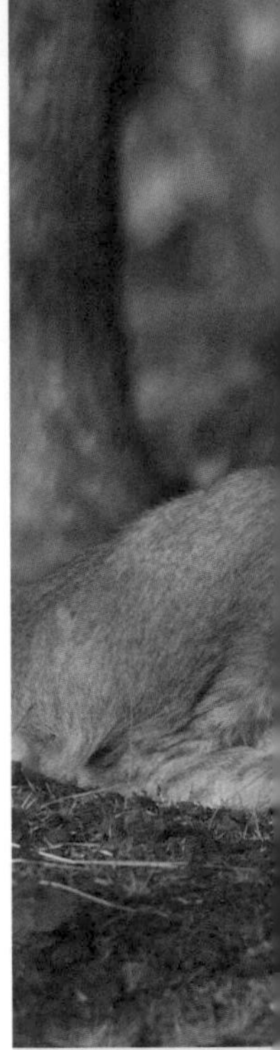

Ahmedabad – Gir National Park

Visit India's Gujarat for its Gandhi heritage (he was born here), exquisite textiles and unique landscapes: arranging a car and driver to get around makes for an easier trip.

FACT BOX

Carbon (kg per person): 149
Distance (km): 1200
Nights: 12
Budget: $-$$
When: Nov-Feb

❶ Ahmedabad

At first glance, Gujarat's bustling capital might seem traffic-choked and unappealing. But take your time exploring and you'll uncover some worthwhile highlights – the superb Calico Museum of Textiles and Gandhi's Ashram headquarters to name two. Throw in some mesmerising temple architecture, glittering night markets and beautifully restored heritage hotels and you'll soon change your mind about the city.

Drive 167km west to Little Rann.

❷ Little Rann

The extraordinary salt flats of Little Rann cover thousands of square kilometres. Famous for their wildlife, they're otherwise seemingly barren, stretching out to the horizon with bare scraps of vegetation appearing like mirages. This is the last home of the peachy-brown Indian wild ass, which roams free and can reach speeds of 30mph. Salt farms dot a desolate landscape in which *agariyas* (salt farmers) make their living extracting salt from shallow mud pans.

Continue 280 west.

❸ Bhuj

Rebuilt after being devastated in a 2001 earthquake, the Kachchh capital has eerily beautiful palaces that deserve a wander, but the main reason to come here is to visit some of the artisanal rural villages around

ABOVE: JAMA MASJID, AHMEDABAD

ABOVE: ASIATIC LIONS, GIR NATIONAL PARK

the town. Here you can see products being made by skilled workers and buy directly from local handicrafts cooperatives, so save your souvenir shopping for Bhuj.

Continue 335km south.

❹ Junagadh

The city of Junagadh sits at the foot of holy Girnar Hill, where pilgrims climb some 10,000 steps to a summit dotted with Jain and Hindu temples. It's an incredible experience to walk up here, starting at dawn and climbing alongside porters who carry those unable (or unwilling) to walk all the way, as well as sadhus, barefoot and even naked.

Drive 50km southeast to Sasan Gir Village, the jumping off point for safaris into the lion sanctuary.

❺ Gir National Park

Gir National Park is the last sanctuary of the Asiatic lion. It's a large, hilly, forested area in which there are now over 400 of the endangered big cats roaming free, as well as deer, crocodiles and leopards. You can take a 4WD safari here from Sasan Gir village; the best chance of spotting the wildlife is at dawn or dusk.

From Sasan Gir it's 368km back to Ahmedabad.

ALTERNATIVES

Extension: Diu

Gujarat's most laidback spot, the island of Diu is a favourite place for Gujaratis to go sink a beer (most of the rest of the state is dry). Much whitewashed architecture remains here from the time it was a Portuguese colony – some local restaurants even serve up Portuguese-influenced dishes.

Trains run to Diu from Sasan Gir (3hr 30min) and Junagadh (6hr 15min).

Diversion: Gondal

To break the long journey from Sasan Gir back to Ahmedabad, stop off at Gondal, once a princely capital run by a family of Rajputs believed to have descended from Krishna. It hugs a greenery-fringed river and still retains several handsome palaces, one of which has been turned into a heritage stay.

The journey from Sasan Gir takes around 3hr 30min.

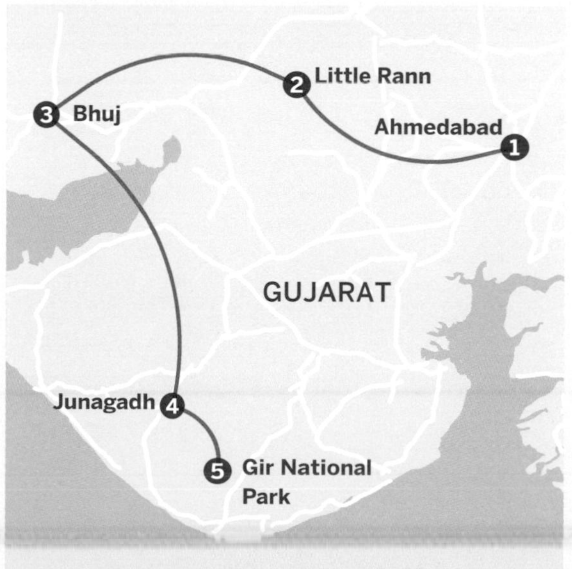

TOUR ROYAL CITIES IN INDIA'S RAJASTHAN

Jaipur – Jodhpur

Rajasthan's princely state is a whimsical wonderland of fortresses, palaces and cities, each with their own signature colour.

FACT BOX

Carbon (kg per person): 48
Distance (km): 643
Nights: 12
Budget: $-$$
When: Nov-Feb

ABOVE: CITY PALACE FROM LAKE PICHOLA, UDAIPUR

❶ Jaipur

Rajasthan's capital is a head-swirling metropolis where the buildings are painted golden pink (giving Jaipur its 'Pink City' nickname). Busy bazaars line colonnaded streets, their stalls laden with the area's famous block-printed textiles and hand-painted ceramics. First stop should be the splendid Hawa Mahal (Palace of Winds), a rose-coloured honeycomb-like mansion with five floors of lattice-work windows. Jaipur centres on the lavish City Palace, now a museum, while palatial in a different way is the city's cinema, Rajmandir, a wonder of 1970s flamingo-pink Art Deco, as fabulously flamboyant inside as out. A rickshaw ride outside the city will take you to another regal relic, Amber Fort, a magnificent clifftop-hugging castle with superlative views of the city.

Multiple trains run to Udaipur every day (7hr 30min), but the best option is the overnight sleeper.

❷ Udaipur

Is this the most romantic of Rajasthan's cities? With its ivory, cupola-topped *havelis* (traditional mansions), temples and icing-pretty City Palace hugging the edge of Lake Pichola, many would say yes. At the centre of the lake is an island where the luxurious Lake Palace Hotel sits in splendid watery isolation,

ABOVE: CITY PALACE, JAIPUR

only accessible via boat. The city might have become overly commercialised in recent years, but sitting on a rooftop watching the sun sink behind the lake as birds swoop overhead is still an essential and hugely charming highlight of a trip to India. Beyond the city limits are the Aravalli Hills, an undulating mauve-tinged mountain range that make a backdrop fit for a Mughal miniature painting.

Hire a taxi to get to Jodhpur (5hr), stopping on the way to see the milk-white marble temples of Ranakpur and Kumbhalgarh's magnificent hilltop fort.

3 Jodhpur

It's easy to guess Jodhpur's nickname when you see that its houses are all painted a heavenly powder blue. Meranghah, the Blue City's regal fort, was built on a 300m-high outcrop in the 15th century, with its vast walls thrusting a further 100m skywards as if growing out of the rock. It dominates the city and you can whizz around it on zipwires. In the streets beneath, even more than in Jaipur, the winding city bazaars feel like a magical link to medieval times, selling spices, incense, leather and fabric.

Head back to Jaipur by overnight train from Jodhpur Junction.

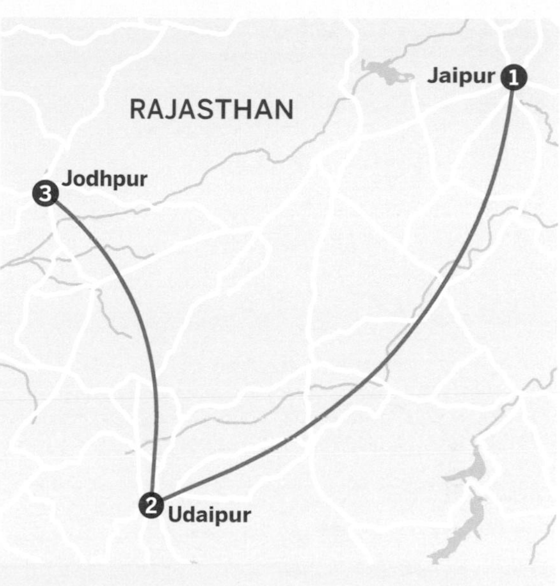

ALTERNATIVES

Extension: Pushkar

Hindu pilgrimage destination Pushkar is a whitewashed small town that's home to over 500 temples, curled around a lake said to have formed when Brahma dropped a lotus petal. It's famous for an extraordinary Camel Fair (Oct/Nov), but is an enchanting place year-round, with a lively market and 52 ghats from which pilgrims bathe in the sacred waters.

Take a bus from Jaipur to Ajmer (every 30min; 2hr 20min), then a taxi to Pushkar (15min).

Diversion: Jaisalmer

Add another colour to your trip is 'the Golden City'. A magnificent sandstone fort-town in the middle of barren desert flats, Jaisalmer was a major staging post on the caravan routes along the Silk Road. Its wealth and importance are carved in stone: the walled city is full of elaborately carved *havelis* and temples. Take a camel safari into the sparse, silky dunes nearby.

There's a daily train from Jodhpur Junction to Jaisalmer (6hr).

SPICES, BOATS AND BACKWATERS IN INDIA'S SOUTH

Mumbai – Kerala's backwaters

Take a trip to the spice-rich Indian south, spotting wild elephants, boating the backwaters and exploring mesmerizing Mumbai.

FACT BOX

Carbon (kg per person): 91
Distance (km): 1477
Nights: 14
Budget: $-$$$
When: Nov-Feb

❶ Mumbai

Home of Bollywood and big business, India's most glamorous city is also its biggest and the gateway to the slower-paced south. Built across multiple islands, Mumbai contains many examples of Art Deco and Indo-Saracenic architecture in the British-era historic centre, including the Gateway to India on the harbour, where boats take visitors across to ancient temples on Elephanta Island. The city's gaping wealth divide is exemplified by Mumbai having one of India's largest slums – and also one of the world's most expensive houses.

Take the overnight train from Chhatrapati Terminus to Margao (11hr); book at least a month ahead. From Margao station take a bus or taxi to Palolem (1hr).

❷ Palolem, Goa

With its arc of silky-sanded, palm-fringed beach, Palolem lets travellers live out the tropical dream, staying in wooden huts by the water and eating coconut-fragrant southern Indian dishes at thatched-roof restaurants. It's one of the safest beaches to swim in Goa, and you can also paddleboard and kayak, visit spice farms inland or practise beachfront yoga.

Take a flight from Goa airport to Kochi (via Bengaluru; 4hr) to continue your trip in Kerala.

❸ Kochi, Kerala

For centuries, the Keralan port of Kochi had traders, colonialists and chancers thriving on the spoils of

ABOVE: HOUSEBOAT ON KERALA'S BACKWATERS

ABOVE: GATEWAY TO INDIA, MUMBAI

the spice coast. The historic centre resonates with the legacy of commerce and settlers, from its Delft tile-lined synagogue to its Chinese fishing nets, and from its colonial-era graveyard to its Portuguese-era bungalows. Today the streets are filled with restaurants, ayurveda practitioners, homestays and galleries, making it a lovely place to explore for a few days.

Take the train south from Kochi to Alappuzha (2hr), jumping-off point for Kerala's famous backwaters.

4 Kerala's backwaters

Kerala's web of rivers, canals and lagoons spreads inland from the coast, meandering for nearly a thousand kilometres. Lined by palm trees and rice paddies, with villages alongside, these water channels are busy with boats, transporting coir (coconut fibre), fish and villagers around this watery world. You can rent a houseboat here – romantic craft adapted from traditional rice barges with driver, chef and staff on board – and take a lazy backwaters trip over several days, one of India's most magical and tranquil experiences.

From Alappuzha, return to Kochi and its international airport for onward flights.

ALTERNATIVES

Extension: Munnar

Inland Munnar is a hill station surrounded by tea plantations, mountains, red splashes of the Flame of the Forest tree, emerald valleys and quicksilver waterfalls. At a cool 1600m it provided respite from the heat during the British Raj – some of the period's veranda-skirted villas remain – and is still a peaceful haven, ideal for walks, tea tasting and discovering fragrant spice gardens.

Munnar is a 4hr taxi or bus ride from Alappuzha.

Diversion: Periyar Wildlife Sanctuary

This vast sanctuary is a patchwork of jungle and hills harbouring around a thousand wild elephants, as well as bison, sambar, wild boar and the occasional tiger. It centres on a huge lake created by the British; there are regular boat cruises. For a better chance of spotting wildlife, take a trek led by local villagers.

The nearest town is Kumilly, accessible by car or bus from Kochi (around 5hr).

ACROSS BHUTAN FROM PARO TO PUNAKHA

Paro – Punakha
This classic itinerary is a perfect introduction to the highlights of quirky Himalayan Bhutan. Private car hire with driver and guide is mandatory, so sit back and enjoy the ride.

FACT BOX

Carbon (kg per person): 35
Distance (km): 280
Nights: 6
Budget: $$$
When: Feb-Apr, Oct-Dec

ABOVE: TAKIN AT MOTITHANG TAKIN PRESERVE, THIMPU

❶ Paro

With the country's only international airport, Paro is the most visited town in Bhutan and, fortunately, also one of its most interesting, so plan to spend your first couple of nights here. The Paro Dzong fortress-monastery is unmissable, as is the nearby National Museum which charts the country's history and traditions, while the beautiful 7th-century Kyichu Lhakhang is one of the kingdom's holiest temples. Most photogenic of all is the iconic Tiger's Nest Monastery, perched miraculously on a cliff-face high above the Paro Valley. Often overlooked is the snail-shell-shaped Dumtse Lhakhang, featuring some of the finest murals in Bhutan. Central Paro's traditional wooden buildings sell some of the country's best handicrafts, so stock up on souvenirs here.

ⓘ Arrange a car and driver in Paro; it's a 60km drive on to Thimpu.

❷ Paro to Thimpu

On the drive to Thimphu there are several excellent opportunities to get off the tour bus route, so take your time and choose a couple of detours here, as well stopping at the traditional incense workshop at Bondey. An excellent three-hour excursion is to the pilgrimage site of Drak Kharpo, with its meditation

ABOVE: TIGER'S NEST MONASTERY

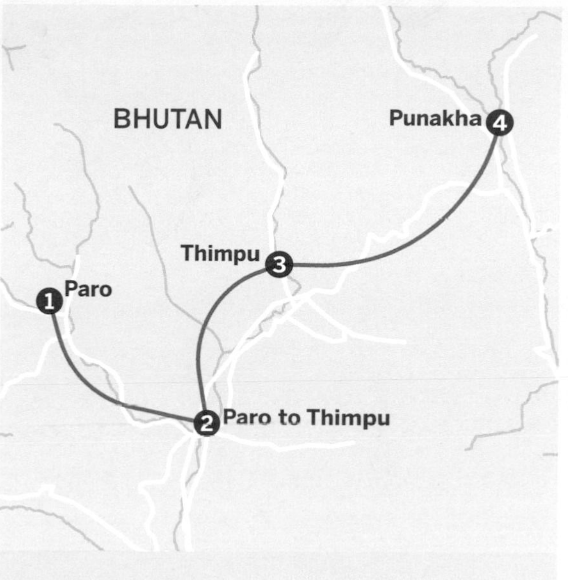

caves and pilgrimage path. Further east, Tamchhog Gompa is a timeless private temple accessed across a 15th-century iron link bridge built by the remarkable Tibetan Renaissance man Tangthong Gyalpo.

4 Thimphu

Bhutan's capital is a fascinating blend of the medieval and modern. Apart from the town's several ancient temples, it's also the best place to see a takin (Bhutan's national animal), catch a traditional archery tournament, and learn about arts and crafts heritage with a visit to a *thangka* painting school or silver jewellery workshop. If you can, time your visit to catch the fascinating weekend farmers market.

It's a 74.5km drive from Thimpu to Punakha.

5 Punakha

Get an early start from Thimphu to reach the Dochula Pass early and revel in the clearest Himalayan views. After snapping photos, descend to the eyebrow-raising temple of Bhutan's 'Divine Madman', whose phallus is painted on houses across the country. Once you've arrived in subtropical Punakha, head for its highlight – a huge *dzong* (fortress-monastery) of historic national importance and the nation's most beautiful building.

Head back to Paro to catch an onward flight.

ALTERNATIVES

Extension: Bumdrak

For a taste of Bhutan trekking, hike half a day from the Paro Valley up to Bumdrak, a 17th-century cliff hermitage at a breathless 3900m. Agencies arrange a night in a tent here (with a real bed) and full camping meals, leaving you to soak up the star-drenched Himalayan sky. The next day, descend early to the Tiger's Nest Monastery, dodging the crowds; temples en route offer fabulous views of the Tiger's Nest.

Arrange your trip through an agency.

Diversion: Haa

If you have some extra time, detour to drive over Bhutan's highest motorable pass, the Chelela, to the remote, forested Haa valley, dotted with shrines, monasteries and traditional villages. En route from Paro, stop off at Kila Nunnery and the cliffside meditation retreat of Dzongdrakha Goemba. The next day, drive on to Thimphu to resume the tour.

Count on three days to make the circuit to Haa.

BOUDHA TO BUDDHA IN NEPAL

Kathmandu – Lumbini

This classic highlights-hitting trip links Nepal's two holiest Buddhist sites with its two largest cities and its most famous national park.

FACT BOX

Carbon (kg per person): 23
Distance (km): 815
Nights: 14
Budget: $
When: Oct-Mar

❶ Kathmandu

Kick things off by heading to the Tibetan centre of Boudha, also called Boudhanath, to follow the maroon-robed monks and Tibetan exiles as they circumambulate Nepal's largest stupa. Boudha is just outside Kathmandu so you can base yourself in the capital and visit the medieval royal architecture of its Durbar Square and the city's endlessly fascinating Old Town, as well as its famously chaotic backpacker district of Thamel, lined with bookstores, mountain-gear shops and craft centres.

Take the morning tourist bus to Pokhara (6hr)

❷ Pokhara

With its traveller-friendly lakeshore cafes and restaurants, Nepal's second city is a great place to unwind. Hike up to the Shanti Stupa for views over the lake of Phewa Tal to the immense Annapurna range looming just 40km away. If you have the time, Pokhara is the perfect place for a short trek in the Annapurna foothills, for anywhere between three and 10 days; consider loops to Ghandruk, Ghachok or Poon Hill for a taster.

Take the daily tourist bus to Chitwan (7hr).

❸ Chitwan

The steamy, jungly plains of the Terai region border

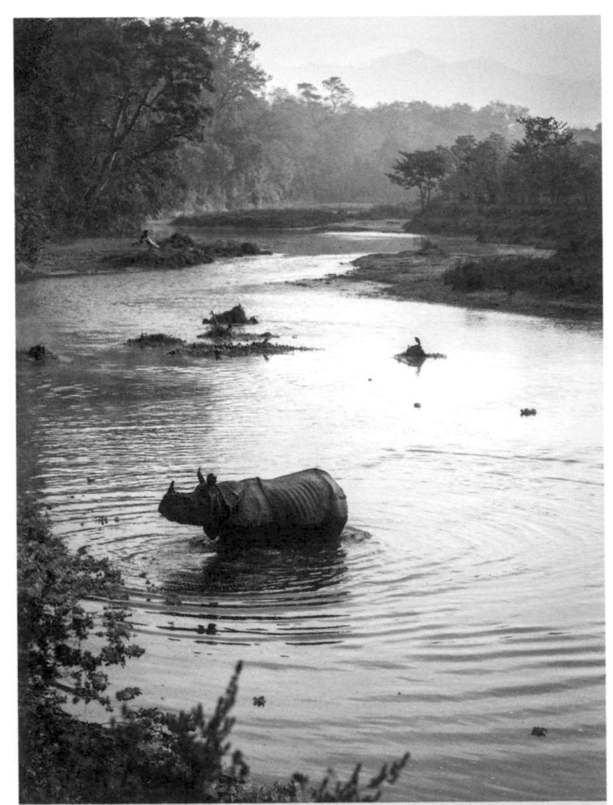

ABOVE: ONE-HORNED RHINO, CHITWAN NATIONAL PARK

ABOVE: DURBAR SQUARE, KATHMANDU

India and reveal a different side to the standard image of Nepal. Chitwan National Park here provides a sanctuary to one-horned rhinos, elephants, gharial crocodiles, leopards, gaur bison and Bengal tigers, amongst others. You can walk with elephants through the forest, go on a dawn jeep or birdwatching safari and overnight in luxurious lodges, enjoying a post-safari sundowner as you listen to primate hoots and howls while dusk falls over the forest.

A morning bus runs to Lumbini via Bhairawa (5hr).

4 Lumbini

This itinerary finishes its pilgrimage at the very spot where Buddha was born, now just the archaeological remains of a small temple and a commemorative pillar erected by the Emperor Ashoka. Rent a bicycle and pedal between dozens of other temples built in the various national styles of Buddhist nations from across the world, then visit the archaeological site of Kapilavastu (Tilaurakot) and walk through the same gateway where Prince Siddhartha renounced the material world and stepped out on his spiritual journey to become the 'Awakened One'.

A daily tourist bus heads back to Kathmandu (8hr) for international connections.

ALTERNATIVES

Extension: Patan

Kathmandu's sister city definitely rewards a day trip. It offers a similarly grandiose Durbar (Royal) Square, the country's best museum, and fascinating backstreets littered with hidden workshops, Buddhist stupas and golden-roofed temples. You'll also find the country's best restaurants and excellent fair-trade handicraft shops, far from Kathmandu's backpacker clamour.

Patan is a 30min taxi ride from Kathmandu.

Diversion: Bandipur

This traditional hilltop Newari town is a great place to break the journey between Kathmandu and Pokhara. Several traditional merchants' houses have been converted into inns and offer stylishly atmospheric places to stay. There's plenty to do here, with a vibrant cultural life in the small village and hikes to nearby mountain viewpoints, temples, caves and even a silkworm farm.

Bandipur is connected by bus to Pokhara (3hr).

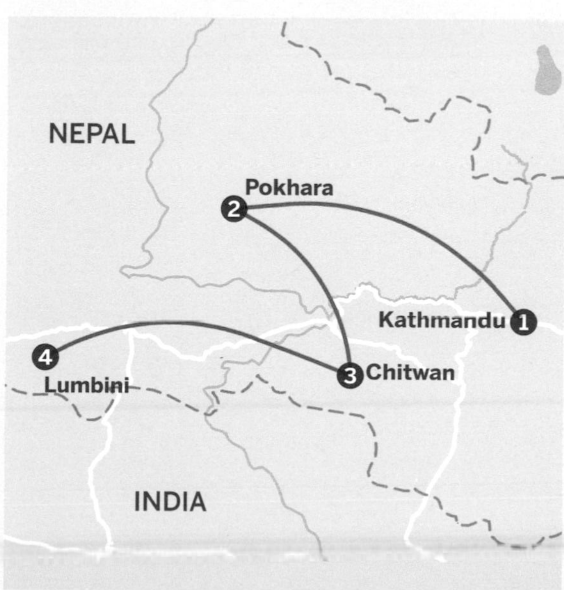

SRI LANKA'S ANCIENT CITIES

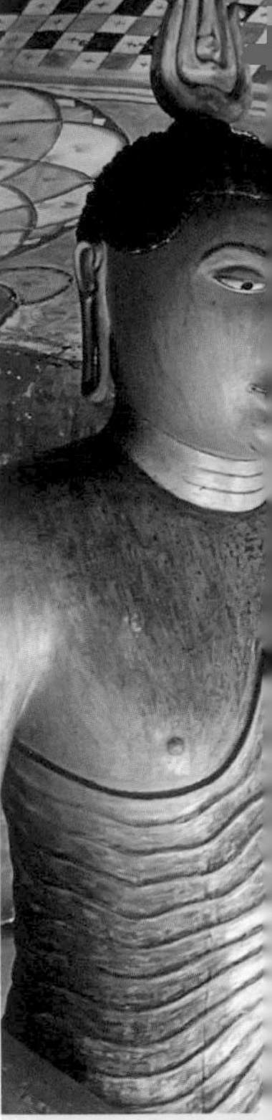

Dambulla – Mihintale

This journey through the heart of Sri Lanka links great Buddhist monuments with ancient capitals and some of the country's most iconic historic sites.

FACT BOX

Carbon (kg per person): 27
Distance (km): 250
Nights: 7
Budget: $$
When: year-round

❶ Dambulla

A two-hour drive from cultural capital Kandy, past a series of fragrant spice gardens, Dambulla is famous for its extravagant Buddhist cave murals. The 2000-year-old paintings and statues spread over five caves burst with colour and character and rank as some of the most beautiful examples of Buddhist art anywhere.

Take the bus to Sigiriya (frequent; 45min).

ABOVE: SIGIRIYA

❷ Sigiriya

When King Kasyapa decided to build a palace atop the 180m-high rock at Sigiriya his architects must have thought him mad, but 1500 years later this royal residence still stands and is one of Sri Lanka's unmissable sights. The hike up passes formal water gardens and famous rock frescoes, climbing ladders between the gigantic paws of a long-vanished stone lion. From the summit, views stretch across a sea of treetops. Catch dawn vistas of the rock from a different angle at nearby Pidurangala rock.

Take the bus to Polonnaruwa, changing at the Dambulla Hwy (frequent; 2hr).

❸ Polonnaruwa

The 800-year-old ruined city of Polonnaruwa was Sri Lanka's capital for over two centuries. Its former importance is seen in the hundreds of stupas, stone palaces, temples, monasteries and water tanks that sprawl across the large site. Elegant stone carvings and sublime rock-carved Buddhas are the artistic

ABOVE: DAMBULLA CAVE TEMPLES

highlights, and there are plenty of hidden corners to explore away from the crowds.

Catch a bus to Anuradhapura (hourly; 3hr).

4 Anuradhapura

For 1300 years Anuradhapura was the epicentre of Buddhism in Sri Lanka. What remains today is a town-sized collection of colossal stupas that stand as some of the largest constructions of the ancient world. The bodhi tree at the Mahabodhi Temple attracts devotees as the most sacred site in the country. Rent a bike or rickshaw to get to the different monasteries, and start early – temperatures can hit 40°C in the afternoon.

There are frequent buses run to Mihintale (30min) or hire an autorickshaw.

5 Mihintale

From Anuradhapura it's an easy day trip to this mountain-top pilgrimage site where Buddhism first took root in Sri Lanka. Climb with pilgrims up stone stairways and past cave dwellings to the dagoba marking the spot where the Emperor Ashoka's son converted the Lankan king.

Back in Anuradhapur, take one of the frequent buses or trains to the capital Colombo (6hr) for onward connections.

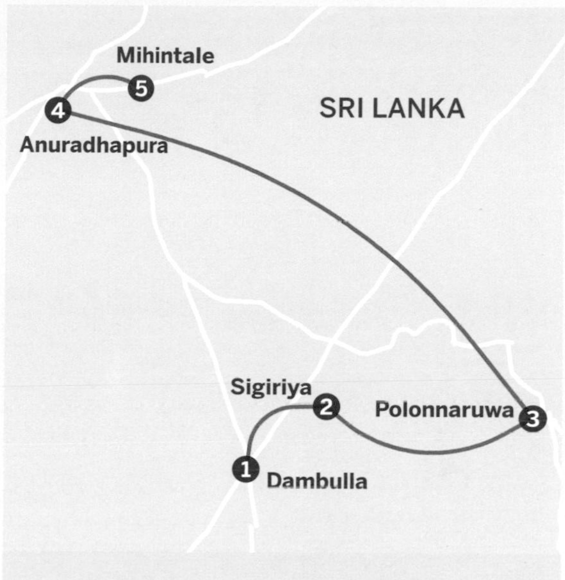

ALTERNATIVES

Extension: Minneriya

Between Polonnaruwa and Anuradhapura/Sigiriya, stop for an early morning or late afternoon jeep safari into this national park, home to sambar deer, leopards, crocodiles and wild elephants. In August and September over 200 of the latter gather here at a single location: a world-renowned natural event known as 'The Gathering'.

Minneriya can be a stopover between buses on the Polonnaruwa-Anuradhapura road.

Diversion: Ritigala

Hidden deep in the forest between Polonnaruwa and Anuradhapura, this half-forgotten monastery lies on the flanks of Ritigala Mountain. Footpaths wind up stone stairways past vine-swathed courtyards, bathhouses and meditation complexes. It's great fun to explore and you'll likely have the site to yourself.

Get here via private car or autorickshaw from Habarana, a 15km return detour off the Polonnaruwa-Anuradhapura road.

ANDAMAN ISLANDS ADVENTURE

Port Blair – Neil Island (Shaheed Dweep)
For a beach escape in a far-flung corner of the world, you can't beat a trip to India's distant, delicious Andaman Islands.

FACT BOX
Carbon (kg per person): 2.4
Distance (km): 130
Nights: 10-15
Budget: \$-\$\$
When: Oct-May

1 Port Blair

With flights to/from mainland India (1370km away), the Andamans' lively coastal capital Port Blair is everyone's first stop. Some intriguing historical sights and great frills-free dining keep you busy while awaiting/arranging onward transport. Explore the Cellular Jail National Memorial (a 20th-century British-built prison for political dissidents) and the jungle-wreathed colonial-era ruins of Ross Island (Netaji Subhas Chandra Bose Dweep), learn about the islands' threatened indigenous communities and, time permitting, day-trip to the 280-sq-km Mahatma Gandhi Marine National Park.

The Andamans' private ferry operators, Makruzz and Green Ocean, have daily services (1hr 30min-2hr 30min) between Port Blair and Havelock Island (Swaraj Dweep); book as far ahead as possible. There are also daily government ferries, only bookable a few days ahead and in person, in this case at Port Blair's Phoenix Bay Jetty.

2 Havelock Island (Swaraj Dweep)

Sugary gold-white beaches, blazing-pink sunsets, sparkling turquoise waves, lushly forested hills and some of Asia's best diving – there are plenty of reasons why heavenly, super-laidback Havelock pulls people to the Andamans. You'll probably want to stay forever, but five days make a great taster. Catch fabulous Radhanagar Beach crowd-free in the

ABOVE: HAVELOCK ISLAND DIVING

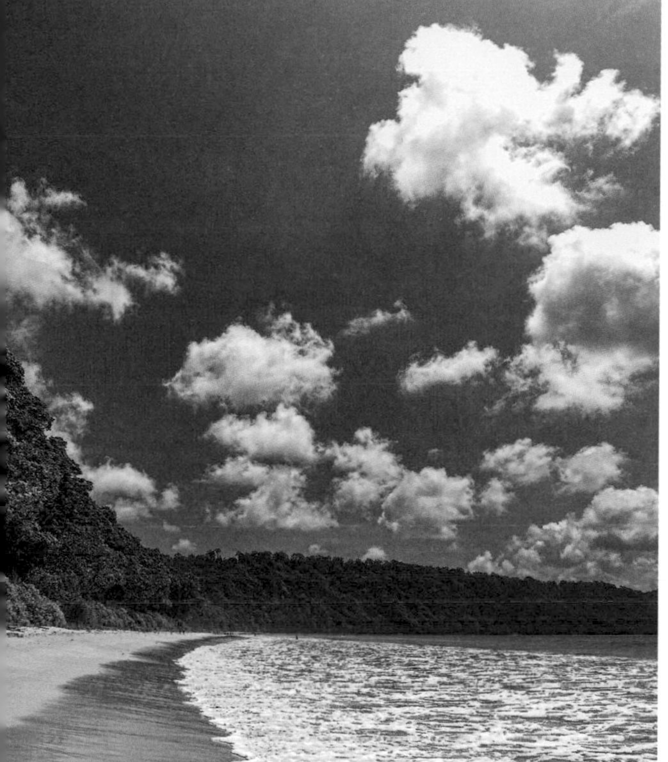

ABOVE: RADHANAGAR BEACH, HAVELOCK ISLAND

early morning, laze on palm-fringed platinum sands all around the island, link up with local dive schools, wander the lively main bazaar and go snorkelling, hiking and kayaking. Down-to-earth cafes and restaurants, bamboo beach huts and a couple of eco-conscious, tropical-inspired boutique hotels also await (beautiful Jalakara is reason alone to visit the Andamans).

Government and private ferries link Havelock with Neil Island (Shaheed Dweep) several times daily (around 1hr).

3 Neil Island (Shaheed Dweep)

More charmingly rustic than Havelock, easy-going Neil unfolds in a vision of electric-green rice paddies, bobbing coconut trees and peaceful little beaches. The heart of the island is its small bazaar and the nearby jetty. Days here (perhaps three or four) are best spent roaming around by bike or scooter, visiting the beaches (sunrise saluting at eastern Beach 5; sunset-gazing at Beach 1) and doing a bit of snorkelling or diving.

There are several daily government and private ferries back to Port Blair (1-2hr), where you'll probably need to stay overnight before onward connections.

ALTERNATIVES

Extension: Middle & North Andaman

Few travellers venture north into the remote jungles of Middle and North Andaman. But rewards include the dazzlingly beautiful twin islets of Ross and Smith, witnessing rare turtles nesting at Kalipur Beach, and family homestays in Mayabunder.

Buses connect Port Blair and Diglipur (several daily; 12hr), some via Mayabunder (10hr), as do a couple of weekly ferries (17hr). Many travellers hire a car and a driver.

Diversion: Little Andaman

For fabulous surf, cream-coloured sand beaches, simple beach-hut accommodation and a wonderfully mellow vibe, travel south from Port Blair to delightful, well-off-the-beaten-track Little Andaman (much of which is an off-limits tribal reserve).

Ferries sail between Port Blair and Little Andaman (4-5 weekly; 5hr 30min to 8hr 30min).

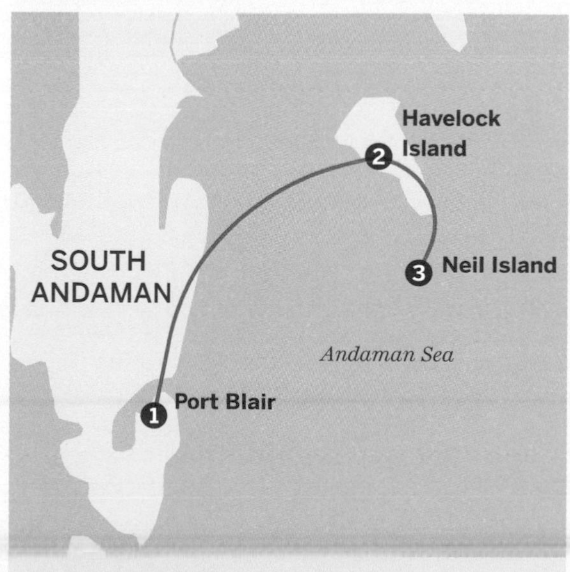

KERALA IN DETAIL, FROM BEACHES TO BACKWATERS

Thiruvananthapuram – Kochi

Dig deep into India's serene southern state to find palm-fringed backwaters, wonderful wildlife and entrancing theatre.

FACT BOX

Carbon (kg per person): 35
Distance (km): 1035
Nights: 23–35
Budget: $-$$
When: Aug-Apr

ABOVE: PADDLING THE ALAPPUZHA BACKWATERS

❶ Thiruvananthapuram (Trivandrum)

Start in Kerala's capital, enjoying its excellent museums, intriguing architecture, fine South Indian cuisine and sandy beaches at Kovalam.

Take the train to Varkala (several daily; 1hr).

❷ Varkala

Catching fiery sunsets from Varkala's rust-red cliffs is a classic Kerala moment. Spend a few days here to make the most of the golden beaches, surf and yoga, mellow guesthouses and cafes and the banyan-shaded temple.

Take another train to Kollam (several daily; 30min-1hr).

❸ Kollam (Quilon) & Munroe Island

Busy Kollam is the gateway to calmer, palm-sprinkled Munroe Island where hushed waterways can be explored by boat.

Hop aboard the State Water Transport canal boat for Alappuzha (daily in season; 8hr), stopping at the famous Amritapuri Ashram, home of Amma, 'the Hugging Mother'.

❹ Alappuzha (Alleppey)

The main base for backwaters houseboat and kayak

ABOVE: MUNNAR TEA PLANTATIONS

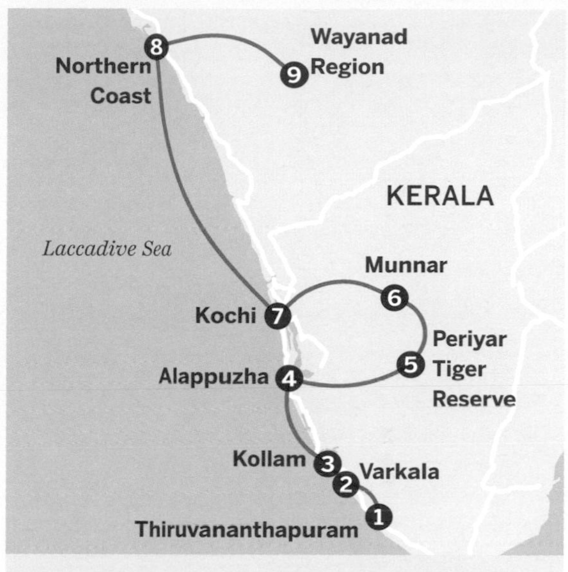

trips. Alappuzha is a lively town with traveller hang-outs and a golden-white beach.

Take a bus to Kumily in Periyar (daily; 6hr).

5 Periyar Tiger Reserve

Rippling across Kerala's Western Ghats, this large, lush wilderness is home to elephants, tigers, deer and more. You'll need three nights minimum for its hiking, spice plantations and cooking classes.

Several buses a day run to Munnar (4-5hr).

6 Munnar

South India's prime tea-growing region shines brightest when seen from an eco-focused homestay among the ludicrously green plantations. Go hiking to appreciate it.

Take a bus to Kochi (regular; 5hr).

7 Kochi (Cochin)

Calm Kochi shows off Kerala's cultural soul, with some of South India's most exciting accommodation, sights, culture and cuisine. Start with historical Fort Cochin and the palaces, synagogues, temples and bazaars.

For the Northern Coast you can take the train to Kannur (frequent; 5hr-6hr 30min); the daily service to Bekal Fort (8hr); or more frequent trains to nearby Kanhangad/Kasaragod.

ALTERNATIVES

Extension: Kerala beach escapes

Just 15km southeast of Trivandrum (and a tempting alternative first stop), Kovalam was once a tranquil fishing village. It's now a pleasant palm-studded beach resort, surrounded by respected ayurvedic retreats. Another blissful beachy area is Marari, with gorgeous accommodation, just north of Alappuzha.

Plentiful buses connect Trivandrum and Kovalam (30min). Northbound buses from Alappuzha stop near Marari (30min). Or take taxis.

Diversion: Mysuru (Mysore)

Over in Karnataka, magical Mysuru is awash with sparkling palaces and grand Wodeyar and British-colonial architecture – you'll need half a day for marvellous Mysuru Palace alone. The city is also a world-renowned yoga centre, and its 10-day Dussehra festival (Sep/Oct) is a highlight.

Buses run to Mysuru from Kannur (6 daily; 6hr), from Mananthavadi (5 daily; 3hr) and from Kalpetta (every 30min; 3hr 30min) in Wayanad.

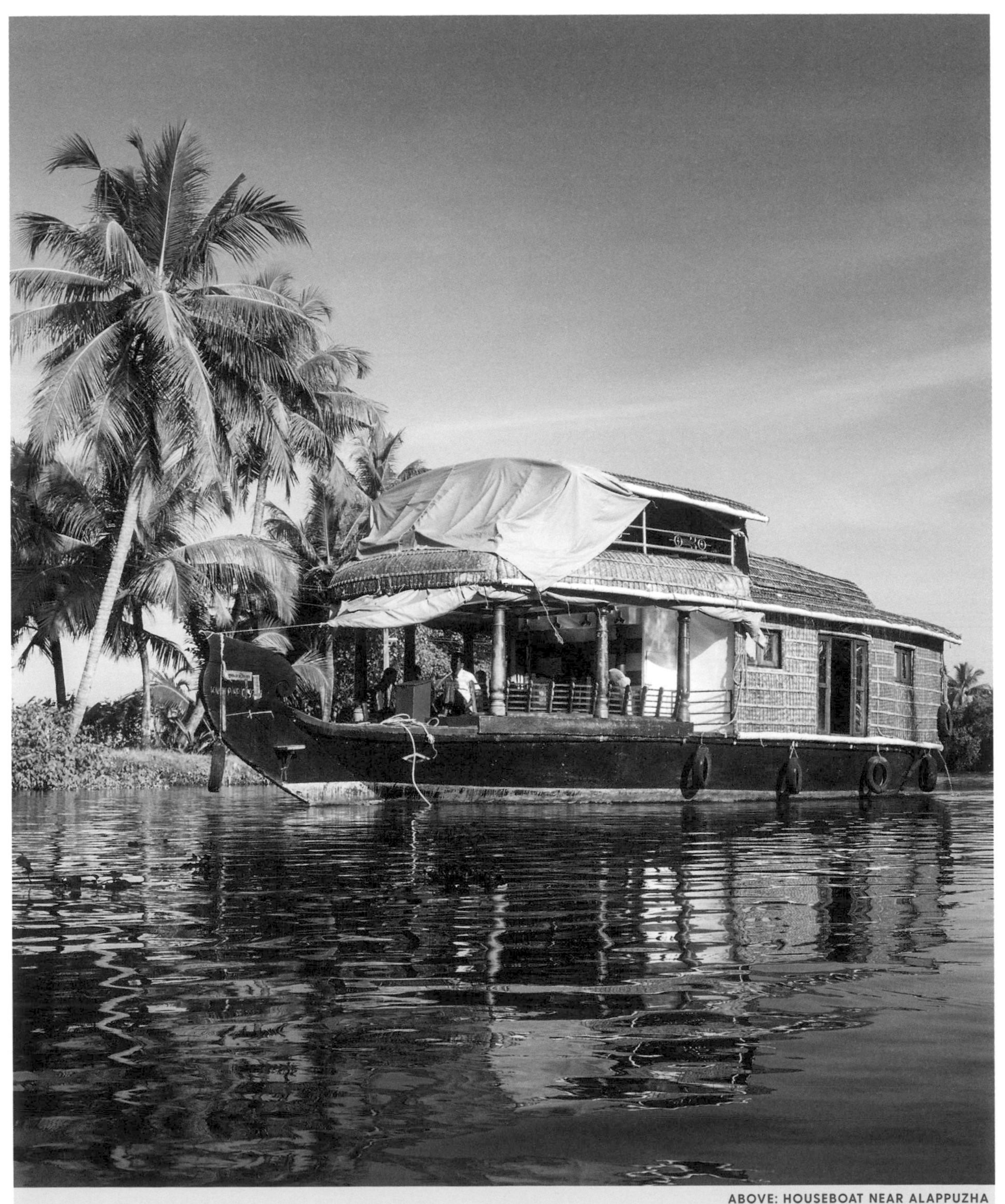

ABOVE: HOUSEBOAT NEAR ALAPPUZHA

8 Northern Coast

Tranquil, untouristed northern Kerala rewards anyone venturing to its coast and backwaters. Kannur is known for its enthralling *theyyam* rituals, gold-sand beaches and low-key homestays, while Bekal has miles of empty sands, a 17th-century fort and exquisite hotels.

Buses run from Kannur to the Wayanad Region town of Mananthavadi (regular; 3hr).

9 Wayanad Region

Take time to explore the unspoilt Wayanad Wildlife Sanctuary. Trek into the hills, relax in homestays and spot wild elephants in Kerala's secluded, mountainous, extraordinarily beautiful northeast corner.

Take a bus back to Kannur or to Kozhikode (very frequent; 3hr), both with airports, or bus back to Kochi (6hr).

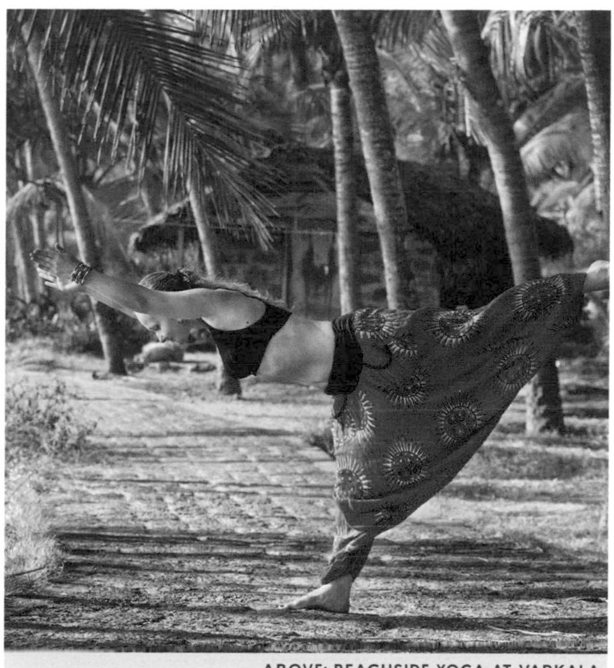

ABOVE: BEACHSIDE YOGA AT VARKALA

ABOVE: VARKALA SEASHORE

HIDDEN HIMALAYAS FROM KASHMIR TO LADAKH

Srinagar – Leh

Take the backdoor to the Himalayas, climbing through the alpine valleys of Islamic Kashmir to the desert moonscape of Buddhist Ladakh.

FACT BOX

Carbon (kg per person): 12
Distance (km): 419
Nights: 7-10
Budget: $$
When: Nov-Apr

❶ Srinagar

Timeless, sometimes troubled, but endlessly fascinating, Srinagar is easy to reach from Delhi (by plane or train and jeep) but is a world away. Papier-mâché patterns paint the interiors of medieval mosques, chinar trees shed their leaves over Mughal gardens and houseboats float serenely on the misty waters of Dal Lake. Check the security situation before you travel as a precaution.

Take a bus or jeep to Kargil (jeeps leave when full; 7-10hr).

❷ Kargil

The trip from Srinagar to Kargil over the Zoji La pass (3528m) leaves behind the idyllic alpine meadows of Kashmir for a lunar landscape of parched deserts in the rain shadow of the Himalaya. There are no must-sees in this busy bazaar town, but treks in the surrounding mountains reveal rock-carved Buddha images and the jaw-dropping scenery of the Suru Valley.

Take another bus or jeep to Mulbekh (jeeps leave when full; 2hr).

❸ Mulbekh

Mulbekh marks the transition from Muslim Kashmir to Buddhist Ladakh, something demonstrated powerfully by the 8m-high Buddha image by the roadside. Hikes cross the barren landscape to the tiny Shargol and Rgyal monasteries, and to the ruined castle high above Mulbekh.

Take the bus to Lamayuru (early morning; 2-3hr).

ABOVE: NISHAT BAGH GARDENS, SRINAGAR

ABOVE: LAMAYURU, LADAKH

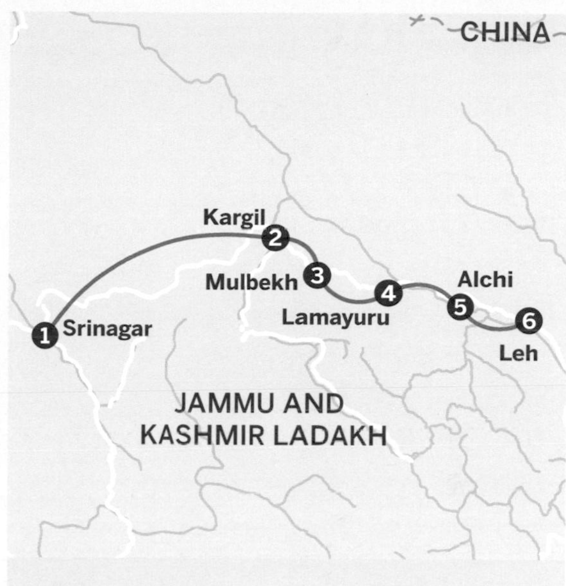

❹ Lamayuru

Crowning an eroded outcrop worm holed with caves, the gompa (monastery) at Lamayuru is a splash of colour against the dust-hued badlands. You can stay in the gompa itself or in village guesthouses, many offering awesome rooftop views.

Take the morning or early afternoon bus to Alchi via Saspol (2hr).

❺ Alchi

Tucked into a side valley off the main road, the crude buildings of Alchi's Choskhor Temple hide what could be Ladakh's most magnificent murals, executed in amazing technicolour – a stunning contrast to the arid, mud-coloured landscape all around.

Take the bus to Leh (early morning; 2-3hr).

❻ Leh

After the meditative emptiness of the stops since Kargil, arriving in Leh feels like a return to big-city living. Well, almost. There are ATMs, boutique hotels, traveller cafes, yoga classes and even places to grab a beer. But the backstreets are unpaved and lead out to medieval monasteries, eroded outcrops and apple orchards.

Take a flight to Delhi (daily; 1hr 30min) for onward connections.

ALTERNATIVES

Extension: Zanskar

With time on your hands, take a jeep along the rugged side-road leading south from Kargil into spectacular Zanskar, a broad plain circled by a crown of saw-tooth mountains. From dusty Padum, clamber up to the spectacular gompas at Karsha, Stongdey, Sani and Phuktal and be struck dumb by the amazing scenery.

Shared jeeps run from Kargil to Padum daily at dawn (10-12hr).

Diversion: Nubra Valley

Why speed south from Leh? North of the Ladakhi capital, a crude road strains across a mountain wall into the wonderfully tranquil Nubra Valley, at one time a busy stop on the Silk Road. Along the way you'll cross the Khardung La (5359m), the second highest motorable pass in the world. Beyond lie apricot orchards, peaceful tented camps and more magnificent monasteries.

Arrange a chartered jeep and permit in Leh (5hr).

MUMBAI TO GOA BY ROAD AND RAIL

Mumbai – Hampi

This rolling train and bus ride from Mumbai to the beaches of Goa adds in detours inland to take in some of India's most spectacular ruins.

FACT BOX

Carbon (kg per person): 67
Distance (km): 1777
Nights: 7
Budget: $$
When: Sep-May

❶ Mumbai

Mumbai is the low-stress gateway to India: less frantic than Delhi or Kolkata, more historic than Bengaluru, with the added bonus of some of India's best fine dining (seek out Trishna or Bombay Canteen to sample some exquisite flavours). The city's complicated history comes alive on the streets of the British-built Fort and on Elephanta Island – take a boat across the bay to see its carving-filled caverns offering a glimpse of the India that existed before Europeans came to town.

Take the Dadar-Madgaon Jan Shatabdi Express to Margao (daily; 8hr 30min). The train passes through some stunning scenery, best viewed in the all-windows premium Vistadome carriage. From Margao take a local bus to Panaji (every 15min; 1hr).

ABOVE: VITTALA TEMPLE CHARIOT, HAMPI

❷ Panaji

Panaji, the relaxed Goan capital, wears its Portuguese influence proudly on its sleeve in the form of terracotta-tiled mansions and gleaming white basilicas, nowhere more so in the Fontainhas neighbourhood.

Take the bus to Old Goa (every 10min; 30min).

❸ Old Goa

A day trip from Panaji to Old Goa is a journey into what might have been if Portugal had retained the reins of power. Timeworn cathedrals, churches and convents spill along the forest-cloaked banks of the Mandovi River, memorials to a city that was bigger than London in its prime.

ABOVE: CHHATRAPATI SHIVAJI MAHARAJ TRAIN STATION, MUMBAI

Take the bus back to Panaji and decide which beach you're going to head to.

4 North coast beaches

Going to Goa without hitting the beach would be preposterous, and on the state's north coast you're spoilt for choice. Local buses zip to Candolim, Calangute and Baga (every 10min; 30min), each offering an idyllic spread of palm-backed sand and palm-thatched, beachside seafood cafes for easy-going lunches and lazy sundowners.

Back in Panaji, take the bus to Margao (every 15min; 1hr), from where you can catch the bus to Hampi (3 daily; 11hr).

5 Hampi

The ruined capital of the Vijayanagar Empire is a place of wonder, with ruined temples spilling across a surreal landscape of eroded outcrops and house-sized boulders. Contrast the intricate, eroded Vittala Temple with the still-used Virupaksha Temple, topped by soaring, deity-covered gopurams (gateway towers).

Round off the trip with a train ride back to Mumbai (2 daily; 19hr) from Hosapete, a 30min bus ride from Hampi (every 30min).

ALTERNATIVES

Extension: Pune

It's a long haul from Hosapete to Mumbai, so consider breaking the journey at Pune, where India's newfound internationalism takes physical form in European-style coffeehouses, craft breweries and flashy fusion restaurants. Work off the calories visiting temples, forts, palaces and, in honour of your train trip, Joshi's Museum Of Miniature Railways.

Trains connect Hosapete and Pune (several daily; 17-18hr).

Diversion: Bengaluru

If you aren't set on looping back to Mumbai, chug on from Hampi to Bengaluru, Karnataka's appealing capital, lauded for its fine food, animated nightlife, ritzy malls and cosmopolitan vibe. Alongside feasting and retail therapy, take trips to palaces, temples and one of India's best art spaces, the National Gallery of Modern Art, before relaxing in popular Cubbon Park.

Trains run from Hosapete to Bengaluru (several daily; 9hr).

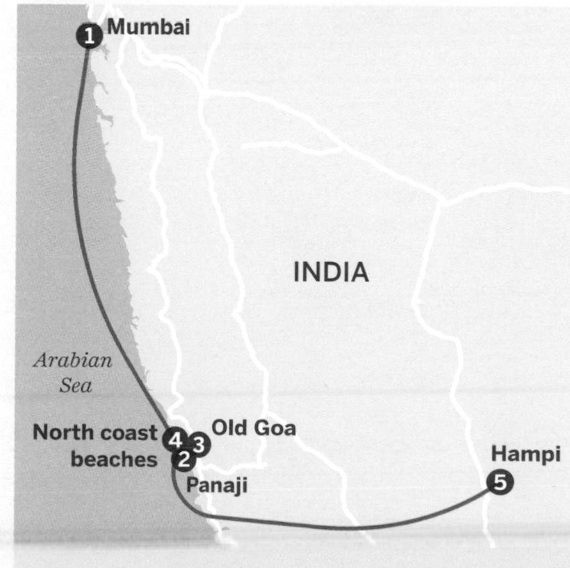

SRI LANKA'S SERENE HILL COUNTRY

Colombo – Badulla

Take a seat on Sri Lanka's best-loved train ride, rattling uphill from Colombo to Kandy and on through tea plantations to the heart of the Hill Country.

FACT BOX

Carbon (kg per person): 13
Distance (km): 252
Nights: 7
Budget: $$
When: Dec-Mar

❶ Colombo

The Sri Lankan capital is a place that grows on you, so linger a few days to feast on fine food and appreciate its contrasts – colonial-era churches huddling between skyscrapers, gleaming malls competing against manic bazaars, and Buddhist and Hindu temples painting the streets with colour.

Take the train from Fort Station to Kandy (10 daily; 3hr).

❷ Kandy

The famous Tooth Temple – home to what may be a genuine tooth of the Buddha – fires up the imagination, but there's much more to see in the capital of the Hill Country. Then there's the stunning journey up here by train, leaving behind the baking heat of the coast for the green cool of the hills.

Take the train to Nanu Oya (5 daily; 4hr), then a three-wheeler to Nuwara Eliya (20min).

❸ Nuwara Eliya

Travelling from Kandy to Nuwara Eliya by train, you can admire the views through the giant windows of the 1st-class observation car, or watch the scenery glide past from the train doorway in second class, with your feet dangling just above the platforms. Once in Nuwara, stay in period hotels, take highland hikes and open a portal from the modern age to British-era Ceylon.

Head back to Nanu Oya (20min) and take the train to Haputale (5 daily; 1hr 30min).

ABOVE: TEA PLANTATIONS ALONG THE KANDY–NUWARA ELIYA LINE

ABOVE: TEMPLE OF THE SACRED TOOTH, KANDY

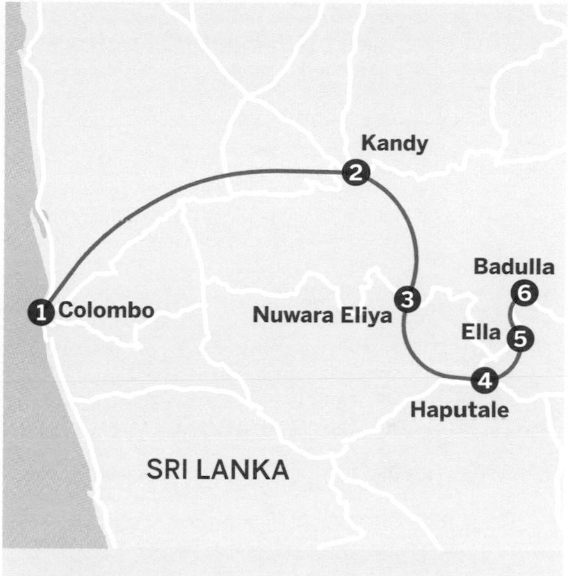

4 Haputale

Clinging to a narrow ridge, Haputale might just be the most stunningly located town in the Hill Country. Take a day to visit the Dambatenne Tea Factory, founded by Thomas Lipton, then catch the views f rom Lipton's Seat, where the tea mogul once surveyed his plantations.

Back on the rails to Ella (5 daily; 1hr 30min).

5 Ella

The train journey to Ella winds through an emerald garden of tea plantations. Ella itself offers up cosy guesthouses (and some of Sri Lanka's best home cooking) and chlorophyll-infused walks to Little Adam's Peak and Ella Rock.

Take the train to Badulla (5 daily; 1hr).

6 Badulla

It's a short hop to the end of the line at Badulla, but the scenery is no less lush and dramatic. Once you get here, you have a choice: board the train for the overnight trip back to Colombo, or plunge downhill to the surf resorts of the East Coast.

Take the train to Colombo (daily; 10hr) or bus to the east coast.

ALTERNATIVES

Extension: Horton Plains National Park

A worthy detour from Nuwara Eliya, this bare highland plateau is a marked contrast to the dense greenery elsewhere in the Hill Country. Hike across its lonely expanse, under the gaze of Sri Lanka's second- and third-highest peaks, Kirigalpotta and Totapola, and the views will take your breath away.

Drivers offer day trips by van from Nuwara Eliya; afterwards, get dropped off at Pattipola and take the Haputale train.

Diversion: Arugam Bay

After the misty cool of the hills, you'll appreciate the change in temperature as you drop down to the east coast. Waste no time: find a guesthouse, rent a surfboard, then test the waters at Baby Point before hitting the more ambitious breaks.

Regular buses run from Badulla to Monaragala (2hr), with connections on to Pottuvil, a short three-wheeler ride from Arugam.

INDIA'S GOLDEN TRIANGLE WITH A DIFFERENCE

Delhi – Jaipur

This take on India's most famous railway circuit offers up a mix of famous sights – Jaipur's palaces, the Taj Mahal, Delhi's Red Fort – and some surprises.

FACT BOX

Carbon (kg per person): 20
Distance (km): 575
Nights: 7
Budget: $
When: Sep-May

❶ Delhi

Chaotic Delhi offers instant sensory overload: just stepping out into the street brings an invigorating jolt of colours, sounds, smells and sensations. Soak it up in the ancient bazaars of Shahjahanabad, following tangled lanes lined with vendors selling wedding cards, saris, steel cookpots, slippers and kites, to reach the magnificent Jama Masjid and the iconic sandstone walls of the Red Fort.

Take the train to Mathura (hourly; 2-3hr).

ABOVE: PURPLE SUNBIRD, KEOLADEO NATIONAL PARK

❷ Mathura and Vrindavan

The birthplace of the Hindu god Krishna is a fascinating detour off the tourist trail, with boat-crowded ghats (ceremonial riverside steps) and backstreets studded with Krishna temples. A thirty-minute bus ride north, Vrindavan – where Krishna frolicked as a child – has more eye-catching temples, including the headquarters of the global Hare Krishna movement.

Get back on the train to Agra (hourly; 1hr).

❸ Agra

Agra is almost as famous for its touts as for the Taj Mahal, but bypass the former and step through the gates to glimpse the latter in all its gleaming glory. Spend as much time as you can here, watching the building change colour throughout the day. Then head to Agra Fort and the nearby ghost city of Fatehpur Sikri, built by Mughal emperor Akbar at lavish expense.

Jump on a bus to Bharatpur (every 30min; 1hr 30min).

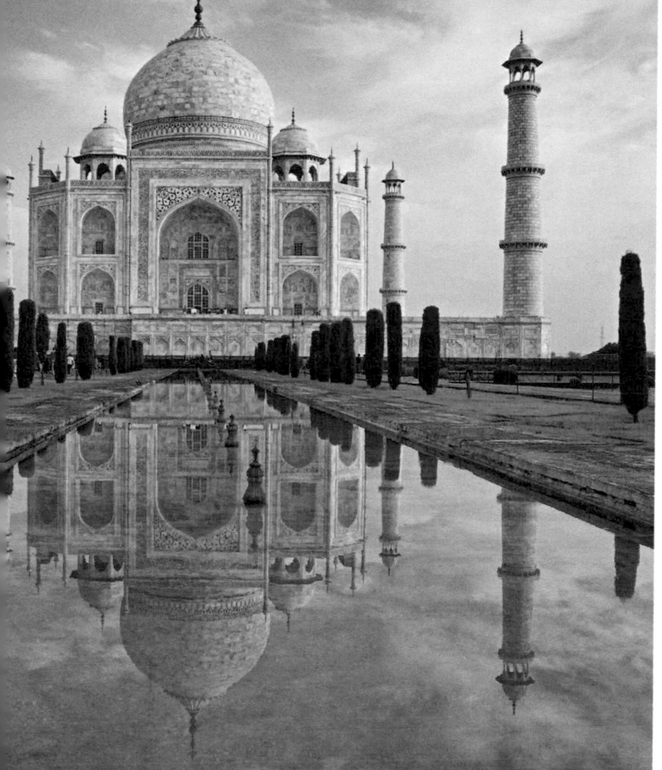

ABOVE: TAJ MAHAL, AGRA

❹ Bharatpur

The dusty plains town of Bharatpur has an interesting tumble down fort, but the main attraction is a waterlogged swamp just south of town, Keoladeo National Park. One of the world's most important wetland bird sanctuaries, it's thronged by some 360 bird species from October to February.

Take the bus to Alwar (hourly; 4hr).

❺ Alwar

Most tourists miss atmospheric Alwar, the oldest of Rajasthan's princely states. A foolish mistake. Swing by the once lavish palace, fort and cenotaph of the local maharaja, then hunt for tigers – with a camera, obviously – in the dry jungles of Sariska Tiger Reserve.

Take the train to Jaipur (hourly; 3-4hr).

❻ Jaipur

The Pink City earned its nickname for the rosy sandstone fortifications and palaces ruled over by generations of Rajput maharajas. Of the myriad sights, make time for elegant City Palace, the filigree-screened Hawa Mahal, nearby Amber fort and Jantar Mantar, the extraordinary observatory built by Jai Singh II.

Complete the 'triangle' with a train ride back to Delhi (9 daily; 4hr 30min-6hr).

ALTERNATIVES

Extension: Gwalior

From Agra, trains rattle south to Gwalior in Madhya Pradesh, described by the Mughal emperor Babur as 'the pearl of India'. After exploring the palaces, temples, museums and monumental gates of Gwalior Fort (built to accommodate royal elephants), peer at the time-worn Jain carvings on the surrounding outcrops, painstakingly restored after being defaced by the Mughals.

Trains run half-hourly to and from Agra (2hr).

Diversion: Ajmer and Pushkar

Having come as far as Jaipur, it's a no-brainer to extend the trip to ancient Ajmer, revered because of its associations with the Sufi saint Khwaja Muin-ud-din Chishti. Nearby lies Pushkar, whose sacred lake is almost engulfed by Hindu temples (come in October or November to catch Pushkar's famous camel fair).

Trains connect to Jaipur (every 30min; 2hr) and local buses zip to Pushkar (30min).

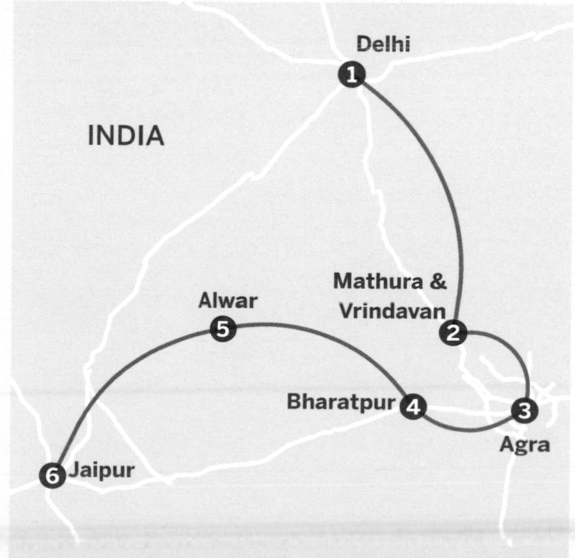

Trip Builder

DRIVE TRIBAL HEARTLANDS IN NORTHEAST INDIA

Guwahati – Kohima

Visit Buddhist and animist tribal villages in the rough, rugged and relatively little-visited Northeast States on this bone-rattling ride by jeep.

FACT BOX

Carbon (kg per person): 283
Distance (km): 2279
Nights: 14
Budget: $$$
When: Sep-May

❶ Guwahati

The Assamese capital is the gateway to the Northeast States, a wild sprawl of jungle-cloaked foothills only tenuously attached to the rest of India by culture, customs and geography. The easiest way to explore this tribal heartland is to arrange permits and a chartered jeep through a travel agency in Guwahati (start the process a month before you plan to arrive). Leave a day free in your schedule to explore Guwahati's flower-filled bazaars and Shakti temples (devoted to female spiritual power).

Drive 182km northeast to Tezpur.

❷ Tezpur

This sleepy country town is unremarkable in itself, but within day-tripping distance is truly remarkable Kaziranga National Park, where sightings of one-horned Indian rhinos are almost guaranteed. Safaris (by jeep, not elephant) are easy to arrange at the park headquarters near Kohora, a two-hour drive from Tezpur.

Back on the road drive 330km northwest to Tawang.

❸ Tawang

A special permit is needed to visit Arunachal Pradesh, but it's worth the effort and cost. The drive north from Tezpur to Tawang struggles high into the mountains, straining over the 4170m Se La Pass. At the end is one of the wonders of the Himalaya: Tawang Gompa, a virtual city of ancient, whitewashed Buddhist monastery buildings crowning a forested ridge.

It's a 2-day, 566km drive to Ziro via Itanagar.

ABOVE: THATCHED HUTS ON STILTS IN NAGALAND

ABOVE: TAWANG GOMPA

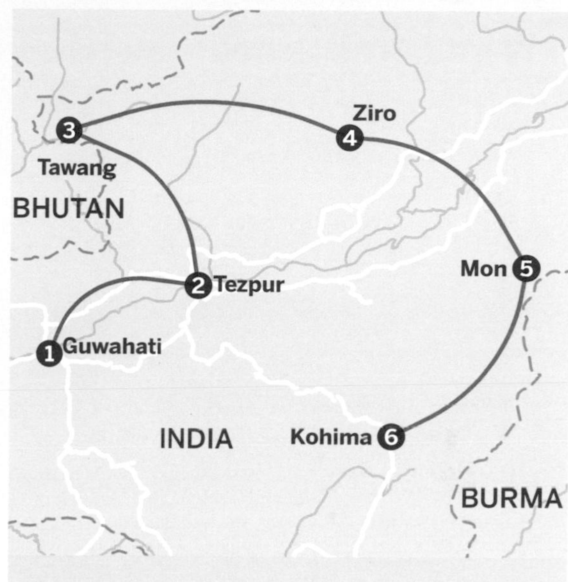

❹ Ziro

After the Himalayan vibe of Tawang, the lush, green foothills of the Ziro Valley feel like another world. Tidy villages of thatched, bamboo stilt-houses mark the homeland of the Apatani people, known for their animist beliefs and facial tattoos and nose-plugs.

Drive 383km southeast to Mon via Dibrugarh.

❺ Mon

Crossing the mighty Brahmaputra River at Dibrugarh, it's a bone-shaking drive to reach the hills of Nagaland. Immaculate villages of Konyak Naga tribespeople spill across the hillsides, decorated with the horns of buffaloes and, in the remotest valleys, with the time-bleached skulls of human head-hunting victims.

Continue for two days/358km to Kohima via Jorhat.

❻ Kohima

Be prepared for another tooth-rattling journey to reach the capital of Nagaland, which snoozes quietly for most of the year but erupts into life each December during the Hornbill Festival, the biggest annual gathering of Naga people.

Head 460km west back to Guwahati.

ALTERNATIVES

Extension: Sivasagar

It's easy to add in a side-trip to Sivasagar – the historic capital of the Ahom Empire – en route from Nagaland back to Guwahati. The city is dotted with timeless temples and the ruins of Ahom palaces. Just north is the sandbar island of Majuli, home to centuries-old *satras* (Hindu monasteries).

Get your jeep driver to divert here on the way from Mon to Kohima.

Diversion: Shillong

The thrill of a visit to Shillong, known in British times as the 'Scotland of the East', isn't just the lush scenery and cool mountain air, it's the journey up here by helicopter from Guwahati – one of the world's most affordable chopper flights. On arrival, wander down to Iew Duh market, where Khasi tribespeople in tartan shawls haggle for veggies, bows and arrows and edible frogs.

Helicopters buzz up and downhill twice daily, Mon-Sat.

TAMIL NADU'S TECHICOLOUR TEMPLES

Chennai – Kanyakumari
Discover India's far south through the kaleidoscopic temple architecture of Tamil Nadu, from lively Chennai's places of worship to peaceful shrines in rural villages.

FACT BOX

Carbon (kg per person): 35
Distance (km): 1010
Nights: 18–25
Budget: $-$$
When: Oct-Mar

❶ Chennai

Dive into Tamil Nadu's dynamic capital and be wowed by its 16th-century Kapaleeshwarar Temple, a rainbow-painted masterpiece. Equally wow-inducing are the superb Tamil cuisine, great shopping and expert-led tours to sights such as Fort St George.

Take the state bus from Chennai Mofussil Bus Terminus (CMBT) to Mamallapuram (very frequent; 2hr-2hr 30min).

❷ Mamallapuram (Mahabalipuram)

Once a great Pallava-kingdom port, mellow 'Mahabs' is home to spectacular Unesco-listed temples and rock carvings. It's also a popular surf hangout, together with nearby Kovalam (Covelong). Well worth a few nights.

Take the bus to Kanchipuram (several daily; 2hr).

❸ Kanchipuram

Chaotic Kanchipuram – the 6th- to 8th-century Pallava capital – is known for its richly adorned temples and handwoven silk saris. It's an easy day trip from Mamallapuram, or soak up the temples at leisure by overnighting.

Take another bus to Tiruvannamalai (frequent; 3hr).

❹ Tiruvannamalai

Spiritual Tiruvannamalai, one of South India's most important temple towns, is home to the 10-hectare

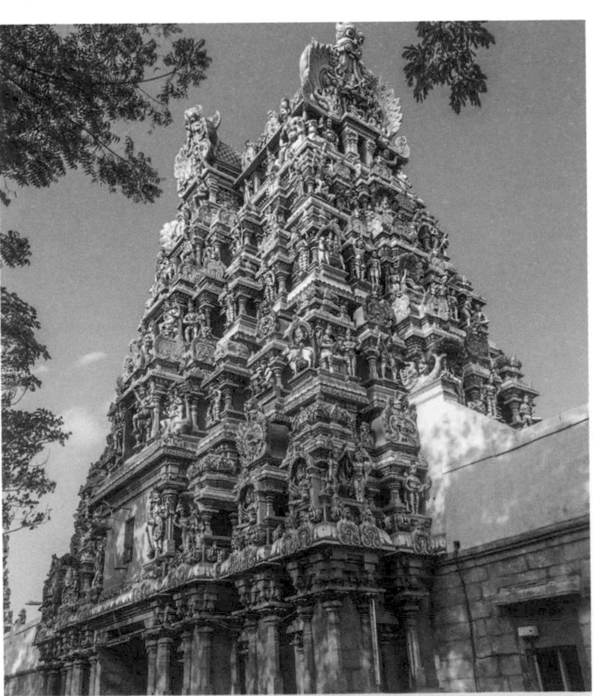

ABOVE: MEENAKSHI AMMAN TEMPLE, MADURAI

ABOVE: SHORE TEMPLE, MAMALLAPURAM

Arunachaleshwar Temple, which honours Shiva in his fire incarnation and rests at the foot of an extinct volcano, surrounded by ashrams.
Take the bus to Puducherry (hourly; 3hr).

5 Puducherry (Pondicherry)
A French colony until 1954, seaside Pondy is a South India highlight, mixing local life and spiritualism with colourful colonial architecture and leafy streets. With stylishly original hotels and outstanding cuisine, allow at least four nights.
Continue on the bus to Kumbakonam (several daily; 3hr).

6 Kumbakonam
Stop overnight in this former medieval power to admire its two Unesco-listed yet still off-the-beaten-track Chola temples, known for their intricate stone carvings.
Take the train to Thanjavur (several daily; 30min-1hr).

7 Thanjavur (Tanjore)
The capital of the great medieval Chola Empire is unmissable for its World Heritage-listed Brihadishwara Temple and maze-like Nayak-era palace.
There are regular trains and buses to Trichy (1hr).

ALTERNATIVES

Extension: Chettinadu
Arid Chettinadu is one of South India's most fascinating, highly rewarding regions, famed for its dramatically opulent mansions (built by wealthy local traders), silk-and-cotton saris and lovely handmade tiles. The tranquil rural towns here host some of Tamil Nadu's most fabulous hotels.
It's best to hire a car and driver in Madurai, Thanjavur or Trichy, though there are buses and trains to the hubs of Pudukkottai and Karaikkudi.

Diversion: Western Ghats
Tamil Nadu's lofty Western Ghats offer misty hill stations, pristine shola forest, twinkling lakes, quiet hiking paths, grand colonial-era mansions and jaw-dropping views. Highlights include the Palani Hills' low-key Kodaikanal, and the Nilgiris' Ooty (Udhagamandalam).
Pre-book the Nilgiri Mountain Railway from Mettupalayam (near Coimbatore) to Ooty (4hr 45min). Buses serve Kodaikanal from Madurai (4hr) and Trichy (6hr).

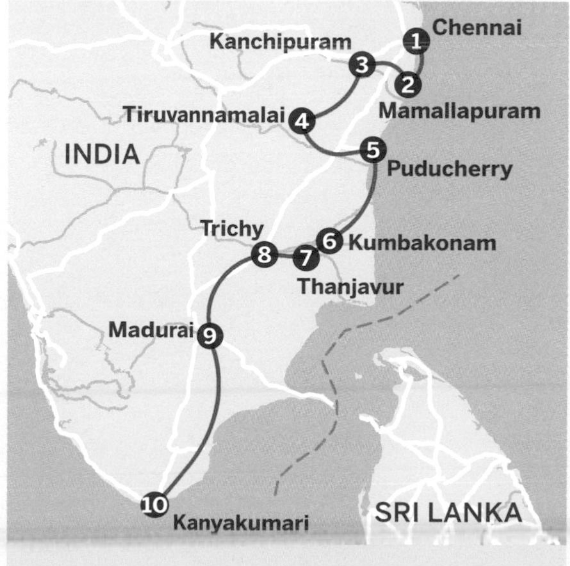

ABOVE: DETAIL OF KAPALEESHWARAR TEMPLE IN CHENNAI

❽ Trichy (Tiruchirappalli)

This hectic transport hub conceals the sprawling, superbly carved Sri Ranganathaswamy Temple and the centuries-old hilltop Rock Fort Temple.

Take the train to Madurai (several daily; 2-3hr).

❾ Madurai

A powerful ancient capital, Madurai harbours one of India's major temples – the dazzling 12-gopuram Meenakshi Amman – and some famously terrific cuisine.

Take the train to Kanyakumari (daily; 5hr).

❿ Kanyakumari

Where better to wrap up than on the subcontinent's southernmost tip, at the temple honouring demon-conquering goddess Kumari? (And don't miss Padmanabhapuram Palace).

Take the train to Madurai, Trichy, Chennai or Thiruvananthapuram for onward airport connections.

ABOVE: SRI AUROBINDO ASHRAM, PUDUCHERRY

ABOVE: BACKSTREET PUDUCHERRY

UP TO THE ANNAPURNAS IN NEPAL

Kathmandu Valley – Poon Hill Loop
Explore the ancient cities of the Kathmandu Valley by public transport, before treading the Himalayan trails that unfurl beyond Pokhara.

FACT BOX
Carbon (kg per person): 13
Distance (km): 530
Nights: 10-12
Budget: $
When: Sep-May

❶ Kathmandu Valley

A hectic, heady and holy metropolis, Kathmandu is Nepal's capital, and a convenient spot to base yourself while exploring other ancient cities in the surrounding Kathmandu Valley. Get your bearings in Durbar Square – flanked by royal palaces, it suffered greatly during the 2015 earthquake but remains the Old City's beating heart. It's a thirty-minute taxi ride from central Kathmandu to the more sedate streets of Bhaktapur – a highlight here is the 18th-century Nyatapola Temple, a lofty pagoda devoted to the goddess Lakshmi. The third city in the trilogy is Patan, half an hour to the south of Kathmandu. It's also known as Lalitpur, which means 'City of Beauty' – an apt description as you wander among its Hindu and Buddhist temples.

To get to Pokhara take a bus (every 30min; 7hr) or flight (very frequent; 30min) from Kathmandu. If flying, sit on the right for the best views of the Himalaya.

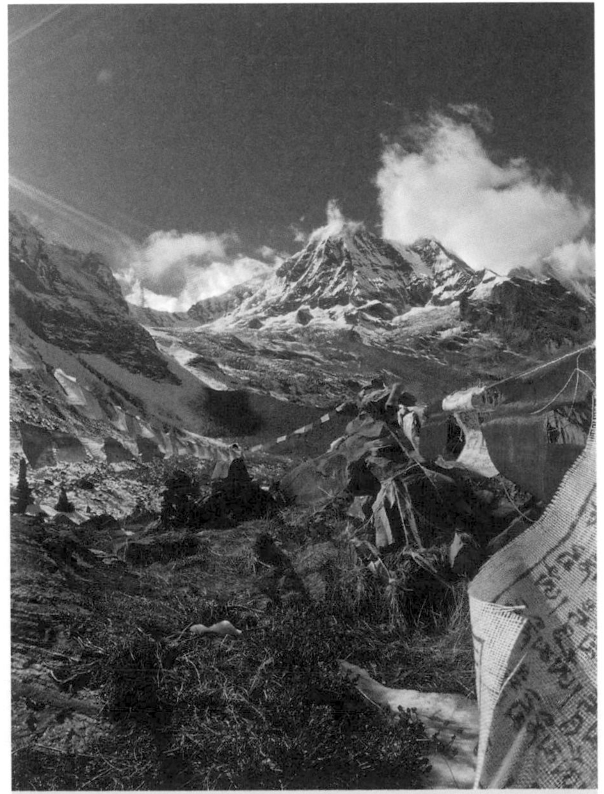

ABOVE: ANNAPURNA BASE CAMP

❷ Pokhara

Nepal's second city is defined by its sublime setting – running to the reedy shores of Phewa Lake, with the unclimbed summit of Machhapuchhre soaring almightily over the rooftops. There's plenty to occupy a few days' stay here: take the easy trek through the

ABOVE: DURBAR SQUARE, KATHMANDU

thick woods to the World Peace Pagoda, and ride the thermals over the lake on paragliding expeditions. For most visitors, however, Pokhara serves as the gateway for epic treks in the Annapurna region – and one last chance to stock up before striking out amongst heavenly summits.

From Pokhara, buses run to Nayapul and the trailhead for the Poon Hill loop (regular; 2hr).

3 The Poon Hill Loop

The Annapurna region is home to two of Nepal's most epic treks: the 11-day Annapurna Sanctuary trek sees walkers travel up to Annapurna Base Camp at 4130m, surrounded by an icy amphitheatre of 8000m peaks; the 21-day Annapurna Circuit loops around the entire namesake range. If you're pressed for time, though, opt for the Poon Hill Loop, which covers around 40km in four days and offers spellbinding panoramas of snow-shrouded summits. Starting in Nayapul, the trail ascends steeply to the hilltop village of Ghorepani and Poon Hill itself, whose grassy peak looks northward to the distant Dhaulagiri massif. Then descend via bucolic Ghandruk, resting weary limbs in cosy tea houses on the way.

From Nayapul, bus back to Pokhara and travel on to Kathmandu by bus or plane.

ALTERNATIVES

Extension: Chitwan National Park

Chitwan National Park offers a drastically different impression of Nepal. Its mosaic of grasslands, forests and marshes extend to the Indian border, and within its boundaries you might spot rhinos, elephants or monkeys among the thick vegetation. Bengal tigers are also resident – although spotting one is akin to a lottery win.

The park is best explored as part of a package safari from Kathmandu.

Diversion: Everest Base Camp Trek

Long and demanding, taking 12-plus days, with 2500m of elevation gain, the rewards of this trek are mighty: the monastery at Tengboche; the Sherpa heartland of Namche Bazaar; and the frozen wastes of Khumbu Icefall.

First take the 30min flight from Kathmandu to Lukla. Book a guided 'EBC' trek through an operator in advance, or travel independently with or without porters.

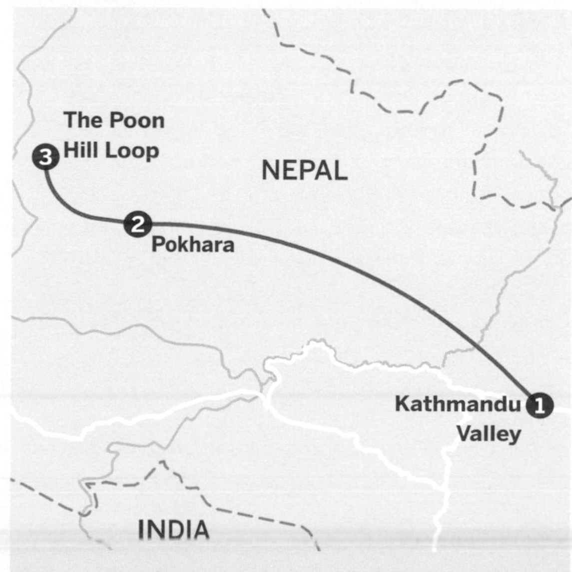

SOUTHEAST ASIA

Rich rainforests, deserted beaches, ancient ruins and marvellously modern cities: Southeast Asia has something for everyone. Gorge on pho and stock up at floating markets in Vietnam; sail Indonsesia's Spice Islands; take a cookery course in Thailand's Chiang Mai; marvel at the temple architecture in Myanmar's Bagan or Cambodia's Siem Reap; and explore techicolour reefs from white-sand shores in Bali and Malaysia.

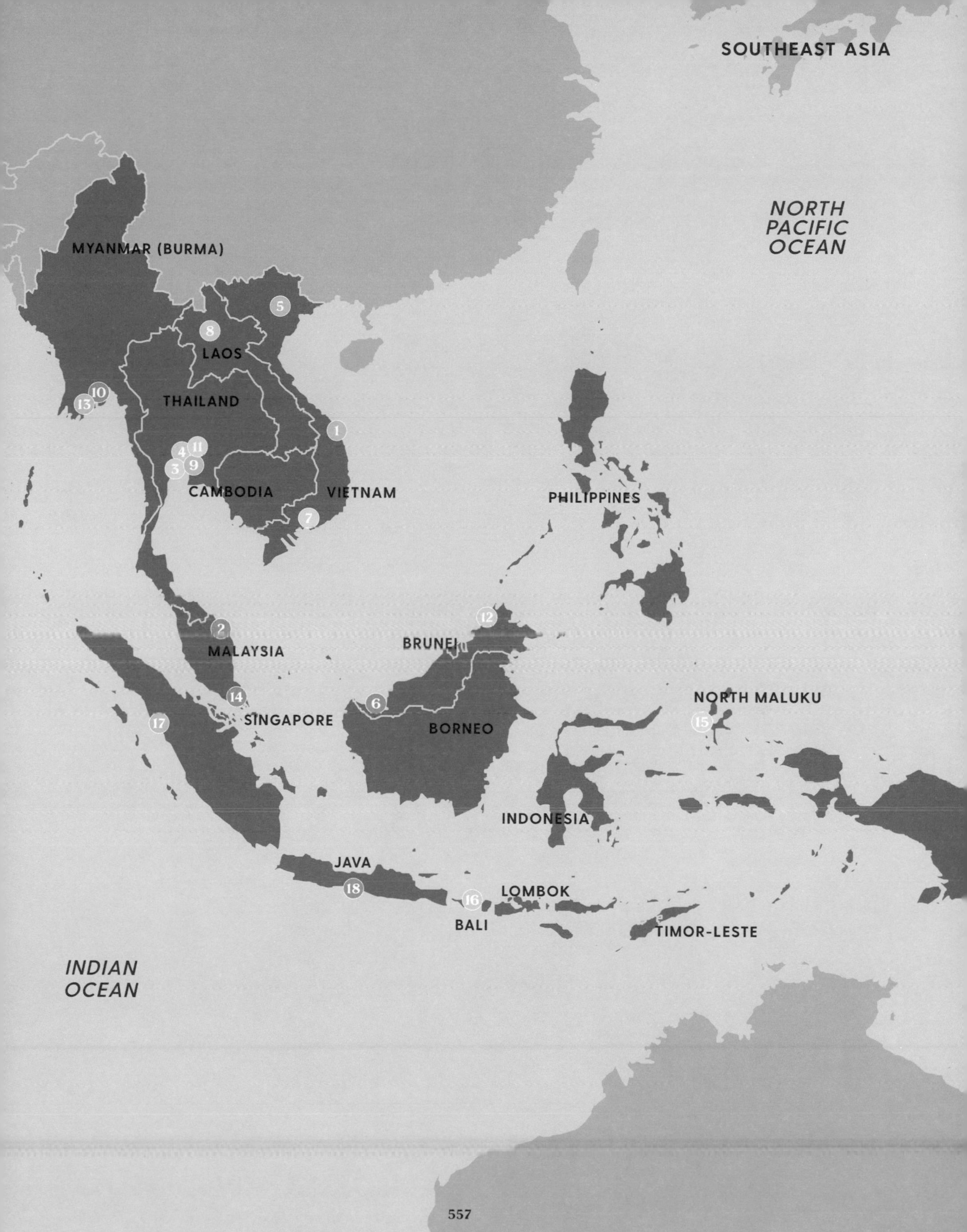

SOUTHEAST ASIA

NORTH PACIFIC OCEAN

MYANMAR (BURMA)

5

8

LAOS

10

13

THAILAND

1

4

11

3

9

CAMBODIA

VIETNAM

PHILIPPINES

7

12

2

BRUNEI

MALAYSIA

14

6

NORTH MALUKU

17

SINGAPORE

BORNEO

15

INDONESIA

JAVA

18

16

LOMBOK

BALI

TIMOR-LESTE

INDIAN OCEAN

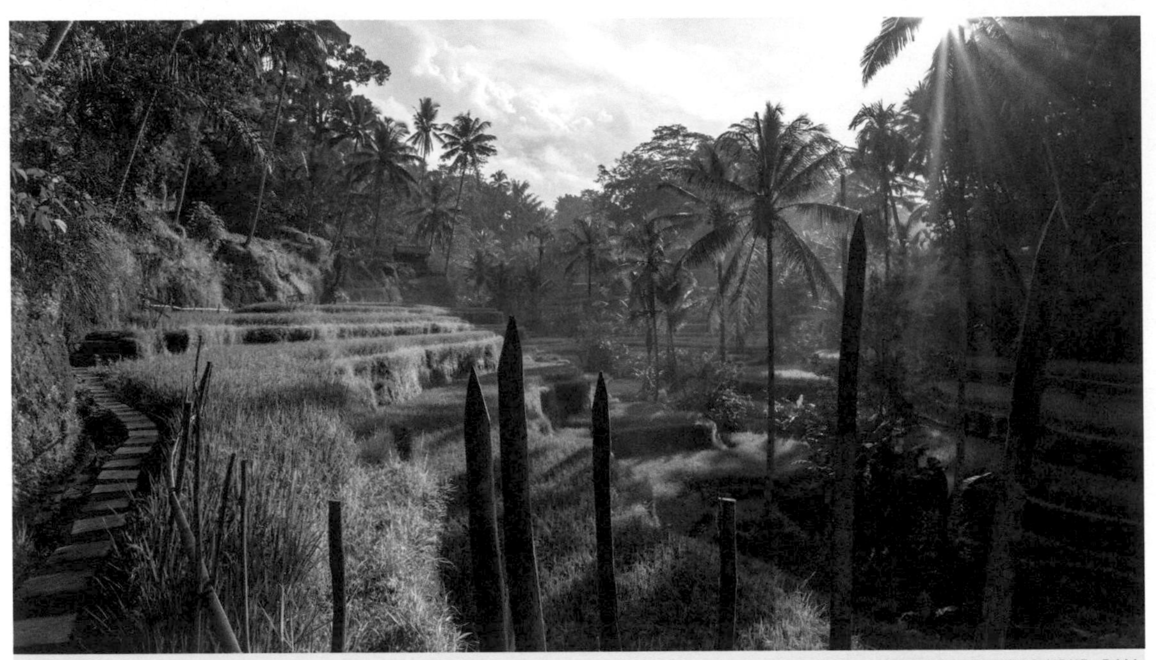

ABOVE: RICE TERRACES NEAR UBUD, BALI

CONTENTS

CENTRAL VIETNAM EXPLORER

Danang – Phong Nha-Ke Bang National Park

Combine heritage, food and activities on a journey north from dynamic Danang and historic Hoi An.

FACT BOX

Carbon (kg per person): 61
Distance (km): 948
Nights: 8-12
Budget: $$
When: Feb-June

❶ Danang

Vietnam's fastest-growing regional city is an energetic place to begin a central coast adventure. Join a street food tour to feast on Danang's famed seafood, secure a front row seat in a Han River bar to view the spectacular fire-breathing Dragon Bridge, and learn about thousand-year-old art in the shaded cloisters of the Museum of Cham Sculpture.

Danang accommodation can book pickups for convenient minibus transfers to Hoi An (1hr). Local buses drop off a 15min walk from central Hoi An.

❷ Hoi An

Relaxed An Bang Beach is a recommended base from which to explore the heritage-filled streets and lanes of Hoi An. Chinese, Japanese and Portuguese trading empires all left their mark on this compact riverside town. Take in Hoi An's ornate Assembly Halls, and explore the rural hinterland on a vintage Vespa or mountain bike; afterwards, enjoy the cosmopolitan cafe and restaurant scene.

Direct buses head to Hue (frequent; 3hr). An exciting alternative is an 'Easy Rider' transfer as a pillion passenger on a motorbike (9hr). The leisurely trip includes lunch, beaches and a stop at Hai Van Pass.

❸ Hue

Hue was the imperial capital of Vietnam from 1802 to 1945 and the mausoleums of the Nguyen dynasty punctuate the banks of the Perfume River south of the city. Explore the sprawling Citadel before journeying downriver by boat, bicycle or motorbike

ABOVE: PHO STALL, HOI AN

ABOVE: CHAOZHOU ASSEMBLY HALL, HOI AN

to the extravagant tombs of the emperors. Don't miss Hue's imperial cuisine – rightly regarded as some of the country's finest.

Take the Reunification Express train to Dong Hoi (5 daily; 3hr 30min).

4 Dong Hoi

Enjoy some oceanside relaxation around Dong Hoi before heading inland for active adventures in Phong Nha-Ke Bang National Park. Try the region's famed *bánh khoai* (shrimp pancakes) at Tu Quy restaurant.

Take the bus to Phong Nha's main town of Son Trach (hourly; 1hr 30min). More convenient is a direct taxi (45min).

5 Phong Nha-Ke Bang National Park

Awarded Unesco World Heritage status in 2003, the Phong Nha region contains the biggest caves on the planet. Book well ahead to explore the world-beating Hang Son Doong, but other subterranean adventures are more readily available. Combine ziplining, kayaking and caving at Hang Toi with overnight camping adventures in Hang En and Hang Va.

Travel from Son Trach back to Dong Hoi to catch an overnight train to Hanoi and its international airport (10hr).

ALTERNATIVES

Day trip: Son Tra Peninsula

Dubbed Monkey Mountain by American soldiers during the Vietnam War, Son Tra Peninsula was a closed military area until relatively recently. Roads best explored in a jeep or on the back of a motorbike wind to the peninsula's summit for brilliant coastal views. Trips usually take in Linh Ung, a colossal Buddha statue, and lunch at local seafood restaurants.

Book an excursion with tour agencies in Danang or Hoi An.

Day trip: Bong Lai Valley

Pick up a hand-drawn map in Son Trach and explore the Bong Lai Valley's quirky collection of locally owned and low-impact attractions. Essential stops include grilled chicken at The Pub With Cold Beer, chilling in a hammock at Moi Moi, and river tubing at the Wild Boar Eco Farm. Feeding the playful ducks at The Duck Stop is also recommended.

Rent a mountain bike in Son Trach town.

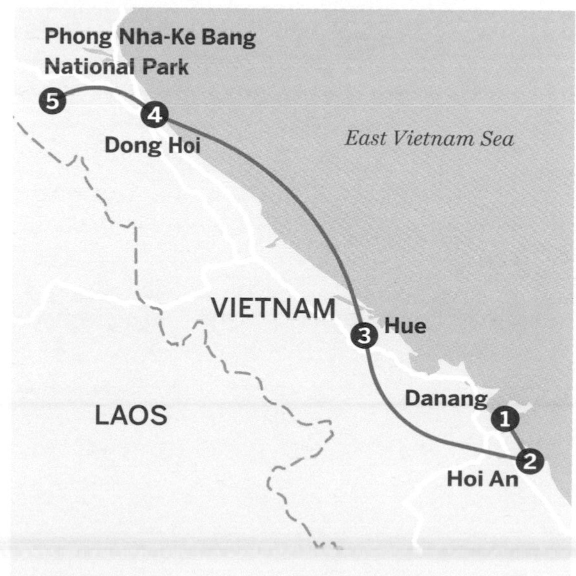

NORTHEAST MALAYSIA ISLAND HOPPER

Kota Bharu – Pulau Kapas

Combine the energy and colour of two less-visited Malaysian cities with offshore detours to relaxing, jungle-clad islands and ravishing reef-studded shores.

FACT BOX

Carbon (kg per person): 10
Distance (km): 228
Nights: 8-12
Budget: $$
When: May-Aug

❶ Kota Bharu

Infused with traditional Malay culture, the state capital of Kelantan is a friendly and easy-going city. Explore the museum precinct around Padang Merdeka – the historic Istana Jahar features a superb collection of Kelantanese crafts – and enjoy the city's favourite breakfast of *nasi kerabu* (coconut-infused blue rice with fish) at Restoran Capital.

ABOVE: KOTA BHARU SPECIALITY, NASI KERABU

Frequent buses from Kota Bharu stop at Kuala Besut (2hr). From there, catch a speedboat to the Perhentian Islands (40min).

❷ Perhentian Islands

This archipelago, mainly comprising Perhentian Besar ('Large') and Perhentian Kecil ('Small'), is a versatile destination combining good diving and snorkelling, hiking on jungle paths and relaxing at beachside resorts ranging from backpacker digs to stylish wooden chalets. Relatively shallow waters, gentle currents and excellent visibility make the Perhentians a good place to learn to dive. A local armada of speedboats facilitates day trips to five smaller islands, and volunteering opportunities include helping to manage the islands' marine resources and supporting the protection of local sea turtle populations.

Catch a speedboat back to the mainland and continue south on frequent buses (2hr).

❸ Kuala Terengganu

Inspired by the economic impetus of offshore oil

ABOVE: PERHENTIAN ISLANDS

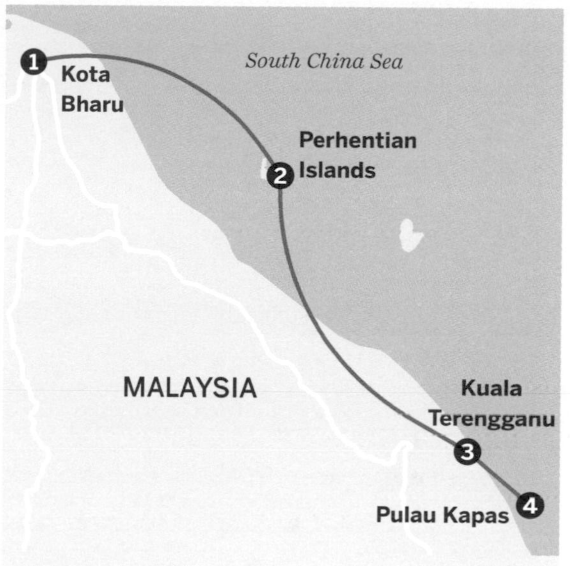

and natural gas exploration, Terengganu's state capital is a beguiling mix of tradition, heritage and modern waterfront parks and promenades. Framed by the wooden balconies of shophouses, the city's Chinatown area combines seafood restaurants and traditional herb dispensaries with hip cafes, bars and the colourful mosaics of Turtle Alley. For an insight into KT's kinetic growth, see the Jambatan Angkat Kuala Terengganu, a modern drawbridge crossing the mouth of the Terengganu River. Visit at sunset for city views from the bridge's upper walkways.

Catch a taxi south to the ferry dock at Marang (30min) and continue by speedboat to Pulau Kapas (15min).

4 Pulau Kapas

Compact Pulau Kapas is a relaxing spot to complete a northeast Malaysian journey. Visit on a weekday for a quieter time and hook up with local dive operators for snorkelling and dive excursions to tiny Pulau Gemia. Located in a hidden rocky cove, Kapas Turtle Valley has stylish batik-adorned bungalows.

Return to the mainland and to Kuala Terengganu, from where frequent buses travel to Kota Bharu (3hr) and flights connect with Kuala Lumpur (1hr).

ALTERNATIVES

Diversion: Dabong

Linked to Kota Bharu by the Jungle Railway, Dabong is an excellent base for exploring Gunung Stong State Park. Ask around Dabong about joining a guided tour that includes hiking, swimming and jungle trekking. Caving and river rafting are other active options.

Take the train from Kota Bharu's Wakaf Bharu station to Dabong (4 daily; 3hr 30min). Hire a taxi in Dabong to continue the 15km to Gunung Stong.

Diversion: Penarik

Stop in Penarik to take a boat trip to the Penarik Firefly Sanctuary. Boats depart as late afternoon merges into an inky, tropical dusk, meandering through mangroves to where thousands of fireflies flicker in lockstep synchronisation. Overnight at Terrapuri Heritage Village, with very comfortable accommodation in restored wooden houses.

Buses south from Kota Bharu (2hr) and north from Kuala Terengganu (1hr) stop in Penarik on request.

THAILAND'S ANCIENT KINGDOMS

Bangkok – Chiang Mai

Embark on a journey into history by bus and train from Bangkok to Chiang Mai, connecting Thailand's most magnificent ancient kingdoms.

FACT BOX

Carbon (kg per person): 27
Distance (km): 744
Nights: 7-10
Budget: $$
When: Nov-Apr

❶ Bangkok

Fly from Bangkok to Chiang Mai and you'll miss some of Thailand's most fascinating ancient sites – so go by bus and train and trace Thailand's history as you go. Start your trip in Ko Ratanakosin, the heart of old Bangkok, where the streets are studded with jewel-like wat (Buddhist monasteries), a testament to the wealth and grandeur of the Chakri kingdom.

Take the train from Hua Lamphong Station to Ayuthuya (hourly; 1hr 30min).

❷ Ayuthaya

Bangkok is just the most recent of a string of capitals founded by former dynasties. Ayuthaya ruled the roost from 1350 to 1767 and its ruined temples and stupas still give a profound impression of lost power. Towering, even in decay, Wat Chai Wattanaram, Wat Ratchaburana and Wat Phra Si Sanphet are some of the most impressive monuments in all of Thailand.

Get back on the train to Phitsanulok (up to 10 daily; 6hr).

❸ Phitsanulok

The provincial capital of Phitsanulok lies off the main tourist trail but you'll probably recognise it from photos thanks to the revered Buddha image enshrined inside Wat Phra Si Ratana Mahathat. Downtown, the Sergeant Major Thawee Folk Museum and Buranathai Buddha Image Foundry offer more insights into the intertwined public and religious life of central Thailand.

Take the bus to Sukhothai (hourly; 1hr).

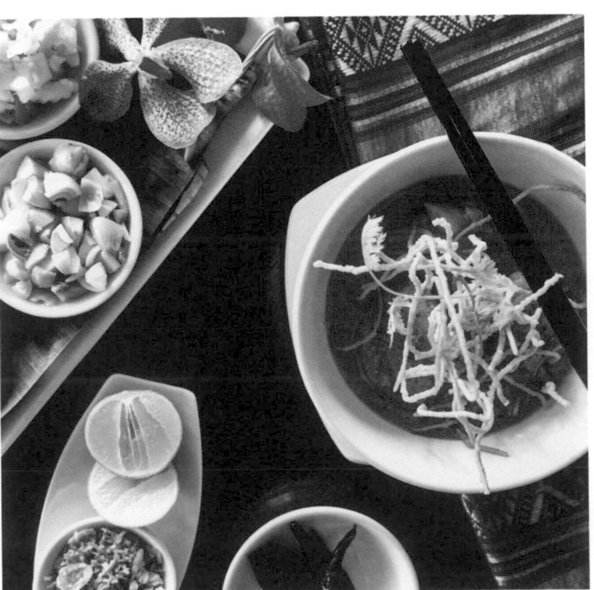

ABOVE: TRY KHAO SOI IN CHIANG MAI

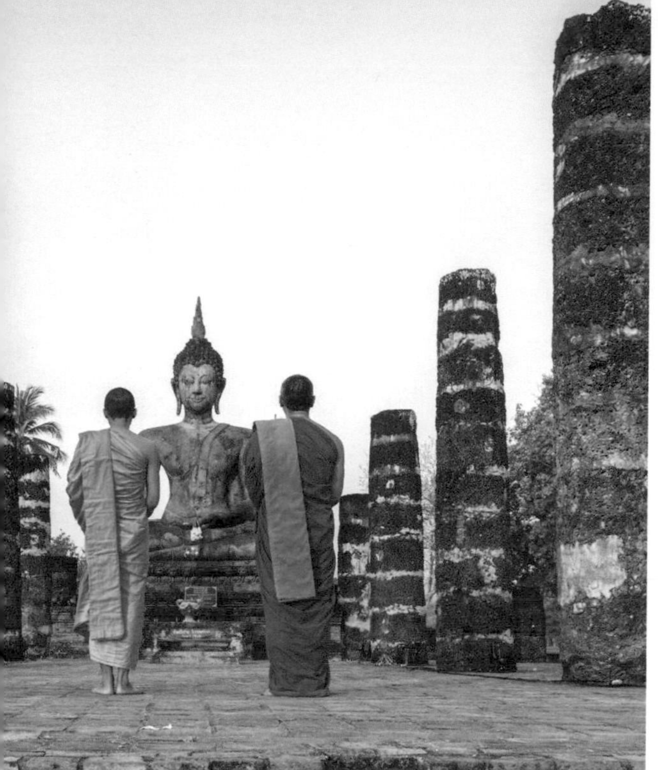
ABOVE: WAT MAHATHAT TEMPLE, SUKHOTHAI

❹ Sukhothai
Some hundred-odd picturesque temples, stupas and ruins mark the extents of Old Sukhothai, which served as the Thai capital until Ayuthaya gained the upper hand. It's a stunning, World Heritage-listed location, worthy of several days of exploring on foot or by rented bicycle.

Take the bus to Lampang (hourly; 3hr).

❺ Lampang
Ignore the travellers charging north and break the trip to Chiang Mai at this gentle provincial capital, loved by those who do visit for its food-filled markets, cool cafes, interesting temples and general lack of stress.

Take the train to Chiang Mai (5 daily; 3hr).

❻ Chiang Mai
Once capital of the northern Lanna kingdom, Chiang Mai is bejewelled with ancient monasteries and night markets serving everything from Chinese-style pork leg and rice (seek out the Cowboy Hat Lady at the Pratu Chang Pheuak market) to *khao soi* (chicken, coconut and noodle soup). Take a cookery course and carry the cuisine home with you.

Flights connect Chiang Mai with cities across Asia.

ALTERNATIVES

Extension: Si Satchanalai-Chaliang Historical Park
Sukhothai isn't the only ancient city between Phitsanulok and Chiang Mai. Off the main road – and the tourist map – Si Satchanalai-Chaliang Historical Park marks the site of two ancient Khmer cities spilling along the banks of the Maenam River. Expect towering temple spires, time-worn Buddha images and a general air of serenity.

Buses run to the historical park from Sukhothai (hourly; 1hr 30min).

Diversion: Lamphun
Between Lampang and Chiang Mai, the riverbank town of Lamphun marks the northern extent of the Mon Dvaravati kingdom, which ruled from 750 to 1281 CE. It's one last chance to experience a quieter version of northern Thailand before the commotion of Chiang Mai, with a scattering of temples and *chedis* (stupas) and a slow, easy pace of life.

Buses roll in regularly from Lampang (1hr 30min) continuing to Chiang Mai (30min).

THE BACK-ROUTE TO ANGKOR: THAILAND TO CAMBODIA

Bangkok – Phnom Penh

Go overland to Angkor, with stops at famous (and outlying) temples and down-time in Bangkok, Siem Reap and Phnom Penh – three of Asia's most engrossing cities.

FACT BOX

Carbon (kg per person): 21
Distance (km): 728
Nights: 7-10
Budget: $$
When: Nov-Apr

❶ Bangkok

Start this journey through the heartland of the former Khmer empire in Bangkok, Southeast Asia's most fun-filled, frenetic metropolis. Fill up on spectacular street food, browse magnificent markets and head to Wat Phra Kaew to stoke your anticipation with a peek at the scale model of Angkor Wat installed by King Rama IV (after a failed attempt to move Angkor Wat stone by stone to Bangkok in 1859).

Take the train to Aranya Prathet (2 daily; 5hr), a *sŏrngtăaou* (pick-up truck) to the Cambodian border at Poipet, then a bus to Sisophon (1-2hr).

❷ Sisophon

Move on quickly from the grimy border town of Poipet to the transport hub of Sisophon. Why visit this dusty junction town? Well, it's the gateway to the temples of Banteay Chhmar, 61km north. You can day-trip easily by shared taxi to these intricately adorned, very Angkor-like Khmer temples, or overnight at one of the charming community homestays close to the ruins.

Take a morning bus or share taxi to Siem Reap (1hr 30min).

❸ Siem Reap

Though it was the famous temples spilling out of the

ABOVE: BAYON TEMPLE, SIEM REAP

ABOVE: ANGKOR WAT, SIEM REAP

jungle that put the town on the map, there's more to the likeable traveller hub of Siem Reap than its location close to Angkor. Drop your bags at a hip hostel or design-mag-friendly boutique hotel, feast on Cambodia's most fabulous food and spend at least a few days being wowed by the temples of Angkor. Everyone makes a beeline for Angkor Wat (sunrise is particularly busy) and the nearby Bayon, but take time to explore the minor temples spilling out from the main complex too.

Take the boat to Phnom Penh (daily; 5-6hr).

4 Phnom Penh

The speedboat ride along the inland sea-sized Tonle Sap to the Cambodian capital used to be a traveller rite of passage, and sitting on deck to watch the waters slide by is still a lot more fun than flying. You'll be dropped off downtown in Phnom Penh, where haunting reminders of the Khmer Rouge years – the killing fields at Choeung Ek, Tuol Sleng prison – are juxtaposed with cool cafes, boisterous bars, a world-class food scene and monasteries overflowing with saffron-robed novices.

Phnom Penh Airport has flights to cities around Asia, including to Bangkok if you want to complete the circle.

ALTERNATIVES

Extension: Banteay Srei

The outlying temple of Banteay Srei is a reminder of what Angkor Wat was like in the early, quieter days of backpacking. Spend a few days being amazed by the grace of the Hindu carvings on the main temple, humbled by the Cambodia Landmine Museum, and surprised by Angkor-era oddities like the 'river of a thousand lingas' at Kbal Spean.

Get there by hired vehicle from Siem Reap (1hr).

Diversion: Koh Rong

After all this inland travel, you've earned some beach time, so slingshot from Phnom Penh through Sihanoukville to board the ferry to Ko Rong. Despite the frenetic backpacker scene at Koh Tuch, the beaches are still stunning, particularly the blindingly white sands around Long Set. The evening frog chorus manages to compete with the booming bass from the bars.

Sihanoukville is 6-10hr by bus from Phnom Penh; the ferry to Ko Rong takes 1hr.

NORTHERN VIETNAM EXPLORER

Hanoi – Ba Be National Park

Journey from Vietnam's exciting capital to explore the country's wild, rugged and remote far northern regions.

FACT BOX

Carbon (kg per person): 40
Distance (km): 1450
Nights: 10-15
Budget: $
When: Mar-May

❶ Hanoi

Negotiate Hanoi's labyrinthine Old Quarter to dine on great street food including *pho bo* (beef noodle soup) followed by *cà phê đá* (iced coffee) around leafy Hoan Kiem Lake. The Temple of Literature and its formal gardens are a perfect antidote if the city's urban buzz gets too much.

Evening sleeper buses run to Sapa from Hanoi's Gia Lam bus station (10hr).

❷ Sapa

Established as a cool-climate hill station by the colonial French in the 1920s, Sapa is now the base to explore northern Vietnam's most popular trekking destination. Book a trek with Sapa O'Chau, a local company run by the region's indigenous Hmong people, to venture into mountain valleys and stay in village homes.

Catch one of the frequent minibuses to Lao Cai (1hr) and then transfer to Ha Giang (6hr).

❸ Ha Giang

Welcome to Vietnam's most fascinating travel frontier, an astonishing landscape of jagged pinnacles and granite outcrops abutting the Chinese border. From the province's namesake capital, head 40km north to the Quan Ba Pass, also known as Heaven's Gate, and a stunning rolling vista of limestone mountains.

Buses depart at 5am and 10am linking Ha Giang to Dong Van (5hr).

❹ Dong Van

The market town of Dong Van is the hub for day-treks to local minority villages, northern Vietnam's most vibrant Sunday market and excursions through jaw-

ABOVE: OLD QUARTER, HANOI

ABOVE: RICE TERRACES NEAR SAPA

dropping scenery to Lung Cu, a massive flag tower on the country's northernmost point.

Book a taxi transfer to Bao Lac (3hr) in time to catch a morning bus from Bao Lac to Cao Bang (4hr).

5 Cao Bang

Plan on a long day's travel to reach Cao Bang, a pleasant provincial capital that's the best transit point for journeying southwest to Ba Be National Park.

Buses from Cao Bang to Cho Ra depart at 9am and noon (5hr). At Cho Ra, arrange a motorbike taxi for the final 18km to the national park.

6 Ba Be National Park

Stay at lakeside homestays and take boat trips around Ba Be's three lakes, framed by forests and towered over by limestone peaks; birdlife is plentiful, there are caves and waterfalls to explore. The area is home to 13 tribal villages, most belonging to the Tay tribal people.

Return to Cho Ra by motorbike and catch a frequent bus (look for 'Hanoi' as a destination on the front) south back to Hanoi's Gia Lam bus station (4hr).

ALTERNATIVES

Extension: Topas Ecolodge

Above a deep and verdant valley 18km from Sapa, Topas Ecolodge makes a fine spot to embark on a few days of hiking and mountain biking. Stylish stone-and-thatch bungalows are designed to take in the stellar views, while textile workshops in local tribal villages, market visits and photography excursions offer a window into local life.

Tapas guests can get free transfers from Sapa (30min).

Diversion: Bac Ha

Bac Ha's Sunday market is a popular destination for daytrippers staying in Sapa – so instead stay in Bac Ha and journey to other nearby weekly markets. Can Cau's Saturday market is popular amongst Flower Hmong and Blue Hmong tribal people, while Sri Cheng on a Wednesday is the trading hub for Nung and Thulao minorities.

Catch a minibus from Sapa to Lao Cai (30min) and transfer to Bac Ga (2hr).

SARAWAK ADVENTURE BY BOAT AND BUS

Kuching – Gunung Mulu National Park
Embark on a Borneo adventure combining exciting river travel, Sarawak's two biggest cities and the natural thrill and spectacle of massive caves.

FACT BOX

Carbon (kg per person): 215
Distance (km): 1584
Nights: 10-12
Budget: $$
When: Apr-Sep

❶ Kuching

Sarawak's riverside capital is an energetic blend of cultures, crafts and cuisines. Cross the Darul Hana pedestrian bridge over the Sungai Sarawak to the colonial garrison of Fort Margherita, dine on tangy Sarawak *laksa* in vintage coffee shops, and wander the shopfront-lined streets and languid lanes of Chinatown.

Frequent flights link Kuching to Sibu (45min).

❷ Sibu

The gateway for boat travel up the remote and forest-clad Batang Rejang, Sibu's riverfront charm is enlivened with the commercial activity of the descendants of migrants from China's Fujian province. Explore the Sibu Heritage Centre and dine amid the stalls of the excellent night market.

Travel by express boat up the Batang Rejang to Kapit (3hr).

ABOVE: THE PINNACLES, GUNUNG MULU NATIONAL PARK

❸ Kapit

Still retaining a frontier spirit, Kapit is the main trading port and upriver town on the Batang Rejang. Sip *kopi* (coffee) at simple riverside cafes, watch the ebb and flow of water-borne commerce and admire Fort Sylvia's colonial wooden architecture.

Continue further upriver on the morning express boat to Belaga (4hr 30min).

❹ Belaga

Blink-and-you'll-miss-it Belaga is another sleepy river settlement even further up the Batang Rejang. Ask around at the town's simple homestays about joining guided treks to tribal longhouses and jungle waterfalls.

Arrange a place on one of the 4WD Land Cruisers that leave around 7.30am heading to Bintulu (4hr).

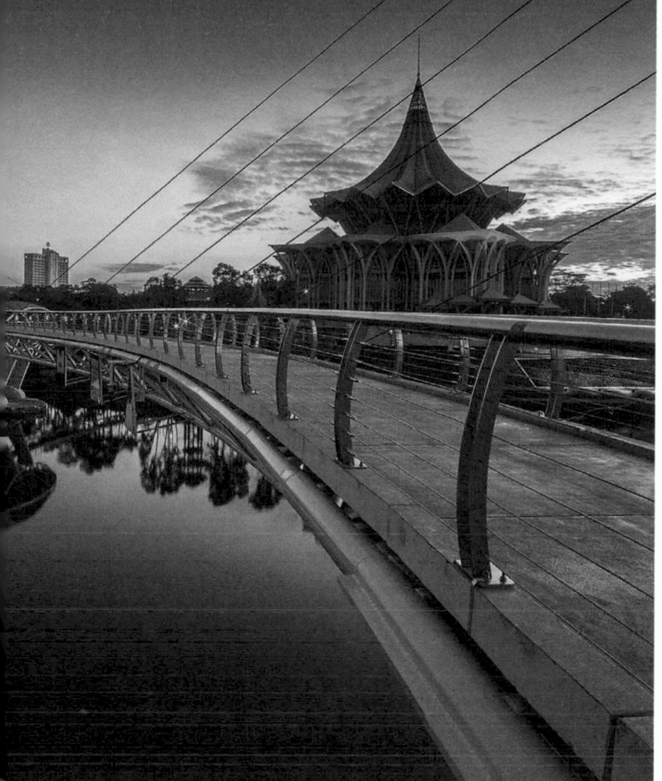

ABOVE: DARUL HANA BRIDGE, KUCHING

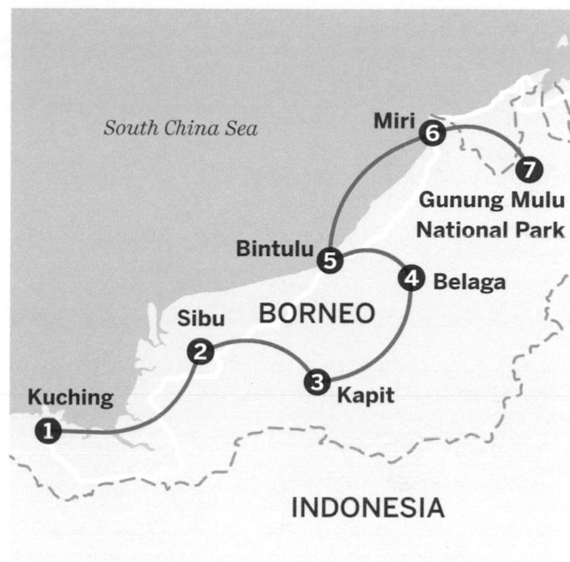

❺ Bintulu

Transformed by natural gas exploration over the last 50 years from a small fishing town into a booming coastal city of 200,000, Bintulu is the perfect place for some urban action after a few days of relaxed river travel. Tuck into cold beers and grilled seafood along Jalan Abang Galau.

Frequent buses link Bintulu to Miri (4hr).

❻ Miri

Sarawak's second city mixes traditional culture with a cosmopolitan energy. Try Kelabit cuisine from Sarawak's remote northeastern highlands at Summit Cafe, and shop for art and textiles at the Miri Handicraft Centre.

MASwings has regular flights from Miri to Gunung Mulu National Park (30min).

❼ Gunung Mulu National Park

Mulu's spectacular Deer Cave is over 2km long and 174m high. Explore on a guided 3km walk before taking in the cathedral-like King's Chamber at the Wind Cave. Adventurous travellers can sign up for the Mulu Canopy Skywalk, a subterranean river walk or jungle trekking.

Fly back to Miri (30min) or Kuching (1hr 40min).

ALTERNATIVES

Diversion: Semenggoh Wildlife Centre

Semenggoh is Sarawak's most accessible place to see orangutans. A semi-wild troop of around 30 of them come down from their natural rainforest habitat to be fed by park rangers at 9am and 3pm every day. Sightings are not guaranteed, but even when there's fruit in the forest, they still usually drop in.

Taxis and local tour companies provide transport from Kuching to Semenggoh (30min).

Diversion: Kelabit Highlands

On Malaysia's northeastern border with Indonesian Kalimantan, the Kelabit Highlands are home to some of Sarawak's most remote communities. The cool mountain climes offer a refreshing environment for day-hikes and longer adventures, guided by knowledgeable villagers. Most accommodation is in comfortable longhouses around the town of Bario.

Take a flight from Miri to Bario (2 daily; 50min).

VIETNAM'S MEKONG DELTA BY SCOOTER

Ho Chi Minh City – Vinh Long

Make like the locals and jump on a scooter in buzzing Ho Chi Minh City before wending your way on a loop through the sultry Mekong Delta.

FACT BOX

Carbon (kg per person): 76
Distance (km): 655
Nights: 6
Budget: $
When: Dec-May

❶ Ho Chi Minh City

The beating heart of southern Vietnam is a vibrant and chaotic city, clogged with traffic but rich with culture. Tick off attractions such as the War Remnants Museum and the Opera House, then head to Tigit Motorbikes to hire a Honda 110cc scooter – the local favourite – and head south. Leave behind the choking outskirts of Ho Chi Minh City for the calmer environment of the Mekong Delta, stopping in My Tho for lunch before continuing to riverside Ben Tre.

Ride 87km from Ho Chi Minh City to Ben Tre.

❷ Ben Tre

Ben Tre is a pleasant town, a riverside hub with a strip of appealing restaurants. This is also your gateway to the sleepy Mekong Delta, away from the tourists and the busy highways, where rice paddies line potholed country lanes and no-frills roadside cafes sell iced coffee with condensed milk.

Back on your scooter head 113km southwest, crossing rivers and skirting paddies, to Can Tho.

❸ Can Tho

The largest city in the delta is home to its largest floating market, Cai Rang, plus restaurants specialising in local dishes such as *sa dec*, a clear soup with rice noodles, and *cong*, a savoury cake with shrimp.

With a full stomach, ride 98km northwest to Long Xuyen, then turn left to Oc Eo.

ABOVE: CAI RANG FLOATING MARKET, MEKONG DELTA

ABOVE: HO CHI MINH CITY

❹ Oc Eo

This tiny village provides access to an archaeological site of the same name, an ancient trading port for Roman merchants. There's also a great local market in which to grab a bowl of *hu tieu*, clear noodle soup topped with meat and herbs.

Head north for 68km to Chau Doc.

❺ Chau Doc

Straddling the Cambodian border, Chau Doc is an historic city that's home to ethnic Vietnamese, Cham and Khmer populations. It's also where you'll find Sam Mountain, the highest point in the delta. Head to the top for incredible views of the surrounding flood plains.

It's a full day's ride – 150km, including several river crossings by ferry – southeast to Vinh Long.

❻ Vinh Long

On the banks of the mighty Mekong, Vinh Long makes a charming final stop, with temples to visit and boat tours to take. Mostly though, you're here to rest up and enjoy the last of the drowsy delta.

Ride back 139km through Ho Chi Minh City's industrial outer suburbs and into the city centre.

ALTERNATIVES

Extension: Dong Sen Thap Muoi

Make the one-hour detour on the road between Chau Doc and Vinh Long to this water-lily covered lake, dotted with small wooden shacks in which groups of locals hang out in hammocks, drinking cheap beer and eating cheap food. There's no-frills accommodation, too, in case you go too hard on those drinks.

Dong Sen Thap Muoi is 94km east of Chau Doc.

Diversion: Soc Trang

If you have time to add a stop between Ben Tre and Can Tho, make it Soc Trang, another riverside hub that in this case is inhabited by the largest population of Khmer people outside Cambodia. It's a quiet town that's the perfect spot to soak up a little Khmer culture in the form of several impressive temples, while also enjoying the riverside idyll.

From Ben Tre, ride 103km southwest to So Trang.

A LEISURELY TRIP THROUGH LAOS

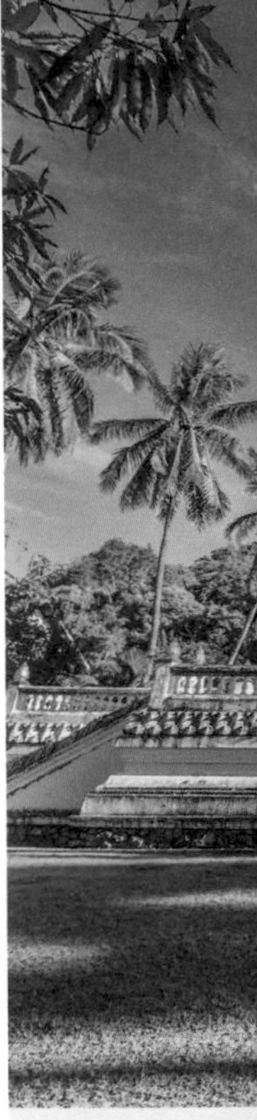

Luang Prabang – Don Det & Don Khon

From ancient capital Luang Prabang take your time covering Laos' highlights, from limestone mountains and secret government hideouts to Khmer ruins and relaxed islands.

FACT BOX

Carbon (kg per person): 204
Distance: 2186
Nights: 16
Budget: $$
When: Oct-Apr

❶ Luang Prabang

Explore gilded Buddhist temples, Lao royal relics and French heritage then spend time in streetside cafes and artisan shops in this magical city, once the seat of a royal kingdom.

Take a minibus to Nong Khiaw from Luang Prabang's bus station (regular; 4hr).

❷ Nong Khiaw

Enjoy a taste of rugged northern Laos with Nong Khiaw's limestone peaks, waterfalls, mountain walks and yoga, all on the picturesque Nam Ou River.

Take the bus for Sam Neua and alight at Nam Et-Phou Louey National Park HQ (daily; 5hr).

ABOVE: WAT PHU, CHAMPASAK

❸ Nam Et-Phou Louey National Park

An epic mountain landscape conceals rare wildlife at this remote park. Spot river creatures on the award-winning Nam Nern Night Safari.

Board the early morning minibus to Sam Neua (6hr), then taxi to Vieng Xai (1hr)

❹ Vieng Xai

Laos' Communist politburo escaped American bombs in the '60s by living in these remote caves. Tour subterranean chambers-turned-offices, a theatre, hospital and elephant stable that supported 20,000 people for 10 years.

Taxi to Sam Neua Airport (1hr) then fly to Vientiane (morning flights Mon-Sat; 1hr 20min).

❺ Vientiane

The smallish, low-slung French-influenced capital is all Buddhist temples, Communist monuments, great food and coffee – providing a fascinating dose of Lao city life away from the tourist trail.

Take a minibus to Vang Vieng (regular; 3hr).

❻ Vang Vieng

Vang Vieng is all about views of sky-hugging limestone peaks curving around the Nam Song River

ABOVE: WAT HO PHA BANG, LUANG PRABANG ROYAL PALACE

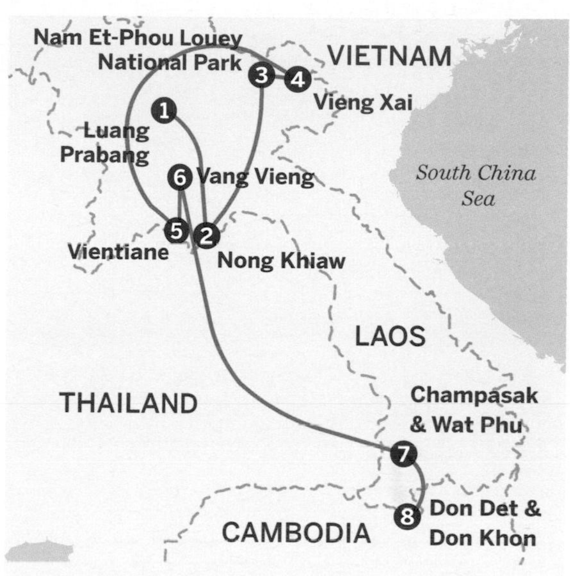

(take a slow tube down it), as well as holy caves and sunset beers.

Head back to Vientiane and take a flight to Pakse (6-7 daily; 1hr 15min). Board a *songthaew* (truck) in Pakse to Ban Muang (1hr) for boats on demand to Champasak (20min).

7 Champasak & Wat Phu

Pretty Mekong-bordering Champasak is the gateway to the Khmer ruins of Wat Phu, built on a holy mountain and wonderfully perfumed by frangipani, Laos' national flower.

Return to Ban Muang by boat and book, through your accommodation, minibus transfers to Don Det and Don Khon (3hr including a boat crossing at Ban Nakasang).

8 Don Det, Don Khon & the 4000 Islands

Kick back in a hammock on the chilled, palm-flecked islands of Don Det and Don Khon. Search for rusting remnants of France's railway dream by bicycle, and try to sight the rare Irrawaddy dolphin by boating on the Mekong.

Book minibus transfers back to Pakse for onward connections from its airport.

ALTERNATIVES

Extension: Elephant Conservation Center

Stroll, gawp, admire and get your hands dirty caring for the last of Laos' diminutive Asian elephants on a three-day trip to the Elephant Conservation Center on Nam Tien Lake. Only 400 *Elephas maximus* survive in the wild in a country once known as the 'Land of a Million Elephants'. The ECC cares for 29 of them.

Take the ECC's no-hassle return transfer from Luang Prabang (daily; 2hr 30min).

Diversion: Bolaven Plateau

The volcanic Bolaven Plateau rises 1300m out of southern Laos with ancient lava cooled across 6000 sq km. Take in coffee farms, single-origin brews, tea and fruit plantations, and dramatic plunging waterfalls. Homestay with locals, check into a coffee resort or zipline into treetop lodgings. Look out for rare hornbills in the forests.

Hire a motorbike from Pakse for the day, or book a tour through your Pakse accommodation.

THAILAND'S NORTHEAST: BANGKOK TO VIENTIANE

Bangkok – Vientiane

Sweep by train and bus from the Thai capital through the country's serene northeast, pausing at temples, natural wonders and Isan enclaves en route to Laos.

FACT BOX

Carbon (kg per person): 31
Distance (km): 800
Nights: 10
Budget: $$
When: Oct-Apr

❶ Bangkok

Overwhelming at first, the Thai capital quickly becomes addictive. After a few days walking its backstreets, ducking into ancient temples and feasting on what could be Asia's finest street food, you'll be itching to return.

Take the train from Hua Lamphong to Pak Chong (every 2hr; 4hr 30min), then a *sŏrngtăaou* (pick-up truck) to Khao Yai National Park (every 30min; 1hr).

❷ Khao Yai National Park

Thailand's oldest national park protects one of the largest intact monsoon forests in the world, sheltering wild elephants, tigers, leopards, gibbons and countless bird species. Hit the hiking trails, then unwind in one of the wineries – yes, wine is made here – just outside the park gates.

Take a *sŏrngtăaou* to Pak Chong (every 30min; 1hr), a train to Nakhon Ratchasima (6 daily; 1hr 30min) and a bus to Phimai (every 30min; 1hr).

❸ Phimai

Laidback Phimai's 11th-century Buddhist temple – topped by towering, time-scarred *prangs* (carved spires) – was the inspiration for Angkor Wat. The site is rarely crowded – and after some peaceful exploration you can spend the night in one of the small, family-run guesthouses that make Phimai feel like a home away from home.

Bus back to Nakhon Ratchasima (every 30min; 1hr) and then take a train to Udon Thani (2 daily; 5hr).

ABOVE: PHA THAT LUANG, VIENTIANE

ABOVE: BANDED KINGFISHER, KHAO YAI NATIONAL PARK

❹ Udon Thani

The Isan town of Udon Thani is noisy and industrious. You'll have a much calmer experience in the surrounding hinterland, which is studded with shrines and beauty spots – don't miss the sacred balancing boulders at Phu Phrabat Historical Park and the Red Lotus Sea with its floating carpet of lotus blooms.

Take the train to Nong Khai (5 daily; 1hr).

❺ Nong Khai

Outwardly, Nong Khai is just a stepping stone to Laos, but the calm mood, fine food and riverside location might just detain you. The glittering wat (monasteries) in town are nothing compared to Sala Kaew Ku, a surreal garden of cement Buddhist sculptures just outside town, created by one eccentric monk.

Take the bus across the border to Vientiane (6 daily; 1hr 30min).

❺ Vientiane

You'll immediately pick up on the French vibe as you reach the Lao capital. Vientiane is a place to absorb slowly, drifting from noodle house to coffee shop, and pausing to ponder in the cloisters of Wat Si Muang and elegant Pha That Luang.

Flights connect Vientiane to cities across Asia.

ALTERNATIVES

Extension: Phanom Rung Historical Park

Crowning a spent volcano, Prasat Phanom Rung is Thailand's most spectacular Khmer-era ruin, a palatial collection of prayer halls, corridors and *prangs* whose name translates as 'Big Mountain Temple'. There are more Khmer relics at Prasat Muang Tam, 8km southeast.

Visit both sites by renting a motorcycle in Nang Rong, accessible by bus from Nakhon Ratchasima (hourly; 2hr).

Diversion: Vang Vieng

The backpacker retreat of Vang Vieng is no longer quite as laidback, nor as hedonistic, as it once was, but the setting – fringed by towering outcrops and paddy fields beside the Nam Song River – is as idyllic as ever. Spend your days caving, kayaking, rock climbing and tubing, and your nights sipping beers over delectable Lao suppers.

Buses run regularly to Vang Vieng from Vientiane (3-5hr).

MYANMAR RIVER-BOAT EXPLORER

Yangon – Bagan

Take this epic sweep along the Ayeyarwady River by boat, train and plane, roaming from Yangon to little-explored Myitkyina and the temples of Mandalay and Bagan.

FACT BOX

Carbon (kg per person): 115
Distance (km): 2520
Nights: 14
Budget: $$
When: Nov-Apr

❶ Yangon

The past grandeur of the former capital of Myanmar is like a faded watercolour, washed out by the monsoon rains. Gigantic gilded stupas soar above a neat grid of colonial-era streets at the fork of the Yangon and Bago rivers. Make time to explore the glimmering Shewdagon Pagoda, which rises over the city like a golden exclamation mark.

Take the train to Myitkyina, changing at Mandalay (daily; 22hr).

❷ Myitkyina

Here in Myanmar's far north you're well off the traveller trail, so you won't have to share Myitkyina's gentle charms with a crowd. Don't miss the riverside town market, where goods arrive piled high on wooden canoes. With the current ban on foreigners travelling by river or road to Bhamo, you'll have to take a short flight to begin the most interesting part of the trip.

Fly to Bhamo (2 weekly; 45min).

❸ Bhamo

Sublime, sleepy Bhamo is a snapshot of what Southeast Asia was like before mass tourism. Old teak houses dot the riverfront, overhung by rain trees that burst into colour during the monsoon. Aside from the gentle pace of life, the main attraction is the chance to board the slow river ferry bound for Mandalay, enjoying front-row views of rural life on the Ayeyarwady.

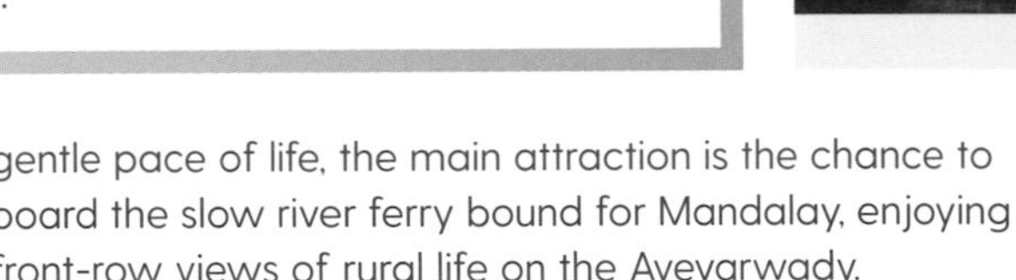

ABOVE: AYEYARWADY RIVERBANK, BAGAN

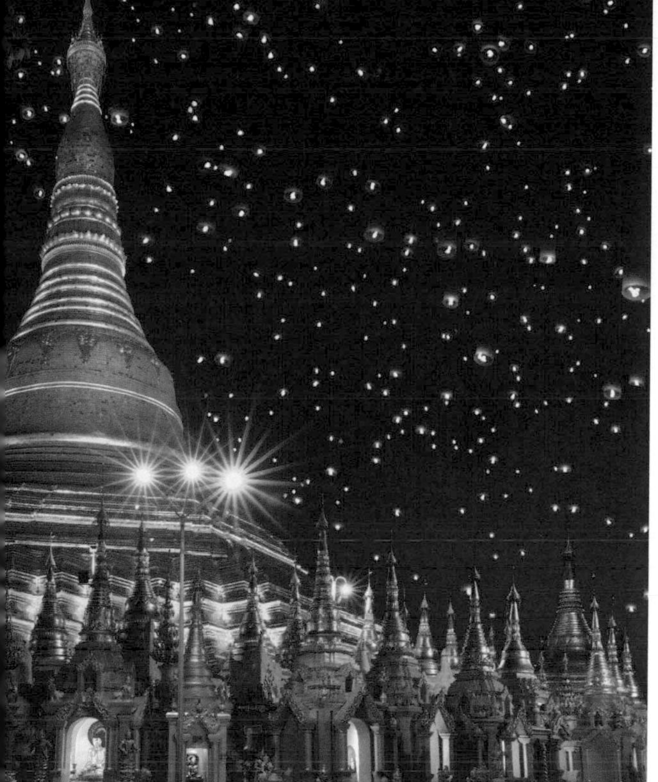

ABOVE: SHEWDAGON PAGODA, YANGON

Take the ferry to Mandalay (1-3 weekly; 40hr).

❹ Mandalay

The modern age has transformed Mandalay into a chaotic sprawl, but there's still much to see: the Buddhist pagodas adorning Mandalay Hill, a full hand of golden *payas* (temples), busy jade markets, marionette theatres and more. Tack on extra time for day trips to the photogenic U-Bein teak bridge at Amarapura and temple towns in the surrounding countryside.

Take the ferry to Nyaung U for Bagan (several weekly, daily Oct-Mar; 10-14hr).

❺ Bagan

The boat ride from Mandalay to Bagan starts with gorgeous vistas of the pagoda-covered hills at Sagaing. But that's nothing compared to the wonders waiting at Bagan itself, where 3000 ruined and restored Buddhist temples stretch to the horizon. Hire a bicycle, e-bike or horse cart and roam at your own pace, kicking off with the towering Shwezigon Pagoda and the temples of Dhammayangyi and Ananda.

Take the train back to Yangon (daily; 16hr) and its onward connections.

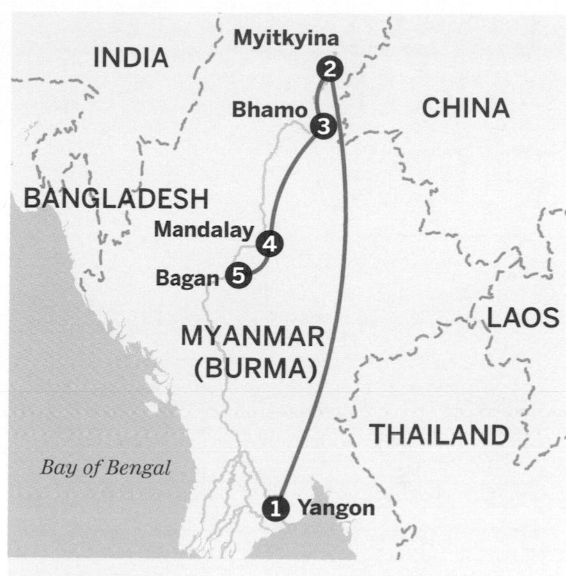

ALTERNATIVES

Extension: Katha

Literature buffs may want to break the journey from Bhamo to Mandalay at somnolent Katha, a forgotten Ayeyarwady township that served as the setting for George Orwell's *Burmese Days*.
The streets are still studded with British-era relics, including the house where Orwell reputedly lived during his time as an officer in the Indian Imperial Police.

Daily ferries run downriver from Bhamo (8-16hr) and on to Mandalay (14-24hr).

Diversion: Bago

With the short distance and easy train connections, it would be remiss not to make the short trip from Yangon to Bago and its sky-piercing pagodas and treasure-filled temples. This was once the capital of southern Burma, as evidenced by looming monuments such as the Shwemawdaw Pagoda, Myanmar's tallest stupa.

You can get to Bago by train from Yangon (6 daily; 1hr 45min).

FROM BANGKOK TO SINGAPORE BY TRAIN

Bangkok – Singapore

Why fly when you can take an epic train ride linking Southeast Asia's mightiest cities, with some beach-time on tropical islands along the way?

FACT BOX

Carbon (kg per person): 87
Distance (km): 1821
Nights: 14
Budget: $$
When: Nov-Mar

❶ Bangkok, Thailand

The busy, captivating Thai capital is Southeast Asia in microcosm. Take the Chao Phraya River Express boat to Wat Arun, wander the gilded courtyards of Wat Phra Kaew, have a traditional Thai massage at Wat Pho, splurge on Thai gastronomy at Michelin-starred Saawaan, then grab a beer in front of the agreeable chaos on Thanon Khao San.

Take the train from Hua Lamphong Station to Chumphon (10 daily; 8hr 30min) then the ferry to Ko Tao (daily; 3hr).

❷ Ko Tao, Thailand

Reef-ringed Ko Tao is one of the cheapest places in the world to learn to scuba dive. Between dives (or learning to dive), trek to island viewpoints, bask on the sands and admire the sunset from the shore every evening.

Take the ferry to Ko Pha-Ngan (daily; 1hr 30min).

❸ Ko Pha-Ngan, Thailand

Ko Pha-Ngan is all about the beach – or more specifically, its infamous full moon beach party. The whole island switches to moon-howling mode once a month at Hat Rin. At other times, just chill on the sand or trek to waterfalls inland.

Take the ferry and bus to Phun Phin train station (2 daily; 4-5hr), a train to Butterworth (changing at the Thai-Malaysia border in Padang Besar; 8hr) and a ferry to Penang (every 30min; 15min).

❹ Penang, Malaysia

More feasts await on cultured Penang Island, where cooks blend Chinese, Malay and Indian cuisine traditions into fabulous fusions. You'll need several days to explore Georgetown's shophouse-lined

ABOVE: KO PHA-NGAN FULL MOON PARTY, THAILAND

ABOVE: KUALA LUMPUR'S PETRONAS TOWERS, MALAYSIA

lanes, historic Chinese clan-houses, and temples and viewpoints in the suburbs.

Ferry back to Butterworth and take the train to Kuala Lumpur (6 daily; 4hr).

❺ Kuala Lumpur, Malaysia

The Malaysian capital is a fast-paced metropolis with a small-town soul, best sampled in food enclaves such as Bangsar Baru and Kampung Baru. Take a culinary journey across Asia in Masjid India and Chinatown, climb the landmark Petronas Towers and mix up theme parks, nature reserves, museums, mosques and temples.

All aboard the train to Singapore, via Gemas and Johor Bahru (3 daily; 9hr).

❻ Singapore

The island at the tip of the peninsula has been consumed by a mega-metropolis: Singapore, part Chinese, part Malay, part Indian, part European and all magnificent. Meander between the skyscrapers to botanic gardens, museums and galleries, with obligatory food stops at Singapore's legendary hawker courts.

Singapore's Changi Airport connects with cities across the globe.

ALTERNATIVES

Extension: Ko Samui

If you don't want to rush to Penang, hop aboard one of the ferries that whoosh daily from Ko Pha-Ngan to Ko Samui. What Samui lacks in raw castaway charm, it compensates for in fine food, fun and wellness options. Mix fine-dining with night-market snacks, and fast, fabulous lunches at informal *kôw gang* (rice and curry) shops with an afternoon of pampering in a spa.

Take a ferry from Ko Pha-Ngan (at least 2 daily; 20-45min).

Diversion: Phetchaburi

Break the journey from Bangkok to Chumphon at lovely Phetchaburi, a one-time Khmer kingdom that today dozes quietly in an ocean of paddy fields. Rising above the temple-studded old centre is Phra Nakhon Khiri Historical Park, mobbed by monkeys and crowned by the elaborate country palace of King Rama IV.

Trains arrive regularly from Bangkok (3hr 30min), continuing on to Chumphon (5hr).

ADVENTUROUS EAST MALAYSIA

Kota Kinabalu – Kuching

Sabah and Sarawak, Malaysia's chunk of the island of Borneo, offer a superb array of national parks, mountain climbing, jungle trekking and wildlife spotting.

FACT BOX

Carbon (kg per person): 146
Distance (km): 1486
Nights: 21
Budget: $$$
When: year-round

❶ Kota Kinabalu

Start in Sabah's capital Kota Kinabalu (KK), eating your way through every Malay dish in the book at the city's authentic, aromatic and bustling Night Market. Take in sunset from the UFO-like Signal Hill Observatory Platform and learn about the state's indigenous cultures at the excellent Sabah Museum and Mari Mari Cultural Village.

Take the bus from either the Inanam or Padang Merdeka bus stations to Kinabalu National Park (frequent; 2hr).

❷ Kinabalu National Park

A park fee, climbing permit, insurance and guide fee are mandatory should you wish to climb 4095m Mt Kinabalu – but Borneo's highest mountain is worth it. You'll also need to set aside at least two days and be in good physical condition as it's all uphill. Thrillseekers will want to descend using the via ferrata route of rungs and rails.

Eastbound buses to Sandakan (4hr 30min) pass by the park entrance. From Sandakan's long-distance bus station, transfer by minibus (20min) to Terminal Bas Sandakan to connect with the shuttle-bus (45min) to Sepilok.

❸ Sepilok

Sepilok is all about encountering 'the man of the forest' – orangutans. You can see these ginger-haired apes in their natural habitat at Sepilok Orangutan Rehabilitation Centre, 25km west of Sandakan. While out here also visit the nearby Bornean Sun Bear

ABOVE: KOTA KINABALU STREET FOOD

ABOVE: KINABALU NATIONAL PARK'S MT KINABALU

Conservation Centre and Rainforest Discovery Centre.
Return to Sandakan from where buses connect with Kota Kinabalu (frequent; 6hr). Fly from KK to Mulu (daily; 55min-4hr), often via Miri.

4 Gunung Mulu National Park

This showcase park is home to caves of mind-boggling proportions, as well as the limestone-shard Pinnacles and old-growth tropical rainforests with 17 vegetative zones. Don't miss the daily exodus of millions of bats from Deer Cave, and be sure to book a slot on the canopy skywalk.
Take a flight from Mulu to Kuching (daily; 1hr 45min).

5 Kuching

Sarawak's riverside capital is a relaxed, historic town sporting colonial and contemporary architecture. The major new Sarawak Museum complex opened in late 2020 with displays on everything from archaeology to zoology; the traditional wood carvings are especially impressive.
Kuching International Airport has connections around the region. The city is also the best place to organise further adventure activities and travels around Sarawak.

ALTERNATIVES

Extension: Semporna Archipelago

One of the world's top diving destinations, these emerald islands surrounded by tropical seas are off Sabah's southeast coast. Only 120 passes are issued daily to dive the dazzling coral reefs of tiny Sipadan, where you share the water with everything from parrotfish to majestic manta rays.
Tawau Airport is closest to the Islands and has direct connections with Kota Kinabalu as well as Kuala Lumpur.

Diversion: Bako National Park

Just 37km northeast of Kuching, this wonderful park is notable for its biodiversity – come here to see rare orchids, proboscis monkeys, bearded pigs and to hike jungle trails to gorgeous beaches. You can visit as a day trip, but you'll get more from the experience if you stay a night or two.
Regular buses run from Kuching to Bako Bazaar (1hr) for motorboat transfers to the park's Teluk Assam jetty (20min).

SOUTHERN MYANMAR DELIGHTS

Yangon – Myeik

Frequent and reliable buses are the best way to travel on this southern Myanmar itinerary, taking in gilded Buddhist temples, ancient cities and idyllic beaches.

FACT BOX

Carbon (kg per person): 170
Distance (km): 1736
Nights: 14
Budget: $$
When: Oct-May

❶ Yangon

The country's largest city, formerly known as Rangoon, is the place to be dazzled by gilded Shwedagon Paya, Myanmar's principal Buddhist temple. Stroll around Kandawgyi Lake and explore downtown Yangon's impressive array of colonial architecture, contemporary art galleries and street-food stalls.

Take a bus from Yangon's Aung Mingalar Bus Terminal to Bago (frequent; 2hr).

ABOVE: MT KYAIKTIYO'S GOLDEN ROCK

❷ Bago

Blissful Buddhas and treasure-filled temples make Bago an appealing and easy first stop on your journey towards Mt Kyaiktiyo. The main sights, including the Shwemawdaw Paya pagoda and the giant reclining Shwethalyaung Buddha, are covered by the Bago Archaeological Zone ticket (K10,000).

Take the bus to Kyaiktiyo (frequent; 3hr).

❸ Mt Kyaiktiyo

A major Buddhist pilgrimage site, Mt Kyaiktiyo's defining feature is an enormous gilded boulder, the Golden Rock, balancing precariously on a cliff edge and crowned with a stupa. The atmosphere during the pilgrimage season (November to March) is especially magical.

From Kyaiktiyo there are frequent buses to Mawlamyine (4hr).

❹ Mawlamyine

On the Thanlwin River, Mawlamyine sports a wealth of crumbling colonial-era buildings, churches, mosques and temples. Kyaikthanlan Paya is the city's tallest stupa, while the Mahamuni (Bahaman) Paya dazzles with its jewel-encrusted central chamber. Possible day trips include Bilu Kyun (Ogre Island) and

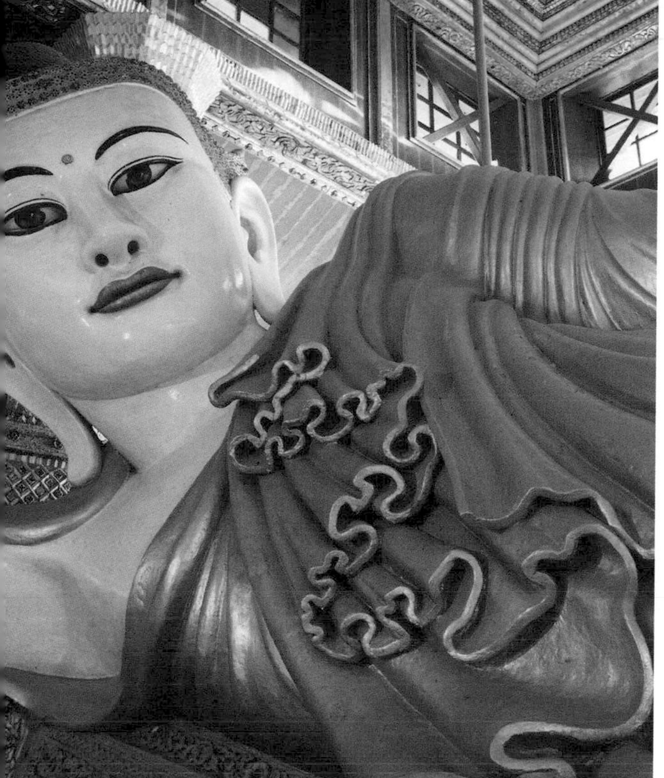

ABOVE: KYAIKTHANLAN PAYA BUDDHA, MAWLAMYINE

the giant reclining Buddha at Win Sein Taw Ya temple.
Take the bus from Mawlamyine to Ye (6 daily; 4hr).

5 Ye

Experience small-town life in attractive Ye, where traditional wooden houses and a large, hectic market sit beside a tree-lined lake. Venture into the countryside to Mon and to Kayin villages such as Kyaing Yar.
Take a minivan to Dawei (4hr).

6 Dawei

This seaside port and low-key travellers' hub sports diverse architecture and offers access to the up-and-coming beach resort area of Maungmagan. Aim for stunning Paradise Beach, about two hours' drive south of Dawei.
Take the bus to Myeik (6 daily; 4hr).

7 Myeik

Grand Sino-Portuguese houses jostle with mosques, churches, wooden homes and colonial-era mansions in this bustling, historic port. It's also a gateway to the 800-odd islands of the Myeik Archipelago – head to Kawthoung to organise a boat trip out to them.
Fly from Myeik back to Yangon (6 weekly; 2hr) for onward international connections.

ALTERNATIVES

Extension: Myeik Archipelago

Accessing these gorgeous, barely-populated islands is not for those on a tight budget – accommodation is either aboard a boat or at one of just four upmarket resorts. Make the investment and you will experience idyllic white-sand beaches sitting in a turquoise sea and some of the best diving in Southeast Asia.
Join a live-aboard cruise or transfer directly to one of the island resorts from Kawthoung.

Diversion: Hpa-An

Kayin State offers a winning combo of superb scenery – forested limestone hills rising above luminous green paddy fields – and a fascinating cultural mix. The logical base is the riverside capital Hpa-An, surrounded by striking caves, rivers and lakes. Climb to the 722m summit of Mt Zwegabin, and marvel at the stadium-size Saddan Cave, its entrance dominated by Buddha statues and pagodas.
Frequent buses connect Hpa-An with Mawlamyine (2hr).

STRAITS SETTLEMENTS: SINGAPORE TO MALAYSIA

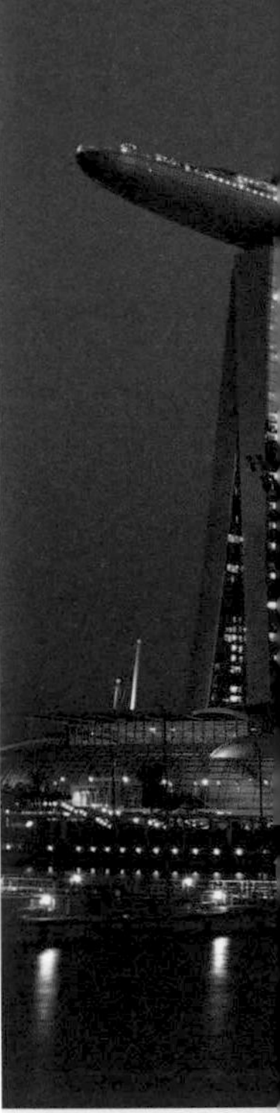

Singapore – George Town

Embark on a two-country bus and train trip through Southeast Asia, focusing on the cultures and colonial history of Singapore and peninsular Malaysia.

FACT BOX

Carbon (kg per person): 122
Distance (km): 1353
Nights: 14
Budget: $$
When: Mar-Nov

❶ Singapore

The city-state of Singapore is the perfect place to start this tropical trip. The astounding architecture of Marina Bay makes a strong contemporary statement but there's also history and heritage to discover in Singapore's excellent museums and galleries – and delicious food to be eaten in its hawker centres, where the region's diverse recipes are cooked to perfection. The MRT transport system makes getting around simple.

Buses run from various central Singapore locations to Melaka Sentral Station (3-4hr), crossing the border into Malaysia.

❷ Melaka

Back in the 15th century, Melaka was the region's richest port and the seat of a sultanate that attracted foreign invaders. Multiple layers of the city's fascinating history are preserved here, from the 16th-century ruins of Portuguese occupation to the many denominations of temples in Chinatown, the heart of the World Heritage zone.

Board another bus from Melaka Sentral to Kuala Lumpur's Terminal Bersepadu Selatan (frequent; 2hr), which is a 13min local train ride south of KL Sentral train station.

❸ Kuala Lumpur

Compared to Singapore and Melaka, Malaysia's capital KL, born from the tin-rush of the mid-19th century, is a teenager. Nonetheless, it's home to gorgeous examples of colonial architecture, such as the buildings surrounding Merdeka Square, plus the Petronas Towers, shiny, soaring symbols of Malay

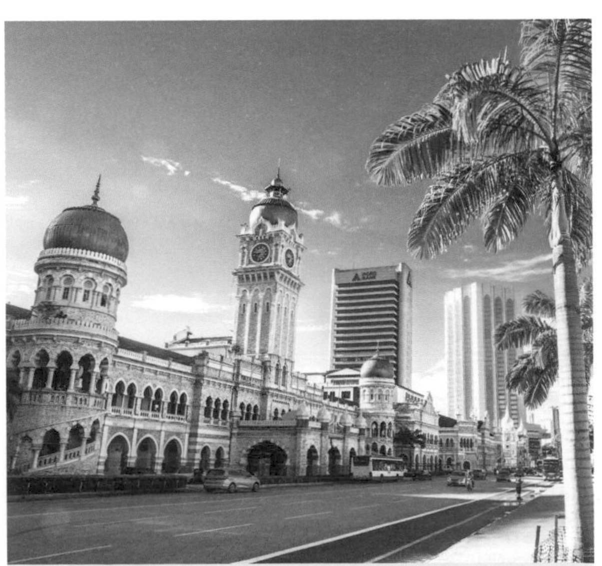

ABOVE: KUALA LUMPUR'S MERDEKA SQUARE, MALAYSIA

ABOVE: MARINA BAY, SINGAPORE

modernity. Other draws include lush botanical gardens and parks, the city's plethora of shopping malls and its high-class food scene.

Take the express train from KL to Ipoh (3 daily; 2hr 30min).

4 Ipoh

Capital of the state of Perak, Ipoh is graced with more fine colonial architecture such as its grand railway station dating from 1917. There's scrumptious street food to be savoured here too, while exploring the street-art decorated lanes of the Old Town. Towering white cliffs flank Ipoh, some with magnificent cave temples pocketed in the limestone.

Take the train to Butterworth (daily; 2hr), and then the ferry to George Town (10min).

5 George Town

Penang's capital has the region's best-preserved World Heritage zone, a swoonworthy mash-up of colonial and local architecture. There's also fantastic street art, Malaysia's best range of street food and easy day trips including up Penang Hill or out to the beaches and jungle of Penang National Park.

Penang International Airport has flights back to Singapore and other international destinations.

ALTERNATIVES

Extension: Cameron Highlands

Sip afternoon tea overlooking the emerald tea plantations and mist-shrouded hills of the Cameron Highlands. This sprawling hill station stays blissfully cool year-round making it the ideal destination for leisurely hiking, nature walks and agro-tourism.

Aim for Tanah Rata, the Highlands' heart, on buses from either Kuala Lumpur (4hr 45min) or Ipoh (2-3hr).

Diversion: Langkawi

If you've felt your journey up Malaysia's west-coast peninsula has been lacking beach time, extend your trip to this lovely archipelago. Beautiful sandy strands are a speciality of the largest island Pulau Langkawi, which also has forest parks, hot springs and waterfalls. Make time to ride the cable car to the 708m summit of Gunung Machinchang for panoramic views.

Fly to Langkawi from Penang (45min) or Singapore (1hr 30min).

SAILING INDONESIA'S SPICE ISLANDS

Ternate – Ambon

Set sail around the Moluccas, a scattering of eastern Indonesian islands now well off the tourist trail but once the centre of the world spice trade.

FACT BOX

Carbon (kg per person): 12
Distance (km): 650
Nights: 25
Budget: $
When: Apr-May, Sep-Oct

1 Ternate

It's hard to believe that in the 16th century this latter-day Indonesian backwater was the heart of the lucrative spice trade. Clinging to the slopes of smoke-puffing Gamalama volcano, Ternate – the wealthiest of the four spice sultanates – remains the regional capital of North Maluku island, though you're unlikely to see many other travellers. The island is accessible by flights from Makassar (Sulawesi) and Jakarta, or by Pelni ferries. Visit the Sultan's Palace and Portuguese forts, peer into the crater lake of Tolire Besar (home, says local legend, to a white crocodile), laze on black-sand beaches and climb Gamalama if it's not rumbling too much.

Frequent speedboats connect Bastiong harbour in Ternate to Rum on Tidore (10min).

2 Tidore

Neighbouring island Tidore is also an historic spice-trading sultanate, with its own perfectly conical volcano (Kiematabu, a worthwhile climb). But unlike Ternate, Tidore has lost its commercial clout, leaving it even sleepier. Here, wander amid traditional whitewashed buildings and crumbling ruins, watch locals harvesting those precious cloves, and shop for Tidorese crafts (pottery, rattan baskets) at Soasio market. Seek out the monument commemorating the arrival of Magellan's circumnavigating crew in 1521 – a nod to busier times.

Take the ferry from Dowora on Tidore to Sofifi on Halmahera (daily; 45min).

ABOVE: WALLACE'S STANDARDWING BIRD OF PARADISE

ABOVE: TERNATE

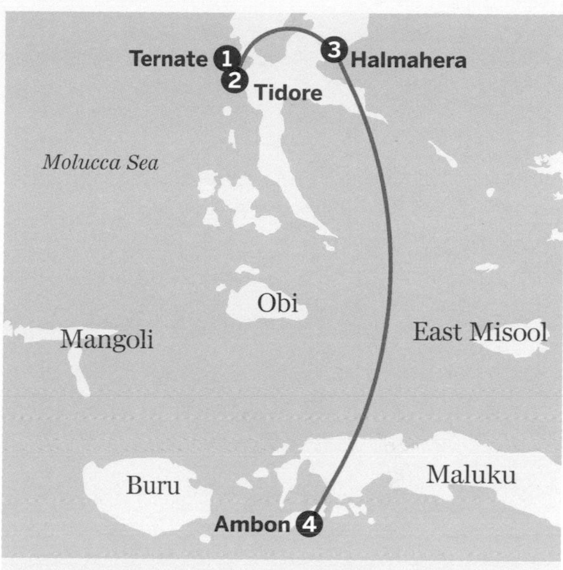

3 Halmahera

Halmahera is huge, mountainous, odd-shaped and under-explored. Allow plenty of time, concentrating on the north. Visit Jailolo (another of the four spice sultanates) and the ethnic Sahu villages to the north, as well as Dodinga (where Alfred Russel Wallace came up with his evolutionary theory around the same time as Darwin) and Aketajawe-Lolobata National Park, a haven for endemic birds, not least the charismatic Wallace's standardwing bird of paradise.
Local boats connect Sofifi, Jailolo and other villages on Halmahera to Ternate. Then take a Pelni ferry from Ternate to Ambon (20hr) – schedules change regularly, delays are likely.

4 Ambon

From Muslim North Maluku, sail into the Christian-majority south. Regional capital Ambon was a former Dutch stronghold – don't miss a visit to Fort Amsterdam (built in 1514) and the ethnographic objects at the Siwalima Museum. Then explore deserted white-sand beaches and head to the mountain villages above Leitimur, where traditional houses and churches sit among fragrant spice trees.
Fly from Ambon's Pattimura Airport to Jakarta (daily; 3hr 30min) for international connections.

ALTERNATIVES

Extension: Bacan

The fourth and final spice-island sultanate is big, offbeat Bacan. Capital Labuha has a fort and the sultan's residence, but better are the jungly hills of the wild interior where Wallace made some of his greatest discoveries, including the Moluccan cuscus (a type of marsupial) and the golden birdwing butterfly; look for parrots, cockatoos, hornbills and endangered black macaques too.
Infrequent ferries run from Ternate to Bacan (6hr) or fly (50min).

Diversion: Banda archipelago

It's not quick or easy to reach this cluster of small, volcanic isles, once the world's only source of nutmeg and mace. But the reward is a mass-tourism-untouched archipelago of colonial architecture and natural good looks, all fringed by magnificent reefs in crystal waters that offer some of the best snorkelling in Indonesia.
Infrequent fast ferries (6hr) and Pelni services (8-14hr) connect Ambon and the Banda islands.

BALI & LOMBOK BEYOND THE CROWDS

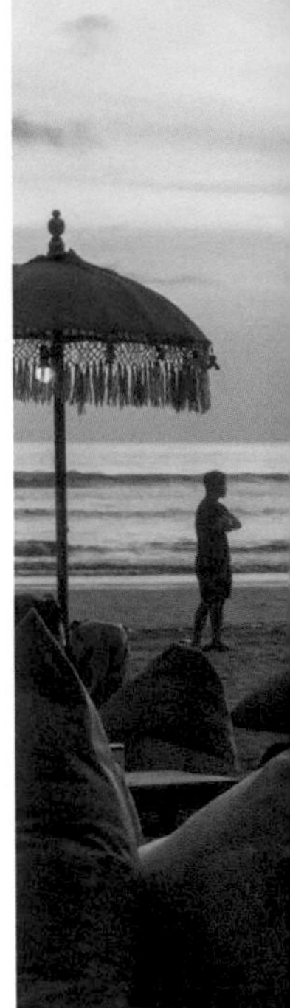

Seminyak – Gunung Rinjani

This two-island hop takes in Bali and Lombok's legendary surf, sand, sights and culture, and journeys inland for epic volcano treks and a vibrant arts scene.

FACT BOX

Carbon (kg per person): 33
Distance (km): 255
Nights: 14
Budget: $
When: year-round

❶ Seminyak

While Kuta has always been the big-name beach drawing tourists to Bali, these days the attention has shifted to the nearby sands of Seminyak, a less sordid version of the original. Here you'll get the magnificent beaches and sublime sunsets, but swap the trashy nightclubs for chic cafes, trendy beach clubs and creative fusion cuisine.

From Seminyak, Bukit Peninsula (25km) is best explored by scooter, which are readily available for hire.

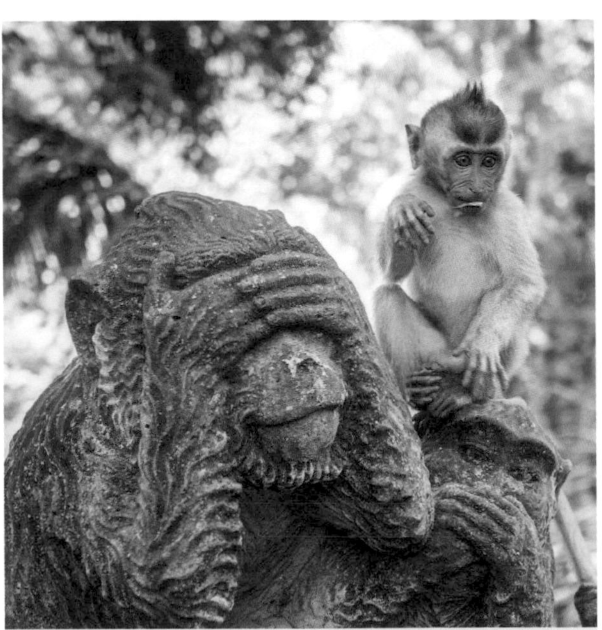

ABOVE: UBUD MONKEY FOREST

❷ Bukit Pensinsula

Long popular with surfers, the Bukit Peninsula has only become popular for backpackers in recent years, as a more laidback alternative to Bali's mainstream beaches. Comprising an attractive string of sandy coves, the region has developed its own little scene with bars and relaxed nightly parties, minus the booming techno.

Take a taxi to Ubud (1hr 30min).

❸ Ubud

Swap the beach towel for some culture by heading inland to Ubud – a cradle for Balinese arts. Set among verdant countryside, this place provides a fascinating insight into Balinese customs, from atmospheric Hindu rituals at its temples to traditional dance and *gamelan* (traditional music) performances. Yoga and wellness are also big draws, attracting health-conscious aspiring yogis from around the world.

ABOVE: SEMINYAK BEACH

Tourist buses (regular; 1hr) head to Padangbai for boats to the Gili Islands (regular; 1hr 30min).

4 Gili Islands

Resting comfortably in the shadow of its more famous neighbour, Lombok may be a short boat trip from Bali, but it's a world away from its mass tourism. The Gili Islands, which sit off Lombok's northwest coast, offer a memorable island escape. Gili Trawangan is the most popular, perfect for lazy days, snorkelling by the beach and and fun nights partying with like-minded travellers.

Take a ferry (regular; 45min) to Bangsal on Lombok. From there take an *ojek* (motorcycle taxi) ride to Senaru for Gunung Rinjani (1hr 30min).

5 Gunung Rinjani

Indonesia has no shortage of active volcanoes (127 to be precise), but few are more popular and sacred than Rinjani. At 3762m, it's the country's second tallest and features astonishing peak vistas and a turquoise crater lake. It takes two to three days to climb, with most summits coinciding with majestic sunrise views.

From the trek's finishing point at Sembalun Lawang, it's a 3hr drive to Lombok's international airport (guided treks typically include transfers).

ALTERNATIVES

Extension: Munduk

If Ubud's feeling too hectic for you, consider heading deeper into Bali's lush interior with a visit to Munduk. This quaint Dutch-colonial summer retreat is set among the foothills of misty mountains and offers plenty of waterfalls and trekking amid attractive jungle.

Munduk is 63km northwest of Ubud, best reached by car or scooter.

Diversion: Kuta (Lombok)

Unlike its namesake beach on Bali, Lombok's version hasn't lost its charm. Hidden away in the south of Lombok, here it's all about turquoise waters and white sands minus the garish nightlife and mass tourism. Whether you're here to surf, or swim and relax by the beach, there's a patch of sand here with your name on it.

Kuta is just 17km south of Lombok's airport – take a taxi to get there.

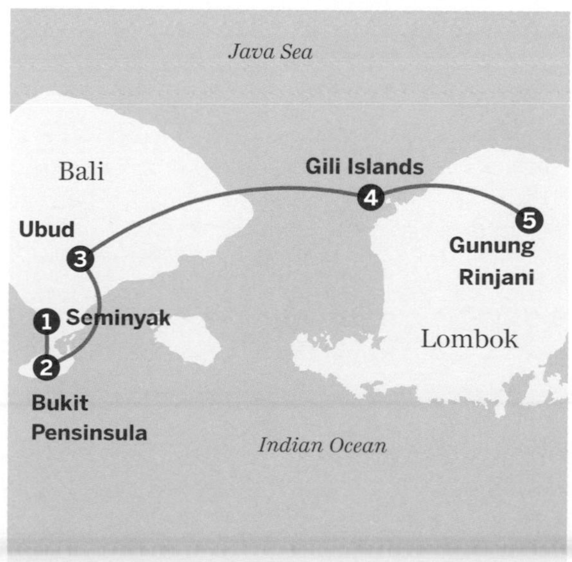

A TASTE OF WEST SUMATRA

Padang – Danau Maninjau

West Sumatra promises Padang cuisine, world-class waves and volcanoes among green landscapes dotted with Minangkabau architecture.

FACT BOX

Carbon (kg per person): 46
Distance (km): 449
Nights: 10
Budget: $
When: year-round

❶ Padang

The West Sumatra capital serves as a popular base for surfers and intrepid backpackers alike. Stroll its crumbling, atmospheric colonial quarter and stop by one of its restaurants specialising in Padang cuisine to sample delectable slow-cooked beef *rendang* and *ayam bakar* (charcoal-grilled chicken), among an assortment of creamy coconut-based curries.

The Mentawai Islands are 150km off mainland Sumatra, accessed from Padang by fast ferry (daily; 6hr). Once there, motorised canoes allow inter-island exploration.

❷ Mentawai Islands

Home to some of the best waves on the planet, these far-flung islands attract surfers from far and wide for their perfect glassy barrels. Others make the boat journey across to the islands to traverse muddy jungles and meet the remote hunter-gatherer tribespeople who live here, learning first-hand about their way of life.

Return to Padang by ferry and catch a bus from Padang to Bukittinggi (regular; 2hr).

❸ Bukittinggi

Bukittinggi is famous for its historical Dutch fort and its yummy West Sumatran cuisine. A short drive from town leads into a bucolic countryside of fluorescent-green rice paddies, palms and looming volcanoes.

ABOVE: MENTAWAI ISLANDS

ABOVE: MINANGKABAU ARCHITECTURE, BUKITTINGGI

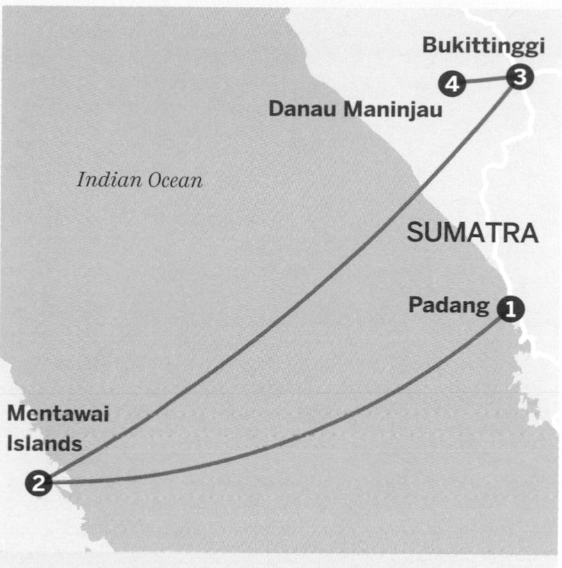

Keep an eye out for the superb homes of the Minangkabau people, featuring multi-tiered spired roofs inspired by the horns of a buffalo. Worthy off-beat excursions include a day at the oxen races, and treks to seek out the rare rafflesia flower in its natural habitat.

Minibuses run between Bukittinggi and Maninjau (regular; 45min); once in Maninjau hire a scooter or taxi to explore the region in more detail as a day trip.

4 Danau Maninjau

This ancient volcanic crater lake offers a wonderful opportunity for some R&R, best spent swimming, canoeing and hiking to waterfalls and traditional villages. Don't miss the local speciality of freshwater crayfish, best accompanied by a cold beer and a sunset view over the primordial lake. The drive here is magnificent, winding you through the verdant countryside of rice paddies, palm trees and smoking volcanoes, before the lake itself is unveiled in dramatic fashion with its panoramic views.

Take a minibus back to Bukittinggi (regular; 45min), from where there are plenty of buses to Padang (regular; 1hr) for flights across Indonesia.

ALTERNATIVES

Extension: Pantai Bungus

If you're seeking an easily-accessible tropical-island beach escape then these low-key islands just south of Padang will do the trick. Spend time island-hopping, lazing in hammocks and sipping frosty sunset drinks on the sand.

Take a taxi ride from Padang to the harbour (1hr), from where you can charter a motorised canoe for the ride to the islands (45min).

Diversion: Kerinci Valley

One of the last strongholds of the endangered Sumatran tiger, this stunning swathe of rainforest is rich in biodiversity and offers wonderful trekking and village homestays. As well as the (small) chance of spotting a tiger, you can also climb to the summit of Southeast Asia's tallest volcano, Gunung Kerinci.

Buses run from Padang to Kerinci's main hub, Sungai Penuh (2 daily; 9hr) where you can arrange local tours and transport

ANCIENT JAVA EXPLORER

Borobudur – Kawah Ijen

Discover Javanese culture, delicious cuisine, World-Heritage Buddhist temples and volcanic landscapes on this trip across Indonesia's most historically important island.

FACT BOX

Carbon (kg per person): 64
Distance (km): 850
Nights: 7
Budget: $
When: year-round

❶ Borobudur

Start your Javanese journey at its centre, where among the serene countryside you'll discover the world's largest Buddhist temple – the monumental World Heritage-listed Borobudur. Dating to the 8th century, its unique circular stupa design astounds, with the intricate craftsmanship featured on its decorative panels taking in scenes relating to Javanese culture.

Take a bus to Yogyakarta (hourly; 1hr 15min).

ABOVE: GUNUNG BROMO

❷ Yogyakarta

At the heart of Javanese culture is the proud and fiercely independent city of Yogyakarta. Take time to explore its palaces and many museums, before taking in traditional music and dance performances along with its thriving contemporary arts scene.

Visit Prambanan as a day trip from Yogyakarta by bus (hourly; 45min).

❸ Prambanan

Dating to the 8th century, Prambanan's temples are as impressive as you'll find anywhere across Asia. While ruins are dotted across its plains, the finest selection are the ones in the main compound, where majestic monuments soar like elegant, ancient skyscrapers.

Take a train from Yogyakarta to Malang (3-4 daily; 9hr).

❹ Malang

Offering an attractive, leafy escape, the Dutch colonial town of Malang is popular among travellers stopping over between Yogyakarta and Bromo volcano. Don't miss its reimagined *kampongs*

ABOVE: BOROBUDUR TEMPLE

(villages), which have undergone a whimsical transformation courtesy of local artists who've painted them entirely in vivid colours and murals.
Take a bus to Probolinggo (hourly; 3hr) then change for another bus to Cemoro Lawang (2 daily; 1hr), the village for Gunung Bromo.

5 Gunung Bromo

The poster child for Java tourism – and arguably Southeast Asia's most famous volcano – Gunung Bromo still brings in the masses for its lunar-like landscape that glows majestically at sunrise.
Arrange a car and driver for the trip to Kawah Ijen (7hr).

6 Kawah Ijen

Making up part of the Ijen volcanic plateau, Kawah Ijen is a big-ticket attraction famed for its unworldly 'blue fire'. It's an effect only visible in the darkness of night, so most visitors begin the climb around 4am to combine seeing its blue sulphuric combustion with a sunrise over its spectacular crater lake.
For onward travel, and maybe a few days on the beach, head to Bali: a 34km drive to Banyuwangi and a 1hr ferry crossing will get you there.

ALTERNATIVES

Extension: Gunung Merapi

On an island filled with volcanoes it's impossible to visit them all, but Merapi is one that you should make an effort for. As one of Asia's most volatile volcanoes it's not always accessible, but if conditions are favourable the summit views up here are magnificent, and the steaming fumaroles dotted across its surface superbly atmospheric.
You can visit Merapi on a day trip from Yogyakarta, 25km south.

Diversion: Solo

Solo (aka Surakata) is a proud town famed for its legacy as a cultural centre for Javanese arts. Here you can experience its rich tradition of batik textiles, visit its palace, many museums (the Sangiran Museum of Ancient Man has the world's largest collection of *Homo erectus* fossils) and sample street food, before shopping for quality handicrafts.
Solo is connected by train to Yogyakarta (hourly; 1hr). The train continues to Malang.

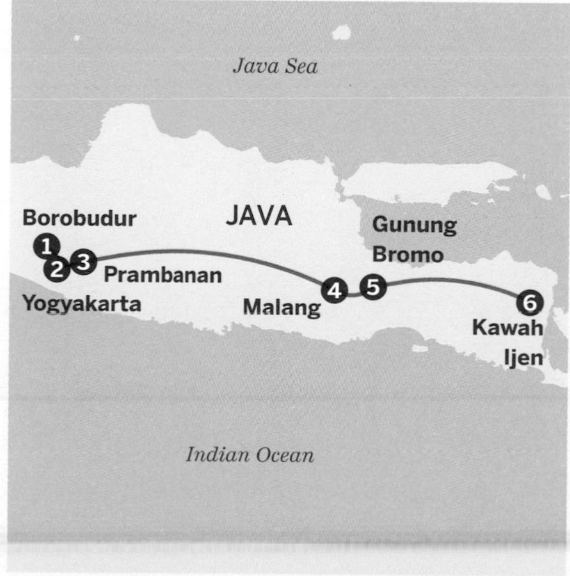

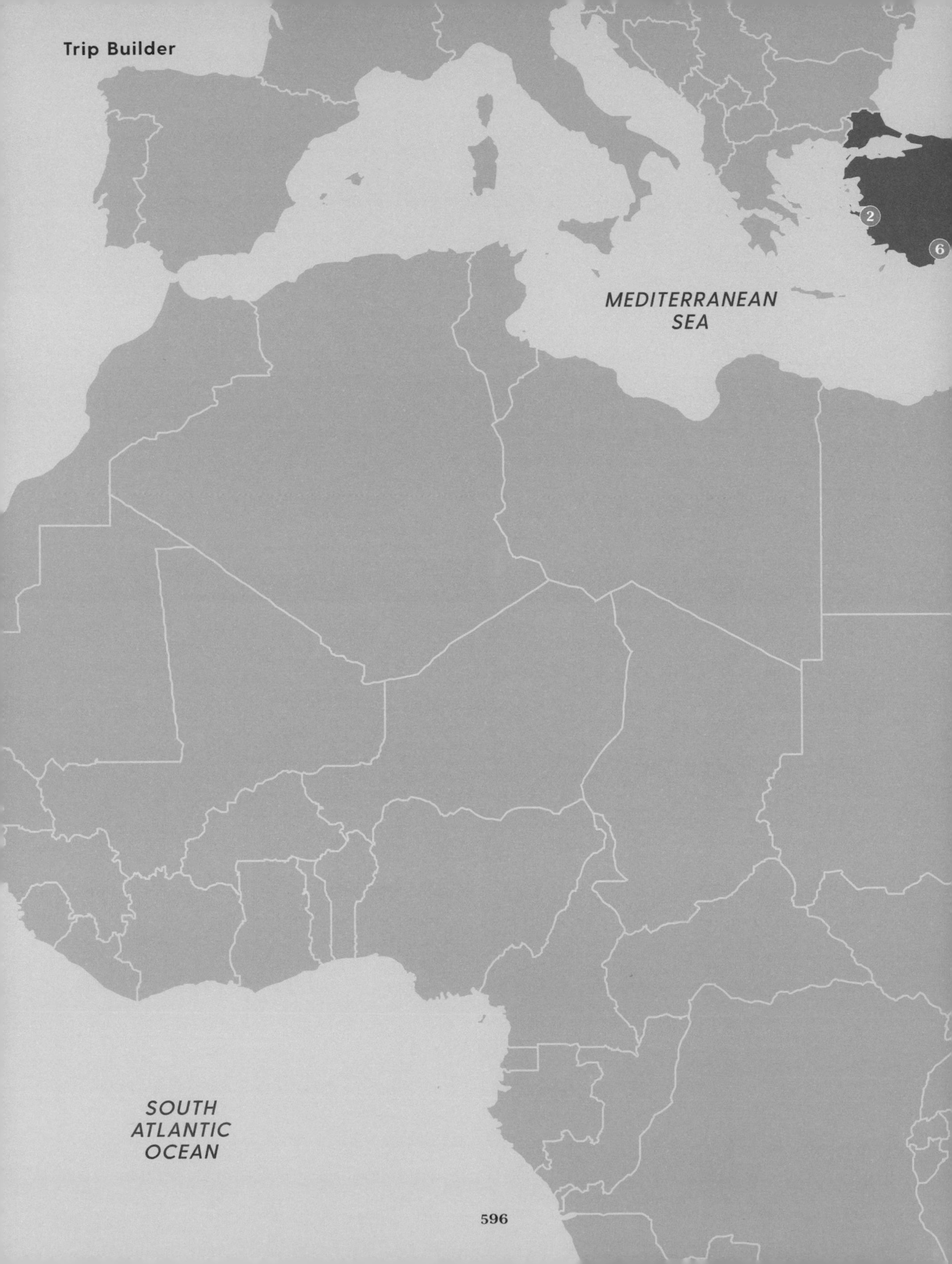

MEDITERRANEAN SEA
SOUTH ATLANTIC OCEAN
2
6

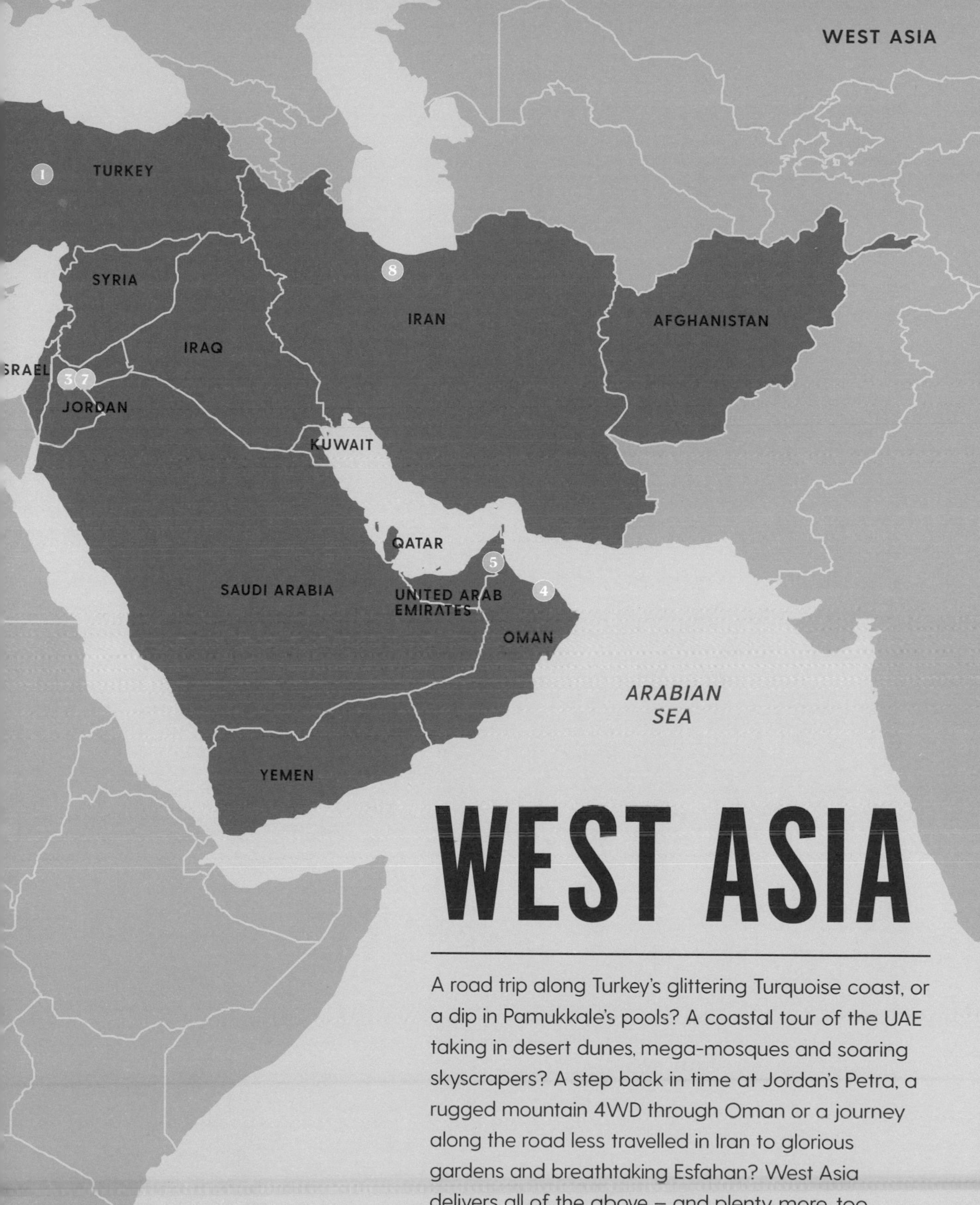

WEST ASIA

A road trip along Turkey's glittering Turquoise coast, or a dip in Pamukkale's pools? A coastal tour of the UAE taking in desert dunes, mega-mosques and soaring skyscrapers? A step back in time at Jordan's Petra, a rugged mountain 4WD through Oman or a journey along the road less travelled in Iran to glorious gardens and breathtaking Esfahan? West Asia delivers all of the above – and plenty more, too.

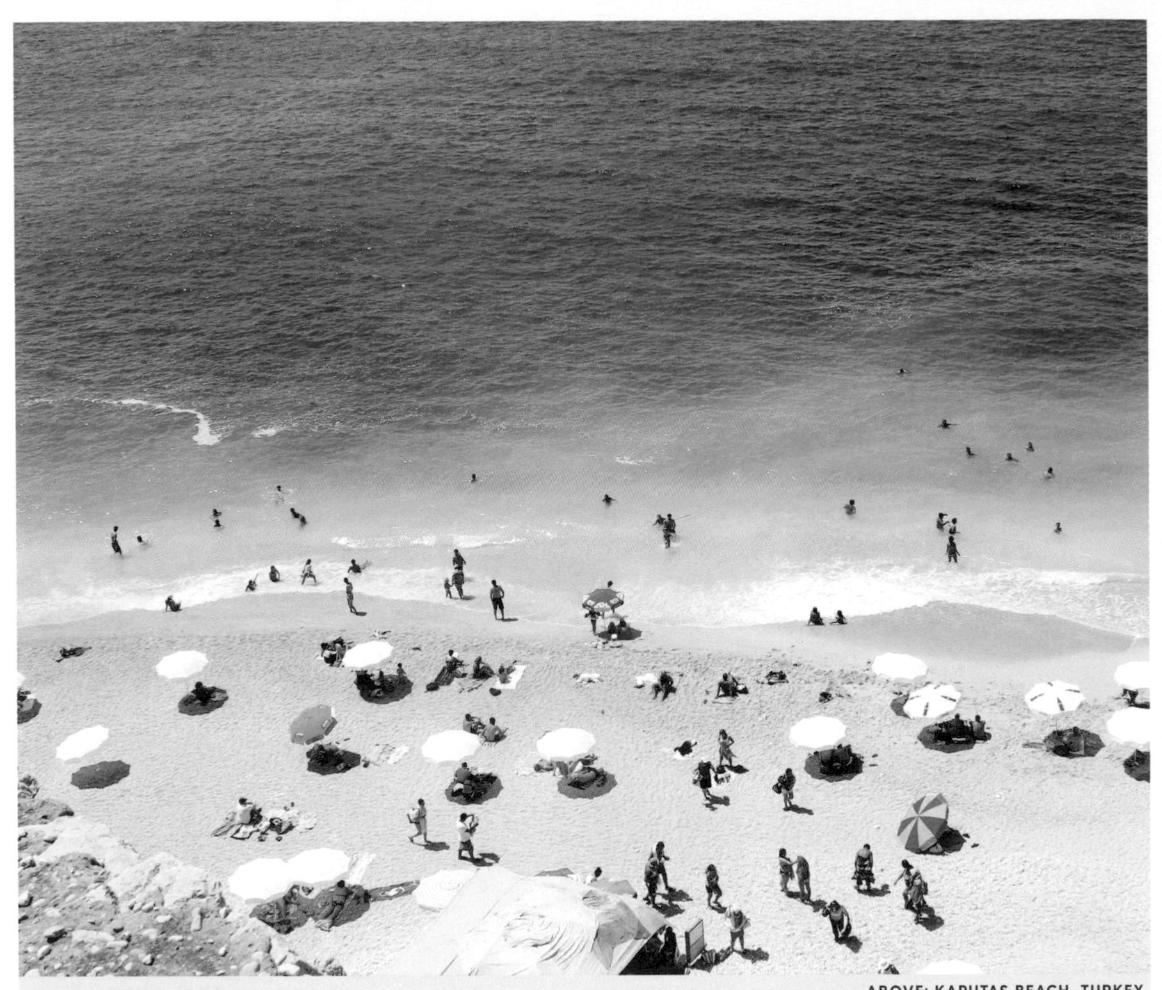

ABOVE: KAPUTAŞ BEACH, TURKEY

CONTENTS

EXPLORING TURKEY'S PAST AND PRESENT

Cappadocia – Mardin

Bounce from fresco-covered cave-churches and glaring mountain-top stone gods to the Neolithic temple where worship may have first begun on this Anatolian bus trip.

FACT BOX

Carbon: (kg per person): 87
Distance (km): 1184
Nights: 8-12
Budget: €€
When: May, Jun, Sep & Oct

❶ Cappadocia

Regular flights from İstanbul land at Cappadocia's two airports, Kayseri and Nevşehir, from where shuttles provide transfers to the region's villages. Once here, a whole host of activities awaits. Hike between fairy-chimney rock pinnacles hidden within valleys of blush-tinged rock. View vibrant Byzantine frescoes in the cave-churches of Göreme Open-Air Museum and Ihlarra Valley. Delve through twisty, claustrophobic tunnels in Kaymaklı Underground City. And hover over the lunar landscape in a hot-air balloon. All that and a great night's sleep in an atmospheric cave-hotel amid otherworldly scenery.

Take the bus from Nevşehir Otogar (bus station) to Gaziantep (3 daily; 7hr 15min).

ABOVE: ZEUGMA MOSAIC MUSEUM'S *GYPSY GIRL*, GAZIANTEP

❷ Gaziantep

Gaze admiringly at world-renowned Roman art at Gaziantep's Zeugma Mosaic Museum, then indulge in the city's star culinary attraction, baklava. Get some of the sticky, syrup-drenched, pistachio goodness to go and enjoy it while strolling the Old Town's skinny bazaar alleys, chock-a-block with copper-workers and historic pastry shops.

Take the bus to Şanlıurfa (hourly; 2hr 30min).

❸ Şanlıurfa

According to local lore, this ancient pilgrimage city was the birthplace of Abraham. Today, it's a place to feed fat sacred carp in Gölbaşı's ponds before losing your bearings in the maze-like bazaar. View a world-class collection in Şanlıurfa Archaeology Museum then head out of town to Göbeklitepe for the T-shaped Neolithic temple pillars that turned modern archaeological theory on its head.

Arrange a tour to Mt Nemrut in Şanlıurfa.

ABOVE: HOT-AIR BALLOONING, CAPPADOCIA

❹ Mt Nemrut

Glowering out across the barren hills below, Mt Nemrut's toppled stone-heads are the handiwork of an upstart Commagene king. After saying hello to King Antiochus and his godly consorts on the summit, wind down the mountain to visit the Commagene capital of Arsameia; stop off at the Karakuş Tümülüsü burial mound along the way.

Back in Şanlıurfa, hop on a bus to Mardin (4 daily; 3hr).

❺ Mardin

This hillside cluster of honey-toned stone houses, home to a rich heritage that harks back to Anatolia's multicultural past, offers panoramas stretching out across the Mesopotamian Plains below. Wriggle your way through the bazaar, take in the vistas from the roof of Zinciriye Medresesi then tour the Syriac monastery of Deyrul Zafaran.

Fly from Mardin Airport back to İstanbul for international connections (2 daily; 2hr).

ALTERNATIVES

Extension: Antakya

Formerly biblical Antioch, where St Paul preached, Antakya is well worth a visit for its Archaeology Museum, which holds world-class Byzantine mosaics; and nearby St Peter's Church, one of the oldest on the planet. Make time, too, to explore the Old Town lanes and feast on Antakya's fusion of Turkish-Arab cuisine.

There's a direct bus from Nevşehir Otogar in Cappadocia (daily; 7hr) or minibuses from Gaziantep (every 30min; 4hr).

Diversion: Van

Extend your trip northeast to Lake Van to see the reliefs in Akdamar Island's church, one of the masterworks of Armenian art. In Van town eat your way through a famed Van *kahvaltı* (breakfast) before visiting the castle and the Urartian fortress of Çavuştepe.

From Mardin, minibuses run to Diyarbakır (hourly; 1hr 15min) from where buses head to Van (several daily; 6hr 30min). Van Airport has daily flights to İstanbul (2hr 15min).

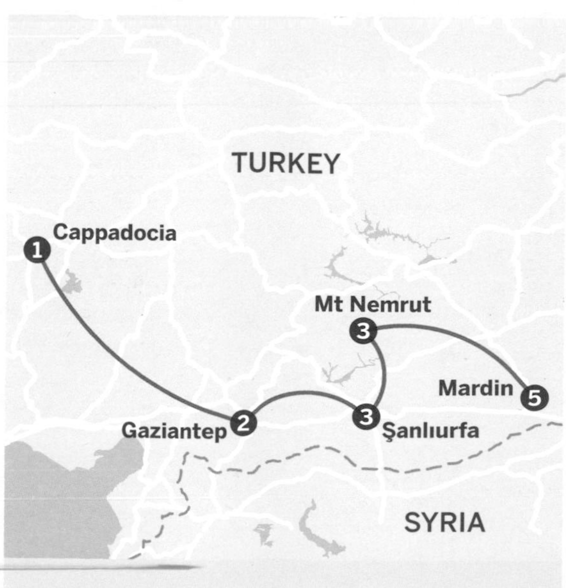

TURKEY'S CLASSICAL-ERA MEDITERRANEAN

İzmir – Antalya

Follow the trail of empires from İzmir to Antalya, down a coastline littered with the ruins of the Classical Era's once triumphant cities.

FACT BOX

Carbon (kg per person): 23
Distance (km): 792
Nights: 10-14
Budget: €€
When: May-Oct

❶ İzmir

Before setting off on the trail of Classical Era archaeological sites, reserve a day for strolling İzmir's seafront *kordon* (promenade) and getting lost amid Kemeraltı Market's alley labyrinth.

Take the train from İzmir's Basmane Gar station to Selçuk (6 daily; 1hr 30min).

❷ Selçuk

Bag Turkey's grandest ruin first and play Roman-for-the-day in Ephesus, once capital of Asia Minor. Afterwards, explore Selçuk where a stork-nest-topped Roman aqueduct marches straight through town, overlooked by a hilltop Byzantine fortress.

Take the bus to Denizli (2 daily; 3hr), then a minibus from Denizli to Pamukkale (every 20min; 20min).

❸ Pamukkale

Pamukkale knits history and natural wonder together into one neat package. Wander the remnants of Roman spa-town Hierapolis at the top of the site then stroll down Pamukkale's hill, wading through white calcite terrace pools of turquoise water along the way.

Return by minibus to Denizli and catch a bus to Fethiye (5 daily; 4hr).

❹ Fethiye

It wasn't all about the Romans round here. Harbour-front Fethiye is a prime base for uncovering the vestiges of the fiercely independent city-states of Ancient Lycia. Day-trip into the surrounding hills to discover the ornate tomb-laden Lycian city ruins of Tlos and Pınara or Xanthos and Letöon.

Hourly buses, taking the coastal road to Antalya, can drop you at the Çıralı turn-off (4hr 45min). Minibuses from the turn-off to Çıralı operate approximately hourly in summer (20min).

ABOVE: EPHESUS' LIBRARY OF CELSUS, SELÇUK

ABOVE: PAMUKKALE

5 Çıralı

It's time for some beach action on Çıralı's 4km-long shore. Swim in the sea, laze on the sand then follow the coast south to explore the rambling remains of the Lycian city of Ancient Olympos. After dark, hike from Çıralı to Mt Olympos where the Chimaera's eternal flames (gas emissions which seep through vents in the rock) have been flickering for at least 2500 years.

Take a minibus back to the turn-off and flag down any Antalya-bound bus on the highway (1hr 30min).

6 Antalya

Bed-down in Antalya's Kaleiçi (Old Town), where cobblestone lanes roll down to a Roman harbour. The next morning, head into the mountainous hinterland to visit the 15,000-seat Roman theatre of Aspendos and the Ancient Pamphylian city of Perge.

Summer charter flights connect to European destinations in summer; regular flights to İstanbul offer more international connections.

ALTERNATIVES

Extension: Blue Cruise

Fethiye is the jumping-off point for Turkey's Blue Cruises, three-night boat trips on which you can soak up the coastal scenery of pine-forest shores backed by craggy mountain peaks. Typical itineraries stop at Butterfly Valley, Kaş and Kekova, before disembarking at Demre, 79km west of Çıralı.

Fethiye tour agencies specialise in Blue Cruises from late April to October. Onward transfers to Çıralı or Olympos are normally included in the package.

Diversion: Kaş

Break up the Fethiye-Çıralı journey leg by stopping in Kaş. This harbour-front town is adventure-activity central for the Turkish Med. Sea-kayak over the sunken city ruins of Kekova Island, and don't miss a day trip to Patara, a Lycian port city whose rambling ruins run down to the beach.

Antalya-bound buses from Fethiye all stop in Kaş (hourly; 2hr 15min).

ISRAEL AND JORDAN COMBINED

Amman – Wadi Rum

See the best of these two countries as part of an epic loop by public transport, from the markets and alleyways of Jerusalem to the windswept deserts of the Arabian peninsula.

FACT BOX

Carbon Cost: 28kg
Distance: 1000
Nights: 10-14
Budget: $$$
When to go: Sep-Apr

❶ Amman, Jordan

The Jordanian capital has an indomitable spirit, best experienced wandering the souks behind the Al-Husseini Mosque, scoffing falafel from Hashem, or hearing the roar of traffic from the cool heights of its Citadel.

Allow a whole day to cross from Amman to Jerusalem via the West Bank due to extensive border checks. JETT buses run from Amman to the King Hussein/Allenby Bridge crossing (daily; 1hr), from where shared taxis travel through the West Bank to Jerusalem (1hr).

ABOVE: PETRA'S MONASTERY, JORDAN

❷ Jerusalem, Israel

Jerusalem's confluence of religions is compelling, captivating and combustible. The highlight is mingling with pilgrims of all faiths: Christians in the shadows of the Church of the Holy Sepulchre; Muslims beneath the Dome of the Rock; Jews beside the Western Wall. The moving Holocaust memorial centre at Yad Vashem offers a modern counterpoint to the city's ancient history.

Regular buses travel from Jerusalem to Tel Aviv (every 30min; 50min).

❸ Tel Aviv, Israel

In many ways the polar opposite of Jerusalem, Tel Aviv is an often hedonistic, beach-facing Mediterranean metropolis. Walk the sequence of sandy beaches southward to the old port of Jaffa, where honey-hued houses rise directly over lapping Mediterranean tides.

Take a bus to Eilat (around 11 daily; 5hr). The Yitzak Rabin/Wadi Araba crossing connects Eilat to Aqaba in Jordan and is a considerably easier undertaking than crossing at King Hussein/ Allenby Bridge.

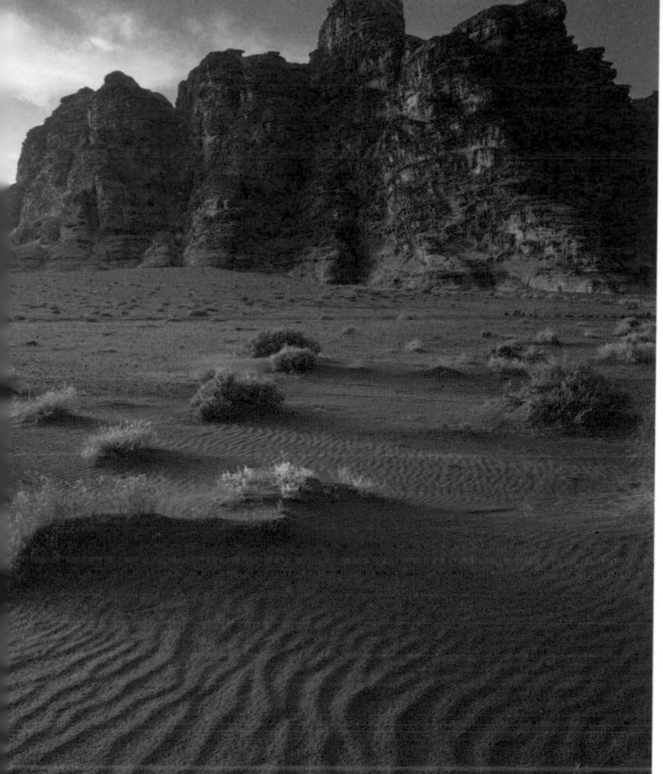
ABOVE: WADI RUM, JORDAN

❹ Eilat, Israel & Aqaba, Jordan

The neighbouring towns of Eilat and Aqaba are separated by a border at the northernmost end of the Red Sea. Both offer opportunities for diving and beachfront lazing; Aqaba also claims a Crusader Castle, once stormed by Lawrence of Arabia.

Arrange a package trip in Aqaba to travel to Wadi Rum (1hr).

❺ Wadi Rum

The cinematic desert of Wadi Rum sees red monoliths rising sheer over sandy wastes. Stay in a tented camp and spend days exploring natural rock arches and springs hidden in the mountains, with nights around the campfire eating earth-oven-cooked Bedouin dishes.

Back in Aqaba, take a bus to Wadi Musa for Petra (regular; 1hr 30min).

❻ Petra

Concealed in narrow canyons beside the Jordan Rift Valley, Petra was once the capital of the Nabatean people. The site is so vast, you could spend many days wandering the craggy pathways that wind among its millennia-old tombs.

From Wadi Musa, buses run to Amman (regular; 4hr) for onward international connections.

ALTERNATIVES

Extension: Bethlehem

Just beyond the Jerusalem suburbs, the little town of Bethlehem is in the West Bank, part of the Palestinian Territories. Most visitors are guided by stars to the Church of Nativity, the Byzantine-era church that allegedly marks the spot where Christ was born. A recent Bethlehem revelation are murals by Banksy, adorning the Separation Wall.

Buses run from Jerusalem to the Wall (regular; 30min), then it's a short taxi ride to the church.

Diversion: Dead Sea

The Dead Sea marks the lowest point on earth and bobbing near-weightless on its salty surface is a high point of any trip to the region. Israel, Jordan and the West Bank all claim stretches of shore where you can wallow in mineral-rich mud up to 420m below sea level.

Amman is well poised for day trips; take a taxi ride to the Amman Beach complex on the northeastern shore (1hr).

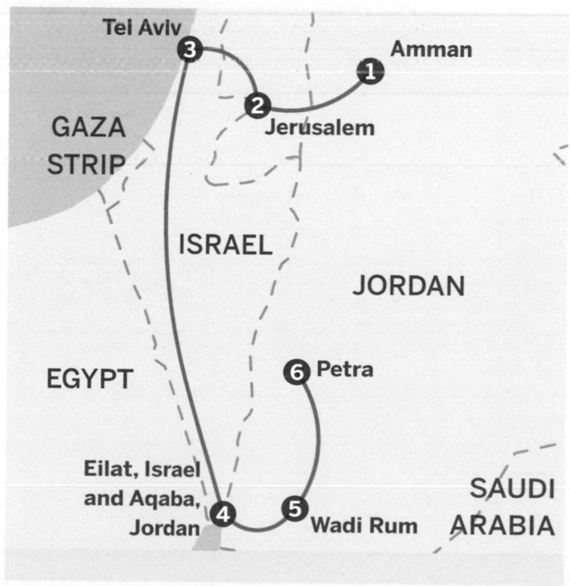

CASTLES AND CANYONS OF OMAN

Muscat – Jebel Shams

Confident off-roaders can steer through one of the Arabian Peninsula's wildest corners, with a self-drive journey from Oman's capital to the summits of the Hajars.

FACT BOX

Carbon (kg per person): 117
Distance (km): 650
Nights: 10
Budget: $$$
When: Oct-Apr

❶ Muscat

With a name that means 'safe anchorage', the capital Muscat reveals both ancient and modern faces of Oman. Mutrah Souq has the aura of the Arabian markets of yore, with merchants selling antiques and wrought-iron lanterns. A counterpoint is the 2011-built Royal Opera House, an opulent marble structure and a legacy of the late opera-loving Sultan.

Drive 90km west from Muscat to Nakhal.

❷ Nakhal

Nakhal offers a first encounter with Oman's famously fearsome fortresses, in the shape of a largely 19th-century bastion that stands guard over palm orchards from its rocky outcrop. After you've stormed the battlements, head to the nearby Al Thowarah hot springs where you can dip into trickling irrigation channels.

Allow an hour to drive the 60km west from Nakhal to Rustaq.

❸ Rustaq

Once the Omani capital, Rustaq is graced by another mighty fortress. It's also a gateway to the Al Hajar Range – the dusty mountain road via Hatt and Wadi Bani Awf is an attraction in its own right, zigzagging up steep cliffs and passing the narrow cleft of Snake Gorge.

Drive 85km south over the mountains to Al Hamra – you'll need a 4WD with supplies and you need to be a confident off-road driver.

ABOVE: NIZWA FORT

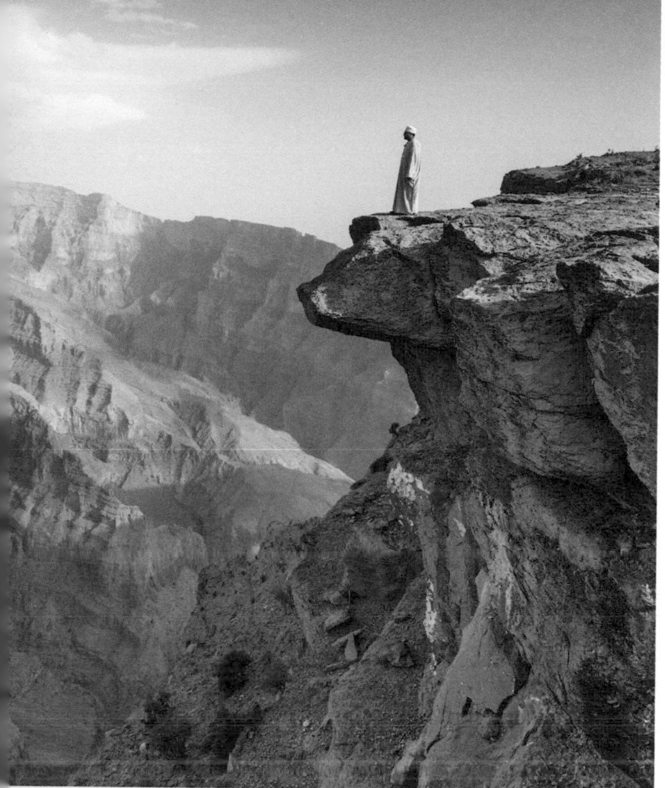
ABOVE: WADI GHUL FROM JEBEL SHAMS

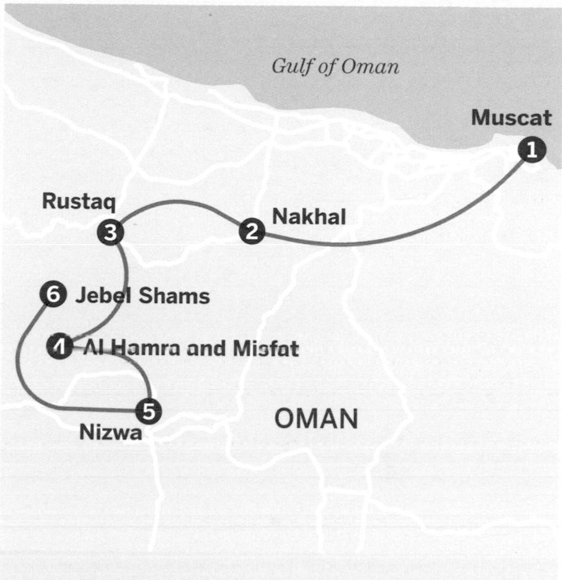

4 Al Hamra and Misfat

You can get a sense of traditional mountain life in Al Hamra and Misfat, neighbouring villages on the southern slopes of the Hajars. In the former, wander among Yemenite mud-brick houses; in the latter, explore lush date plantations that spill down the mountainside.

Continue 50km south to Nizwa

5 Nizwa

The attractive city of Nizwa is a staging post for adventures in the canyons and summits of the Hajars. You can't miss Nizwa Fort, graced by soaring towers from which would-be invaders had boiling-hot date oil dropped on their heads.

Drive 95km west to the summit of Jebel Shams.

6 Jebel Shams

Jebel Shams, 'Mountain of the Sun', is ironically a good place to escape the furnace-hot desert, with cooling breezes circulating around its 3000m summit. Drive up its slopes for views down into Wadi Ghul, the 'Grand Canyon of Arabia', as desolate and awe-inspiring as its American cousin.

Return to Nizwa, from where it's 270km north to Muscat and onward connections.

ALTERNATIVES

Extension: Sharqiya Sands

In the eastern edge of Oman, Sharqiya Sands (formerly Wahiba Sands) is a textbook desert of undulating dunes and blood-red sunsets. Joining a tour is essential to navigate the treacherous sands and reach remote campsites – accommodation generally takes the form of canvas tents, pitched beside Bedouin campfires under clear, star-filled Arabian skies.

Arrange a tour in advance for the 200km trip to the sands.

Diversion: Dhofar

Separated from the north of the country by a near-endless gravel desert, the Dhofar region sees low-slung hills rising by the azure waters of the Indian Ocean, and is famous for its ancient frankincense trade. Salalah is the entry point for the surrounding coast – the Tomb of Job, the fishing town of Mirbat and the sea cliffs at Mughshail are all within an hour's drive.

You can fly to Salalah from Muscat (2 daily; 1hr).

A COASTAL ROAD TRIP ALONG THE ARABIAN GULF

Abu Dhabi – Musandam Peninsula

Hire a car and travel by road through the megacities and deserts of the UAE, before boarding a dhow and sailing the fjord-like inlets of Oman.

FACT BOX

Carbon (kg per person): 118
Distance (km): 1050
Nights: 10-12
Budget: $$$
When: Oct-Apr

❶ Abu Dhabi, UAE

Abu Dhabi is the capital of the UAE and more traditional than its famous coastal sibling, Dubai, though it has a similar appetite for superlative architecture. This reaches its most beautiful expression in the Sheikh Zayed Grand Mosque – marvel at the twinkling chandeliers in the prayer hall (unusually, it is open to non-Muslims).

From Abu Dhabi drive 230km south to Liwa.

ABOVE: EMPTY QUARTER DUNES, UAE

❷ Liwa, UAE

The UAE claims a sizable chunk of the Empty Quarter – the world's largest sand desert, extending south into the Arabian Peninsula. Liwa is a gateway for adventures among the dunes, including camel trekking and 4WD trips in the shadow of the 300m-high Moreeb Hill (one of the world's tallest dunes).

It's a dusty 320km drive north to Dubai.

❸ Dubai, UAE

Dubai's landmarks know no limits, from the futuristic sail-like silhouette of the Burj Al Arab to the artificial islands of the Palm Jumeirah. None, however, is as epic as the Burj Khalifa, the world's highest building, reaching a giddy 828m. A lightning-fast, ear-popping lift shuttles you up to the viewing deck for views over the Gulf.

Drive 120km northeast.

❹ Ras Al Khaimah, UAE

The northernmost emirate in the UAE, Ras Al Khaimah has also seen a development explosion. A refreshing counterpoint to this is the old fishing village of Jazirat Al Hamra to the southwest of Ras Al Khaimah city, a picture of life before oil money, with coral-stone houses and wind towers.

ABOVE: DUBAI'S GRAND MOSQUE, UAE

Cross the border into Oman and drive 80km to Khasab (cross-border car rentals are possible with the right insurance).

5 Khasab, Oman

The low-key capital of the Omani exclave of Musandam, Khasab sits in the middle of a jagged coastline that conjures up storybook associations with *Sinbad the Sailor*. You might hear whispers of the *Arabian Nights* in 17th-century Khasab Fort, whose towers rise up against a backdrop of parched escarpments.

From Khasab port, numerous outfits offer dhow excursions around Musandam.

6 The Musandam Peninsula, Oman

The Musandam Peninsula is Arabia's (warmer) answer to the Norwegian fjords, with blue inlets travelling deep between dusty summits. The way to travel here is by dhow – Arabian sailboats. Join a departure from Khasab to sail in the company of leaping dolphins, laying anchor to explore fishing villages accessible only from the sea.

Drive 200km southwest back to Dubai, whose airport has international connections.

ALTERNATIVES

Extension: Sir Bani Yas

The island of Sir Bani Yas was once the private project of Sheikh Zayed, founder of the UAE. A kind of real-life Jurassic Park, Sir Bani Yas is home to a curious array of native species – the endangered Arabian oryx and sand gazelles – along with introduced creatures such as cheetahs and giraffes.

From Abu Dhabi it's 270km to the Sir Bani Yas ferry dock.

Diversion: Sharjah

The third biggest city in the UAE, Sharjah lives in the shadow of next-door Dubai. It nonetheless has a proud backstory of its own. Find out more at its Heritage Museum, set in a former pearl merchant's house, then practise your bartering in Souk Al Arsah which sells everything from pashminas to *khanjar* daggers.

Sharjah begins at the eastern city limits of Dubai: it's about 25km between the two centres.

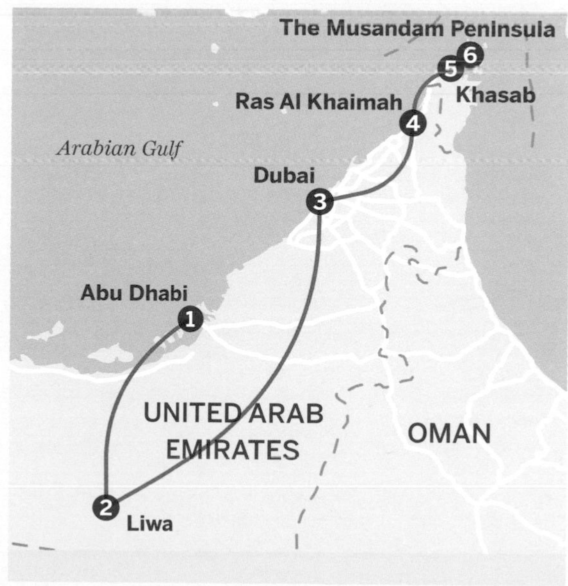

TURKEY'S TURQUOISE COAST

Antalya – Kabak Beach

Rent a car and take an historical road trip along Anatolia's most beguiling shoreline, stopping at a Greek island, storied castles and secluded coves of turquoise shallows.

FACT BOX

Carbon (kg per person): 63
Distance (km): 506
Nights: 10-14
Budget: $$
When: Mar-Jun, Sep-Nov

❶ Antalya, Turkey

Unjustly regarded as just a gateway to coastal resorts nearby, the city of Antalya rewards anyone who lingers. Stroll the back alleys of Kaleiçi, the old quarter of Ottoman mansions, bazaars and mosques. Set a course for the Roman harbour. Or immerse yourself in the baths of Sefa Hamam, to be steamed and scrubbed to contentment amidst 13th-century architecture.

Drive 60km south along the coast to Yanartaş.

❷ Yanartaş, Turkey

Yanartaş offers one of the strangest sights in the Mediterranean – naturally occurring flames bursting forth from the slopes of Mt Olympos. In classical mythology, this was believed to be home of the fire-breathing monster, the Chimaera (who these days obligingly toasts your marshmallows). Park at the seaside hamlet of Çıralı, from where it's a 30min walk to the fires.

It's a beautiful 170km drive west to Uçağız for the boat to Kekova.

❸ Kekova, Turkey

Kekova is a pocket paradise on the Turquoise Coast, a beautiful muddle of islets and peninsulas, a sunken Lycian city (where turtles can sometimes be spotted swimming about the ruins) and a Crusader castle guarding the whole scene. To get to the island, take a glorious 10min motorboat ride from the fishing village of Uçağız.

ABOVE: KEKOVA COASTLINE, TURKEY

ABOVE: GREEK ISLAND-OUTPOST, KASTELLORIZO

From Uçağız, drive west for 30km to Kaş; park here and take the ferry to Kastellorizo (2 daily in summer, daily in winter; 25min).

4 Kastellorizo, Greece

Just over a kilometre from the Turkish mainland, the miniature island of Kastellorizo is a speck of Greek territory adrift amid the bays and headlands of the Turquoise Coast. Visit on a day trip (or longer) from the port of Kaş to wander among colourful harbourside houses and hike dusty trails in search of lonely Orthodox chapels.

Return to Kaş and drive 150km to Kabak; go via the resort of Ölüdeniz if you want to avoid the hair-raising, if spectacular, mountain road.

5 Kabak Beach, Turkey

Kabak Beach is the polar opposite of the thronging resort towns to the north – a secluded strip of sand guarded by sprawling pine forests, accessed only via a perilously steep road from the cliffs. The bohemian hamlet beside the beach is a launchpad for snorkelling adventures and yoga.

It's a beautiful 80km coastal drive from Kabak to Dalaman Airport, which has flights to European destinations.

ALTERNATIVES

Extension: The Lycian Way

The 540km Lycian Way is Turkey's most hallowed long-distance footpath, following old mule tracks, country lanes and Roman roads along the shore from Fethiye to Antalya. It also links the lost cities of the Lycian civilisation which once ruled this area. You'll need a month to tackle it all – or walk the stretch from Ölüdeniz to Kınık in five days.

Ölüdeniz is a 20km (40min) drive north from Kabak.

Diversion: Datca and Bozburun Peninsulas

Continue eastward from Ölüdeniz and another glorious coastal drive beckons in the Datça and Bozburun Peninsulas, which together reach like an outstretched hand into the Aegean Sea. The former offers little windmills, rustic villages and the evocative ruins of Kndios; the latter feels wilder, with scrubby hills dipping down to balmy coves.

Reach the two peninsulas on a 150km drive from Kabak

DRIVE JORDAN'S HISTORIC KING'S HIGHWAY

Amman – Petra

Jordan's classic road trip is Route 35, more poetically known as the King's Highway, a journey of souks, Crusader castles and ancient desert cities.

FACT BOX

Carbon (kg per person): 62
Distance (km): 499
Nights: 6
Budget: $$
When: Mar-May

❶ Amman

Jordan's capital is mostly 20th century, but it has some diverting older sights. Top of the list is the walled Citadel, which sits on Jebel Al Qala'a (850m), Amman's highest hill. The city's restored Roman theatre is another atmospheric must-see, as is the Jordan Museum – look out for the 9500-year-old plaster mannequins of Ain Ghazal and Jordan's share of the famous Dead Sea Scrolls.

Rent a car and head 35km south on the first leg of the King's Hwy.

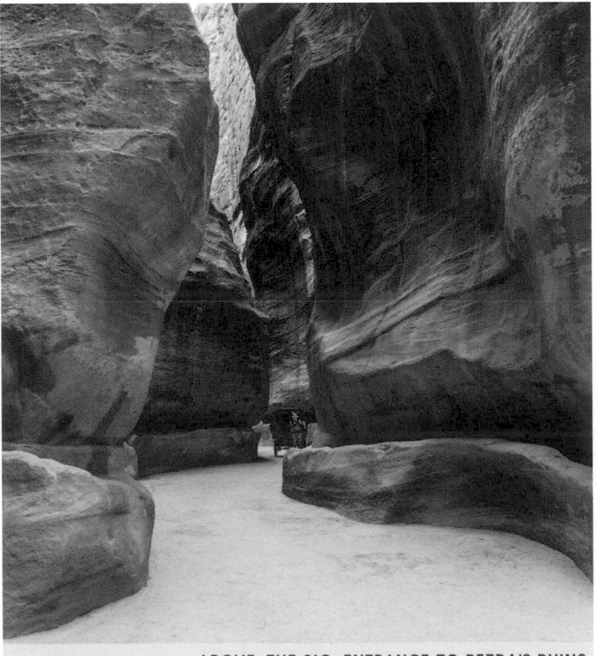

ABOVE: THE SIQ, ENTRANCE TO PETRA'S RUINS

❷ Madaba & Mt Nebo

The market town of Madaba is best known for its Byzantine-era mosaics: the most famous of these is the 6th-century map of the Middle East on the floor of St George's Church. Mt Nebo, 9km northwest of town, is where Moses supposedly saw the Promised Land (it's said he died at the ripe old age of 120 and was buried nearby). The Moses Memorial Church on the summit commands far-reaching views of the Dead Sea, Israel and the Palestinian Territories beyond.

Back on the King's Hwy head 88km south.

❸ Karak

The Crusader stronghold of Karak (or Kerak) was a place of legend in the battles between the Christian Crusaders of the Franks and the Islamic armies of Saladin. Reconstruction and excavation work is

ABOVE: KARAK'S CRUSADER STRONGHOLD

ongoing but you can visit the castle's church, barracks, prison cells and defensive keep.

Continue 111km south.

4 Shobak

For more insights into Jordan's turbulent medieval period, stop at another dramatic Crusader stronghold, Shobak. Perched on a hill called Mons Realis (Royal Mountain), it's less intact than Karak. Built by the Crusader King Baldwin I in 1115, it withstood numerous attacks from Saladin's armies before succumbing in 1189.

It's 32km south from Shobak to Wadi Musa, the gateway to Petra.

5 Petra

The ancient city of Petra looks like it's been made for a movie set – and indeed it's been used as a ready-made backdrop in films from *Indiana Jones* to *The Mummy*. It was built in the 3rd century BCE by the Nabataeans, who carved magnificent palaces, temples, tombs, storerooms and stables into the soft stone cliffs. It's approached through the anticipation-building 1.2km-long Siq – a high-walled cleft in the cliffs caused by tectonic movement.

It's 233km back to Amman along the King's Hwy.

ALTERNATIVES

Extension: Wadi Rum

If you've dreamed of experiencing the desert, Wadi Rum offers a glimpse you'll never forget. Inhabited by nomadic Bedouin, the village of Rum serves as a gateway to the dunes; explore by 4WD, quadbike, camel or on foot. The most striking dunes are at Al Hasany; for full desert immersion, stay overnight at a Bedouin desert camp. You'll rarely see skies so clear.

Wadi Rum is 114km south of Petra. Arrange desert expeditions in Rum.

Diversion: The Dead Sea

At up to 434m below sea level, the Dead Sea is the lowest point on earth. Its salts and minerals have long been exploited for their skin-friendly properties, and its shores are lined with spa resorts. You don't actually swim in the Dead Sea – the salinity means you just float on the surface.

Resorts along the sea are an easy day trip from Amman or Madaba.

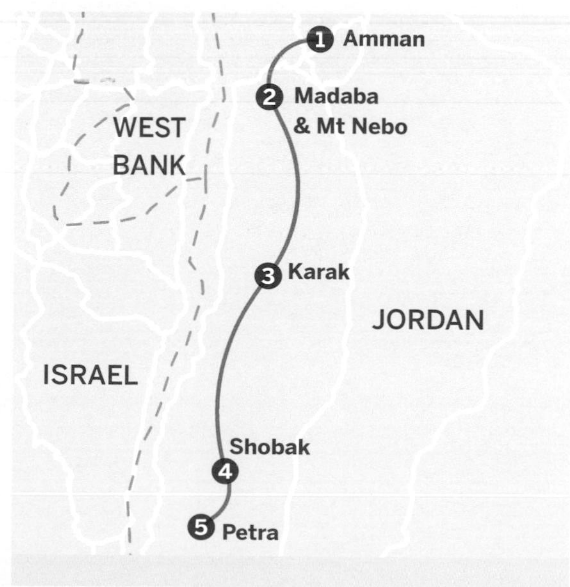

A TAILOR-MADE TOUR FROM TEHRAN TO SHIRAZ

Tehran – Shiraz

Less stressful than the bus system, and more affordable than an organised group tour, a private guided road trip is the best way to tour Iran's highlights.

FACT BOX

Carbon (kg per person): 117
Distance (km): 940
Nights: 10
Budget: $$
When: year-round

❶ Tehran

Arrange to meet your local guide (try Sufi Tavafi, one of Iran's only female driver-guides; synotrip.com) in Tehran and allow several days to explore the nation's beating heart, which hugs the lower slopes of the snow-capped Alborz Mountains. Get an insight into the luxurious lives of the former shahs at Golestan Palace, the Sa'd Abad Museum Complex and the Niyavaran Cultural-Historic Complex before seeking out Haj Ali Darvish tea house in the Grand Bazaar for one of the most memorable teas you'll enjoy in the country. Nearby, perennially busy Moslem restaurant will satisfy your deepest Persian food cravings.

With your guide, drive 250km (around 3hr) south to Kashan.

❷ Kashan

This delightful, laidback oasis city is one of Iran's most alluring destinations. Take a turn through the Unesco-recognised Bagh-e Fin garden, tour grand 19th-century Persian homes-turned-museums (including the resplendent Khan-e Boroujerdi with its ornately decorated courtyard), and check out the 500-year-old Sultan Mir Ahmad bath-house before spending the night at an atmospheric hotel such as the beautifully restored Saraye Ameriha, which has a charming cafe.

From Kashan, it's a 210km (2hr 30min) drive south to Esfahan.

ABOVE: IRANIAN TEA

ABOVE: MASJED-E NASIR AL MOLK, SHIRAZ

3 Esfahan

Wandering among its tree-lined boulevards and historic buildings, it's easy to understand why Esfahan is Iran's top tourist destination. A city of artisans, this is *the* place to purchase a Persian carpet – but your first stop should be the beautiful Naqsh-e Jahan (Imam) Square, dominated by the breathtaking blue-tiled Masjed-e Shah. Come nightfall, head to Khajoo Bridge to listen to locals singing Persian songs together under its historic arches.

Continue south for 480km (6hr) to Shiraz.

4 Shiraz

Shiraz, the city of poets, is a fittingly atmospheric place to end your Iran adventure. It's home to splendid gardens (don't miss Bagh-e Naranjestan) and exquisite mosques (such as the famous Masjed-e Nasir Al Molk or Pink Mosque; go early to snap the classic shot of the morning sun filtering through the colourful stained glass); it's also within easy reach (a 1hr drive) of the still-spectacular ruins of Persepolis, destroyed by Alexander the Great in 330 BCE. Leave time for last-minute shopping at Shiraz's historic Vakil Bazaar, where you can pick up carpets, spices, jewellery and more.

Shiraz's airport has onward international flights, or drive back to Tehran with your guide.

ALTERNATIVES

Extension: Qom

If you can't get enough mosques, factor in a few hours during the trip from Tehran to Kashan to visit the holy city of Qom. Highlights include enormous Imam Hassan Mosque and the Hazrat-e Masumeh, a magnificent shrine marking Qom's physical and spiritual centre. Grab some local kebab (barbecued lamb *kofta* and vegetables wrapped in flatbread) before you leave.

It's a 147km drive from Tehran to Qom.

Diversion: Yazd

From Esfahan, head southeast to the desert city of Yazd. Get lost in its maze of winding lanes lined with mudbrick houses, stop in hidden tea houses and marvel at exquisite mosques, including the vision of turquoise tiles that is Masjed-e Jameh. Bed down in a traditional guesthouse such as the elegantly restored Narenjestan Traditional House.

Yazd is 310km (4hrs) southeast of Esfahan.

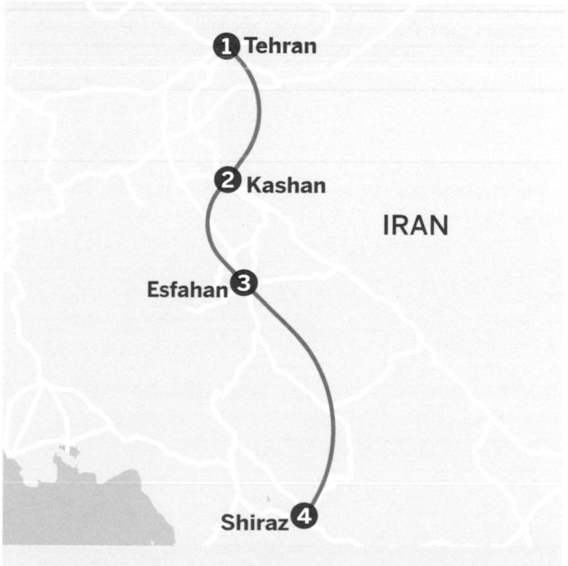

INDEX

A

B

C

D

E

F

Lonely Planet's Trip Builder
October 2021
Published by Lonely Planet Global Limited
ABN 36 005 607 983
www.lonelyplanet.com
1 2 3 4 5 6 7 8 9 10

Printed in Malaysia
ISBN 978 18386 9334 3

Written by Abigail Blasi, Adam Karlin, Alex Crevar, Amy Balfour, Andrew Bender, Ann Abel, Bailey Freeman, Ben Groundwater, Bradley Mayhew, Brendan Sainsbury, Brett Atkinson, Bridget Gleeson, Carolyn Heller, Charles Rawlings-Way, Claire Boobbyer, Cliff Wilkinson, Etain O'Carroll, Helen Ranger, Isabella Noble, James Bainbridge, James Smart, Jess Lee, Joe Bindloss, John Brunton, Kate Armstrong, Kate Morgan, Kerry Walker, Kevin Raub, Lorna Parkes, Luke Waterson, Monique Perrin, Nicola Williams, Oliver Berry, Oliver Smith, Peter Dragicevich, Philip Tang, Regis St Louis, Sam Haddad, Sarah Baxter, Simon Richmond, Stephen Lioy, Tamara Sheward, Trent Holden, Virginia Maxwell, Valerie Stimac, Yolanda Zappaterra

General Manager, Publishing Piers Pickard
Commissioning Editor Robin Barton
Editors Cliff Wilkinson, Polly Thomas
Art Director Daniel Di Paolo
Layout Designer Jo Dovey
Cartography Katerina Pavkova
Print Production Nigel Longuet

STAY IN TOUCH
lonelyplanet.com/contact

Lonely Planet Global Limited, Ireland
Digital Depot, Roe Lane (off Thomas St),
Digital Hub, Dublin 8, D08 TCV4

Paper in this book is certified against the Forest Stewardship Council™ standards. FSC™ promotes environmentally responsible, socially beneficial and economically viable management of the world's forests.